ELECTRONIC PRINCIPLES

Other Books by Albert P. Malvino

Digital Computer Electronics, An Introduction to Microcomputers

Transistor Circuit Approximations

Experiments for Transistor Circuit Approximations

Resistive and Reactive Circuits

Electronic Instrumentation Fundamentals

Basic Electronics: A Text-lab Manual (with P. Zbar)

Digital Principles and Applications (with D. Leach)

Experiments for Electronic Principles

ELECTRONIC PRINCIPLES Third Edition

Albert Paul Malvino, Ph.D.

McGraw-Hill Book Company

NEW YORK ATLANTA DALLAS ST. LOUIS SAN FRANCISCO
AUCKLAND BOGOTÁ GUATEMALA HAMBURG JOHANNESBURG LISBON
LONDON MADRID MEXICO MONTREAL NEW DELHI PANAMA PARIS
SAN JUAN SÃO PAULO SINGAPORE SYDNEY TOKYO TORONTO

Sponsoring Editor: Paul Berk
Editing Supervisor: Pat Nolan
Design Supervisor/Cover Designer: Judith Yourman
Production Supervisor: S. Steven Canaris

Text Designer: Suzanne Bennett and Associates
Cover Photographer: Richard Megna/Fundamental Photographs

Library of Congress Cataloging in Publication Data

Malvino, Albert Paul.
  Electronic principles.

  Includes index.
  1. Electronics.   I. Title.
TK7816.M25   1983   621.381   83-19987
ISBN 0-07-039912-3

Electronic Principles, Third Edition

1 2 3 4 5 6 7 8 9 0   DOCDOC   8 9 0 9 8 7 6 5 4 3

ISBN 0-07-039912-3

To Joanna,
My brilliant and beautiful wife
without whom I would be nothing.
She always comforts and consoles,
never complains or interferes,
asks nothing and endures all,
and writes my dedications.

CONTENTS

Preface

This third edition is not a skin-deep revision. Almost every chapter has been rewritten to reflect the changes that have taken place in industry. Some of the earlier chapters on discrete devices have been combined to make room for new material. Although the discussion of discrete devices has been streamlined, you will still find a complete treatment of diodes and transistors because an understanding of what these components are and how they function is the foundation needed to understand ICs.

In revising the book, I discovered several areas that needed expansion. Now, you will find new discussions of troubleshooting, optoelectronic devices, power-supply filtering, load lines, graphical analysis, cascaded stages, *h* parameters, classes D through S, JFET switches, JFET voltage-variable resistances, dual-gate MOSFETs, VMOS interface circuits, diff-amp analysis, negative-feedback circuits, foldback current limiting, parasitic oscillations, and phase-locked loops.

Besides the foregoing changes, I wrote many new chapters to cover topics such as voltage and current feedback, JFET-controlled op-amp circuits, voltage-controlled current sources, current boosters, active Butterworth filters, comparators with hysteresis, window comparators, Schmitt triggers, integrators, differentiators, waveshaping circuits, dc-to-dc converters, switching regulators, 555 timers, and thyristors.

In addition to the changes in material, this new edition contains two major changes in format. First, I have rewritten the book to allow you to use either conventional or electron flow. Since either approach is valid, there is no reason why I should saddle you with one type of flow when you prefer the other. Chapter 1 discusses both types of flow and indicates how either can be used in subsequent chapters.

The second major change is in the homework problems. Because of many requests, I have expanded the problems at the end of each chapter to include five categories: straightforward, troubleshooting, design, challenging, and computer. The straightforward section is similar to the problems of earlier editions. The troubleshooting, design, challenging, and computer sections are new with this edition. Consider these new problems optional. If they fit your program, fine. If not, ignore them. For example, most schools will want to include the straightforward and troubleshooting problems because they are basic to any technician program. Other schools may use the design and challenging problems as well. And finally, schools where computers are available for students may want to include the computer problems to round out their programs. Chapter 1 describes the new homework problems in more detail. In my opinion, the problems in this book will enhance any program and bring a new dimension to electronics education.

As before, this book is for a student taking a first course in linear electronics. The prerequisites are a dc-ac course, algebra, and some trigonometry (at least enough to work with sine waves). In many schools it will be possible to take the ac and trigonometry courses concurrently.

A final point. In addition to this textbook, a correlated laboratory manual *Experiments for Electronic Principles* is available. It contains over 50 experiments including optional exercises in troubleshooting and design. An extensive instructor's guide is also available.

Albert Paul Malvino

Acknowledgments

First, I would like to thank everyone who returned a questionnaire for *Electronic Principles* in the Spring of 1982. I know it took a lot of time to answer the questions and to add comments on what to delete, expand, and include. I want you to know I read each questionnaire carefully. Without them to guide me and the many excellent suggestions given, I could not have attained the perspective and quality found in this third edition.

Second, I want to thank each instructor who allowed me to interview him. My visit to your school and talking to you gave me great insight into what had to be done. Our exchange of ideas led to excellent new ideas that I could never have come up with alone. I found the interviews an invaluable asset during the months of rewrite and revision.

Third, I want to thank all the reviewers who read some or all of the revised manuscript. Several times, you caught me wandering off the true path. Furthermore, you were constantly suggesting better ways to discuss ideas, word passages, and draw figures. Your help has raised the quality of the final product to a level that would have been impossible working alone.

Finally, I want to acknowledge and thank each of the following colleagues who made significant contributions to this third edition:

Richard Berg, West Valley College
Adrian Berthiaume, Northern Essex Community
    College
Alfred E. Black, ITT Technical and Business Institute
Marvin Chodes, New Hampshire Vocational-Technical
    College
Charles J. Cochran, ITT Technical and Business
    Institute
Daniel Courtney, Springfield Technical Community
    College

James D. Feeney, Southern Maine Vocational Technical Institute

James Fisk, Northern Essex Community College

Lawrence Fryda, Larimer County Vocational-Technical Center

Jack Hain, Shore Community College

Herbert N. Hall Jr., Lakeland College

Donald C. Jameson, Santa Monica College

Theodore Johnson, Berkshire Community College

Joseph Kittel, Vermont Technical College

Karl R. Laurin, Greensburg Institute of Technology

William H. Lauzon and staff, Technical Careers Institute

Joseph J. Macko, Heald College

Leo D. Martin, Iowa Western Community College

Michael A. Miller, DeVry Institute of Technology

Gary Mullet, Springfield Technical Community College

David L. Newton, ITT Technical and Business Institute

James O'Malley Jr., Security Systems

John R. Paris, Madisonville State Vocational-Technical School

Roger J. Pines, George C. Wallace State Community College

Peter Rasmussen, Vermont Technical College

Marvin Rogers, Vermont Technical College

Wayne Roy, New Hampshire Vocational-Technical College

Bernard Rudin, Community College of Philadelphia

Roy M. Sutcliffe, Idaho State University

James E. Teza, Butler County Community College

Ernest W. Trettel, Minneapolis Technical Institute

Edward V. Tuba, Heald Institute of Technology

James W. Waddell, ITT Technical and Business Institute

Frank Wang, New Hampshire Vocational-Technical College

Frank M. Wiesenmeyer, Richland Community College

Introduction

One of the prerequisites for reading this book is a course in dc circuit theory in which topics like Ohm's law, Kirchhoff's laws, and other circuit theorems have been discussed. This first chapter reviews a few basic ideas and introduces some viewpoints that you might have missed the first time through basic dc theory.

1-1 CONVENTIONAL AND ELECTRON FLOW

Which way do electric charges flow? Murphy's law says that the number of deeply held beliefs is equal to the number of possibilities, no matter how ridiculous. Fortunately, there are only two possible directions for current: plus to minus, or minus to plus.

THE FLUID THEORY

Franklin (1750) made an outstanding contribution with his fluid theory of electricity. He visualized electricity as an invisible fluid. If a body had more than its normal share of this fluid, he said it had a positive charge; if the body had less than a normal share, its charge was considered negative. On the basis of this theory, Franklin concluded that electric fluid flowed from positive (excess) to negative (deficiency).

The fluid theory was easy to visualize and agreed with all experiments conducted in the eighteenth and nineteenth centuries. As a result, everybody accepted the notion that charges were flowing from positive to negative (now called *conventional* flow). Between 1750 and 1897, a large number of concepts and formulas based on the conventional flow came into existence. During this period, the scientific community became committed to conventional flow as a way of life.

Even today, the bulk of engineering literature continues to use conventional flow. Somebody (usually an engineer or scientist) who invents a new device tends to insert arrows on the device that point in the direction of conventional current.

THE ELECTRON

In 1897, Thomson discovered the electron and proved that it had a negative charge. Nowadays, the planetary concept of matter is well known. Matter is made up of atoms. Each atom is a positively charged nucleus surrounded by orbiting electrons. The outward push of centrifugal force on each electron is exactly balanced by the inward pull of the nucleus. Therefore, electrons travel in stable orbits, in a manner similar to the motion of the planets around the sun.

A copper atom has 29 protons and 29 electrons. Of the 29 electrons, 28 travel in tight orbits around the nucleus; because of their small orbits, these electrons are locked into the atom by the strong pull of the nucleus. But the 29th electron travels in a very large orbit. Since it is relatively far from the nucleus, this electron feels almost no nuclear attraction. As a result, it is called a *free electron* because it can easily wander from one copper atom to the next.

ELECTRON FLOW

In a piece of copper wire, the only charges that flow are the free electrons. Under the influence of an electric field, these free electrons flow out of the negative terminal of a battery through the wire to the positive terminal. This is exactly opposite to conventional flow, which creates a problem. Everybody now agrees that charges actually flow from negative to positive in a piece of copper wire, but not everyone is willing to discard the use of conventional flow.

Why the resistance to change? Because once you get above the atomic level, it makes no difference whether you visualize charges flowing from negative to positive or vice versa. Mathematically, you get the same answers either way. Therefore, even though electron flow is the truth, the whole truth, and nothing but the truth, conventional flow preserves the mathematical foundations of almost 200 years of circuit theory.

What it comes down to is this: It is convenient for engineers to use both conventional and electron flow, rather than choosing one or the other. At the atomic level, they use electron flow to explain what is actually happening. Above the atomic level, they pretend that a hypothetical positive charge flows, rather than an electron. Maybe someday the engineering community will change to electron flow when analyzing circuits mathematically, but at this time the consensus is that such a change is not worth the hassle.

EITHER FLOW VALID

When a device is discussed for the first time, both types of flow will be shown, with a solid arrow for conventional flow and a dashed arrow for electron flow. You can use either type of flow, so ignore the one you don't want. As an example, Fig. 1-1*a* shows a circuit with conventional current, and Fig. 1-1*b* shows the same circuit with electron flow. In using this book, you should settle on either conventional flow or electron flow; either is valid. Furthermore, occasionally seeing both types of current is probably good training because you will encounter both types in industry.

After we introduce a device, we will drop the use of current arrows. Again, this is good practice because industrial schematics show voltage polarities but not current

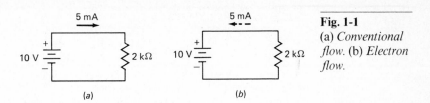

directions. It's up to you to know that charges flow from positive to negative (if you use conventional flow) or from negative to positive (if you prefer electron flow).

1-2 VOLTAGE SOURCES

For any electronic circuit to work, there has to be a source of energy. An energy source is either a voltage source or a current source. This section discusses the voltage source, and the next section is about the current source.

IDEAL VOLTAGE SOURCE

An *ideal* or perfect voltage source produces an output voltage that does not depend on the value of load resistance. The simplest example of an ideal voltage source is a perfect battery, one whose *internal resistance* is zero. For instance, the battery of Fig. 1-2a produces an output voltage of 12 V across a load resistance of 10 kΩ; Ohm's law tells us that the load current is 1.2 mA. If we reduce the load resistance to 30 Ω, as shown in Fig. 1-2b, the load voltage is still 12 V; the load current, however, increases to 0.4 A. (Don't reach for your calculator; all calculations in this section should be done mentally.)

Figure 1-2c shows an adjustable load resistance (rheostat). The ideal voltage source will always produce 12 V across the load resistance, regardless of what value it is adjusted to. Therefore, the load voltage is constant; only the load current changes.

REAL VOLTAGE SOURCE

An ideal voltage source cannot exist in nature; it can exist only in our minds as a theoretical device. It is not hard to understand why. Suppose the load resistance of Fig. 1-2c approaches zero; then the load current approaches infinity. No real voltage source can produce infinite current because every real voltage source has some internal resistance. This resistance is typically less than 1 Ω. For instance, a flashlight battery has an internal resistance of less than 1 Ω, a car battery has an internal resistance of less than 0.1 Ω, and an electronic voltage source may have an internal resistance of less than 0.01 Ω.

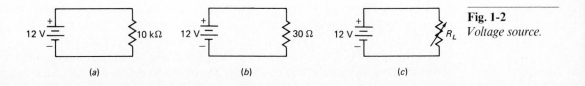

Fig. 1-3
Load current.

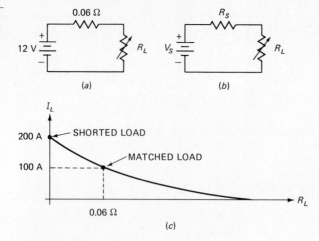

(a)   (b)

(c)

SHORTED-LOAD CURRENT

The internal resistance of a real voltage source appears in series with the load resistance. For instance, Fig. 1-3*a* shows a 12-V source with an internal resistance of 0.06 Ω. If we reduce the load resistance to zero, Ohm's law gives

$$I = \frac{12 \text{ V}}{0.06 \text{ } \Omega} = 200 \text{ A}$$

This is the maximum load current that the real voltage source can deliver; this maximum load current is called the *shorted-load current.*

GRAPH OF LOAD CURRENT

You can visualize any real voltage source as shown in Fig. 1-3*b*: An ideal voltage source V_S is in series with an internal resistance R_S. With Ohm's law,

$$I_L = \frac{V_S}{R_S + R_L} \tag{1-1}$$

When the load resistance increases, the load current decreases. Plotting load current versus load resistance for Fig. 1-3*a*, we get the graph of Fig. 1-3*c*. There are no surprises here. The load current is 200 A for zero load resistance. Then, as load resistance increases toward infinity, the load current decreases toward zero. Note the intermediate point where the load resistance matches the internal resistance; at this point the load current is half the shorted-load current.

Figure 1-4 shows I_L versus R_L for any circuit. When R_L is zero, I_L is maximum and equal to V_S/R_S. When R_L equals R_S, I_L is half the maximum value and equal to $V_S/2R_S$. Further increases in R_L cause I_L to decrease toward zero.

LOAD VOLTAGE

When the load resistance increases to infinity in Fig. 1-3*b*, the load voltage approaches the ideal source voltage. We can prove this as follows. The load voltage

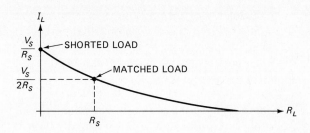

Fig. 1-4
*Load current
versus load
resistance.*

equals

$$V_L = I_L R_L$$

Since $I_L = V_S/(R_S + R_L)$, we can write

$$V_L = \frac{V_S}{R_S + R_L} R_L$$

or

$$V_L = \frac{R_L}{R_S + R_L} V_S \qquad (1\text{-}2)$$

Look at the denominator. When R_L approaches infinity, it *swamps out* (overpowers or makes negligible) the internal resistance. For instance, when R_L is 100 times greater than R_S, the load voltage is approximately 99 percent of the source voltage. When R_L equals infinity (open load), the load voltage equals the ideal voltage.

Figure 1-5*a* illustrates the swamping effect of large load resistance. This is a graph of Eq. (1-2) for a source voltage of 12 V and an internal resistance of 0.06 Ω. When the load resistance is zero (shorted), the load voltage is zero. When the load resistance equals the internal resistance (0.06 Ω), V_L equals 6 V because half the source voltage is dropped across R_S.

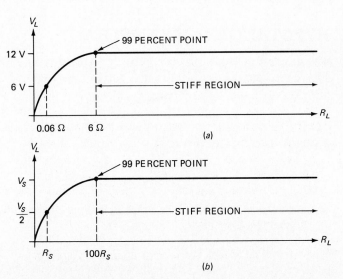

Fig. 1-5
*Load voltage
versus load
resistance.*

As the load resistance continues to increase, the load voltage begins to plateau at 12 V. The load voltage asymptotically approaches 12 V, the value of the ideal or open-load voltage. Notice the 99 percent point. At this point, the load resistance is 100 times greater than the internal resistance and the load voltage is approximately 99 percent of the source voltage.

Figure 1-5*b* shows V_L versus R_L for any circuit. When R_L is zero, V_L is zero. When R_L equals R_S, V_L is half of V_S. For larger load resistances, the load voltage approaches the ideal source voltage. When R_L is greater than $100R_S$, V_L is more than 99 percent of V_S.

STIFF VOLTAGE SOURCE

Often, the load resistance is much larger than the internal resistance of a voltage source. This is equivalent to saying that the internal resistance is much smaller than the load resistance. In this book, a *stiff voltage source* is one whose internal resistance is at least 100 times smaller than the load resistance:

$$R_S \leq 0.01R_L$$

This is equivalent to saying that R_L is at least 100 times greater than R_S. The important thing about a stiff voltage source is this: It produces a load voltage that is between 99 and 100 percent of the ideal source voltage.

With a stiff voltage source the difference between the load voltage and the ideal or open-load voltage is less than 1 percent, small enough to ignore for most trouble-shooting, analysis, and design. The word "stiff" reminds us that the source is delivering an almost ideal voltage to the load resistance.

1-3 CURRENT SOURCES

A voltage source has a very small internal resistance. A *current source* is different; it has a very large internal resistance. Furthermore, a current source produces an output current that does not depend on the value of load resistance.

HYPOTHETICAL EXAMPLE

The simplest example of a current source is the combination of a battery and a large source resistance, as shown in Fig. 1-6*a*. In this circuit, the load current is

$$I_L = \frac{V_S}{R_S + R_L}$$

When R_L is zero, the current is

$$I_L = \frac{12 \text{ V}}{10 \text{ M}\Omega} = 1.2 \ \mu\text{A}$$

Because R_S is so large, the load current is approximately 1.2 μA for a large range of R_L. For instance, when R_L is 10 kΩ,

$$I_L = \frac{12 \text{ V}}{10.01 \text{ M}\Omega} = 1.1988 \ \mu\text{A}$$

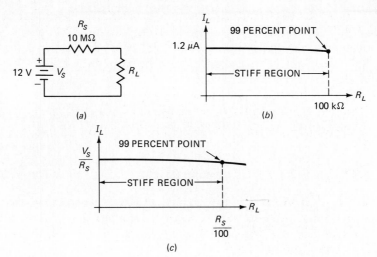

Fig. 1-6
Current source.

When R_L is 100 kΩ,

$$I_L = \frac{12 \text{ V}}{10.1 \text{ M}\Omega} = 1.188 \; \mu\text{A}$$

which is still very close to 1.2 μA.

Figure 1-6*b* shows a graph of load current versus load resistance. As you can see, the load current is approximately constant. When the load resistance equals 100 kΩ, the load current is 99 percent of the ideal value. Figure 1-6*c* is a graph for any circuit. The important thing to realize is that load current is approximately constant as long as R_L is less than $R_S/100$. This is equivalent to saying that R_S is at least 100 times greater than R_L.

For future discussions, a *stiff current source* is one whose internal resistance is at least 100 times greater than the load resistance:

$$R_S \geq 100 R_L$$

As indicated in Fig. 1-6*c*, a stiff current source produces a load current that is between 99 and 100 percent of the ideal or shorted-load value.

PRACTICAL CURRENT SOURCES

Current sources are rarely built with a battery-resistor combination because the large series resistance makes getting enough current for practical applications impossible. Instead, transistors and other devices are connected in such a way as to produce stiff current sources that can deliver amperes of load current. You will learn more about practical current sources in later chapters.

EXAMPLE

Norton's theorem (discussed in basic circuits books) uses the symbol of Fig. 1-7*a* for an ideal current source, one whose internal resistance is infinite. Such a device can

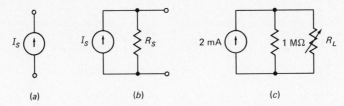

Fig. 1-7 (a) *Symbol for current source.* (b) *Internal resistance is in parallel.* (c) *Load resistance.*

exist only in our minds because any real current source has an internal resistance. This internal resistance is very high, but never infinite.

The internal resistance of a real current source is always shown in parallel with an ideal current source (see Fig. 1-7*b*). This parallel connection is necessary to shunt the source current during open-load conditions. For instance, suppose that the load resistance of Fig. 1-7*c* is shorted (0 Ω). Then all the charges flow through the shorted load, and the load current is 2 mA. If the load resistance increases, the current source remains stiff up to a load resistance of 10 kΩ; for this value of load resistance, approximately 99 percent of the charges pass through the load resistance.

With further increases in load resistance, more of the charges are shunted through the internal resistance. When the load resistance is 1 MΩ, the source current divides equally, with 1 mA through the internal resistance and 1 mA through the load resistance.

As the load resistance goes to infinity or an open-load condition, the load current drops to zero and all of the 2 mA passes through the internal source resistance.

1-4 THEVENIN'S THEOREM

Every once in a while, somebody makes a big breakthrough in engineering and carries all of us to a new high. M. L. Thevenin made one of these quantum jumps when he discovered a circuit theorem now called *Thevenin's theorem.* You probably covered Thevenin's theorem in your basic dc circuits course and no doubt were told it was important. But it's impossible to tell a beginner how valuable the theorem is. If the statement

THEVENIN'S THEOREM IS VITALLY IMPORTANT

were repeated 1000 times, it still would not indicate how useful the theorem is to anyone who troubleshoots, analyzes, or designs electronic circuits.

BASIC IDEA

Suppose someone hands you the schematic diagram of Fig. 1-8*a* and asks you to calculate the load current for each of these values of R_L: 1.5 kΩ, 3 kΩ, and 4.5 kΩ. One solution is based on combining series and parallel resistances to get the total resistance seen by the source; then you calculate the total current and work back toward the load, dividing the current until you find the load current. After you work

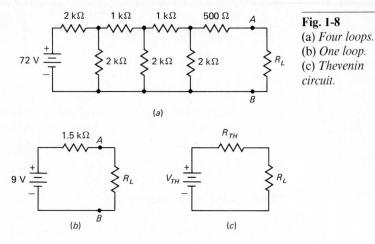

Fig. 1-8
(a) *Four loops.*
(b) *One loop.*
(c) *Thevenin circuit.*

out the load current for 1.5 kΩ, you can repeat the whole boring process for 3 kΩ and 4.5 kΩ.

Another approach is simultaneous solution of Kirchhoff loop equations. Assuming that you know how to solve four simultaneous loop equations, you can grind your way to the answer for a load resistance of 1.5 kΩ. Then you have to repeat the process for 3 kΩ and for 4.5 kΩ. Half an hour later (more or less), you may have worked out all three load currents.

On the other hand, suppose someone asks you to solve for the load currents in Fig. 1-8b, given load resistances of 1.5 kΩ, 3 kΩ, and 4.5 kΩ. Faster than anyone can pull out a calculator, you can mentally work out a load current of

$$I_L = \frac{9\ \text{V}}{3\ \text{k}\Omega} = 3\ \text{mA}$$

for a load resistance of 1.5 kΩ. You can also calculate load currents of 2 mA for 3 kΩ and 1.5 mA for 4.5 kΩ.

Why is the second circuit so much easier to analyze than the first? Because the second circuit has only one loop, compared with four loops in the first circuit. Anybody can solve a one-loop problem because all it takes is Ohm's law.

This is where Thevenin's theorem comes in. Thevenin discovered that any multiloop circuit like Fig. 1-8a can be reduced to a single-loop circuit like Fig. 1-8b. You can have a nightmare of a circuit, but it still can be reduced to a single-loop circuit (see Prob. 1-17 if you like nightmares). This is why experienced engineers and technicians like Thevenin's theorem so much: It takes big, complicated circuits and turns them into simple one-loopers like the equivalent circuit of Fig. 1-8c.

The general idea is this: Whenever you're after the load current in a circuit with more than one loop, *think Thevenin,* or at least consider it as a possible way to go. More often than not, Thevenin's theorem will offer the most efficient way to solve the problem, especially if the load resistance takes on several values.

In this book, to *thevenize* means to apply Thevenin's theorem to a circuit, that is, to reduce a multiloop circuit with a load resistance to an equivalent single-loop circuit with the same load resistance. In the Thevenin equivalent circuit, the load resistor sees a single source resistance in series with a voltage source. How much easier can life be than this?

THEVENIN VOLTAGE

Recall the following ideas about Thevenin's theorem. The *Thevenin voltage* is the voltage that appears across the load terminals when you open the load resistor. Because of this, the Thevenin voltage is sometimes called the open-circuit or open-load voltage.

THEVENIN RESISTANCE

The *Thevenin resistance* is the resistance looking back into the load terminals when all sources have been reduced to zero. This means replacing voltage sources by short circuits and current sources by open circuits.

ANALYZING BUILT-UP CIRCUITS

When a multiloop circuit is already built, you can measure the Thevenin voltage as follows. Physically open the load resistor by disconnecting one end of it, or remove it altogether from the circuit. Then use a voltmeter to measure the voltage across the load terminals. The reading you get (assuming no voltmeter loading error) is the Thevenin voltage.

Next, measure the Thevenin resistance as follows. Reduce all sources to zero. This means physically replacing voltage sources with short circuits and physically opening or removing current sources. Then use an ohmmeter to measure the resistance between the load terminals. This is the Thevenin resistance.

As an example, suppose you have breadboarded the unbalanced Wheatstone bridge shown in Fig. 1-9a. To thevenize the circuit, you physically open the load resistance and measure the voltage between *A* and *B* (the load terminals). Assuming no measurement error, you will read 2 V. Next, you replace the 12-V battery by a short circuit and measure the resistance between *A* and *B*; you should read 4.5 kΩ. Now you can draw the Thevenin equivalent of Fig. 1-9b. With it, you can easily and quickly calculate the load current for any value of load resistance.

Fig. 1-9
(a) *Wheatstone bridge.* (b) *Thevenin equivalent.*

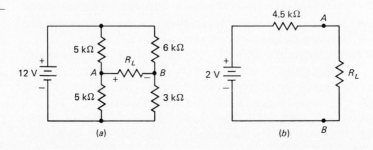

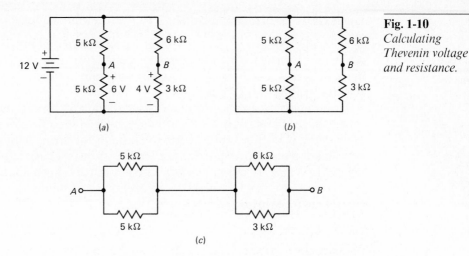

Fig. 1-10
*Calculating
Thevenin voltage
and resistance.*

(a) (b)

(c)

ANALYZING SCHEMATICS

If the circuit is not built, you are going to have to use your head instead of a VOM to find the Thevenin voltage and resistance. Given the unbalanced Wheatstone bridge of Fig. 1-9*a*, you mentally open the load resistor. If you are visualizing correctly, you can then see a voltage divider on the left side and a voltage divider on the right side. The one on the left produces 6 V, and the one on the right produces 4 V, as shown in Fig. 1-10*a*. The Thevenin voltage is the difference of these two voltages, which is 2 V.

Next, mentally replace the 12-V battery with a short circuit to get Fig. 1-10*b*. By redrawing the circuit, you get the two parallel circuits shown in Fig. 1-10*c*. Now it's easy to mentally calculate a Thevenin resistance of 4.5 kΩ.

1-5 NORTON'S THEOREM

The Norton theorem takes only a few minutes to review because it's closely related to the Thevenin theorem. Given a Thevenin circuit like Fig. 1-11*a*, the Norton theorem says that you can replace it by the equivalent circuit of Fig. 1-11*b*. The Norton equivalent has an ideal current source in parallel with a source resistance. Notice that the current source pumps a fixed current V_{TH}/R_{TH}; also notice that the source resistance has the same value as the Thevenin resistance.

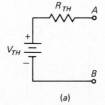
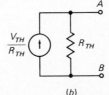

Fig. 1-11
(a) *Thevenin
circuit.* (b) *Norton
circuit.*

(a) (b) (c)

CURRENT DIRECTION

Incidentally, the arrow in the Norton current source points in the direction of conventional current; this is because the inventor was an engineer. This is the first of many devices where the schematic arrow will point in the direction of conventional current.

If you prefer to use electron flow, your training in reverse thinking starts now. When you see the arrow in a current source, it means that electrons go the other way. To ease the pain, you can visualize the head of the arrow pointing at the positive terminal of the source, while the tail represents the negative terminal, as shown in Fig. 1-11c. The plus and minus signs are not usually shown on schematics, so that you will have to mentally supply your own plus and minus signs when you see the current-source symbol.

NORTON RESISTANCE

It's easy to remember the Norton resistance because it has the same value as the Thevenin resistance. For instance, if the Thevenin resistance is 2 kΩ, then the Norton resistance is 2 kΩ; the only difference is that the Norton resistance appears in parallel with the source.

AN EXAMPLE

Here is an example of replacing the Thevenin circuit of Fig. 1-12a by an equivalent Norton circuit. First, short the load terminals as shown in Fig. 1-12b and calculate the load current, which is 5 mA. This shorted-load current equals the Norton current. Second, draw the Norton equivalent circuit of Fig. 1-12c. Notice that the Norton current equals 5 mA and the Norton resistance is 2 kΩ, identical to the Thevenin resistance.

1-6 TROUBLESHOOTING

Troubleshooting is an art and a science; it means finding out why a circuit is not doing what it is supposed to do. Every engineer and technician does a lot of troubleshooting when designing or testing electronic circuits. In this book we will mentally troubleshoot circuits by posing questions like, "What happens to the load voltage if such-and-such is shorted?" or, "What happens to the load current if such-and-such is open?"

Fig. 1-12
Deriving Norton circuit from Thevenin circuit.

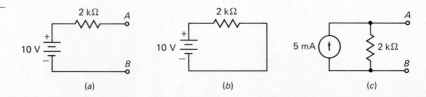

(a)   (b)   (c)

SHORTS AND OPENS

Devices like resistors and transistors can *short* or *open* in a number of ways. One way to destroy any device is by exceeding its maximum-power rating. On data sheets for various devices, manufacturers carefully list the *absolute maximum ratings* for their devices. Do not take these ratings lightly; if you exceed the maximum-power rating, for example, you will almost always have to buy a new device. Sometimes, exceeding the maximum-power rating burns out the interior of the device, leaving an empty space called an open circuit. At other times, excessive power dissipation melts the interior of the device, producing an expensive short circuit.

Another interesting way to create unwanted short and open circuits is during the stuffing and soldering of printed-circuit boards. An undesirable splash of solder may connect two nearby conducting lines; known as a *solder bridge*, this effectively shorts any device between the two conducting lines. On the other hand, a poor solder connection usually means no connection at all; known as a *cold-solder joint*, this implies that the device appears to be open.

SUBTLE TROUBLE

A shorted or open device is the most common trouble you can run into. But sometimes the changes are more subtle. For instance, temporarily applying too much heat to a resistor may permanently change the resistance by several percent. If the value of resistance is critical, the circuit may not work properly after the heat shock. In this case, it may be a bit more difficult to isolate the faulty resistor in a built-up circuit.

EXAMPLE

Let's start our mental troubleshooting practice now. Figure 1-13 shows a voltage divider. If there are no troubles, the voltage between point A and ground is 3 V. Now let's work through some possible troubles.

If R_1 is shorted, what is the voltage between point A and ground? In this case, all the source voltage appears between point A and ground; therefore, the answer is 9 V.

How much voltage is there between point A and ground if R_1 is open? In this case, there is no current through R_2. Ohm's law tells us there is no voltage across R_2. As a result, the answer is 0 V.

Suppose R_2 is shorted. What happens to the voltage from point A to ground? There will be current through R_2, but the zero resistance means there is zero voltage.

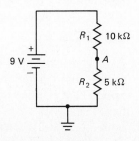

Fig. 1-13
Troubleshooting.

In other words, a shorted R_2 forces point A to be at ground potential. Therefore, the answer is 0 V.

Finally, what is the voltage between point A and ground when R_2 is open? There are two ways to see this. First, the logical way: Since there is no current in R_1, there is no voltage drop across R_1; therefore, all the source voltage must appear between point A and ground, and so the answer is 9 V.

If the foregoing logic escapes you, then try the second method based on Kirchhoff's voltage law. Write a loop equation to get

$$IR_1 + V_2 - 9 = 0$$

where I is the loop current. Rearrange the equation to get

$$V_2 = 9 - IR_1$$

Since R_2 is open, there is no loop current in Fig. 1-13, and the equation simplifies to

$$V_2 = 9$$

Therefore, the answer is 9 V, the same as found by logic.

1-7 STUDY AIDS

At the end of each chapter you will find a variety of problems to help reinforce and expand your understanding of the topics discussed in this book. Here is a brief description of the different types of problems and how to use them.

STRAIGHTFORWARD

This is a preliminary set of homework problems designed to help you interact with the material in a quantitative way. You will be asked to solve for voltages, currents, and other quantities. Often, these problems will be easy, done in minutes or even seconds. Sometimes they will take a little longer, but in all cases, you will be analyzing topics discussed in the text.

TROUBLESHOOTING

Here you will be asked to figure out what voltages, currents, and other quantities would be if troubles occurred in a circuit. Most of the time, the troubles are shorted or open devices. Sometimes the problems start by giving you the trouble and asking you to list the possible causes of such a trouble. You may find these problems entertaining, even fun.

DESIGN

Here you will find problems that ask you to design a circuit that meets certain specifications. This level of work combines the creative and critical functions of your mind. That is, you have to dream up some circuit that might do the job, then you have to analyze the circuit to see if it does the job. This is where electronics becomes both an art and a science. In other words, there may be many right ways to do the same job. If you can find one of them, you've got a design. If you can find two or more, you can compare them for simplicity, cost, elegance, etc., and decide which is the better all-around design.

Incidentally, after you calculate values of resistance and capacitance, check the standard values shown in Appendixes 2 and 3; these are commercially available sizes. Select the nearest standard value or the next higher standard value, depending on the kind of circuit you are designing.

CHALLENGING

If you have a strong grasp of the material, you may want to take a look at these problems. Here you will find a couple of difficult problems, requiring a fairly high level of competence. You should be able to solve most of these problems, although they will take more time and may stretch you to your limits.

COMPUTER

Sooner or later you will encounter a computer. It can show up anywhere: on an engineer's desk, in a friend's home, on a production line, etc. At the rate prices are dropping, you may even buy your own computer someday (if you don't have one already).

A computer is a tool, similar to a pocket calculator. The main difference is that you can type instructions into a computer telling it how to solve a problem. The list of instructions you type in is called the *program.* By changing the program, you change the operation of the computer.

BASIC is the name of the most widely used language for telling a computer what to do. At the end of each chapter you will find a couple of self-explanatory problems that will help you learn BASIC and how to program a computer. The problems are kept simple and few in number so that you don't spend too much time on them. The intention is to expose you to the rudiments of BASIC, not to make you an expert in BASIC programming.

Consider the computer problems optional. If you already know BASIC, you can skip some of the earlier problems. Or if you're just not interested in learning BASIC at this time, ignore this final set of problems.

PROBLEMS

Straightforward

1-1. The internal resistance of a voltage source equals 0.05 Ω. How much voltage is dropped across this internal resistance when the current through it equals 2 A?

1-2. A voltage source is temporarily shorted. If the ideal source voltage is 6 V and the shorted-load current is 150 A, what is the internal resistance of the source?

1-3. In Fig. 1-14, the ideal source voltage is 9 V and the internal resistance is 0.4 Ω. If the load resistance is zero, what is the load current?

1-4. The ideal source voltage is 12 V in Fig. 1-14. If the internal resistance is 0.5 Ω, what is the load current when the load resistance equals 50 Ω? The load voltage?

1-5. Repeat Prob. 1-4 for R_S of 5 Ω.

Fig. 1-14

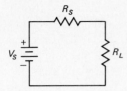

1-6. In Fig. 1-14, the ideal source voltage is 10 V and the load resistance is 75 Ω. If the load voltage equals 9 V, what does the internal resistance equal? Is the voltage source stiff?

1-7. A voltage source has an internal resistance of 2 Ω. For the source to be stiff, what is the minimum allowable load resistance?

1-8. The ideal source voltage is 15 V and the internal resistance is 0.03 Ω in Fig. 1-14. For what range of load resistance is the voltage source stiff?

1-9. A load resistance can be adjusted from 20 Ω to 200 kΩ. If a voltage source acts stiff for the entire range of load resistance, what can you say about the internal resistance?

1-10. In Fig. 1-15, the ideal source current is 10 mA and the internal resistance is 100 kΩ. If the load resistance equals zero, what does the load current equal?

Fig. 1-15

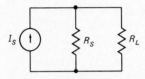

1-11. The ideal source current is 5 mA and the internal resistance is 250 kΩ in Fig. 1-15. If the load resistance is 10 kΩ, what is the load current? Is this a stiff current source?

1-12. A current source has an internal resistance of 150 kΩ. For what range of load resistance is the current source stiff?

1-13. Somebody hands you a black box with a 2-kΩ resistor connected across its load terminals. How can you measure its Thevenin voltage?

1-14. The black box in Prob. 1-13 has a knob on it that allows you to reduce all internal voltage and current sources to zero. How would you measure the Thevenin resistance?

1-15. Calculate the load current in Fig. 1-16 for each of these load resistances: 0, 1 kΩ, 2 kΩ, 3 kΩ, 4 kΩ, 5 kΩ, and 6 kΩ. (Use Thevenin's theorem.)

Fig. 1-16

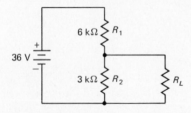

1-16. Solve Prob. 1-15. Then solve the same problem without using Thevenin's theorem; that is, use ordinary methods involving Ohm's and Kirchhoff's laws, etc. After you are finished, comment on what you have learned about the Thevenin theorem.

1-17. You are in the laboratory looking at a circuit like Fig. 1-17. Somebody challenges you to find the Thevenin circuit driving the load resistor. Describe an experimental procedure for measuring the Thevenin voltage and the Thevenin resistance.

1-18. A Thevenin circuit has a Thevenin voltage of 10 V and a Thevenin resistance of 5 kΩ. What is the Norton current? The Norton resistance?

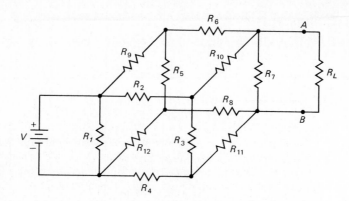

Fig. 1-17

1-19. Draw the Norton equivalent for Fig. 1-16. If the load resistance is 16 kΩ, how much current is there through the Norton resistance?

Troubleshooting

1-20. Suppose the load voltage of Fig. 1-16 is 36 V. What is wrong with R_1?

1-21. The load voltage of Fig. 1-16 is zero. The battery and the load resistance are all right. Suggest two possible troubles.

1-22. If the load voltage is zero in Fig. 1-16 and all resistors are normal, where does the trouble lie?

1-23. The load voltage is 12 V in Fig. 1-16. What is the probable trouble?

Design

1-24. Design a hypothetical current source using the battery-resistor combination described in Sec. 1-3. The current source must meet the following specifications: It must supply a stiff 1 mA of current to any load resistance between 0 and 10 kΩ.

1-25. Design a voltage divider (similar to Fig. 1-16) that meets these specifications: Ideal source voltage is 30 V, open-load voltage is 15 V, and Thevenin resistance is equal to or less than 2 kΩ.

1-26. Design a voltage divider like Fig. 1-16 so that it produces a stiff 10 V to all load resistances greater than 1 MΩ. Use an ideal source voltage of 30 V.

Challenging

1-27. Somebody hands you a D-cell flashlight battery and a volt-ohm-milliammeter (VOM). You have nothing else to work with. Describe an experimental method for finding the Thevenin equivalent circuit of the flashlight battery.

1-28. You have a D-cell flashlight battery, a VOM, and a box of various resistors. Describe a method that uses one of the resistors to find the Thevenin resistance of the battery.

1-29. Calculate the load current in Fig. 1-18 for each of these load resistances: 0, 1 kΩ, 2 kΩ, 3 kΩ, 4 kΩ, 5 kΩ, and 6 kΩ.

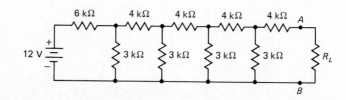

Fig. 1-18

Computer

1-30. The PRINT instruction is the first BASIC command we discuss. If a computer executes a statement like

PRINT "TODAY IS SUNNY."

then the computer screen will display

TODAY IS SUNNY.

In other words, the computer prints the words following the PRINT command; these words must be inside quotation marks.

What does the computer screen show when the computer executes this statement:

PRINT "PROGRAMMING IS EASY."

1-31. When you program a computer, you have to assign a line number to each command to keep track of the different commands. These line numbers are usually spaced 10 apart. For instance, here is a program with line numbers:

10 PRINT "TODAY IS SUNNY."
20 PRINT "PROGRAMMING IS EASY."
30 PRINT "MAYBE I CAN LEARN THIS STUFF."

If this program is run on a computer, it will execute three PRINT commands. What do you think will appear on the computer screen?

1-32. Write a program with three PRINT statements using line numbers 10, 20, and 30 that prints your name on the first line, your address on the second line, and the city and state and zip code on the third line.

1-33. The END statement tells a computer to stop processing data. For instance,

10 PRINT "IRELAND IS GREEN"
20 PRINT "CALIFORNIA IS GOLDEN"
30 END

In this program, the END statement stops the execution. Theoretically, an END statement should be the last instruction in your program because all computers recognize the END statement. But most microcomputers do not require the END statement; they automatically stop execution when they run out of instructions. In this book, the END statement is optional.

Write a program that prints out the name and address of your school (or place of work).

Diode Theory

The chapter begins by discussing semiconductor theory. The treatment is simple, sometimes idealized; otherwise, we would be caught up in highly advanced mathematics and physics. The middle part of the chapter is about *pn* junctions, the key idea behind a semiconductor diode. The word "diode" is a compound meaning "two electrode," where "di" stands for two and "ode" for electrode. This discussion tells how a diode works and prepares you for the transistor, which combines two diodes in a single device. The chapter ends with approximations for the diode. These approximations simplify the troubleshooting, analysis, and design of diode circuits.

2-1 SEMICONDUCTOR THEORY

You already know something about atoms, electrons, and protons. This section will add to your knowledge. You will see how silicon atoms combine to form crystals (the backbone of semiconductor devices).

ATOMIC STRUCTURE

Bohr idealized the atom. He saw it as a nucleus surrounded by orbiting electrons (Fig. 2-1*a*). The nucleus has a positive charge and attracts the electrons. The electrons would fall into the nucleus without the centrifugal force of their motion. When an electron travels in a stable orbit, it has just the right velocity for centrifugal force to balance nuclear attraction.

Three-dimensional drawings like Fig. 2-1*a* are difficult to show for complicated atoms. Because of this, we will symbolize the atom in two dimensions. For example, an isolated silicon atom (Fig. 2-1*b*) has 14 protons in its nucleus. Two electrons travel in the first orbit, eight electrons in the second, and four in the outer or *valence* orbit. The 14 revolving electrons neutralize the charge of the 14 protons, so that from a distance the atom acts electrically neutral.

Especially important, the outer orbit contains four electrons. Because of this, silicon is called *tetravalent* ("tetra" is Greek for "four"). Incidentally, the nucleus and inner electrons are called the *core* of an atom. The core is a bore; nothing of

Fig. 2-1
(a) *Bohr model.*
(b) *Silicon atom.*

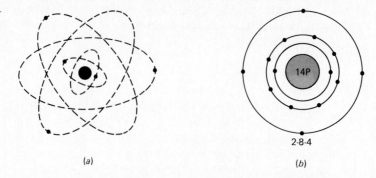

(a)   (b)

interest happens inside the core. Our focus is almost always on the valence orbit; this is where all the action is in semiconductors.

ENERGY LEVELS

You might think that an electron could travel in an orbit of any radius, provided its velocity had the right value. Modern physics says otherwise; only certain orbit sizes are permitted. For instance, in Fig. 2-2a, electrons can travel in the first, second, and third orbits, but they cannot travel in intermediate orbits. That is, all radii between r_1 and r_2 are forbidden; similarly, all radii between r_2 and r_3 are forbidden. (If you want to know why, you will have to study quantum physics.)

In Fig. 2-2a it takes energy to move an electron from a smaller to a larger orbit, because work has to be done to overcome the attraction of the nucleus. Therefore, the larger the orbit of an electron, the higher its energy level or potential energy with respect to the nucleus.

For convenience in drawing, everybody visualizes the curved orbits like the horizontal lines of Fig. 2-2b. The first orbit represents the first energy level, the second orbit represents the second energy level, and so on. The higher the energy level, the larger the orbit of the electron. In other words, an energy level is just another way of saying orbital radius.

If external energy like heat, light, or other radiation bombards an atom, it can add energy to an electron and lift it to a higher energy level (larger orbit). The atom is then said to be in a state of *excitation*. This state does not last long, because the energized

Fig. 2-2
(a) *Magnified view of atom.*
(b) *Energy levels.*

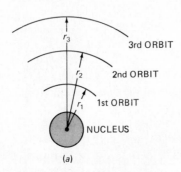

(a)

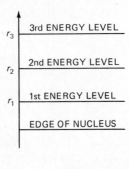

(b)

electron soon falls back to its original energy level. As it falls, it gives back the acquired energy in the form of heat, light, or other radiation.

CRYSTALS

An isolated silicon atom has four electrons in its valence orbit. To be chemically stable, a silicon atom needs eight electrons in its valence orbit. Because of this, a silicon atom will combine with other atoms in such a way as to produce eight electrons in its valence orbit.

When silicon atoms combine to form a solid, they arrange themselves in an orderly pattern called a *crystal*. The forces that hold the atoms together are known as *covalent bonds*. Figure 2-3*a* illustrates the idea. The silicon atom positions itself between four other silicon atoms. Each neighbor then shares an electron with the central atom. In this way, the central atom picks up four electrons, giving it a total of eight in its valence orbit.

The eight electrons do not belong exclusively to the central atom; they are shared by the four surrounding atoms. Since the adjacent cores have a net positive charge, they attract the shared electrons, creating equal and opposite forces. This pulling in opposite directions is the covalent bond, the glue that holds atoms together. (The situation is analogous to tug-of-war teams pulling on a rope. As long as the teams pull with equal and opposite forces, they remain immobile and bonded together.)

HOLES

When outside energy lifts a valence electron to a higher energy level (larger orbit), the departing electron leaves a vacancy in the outer orbit (see Fig. 2-3*b*). We call this vacancy a *hole*. These holes are one of the reasons diodes and transistors work the way they do. Later sections will tell you more about holes.

ENERGY BANDS

When a silicon atom is isolated, the orbit of an electron is controlled by the charges of the isolated atom. But when silicon atoms combine into a crystal, the orbit of an electron is influenced by the charges of many adjacent atoms. Since each electron has a different position inside the crystal, no two electrons see exactly the same pattern of surrounding charges. Because of this, the orbit of each electron is different.

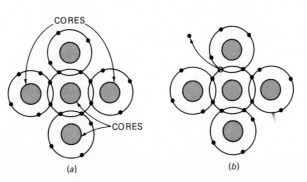

(a)   (b)

Fig. 2-3
(a) *Covalent bonds.* (b) *Hole.*

Fig. 2-4
Energy bands.

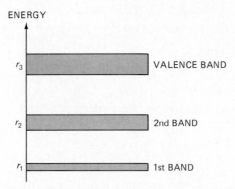

Figure 2-4 shows what happens to energy levels. All electrons traveling in first orbits have slightly different energy levels because no two see exactly the same charge environment. Since there are billions of first-orbit electrons, the slightly different energy levels form a cluster or *band.* Similarly, the billions of second-orbit electrons, all with slightly different energy levels, form the second energy band shown. And all third-orbit electrons form the third band.

In Fig. 2-4 the energy bands are dark. This will be our way of indicating filled or saturated bands, that is, those in which all available orbits are already occupied by electrons. For instance, the valence band is full because the valence orbit of each atom has eight electrons in it.

Figure 2-4 shows the energy bands in a silicon crystal at absolute zero temperature (−273°C); this is where molecular vibration ceases, meaning that it is as cold as it can get. There is absolutely no current at absolute zero.

2-2 CONDUCTION IN CRYSTALS

Each copper atom has one free electron. Since this electron travels in an extremely large orbit (high energy level), the electron can barely feel the attraction of the nucleus. In a piece of copper wire the free electrons are contained in an energy band called the *conduction band.* These free electrons can produce large currents.

A piece of silicon is different. Figure 2-5a shows a bar of silicon with metal end surfaces. An external voltage sets up an electric field between the ends of the crystal. Does current flow? It depends. On what? On whether there are any movable electrons inside the crystal.

ABSOLUTE ZERO

At absolute zero temperatures, electrons cannot move through the crystal. All valence electrons are tightly held by the silicon atoms because they are part of the covalent bonds between atoms. Figure 2-5b shows the energy-band diagram. When the first three bands are filled, electrons in these bands cannot move easily because there are no empty orbits. But beyond the valence band is the conduction band. If a valence electron could be lifted into the conduction band, it would be free to move from one atom to the next. At absolute zero temperature, however, the conduction band is empty; this means that no current can exist in the silicon crystal.

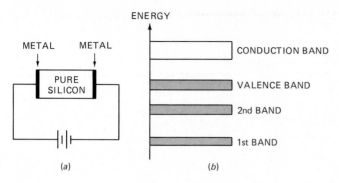

Fig. 2-5   (a) *Circuit.* (b) *Energy bands at absolute zero temperature.*

ABOVE ABSOLUTE ZERO

Raise the temperature above absolute zero and things change. The incoming heat energy breaks some covalent bonds; that is, it knocks valence electrons into the conduction band. In this way, we can get a limited number of conduction-band electrons, symbolized by the minus signs of Fig. 2-6*a*. Under the influence of the electric field, these free electrons move to the left and set up a current.

Above absolute zero, we visualize the energy bands as shown in Fig. 2-6*b*. Heat energy has lifted some electrons into the conduction band, where they travel in very large orbits. In these conduction-band orbits, the electrons are only loosely held by the atoms and can easily move from one atom to the next.

In Fig. 2-6*b,* each time an electron is bumped up to the conduction band, a hole is created in the valence band. Therefore, the valence band is no longer saturated or filled; each hole in the valence band represents an available orbit of rotation.

The higher the temperature, the greater the number of valence electrons kicked up to the conduction band and the larger the current in Fig. 2-6*a*. At room temperature (around 25°C), the current is too small to be useful. At this temperature, a piece of silicon is neither a good insulator nor a good conductor. For this reason, it is called a *semiconductor.*

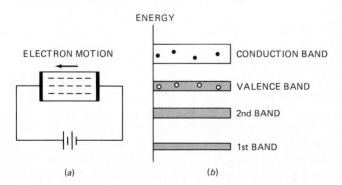

Fig. 2-6   (a) *Electron flow.* (b) *Energy bands at room temperature.*

SILICON VERSUS GERMANIUM

Germanium, another tetravalent element, was widely used in the early days of semiconductors. But nowadays it's rarely used in new designs. At room temperature a silicon crystal has almost no free electrons compared with a germanium crystal. This is the key reason why silicon has totally overshadowed germanium in the fabrication of diodes, transistors, and other semiconductor devices. Later sections will explain why this lack of free electrons in pure silicon is a tremendous advantage.

HOLE CURRENT

The holes in semiconductors also produce a current. This is what makes semiconductors distinctly different from copper wire. In other words, a semiconductor offers two paths for current, one through the conduction band (large orbits) and another through the valence band (smaller orbits). As an analogy, copper wire is like a one-lane road and a semiconductor is like a two-lane road.

The conduction-band current in a semiconductor is similar to that in copper wire. But the valence-band current is brand new. Look at the hole at the extreme right of Fig. 2-7a. This hole attracts the valence electron at *A*. With only a slight change in energy, the valence electron at *A* can move into the hole. When this happens, the original hole vanishes and a new one appears at position *A*. The new hole at *A* can attract and capture the valence electron at *B*. When the valence electron moves from *B* to *A*, the hole moves from *A* to *B*. The valence electrons can continue to move along the path shown by the arrows; the holes move in the opposite direction. What it boils down to is this: Because there are holes in valence orbits, there is a second path along which electrons can move through a crystal. (Not shown in this drawing are the much larger conduction-band orbits in which the free electrons are traveling.)

Here is what happens in terms of energy levels. To begin with, thermal energy (the same as heat energy) bumps an electron from the valence band into the conduction band. This leaves a hole in the valence band, as shown in Fig. 2-7b. With a slight energy change, the valence electron at *A* can move into the hole. When this happens, the original hole disappears and a new one appears at *A*. Next, the valence electron at *B* can move into the new hole with a slight change in energy. In this way, with minor changes in energy, valence electrons can move along the path shown by the arrows.

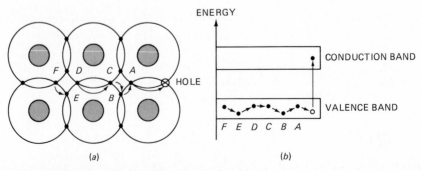

(a)   (b)

Fig. 2-7   (a) *Hole current.* (b) *Energy diagram of hole current.*

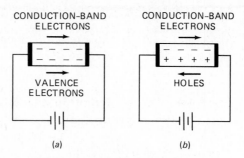

Fig. 2-8
Two paths for current.

This is equivalent to the hole moving through the valence band along the *ABCDEF* path.

ELECTRON-HOLE PAIRS

When we apply an external voltage across a crystal, it forces the electrons to move. In Fig. 2-8*a*, two kinds of movable electrons exist, conduction-band electrons and valence electrons. The movement to the right of valence electrons means that holes are moving to the left.

Most engineers talk about the motion of holes rather than the motion of valence electrons. In a pure semiconductor, the existence of each conduction-band electron means that there is a hole in the valence orbit of some atom. In other words, thermal energy produces electron-hole pairs. The holes act like positive charges and for this reason are shown as plus signs in Fig. 2-8*b*. (A phenomenon known as the Hall effect confirms that holes act like positive charges.) As before, we visualize conduction-band electrons moving to the right. But now, we think of holes (positive charges) moving to the left.

RECOMBINATION

In Fig. 2-8*b*, each minus sign is a conduction-band electron in a large orbit, and each plus sign is a hole in a smaller orbit. Occasionally, the conduction-band orbit of one atom may intersect the hole orbit of another. Because of this, every so often a conduction-band electron can fall into a hole. This merging of a free electron and a hole is called *recombination*. When recombination takes place, the hole does not move elsewhere, it disappears.

Recombination occurs continuously in a semiconductor. This would eventually fill every hole except for one thing: Incoming heat energy keeps producing new holes by lifting valence electrons up to the conduction band. *Lifetime* is the name given to the average time between the creation and the disappearance of an electron-hole pair. Lifetime varies from a few nanoseconds to several microseconds, depending on how perfect the crystal structure is, and on other factors.

2-3 DOPING

A pure silicon crystal (one in which every atom is a silicon atom) is known as an *intrinsic* semiconductor. For most applications, there are not enough free electrons and holes in an intrinsic semiconductor to produce a usable current. *Doping* means adding impurity atoms to a crystal to increase either the number of free electrons or the number of holes. When a crystal has been doped, it is called an *extrinsic* semiconductor.

n-TYPE SEMICONDUCTOR

To get extra conduction-band electrons, we can add pentavalent atoms; these have five electrons in the valence orbit. After we add pentavalent atoms to a pure silicon crystal, we still have mostly silicon atoms. But every now and then we find a pentavalent atom between four neighbors, as shown in Fig. 2-9*a*. The pentavalent atom originally had five electrons in its valence orbit. After forming covalent bonds with four neighbors, this central atom has an extra electron left over. Since the valence orbit can hold no more than eight electrons, the extra electron must travel in a conduction-band orbit.

Figure 2-9*b* shows the energy bands of a crystal that has been doped by a pentavalent impurity. We have a large number of conduction-band electrons, produced mostly by doping. Only a few holes exist, created by thermal energy. We call the electrons the *majority* carriers and the holes the *minority* carriers. Doped silicon of this kind is known as an *n-type* semiconductor, where *n* stands for negative.

A final point: Pentavalent atoms are often called *donor* atoms because they produce conduction-band electrons. Examples of donor impurities are arsenic, antimony, and phosphorus.

p-TYPE SEMICONDUCTOR

How can we dope a crystal to get extra holes? By using a trivalent impurity (one with three electrons in the outer orbit). After we add the impurity, we find each trivalent atom between four neighbors, as shown in Fig. 2-10*a*. Since each trivalent atom brought only three valence-orbit electrons with it, only seven electrons will travel in its valence orbit. In other words, one hole appears in each trivalent atom. By controlling the amount of impurity added, we can control the number of holes in the doped crystal.

Fig. 2-9
Doping with donor impurity.

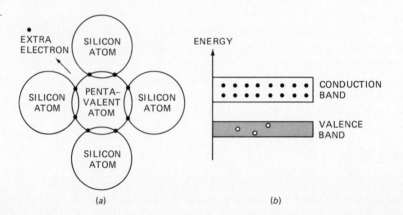

(a)   (b)

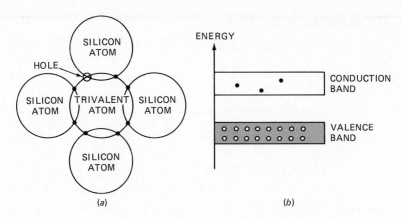

Fig. 2-10
*Doping with
acceptor impurity.*

A semiconductor doped by a trivalent impurity is known as a *p-type* semiconductor; the *p* stands for positive. As shown in Fig. 2-10*b*, the holes of a *p*-type semiconductor far outnumber the conduction-band electrons. For this reason, the holes are the majority carriers in a *p*-type semiconductor, while the conduction-band electrons are the minority carriers.

Trivalent atoms are also known as *acceptor* atoms because each hole they contribute may accept an electron during recombination. Examples of acceptor impurities are aluminum, boron, and gallium.

BULK RESISTANCE

A doped semiconductor still has resistance. We call this resistance the *bulk* resistance. A lightly doped semiconductor has a high bulk resistance. As the doping increases, the bulk resistance decreases. Bulk resistance is also called *ohmic* resistance because it obeys Ohm's law; that is, the voltage across it is proportional to the current through it.

2-4 THE UNBIASED DIODE

It is possible to produce a crystal like Fig. 2-11*a* that is half *p*-type and half *n*-type. The junction is where the *p*-type and *n*-type regions meet. A *pn* crystal like this is commonly called a *diode*.

Figure 2-11*a* shows the *pn* crystal at the instant of formation. The *p* side has many holes (majority carriers), and the *n* side has many free electrons (also majority carriers). The diode of Fig. 2-11*a* is unbiased, which means that no external voltage is applied to it.

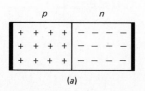

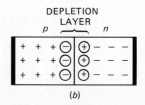

Fig. 2-11
(a) *Before
diffusion.* (b) *After
diffusion.*

DEPLETION LAYER

Because of their mutual repulsion, the free electrons on the *n* side diffuse or spread in all directions. Some diffuse across the junction. When a free electron leaves the *n* region, its departure creates a positively charged atom (a positive ion) in the *n* region. Furthermore, as it enters the *p* region, the free electron becomes a minority carrier. With so many holes around it, this minority carrier has a short lifetime; soon after entering the *p* region, the free electron will fall into a hole. When this happens, the hole disappears and the associated atom becomes negatively charged (a negative ion).

Each time an electron diffuses across the junction, it creates a pair of ions. Figure 2-11*b* shows these ions on each side of the junction. The circled plus signs are the positive ions, and the circled minus signs are the negative ions. The ions are fixed in the crystal structure because of covalent bonding and cannot move around like the free electrons and holes. As the number of ions builds up, the region near the junction is depleted of free electrons and holes. We call this region the *depletion layer*.

BARRIER POTENTIAL

Beyond a certain point, the depletion layer acts like a barrier to further diffusion of free electrons across the junction. For instance, visualize a free electron in the *n* region diffusing left into the interior of the depletion layer (see Fig. 2-11*b*). Here it encounters a negative wall of ions repelling it back to the right. If the free electron has enough energy, it can break through the wall and enter the *p* region, where it falls into a hole and creates another negative ion.

The strength of the depletion layer keeps increasing with each crossing electron until an equilibrium is reached; at this point the internal repulsion of the depletion layer stops further diffusion of free electrons across the junction.

The difference of potential across the depletion layer is called the *barrier potential*. At 25°C, this barrier potential approximately equals 0.7 V for silicon diodes. (The germanium diode has a barrier potential of 0.3 V.)

2-5 FORWARD BIAS

Figure 2-12*a* shows a dc source across a diode. The positive source terminal is connected to the *p*-type material, and the negative terminal to the *n*-type material. We call this connection *forward bias*. As a memory aid, notice that the + connects to the *p* side and the − to the *n* side.

LARGE FORWARD CURRENT

Forward bias can produce a large forward current. Why? The negative terminal of the source repels the free electrons in the *n* region toward the junction. These extra-energy electrons can then cross the junction and fall into holes. The recombinations occur at varying distances from the junction, depending on how long a free

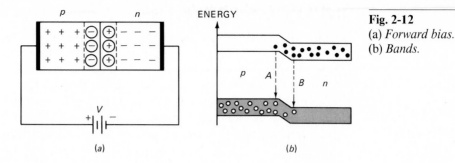

Fig. 2-12
(a) *Forward bias.*
(b) *Bands.*

electron can avoid falling into a hole. The odds are high that recombinations occur close to the junction.

As the free electrons fall into the holes, they become valence electrons. Then, traveling as valence electrons, they continue to move to the left through the holes in the *p* material. When valence electrons reach the left end of the crystal, they leave the crystal and flow into the positive terminal of the source.

Here is the life story of a single electron in Fig. 2-12*a* as it moves from the negative terminal of the battery to the positive terminal:

1. After leaving the negative terminal, it enters the right end of the crystal.
2. It travels through the *n* region as a free electron.
3. Near the junction it recombines and becomes a valence electron.
4. It travels through the *p* region as a valence electron.
5. After leaving the left end of the crystal, it flows into the positive source terminal.

ENERGY BANDS

Figure 2-12*b* shows how to visualize the flow in terms of energy bands. To begin with, the barrier potential gives the *p* bands slightly more energy than the *n* bands; this is why the *p* bands are higher than the *n* bands. The forward bias pushes the conduction-band electrons in the *n* region toward the junction. Soon after entering the *p* region, each electron falls into a hole (path *A*). As a valence electron, it can continue its journey toward the left end of the crystal.

Sometimes a conduction-band electron can fall into a hole before it crosses the junction. In Fig. 2-12*b* a valence electron may cross the junction from right to left; this leaves a hole just to the right of the junction. This hole does not live long. A conduction-band electron soon falls into it (path *B*).

No matter where the recombination takes place, the result is the same. A steady stream of conduction-band electrons moves toward the junction and falls into the holes near the junction. The captured electrons (now valence electrons) move left in a steady stream through the holes in the *p* region. In this way, we get a continuous flow of electrons through the diode.

Incidentally, as the free electrons fall along paths *A* and *B,* they fall from a higher energy level to a lower one. As they fall, they radiate energy in the form of heat and light. Remember this. It will come in handy later when we discuss light-emitting diodes.

2-6 REVERSE BIAS

If you reverse the polarity of the dc source, you *reverse-bias* the diode, as shown in Fig. 2-13a. Now the + connects to the *n* side and the − to the *p* side. What effects does reverse bias have?

DEPLETION LAYER

The reverse bias of Fig. 2-13a forces free electrons in the *n* region away from the junction toward the positive source terminal; also, holes in the *p* region move away from the junction toward the negative terminal. The departing electrons leave more positive ions near the junction, and the departing holes leave more negative ions. Therefore, the depletion layer gets wider. The greater the reverse bias, the wider the depletion layer becomes. The depletion layer stops growing when its difference of potential equals the source voltage.

Figure 2-13b is an alternative way of seeing the same idea. When reverse bias is first applied, conduction-band electrons and holes move away from the junction. The depletion layer gets wider until its difference of potential equals the source voltage. When this happens, free electrons and holes stop moving.

MINORITY-CARRIER CURRENT

Is there any current at all after the depletion layer adjusts to its new width? Yes. A very small current does exist. Here's why. Thermal energy continuously creates a limited number of free electrons and holes on both sides of the junction. Because of the minority carriers, a small current exists in the circuit.

The reverse current caused by the minority carriers is called the *saturation current*, designated I_S. The name saturation reminds us that we cannot get more of this reverse current than is produced by thermal energy. In other words, increasing the reverse voltage will not increase the number of thermally created minority carriers. Only an increase in temperature can increase I_S. Based on experiments, a practical rule for all silicon diodes is this: I_S approximately doubles for each 10°C rise in temperature. For example, if I_S is 5 nA (nanoamperes) at 25°C, it is approximately 10 nA at 35°C, 20 nA at 45°C, 40 nA at 55°C, and so on.

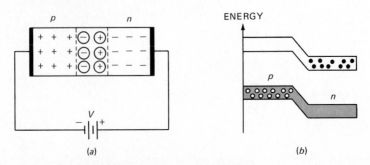

Fig. 2-13 (a) *Replace by reverse bias.* (b) *Replace by energy bands.*

A silicon diode has a much smaller I_S than a germanium diode. This is one of the reasons why silicon dominates the field of semiconductor devices.

SURFACE-LEAKAGE CURRENT

Besides the reverse current through the crystal, there is a small current on the surface of the crystal. This other component of reverse current is called *surface-leakage current,* which we will symbolize by I_{SL}. It is caused by surface impurities that set up ohmic paths for current. Like thermally produced current, surface-leakage current is extremely small.

REVERSE CURRENT

Data sheets for diodes lump I_S and I_{SL} into a single current called the reverse current I_R; it is usually specified at a particular value of reverse voltage V_R and ambient temperature T_A. Since I_S is temperature-sensitive and I_{SL} is voltage sensitive, I_R is both temperature- and voltage-sensitive. For instance, the I_R of a 1N914 (a widely used diode) is 25 nA for a reverse voltage V_R of 20 V and an ambient temperature T_A of 25°C. If either the voltage or the temperature increases, the reverse current increases. As a rule, a designer selects a diode whose reverse current is small enough to ignore in the particular application.

BREAKDOWN VOLTAGE

If you increase the reverse voltage, you literally will reach a breaking point, called the *breakdown* voltage of the diode. For rectifier diodes (those optimized to conduct better one way than the other), the breakdown voltage is usually greater than 50 V. Once the breakdown voltage is reached, the diode can conduct heavily. Where do the carriers suddenly come from? Figure 2-14a shows a thermally produced free electron and hole inside the depletion layer. Because of the reverse bias, the free electron is pushed to the right. As it moves, it gains speed. The greater the reverse bias, the faster the electron moves (which is equivalent to saying that it gains more energy). After a while, the free electron may collide with a valence electron, as shown in Fig. 2-14b. If the free electron has enough energy, it can dislodge the valence electron, so that there

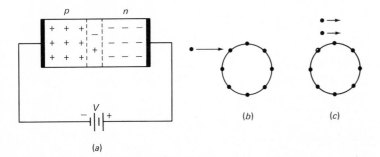

Fig. 2-14 *Breakdown.* (a) *Minority carriers in depletion layer.* (b) *Free electron hits valence electron.* (c) *Two free electrons.*

are two free electrons (see Fig. 12-14c). Both of these can now accelerate and dislodge other valence electrons until an all-out avalanche occurs. Because of the huge number of free electrons, the diode of Fig. 2-14a will conduct heavily and be destroyed by excessive power dissipation.

Most diodes are not allowed to break down. In other words, by deliberate design the reverse voltage across a rectifier diode is always kept less than its breakdown voltage. There is no standard symbol for breakdown voltage. It has been symbolized on various data sheets as follows:

$V_{(BR)}$: voltage breakdown

BV: breakdown voltage

PRV: peak reverse voltage

PIV: peak inverse voltage

V_{RWM}: voltage reverse working maximum

V_{RM}: voltage reverse maximum

and others. Some of these are dc ratings and some are ac ratings. You will have to consult individual data sheets for the conditions attached to these breakdown ratings.

2-7 LINEAR DEVICES

Ohm's law tells us that the current through an ordinary resistor is proportional to the voltage across the resistor. This means that a graph of resistor current versus resistor voltage is linear. For instance, given a resistor of 500 Ω, its *I-V* graph looks like Fig. 2-15. Notice the sample points. The current is 1 mA for a voltage of 0.5 V and 2 mA for 1 V. In either case, the ratio of voltage to current equals 500 Ω. Reversing the voltage has no effect on the linearity of the graph. A reverse current of −1 mA exists for a reverse voltage of −0.5 V; the current increases to −2 mA for −1 V.

An ordinary resistor is often called a *linear* device because its *I-V* characteristic is linear. A resistor is only one example of a linear component. You will see others in later chapters. An ordinary resistor is also referred to as a *passive* device because all it

Fig. 2-15
Current versus voltage in a linear resistance.

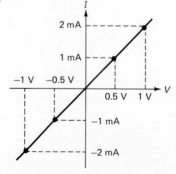

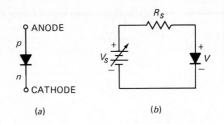

Fig. 2-16
(a) *Diode symbol.*
(b) *Diode circuit.*

does is dissipate power; it cannot generate power. A battery, on the other hand, is an *active* device because it can generate power for a passive device like a resistor. Chapter 5 introduces the transistor, another active device.

2-8 THE DIODE GRAPH

Figure 2-16a shows the schematic symbol of a rectifier diode. The *p* side is called the *anode,* and the *n* side the *cathode.* As previously discussed, forward bias can produce a large electron flow from the *n* side to the *p* side; this is equivalent to having a large conventional flow from the *p* side to the *n* side. As a reminder, the arrow on the diode symbol points in the easy direction for conventional current.

If you prefer electron flow, you will have to use reverse thinking; the easy direction for electron flow is against the arrow. It may help to remember that the arrow points in the direction from which the free electrons come.

EXPERIMENTAL DATA

Figure 2-16b shows a circuit you can set up in the laboratory to measure the current and voltage of a diode. With the source polarity shown, the diode is forward-biased. The greater the source voltage, the larger the diode current. By varying the source voltage, you can measure the diode current (connect an ammeter in series) and the diode voltage (connect a voltmeter in parallel with the diode). By plotting the corresponding currents and voltages, you get the graph of the forward region shown in Fig. 2-17. If you reverse the source voltage, you can get readings for the reverse region. These readings will be extremely small below the breakdown point.

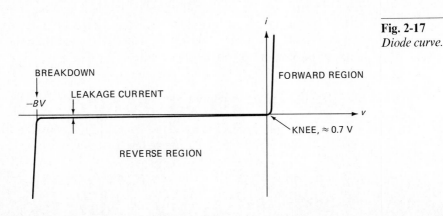

Fig. 2-17
Diode curve.

The first thing to notice is the *nonlinearity* of a diode. No longer do we see a nice straight line like the one given for a resistor. Instead, we get a highly nonlinear *I-V* graph. Stated another way, current is no longer proportional to voltage. As will be discussed later, this nonlinear characteristic is what makes the diode such a useful device.

KNEE VOLTAGE

What does the graph of Fig. 2-17 tell us? When forward bias is applied, the diode doesn't conduct heavily until we overcome the barrier potential. This is why the current is small for the first few tenths of a volt. As we approach the barrier potential (around 0.7 V for a silicon diode), free electrons and holes start crossing the junction in larger numbers. This is the reason the current starts increasing rapidly. Above 0.7 V, small increases in voltage produce a large increase in current.

The voltage at which the current starts to increase rapidly is called the *knee* or *offset* voltage. For a silicon diode, this voltage equals the barrier potential, around 0.7 V. (The germanium diode has an offset voltage of 0.3 V.)

BULK RESISTANCE

Above the knee voltage, the diode current increases rapidly; small increases in diode voltage cause large increases in diode current. The reason is this: After the barrier potential has been overcome, all that impedes current is the bulk or ohmic resistance of the *p* and *n* regions. This resistance is linear. In other words, a diode combines a highly nonlinear resistance (the junction) and a linear bulk resistance (the *p* and *n* regions outside the depletion layer). Below 0.7 V the nonlinearity of the junction predominates; above 0.7 V the linearity of the bulk resistance takes over.

REVERSE REGION

When you reverse-bias the diode of Fig. 2-17, you get an extremely small reverse current (sometimes called the *leakage* current). If you increase the reverse voltage enough, you eventually will reach the breakdown voltage of the diode (some diodes have breakdown voltage of hundreds of volts). As you know, a rectifier diode should always operate below the breakdown voltage. To ensure this, a designer deliberately selects a diode type whose breakdown voltage is larger than the maximum reverse voltage expected during normal operation.

POWER AND CURRENT RATINGS

A rectifier diode is optimized for its unilateral action. You can think of a diode as a one-way conductor because it has a low forward resistance and a high reverse resistance. One way to destroy a diode is to exceed its reverse breakdown voltage. Another way to ruin a diode is to exceed its *maximum power rating*. Any device has a power dissipation, the product of its voltage and current. If this power dissipation is too high, the device will burn out, leaving you with an open or a short. Typically, the voltage at the destruct point of a diode is well above the knee voltage, a volt or more. The product of this voltage and the current produces so much heat that the diode is wiped out.

Manufacturers sometimes specify the power rating of a diode on data sheets. For instance, the 1N914 has a maximum power rating of 250 mW. More often, the data sheet lists only the *maximum current* that the diode can handle. This is more convenient to measure and to design with.

As an example, the data sheet of a 1N4003 does not include a maximum power rating, but it does list a maximum dc forward current of 1 A. In effect, we are being told that if we allow more than 1 A of steady current through a 1N4003, it may be destroyed or its life shortened.

Incidentally, data sheets define two classes of rectifier diodes: *small-signal* diodes (those with a power rating less than 0.5 W) and *rectifiers* (those with a power rating more than 0.5 W). The 1N914 is a small-signal diode because its power rating is 0.25 W; the 1N4003 is a rectifier because its power rating is 1 W.

CURRENT-LIMITING RESISTOR

This brings us to why a resistor is almost always used in series with a diode. Back in Fig. 2-16*b*, R_S is called a *current-limiting* resistor. The larger we make R_S, the smaller the diode current. In other diode circuits to be discussed, there will always be a current-limiting resistor in series with the diode. A designer selects a value of R_S that keeps the maximum forward current below the maximum current rating of the diode.

Even in circuits where you cannot see a resistor (like a black box driving a diode), the Thevenin resistance facing the diode may be enough to hold the current under the maximum current rating of the diode. The point is that there must always be sufficient resistance in series with the diode to limit the current to less than the maximum current rating.

2-9 LOAD LINES

This section introduces you to the *load line,* a tool used to find the exact value of diode current and voltage. Later sections will show you how to apply the load-line method to transistors and other semiconductor devices.

EQUATION FOR THE LOAD LINE

How can we find the exact diode current and voltage in Fig. 2-18? In this series circuit a voltage source V_S forward-biases the diode through a current-limiting resistor R_S. The voltage from the left end of the resistor to ground is V_S, the source voltage. The voltage from the right end of the resistor to ground is V, the diode voltage. Therefore,

Fig. 2-18
*Analyzing circuit
with load line.*

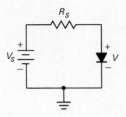

the difference of potential across the resistor is $V_S - V$, and the current is

$$I = \frac{V_S - V}{R_S} \tag{2-1}$$

Because of the series circuit, this current is the same throughout the circuit.

AN EXAMPLE

If the source voltage and current-limiting resistance are given, then only the diode current and voltage are unknown. For instance, if the source voltage is 2 V and the current-limiting resistance is 100 Ω, then Eq. (2-1) becomes

$$I = \frac{2 - V}{100} \tag{2-2}$$

Equation (2-2) is a linear relation between current and voltage. If we plot this equation, we will get a straight line. For instance, let V equal zero. Then

$$I = \frac{2\text{ V} - 0\text{ V}}{100\ \Omega} = 20\text{ mA}$$

Plotting this point ($I = 20$ mA, $V = 0$) gives the point on the vertical axis of Fig. 2-19. This point is called *saturation* because it represents maximum current.
    Here's how to get another point. Let V equal 2 V. Then Eq. (2-2) gives

$$I = \frac{2\text{ V} - 2\text{ V}}{100\ \Omega} = 0$$

When we plot this point ($I = 0$, $V = 2$ V), we get the point shown on the horizontal axis (Fig. 2-19). This point is called *cutoff* because it represents minimum current.
    By selecting other voltages, we can calculate and plot additional points. Because Eq. (2-2) is linear, all points will lie on the straight line shown in Fig. 2-19. (Try plotting other points if you don't believe it.) The straight line is called the *load line.*

THE *Q* POINT

Figure 2-19 shows the graph of a load line and a diode curve. The point of intersection represents a simultaneous solution. In other words, the coordinates of point *Q* are the values of diode current and voltage for a source voltage of 2 V and a current-limiting resistance of 100 Ω. Reading the coordinates of the *Q* point, we get a diode current of approximately 12.5 mA and a diode voltage of 0.75 V. The *Q* point

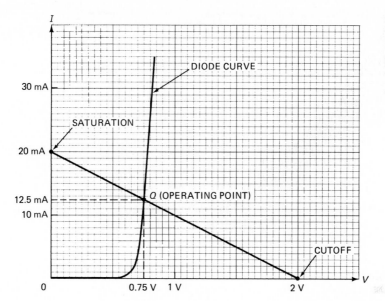

Fig. 2-19
*Load line
intersects diode
curve at operating
point.*

is referred to as the *operating point* because it represents the current through the resistor and the diode.

OTHER LOAD LINES

Equation (2-1) is linear for any source voltage and series resistance. Two points determine a straight line; therefore, we can always plot the load line by drawing a straight line through the vertical and horizontal intercepts (the points where the line crosses the vertical and horizontal axes).

Since the voltage is zero at the upper end of the load line, Eq. (2-1) gives a current of

$$I = \frac{V_S}{R_S} \qquad \text{(saturation)} \qquad (2\text{-}3)$$

Similarly, because the current is zero at the lower end of the load line, Eq. (2-1) yields

$$V = V_S \qquad \text{(cutoff)} \qquad (2\text{-}4)$$

REMEMBERING THE ENDS OF THE LOAD LINE

A way to remember the ends of the load line is as follows. Refer back to Fig. 2-18, where all this started. Letting V equal zero is equivalent to shorting the diode and asking how much current there is in the circuit. If you're visualizing this correctly, the answer is V_S/R_S; this is the current at saturation, the upper end of the load line. Likewise, letting I equal zero is equivalent to opening the diode and asking how much voltage there is across the diode. The answer is V_S; this is the voltage at cutoff, the lower end of the load line.

2-10 DIODE APPROXIMATIONS

Resistors typically have tolerances of ±5 percent; diode knee voltages may have tolerances of ±10 percent; other devices we will discuss have tolerances from ±20 to ±50 percent, or more. Let us be sensible and realize that exact mathematical answers to circuit problems have little value if component tolerances are running 5, 10, 20 percent or more. What makes sense in the real world of everyday electronics is approximate answers.

IDEAL DIODE

So let's approximate diode behavior. What does a diode do? It conducts well in the forward direction and poorly in the reverse direction. Boil this down to its essence, and this is what you get: An *ideal diode* acts like a perfect conductor (zero voltage) when forward-biased and like a perfect insulator (zero current) when reverse-biased, as shown in Fig. 2-20a.

In circuit terms, an ideal diode acts like a switch. When the diode is forward-biased, it is like a closed switch (Fig. 2-20b). If the diode is reverse-biased, the switch opens.

Extreme as the ideal-diode approximation seems at first, it gets you started in seeing how diode circuits work; you don't have to worry about the effects of offset voltage or bulk resistance. There will be times when the ideal approximation will be too inaccurate; for this reason, we need a second and third approximation.

SECOND APPROXIMATION

We need about 0.7 V of offset voltage before a silicon diode will really conduct well. When the source voltage is large, this 0.7 V is too small to matter. But when the source voltage is not large, we may want to take the knee voltage into account.

Figure 2-21a shows the graph for the *second approximation.* The graph says that no current flows until 0.7 V appears across the diode. At this point the diode turns on. No matter how large the forward current, we assume that the diode voltage is 0.7 V.

Figure 2-21b is the equivalent circuit for the second approximation. The idea is to think of the diode as a switch in series with a battery of 0.7 V. If the source voltage is

Fig. 2-20
(a) *Ideal diode.*
(b) *Switch is equivalent to ideal diode.*

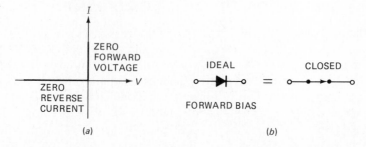

(a)   (b)

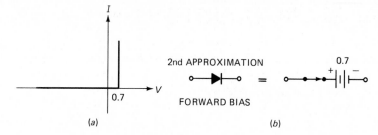

Fig. 2-21   (a) *Second approximation.* (b) *Switch and battery is equivalent circuit.*

greater than 0.7 V, the switch closes and the diode voltage is 0.7 V. If the source voltage is less than 0.7 V or if the source voltage is negative, the switch opens.

THIRD APPROXIMATION

In the *third approximation* of a diode, we include bulk resistance r_B. As before, the diode turns on at 0.7 V. Then additional voltage appears across the bulk resistance, so that the total diode voltage is greater than 0.7 V. Figure 2-22a shows the effects of r_B. After the silicon diode turns on, the current produces a voltage across r_B. The greater the current, the larger the voltage. Because r_B is linear, the voltage increases linearly as the current increases.

The equivalent circuit for the third approximation is a switch in series with a battery of 0.7 V and a resistance of r_B (see Fig. 2-22b). After the external circuit has overcome the barrier potential, the diode current produces an *IR* drop across the bulk resistance. Therefore, the total voltage across the silicon diode equals

$$V_F = 0.7 + I_F r_B \qquad (2\text{-}5)$$

WHICH ONE TO USE

For most practical work, the second approximation is the best compromise; this is the one we will use throughout this book unless otherwise indicated. Furthermore, when you do the problems at the end of each chapter, use the second approximation unless you are specifically told otherwise.

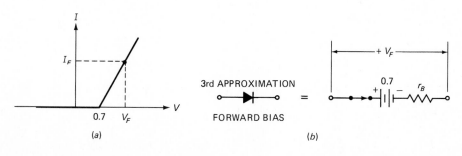

Fig. 2-22
(a) *Third approximation.* (b) *Equivalent circuit.*

Fig. 2-23

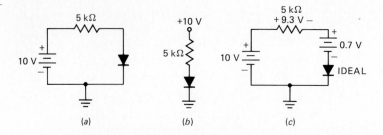

(a) (b) (c)

EXAMPLE 2-1

Use the second approximation to find the diode current in Fig. 2-23a.

SOLUTION

The diode is forward-biased; therefore, it drops 0.7 V. The voltage across the resistor is the difference between the source voltage and the diode voltage (10 V − 0.7 V), which equals 9.3 V. Therefore, the diode current is

$$I = \frac{9.3 \text{ V}}{5 \text{ k}\Omega} = 1.86 \text{ mA}$$

By the way, industrial schematics don't usually show complete circuits like Fig. 2-23a. Instead, you get an abbreviated drawing like Fig. 2-23b. Since the source is grounded on one side, the usual practice is to show only the potential of the other side with respect to ground. In Fig. 2-23a, the potential of the positive side with respect to ground is +10 V. When you see an industrial schematic like Fig. 2-23b with voltage at a point, always remember that the potential is with respect to ground.

Figure 2-23c shows an equivalent circuit that emphasizes the second approximation and its effects. As indicated, 9.3 V is dropped across the resistor and 0.7 V across the diode.

Fig. 2-24

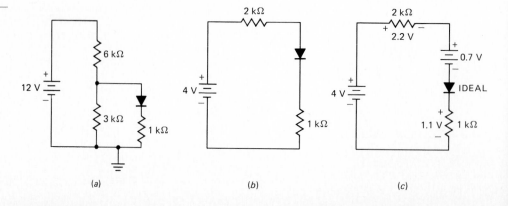

(a) (b) (c)

EXAMPLE 2-2

Calculate the current through the 1-kΩ resistor of Fig. 2-24*a*.

SOLUTION

What are you supposed to do whenever you see more than one loop? Right! Think Thevenin, or at least consider using the Thevenin theorem. The easiest way to solve this problem is to thevenize the voltage divider to get the equivalent circuit of Fig. 2-24*b*. Since the diode is forward-biased, it drops 0.7 V. The rest of the source voltage appears across the total resistance of 3 kΩ. Therefore, the current in the series circuit is

$$I = \frac{4 \text{ V} - 0.7 \text{ V}}{3 \text{ k}\Omega} = 1.1 \text{ mA}$$

If you don't understand the foregoing logic, write a Kirchhoff voltage equation for Fig. 2-24*b* to get

$$-4 + 2000I + 0.7 + 1000I = 0$$

If you solve this for *I*, you get 1.1 mA.

Figure 2-24*c* shows an equivalent circuit. The upper resistor has a voltage drop of 2.2 V, the diode a drop of 0.7 V, and the lower resistor a drop of 1.1 V.

2-11 DC RESISTANCE OF A DIODE

If you take the ratio of total diode voltage to total diode current, you get the *dc resistance* of the diode. In the forward direction this dc resistance is symbolized by R_F; in the reverse direction, it is designated R_R.

FORWARD RESISTANCE

Because the diode is a nonlinear resistance, its dc resistance varies with the current through it. For example, here are some pairs of forward current and voltage for a typical 1N914: 10 mA at 0.65 V, 30 mA at 0.75 V, and 50 mA at 0.85 V. At the first point the dc resistance is

$$R_F = \frac{0.65 \text{ V}}{10 \text{ mA}} = 65 \text{ }\Omega$$

At the second point,

$$R_F = \frac{0.75 \text{ V}}{30 \text{ mA}} = 25 \text{ }\Omega$$

And at the third point,

$$R_F = \frac{0.85 \text{ V}}{50 \text{ mA}} = 17 \text{ }\Omega$$

Notice how the dc resistance decreases as the current increases. In any case, the forward resistance is low.

REVERSE RESISTANCE

Similarly, here are two sets of reverse current and voltage for a 1N914: 25 nA at 20 V; 5 μA at 75 V. At the first point, the dc resistance is

$$R_R = \frac{20 \text{ V}}{25 \text{ nA}} = 800 \text{ M}\Omega$$

At the second point,

$$R_R = \frac{75 \text{ V}}{5 \text{ }\mu\text{A}} = 15 \text{ M}\Omega$$

Notice how the dc resistance decreases as we approach the breakdown voltage (75 V). Nevertheless, the reverse resistance of the diode is still high, well into the megohms.

TROUBLESHOOTING

You can quickly check the condition of a diode with an ohmmeter. Measure the dc resistance of the diode in either direction, then reverse the leads and measure the dc resistance again. The forward current will depend on which ohmmeter range is used, which means you will get different readings on different ranges. The main thing to look for, however, is a high ratio of reverse to forward resistance. How high is high? For typical silicon diodes used in electronics work, the ratio should be higher than 1000:1.

Using an ohmmeter to check diodes is an example of go/no-go testing. You're really not interested in the exact dc resistance of the diode; all you want to know is whether the diode is acting approximately like a one-way conductor or not—that is, if it has a low resistance in the forward direction and a high resistance in the reverse direction. Diode troubles are indicated if you find any of the following: extremely low resistance in both directions (diode shorted); high resistance in both directions (diode open); somewhat low resistance in the reverse direction (called a leaky diode).

The test is usually done with the diode out of a circuit. But even when the diode is in a circuit, an ohmmeter check (turn the circuit power off first) should indicate lower resistance one way than the other.

A final point: Some ohmmeters can produce enough current on low ranges to destroy a small-signal diode. For this reason, you should test small-signal diodes on ranges greater than $R \times 10$. On these higher scales, the internal resistance of the ohmmeter prevents excessive diode current.

PROBLEMS

Straightforward

2-1. A silicon diode has a saturation current of 2 nA at 25°C. What is the value of I_S at 75°C? At 125°C?

2-2. At 25°C, a silicon diode has a reverse current of 25 nA. The surface-leakage component equals 20 nA. If the surface current still equals 20 nA at 75°C, what does the total reverse current equal?

2-3. When forward-biased in a particular circuit, a diode has a current of 50 mA. When reverse-biased, the current drops to 20 nA. What is the ratio of forward to reverse current?

2-4. What is the power dissipation in a forward-biased silicon diode if the diode voltage is 0.7 V and the current is 100 mA?

2-5. Draw the *I-V* graph of a 2-kΩ resistor. Label the point where the current is 4 mA.

2-6. Sketch the *I-V* graph of a silicon diode with an offset 0.7 V and a PIV of 50 V. In your own words, explain each part of the graph.

2-7. In Fig. 2-25, what is the approximate power dissipation of the diode if the current is 50 mA? How much is it when the current is 100 mA?

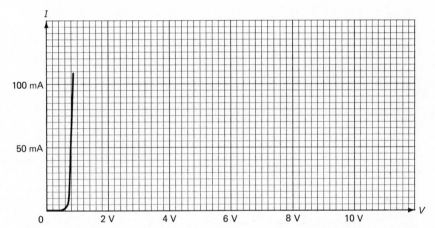

Fig. 2-25

2-8. A source voltage of 8 V drives a diode through a current-limiting resistor of 100 Ω. If the diode has the *I-V* characteristic of Fig. 2-25, what is the current at the upper end of the load line? The voltage at the lower end of the load line? What are the approximate current and voltage at the *Q* point? The power dissipation of the diode?

2-9. Repeat Prob. 2-8 for a resistance of 200 Ω. Describe what happens to the load line.

2-10. Repeat Prob. 2-8 for a source voltage of 2 V. What happens to the load line?

2-11. The source voltage is 9 V and the source resistance is 1 kΩ in Fig. 2-26a. Calculate the current through the diode.

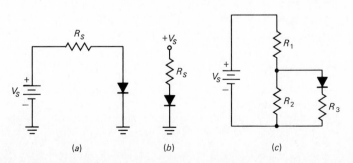

Fig. 2-26

2-12. V_S is 12 V and R_S is 47 kΩ in Fig. 2-26*b*. What is the diode current?

2-13. In Fig. 2-26*c*, $V_S = 15$ V, $R_1 = 68$ kΩ, $R_2 = 15$ kΩ, and $R_3 = 33$ kΩ. Calculate the diode current. Is the voltage divider stiff? (The word "stiff" is defined in Sec. 1-2.)

2-14. Suppose the source voltage is reversed in Fig. 2-26*c*. If $V_S = 100$ V, what does the reverse voltage across the diode equal? (Use the resistance values given in Prob. 2-13.)

2-15. Here are some diodes, their breakdown voltages, and their current ratings:

Diode	PRV rating	I_{max}
1N914	75 V	200 mA
1N4001	50 V	1 A
1N1185	120 V	35 A

In Fig. 2-26*c* the polarity of the source is reversed. If $V_S = 200$ V, $R_1 = 10$ kΩ, and $R_2 = 10$ kΩ, which of the foregoing diodes breaks down when used in the circuit?

2-16. The source voltage is 100 V and the source resistance is 220 Ω in Fig. 2-26*b*. Which of the diodes listed in Prob. 2-15 are okay to use in the circuit?

2-17. Calculate the approximate forward resistance in Fig. 2-25 for each of the following currents: 10 mA, 50 mA, and 100 mA.

2-18. Here are some diodes and their worst-case specifications:

Diode	I_F	I_R
1N914	10 mA at 1 V	25 nA at 20 V
1N4001	1 A at 1.1 V	10 μA at 50 V
1N1185	10 A at 0.95 V	4.6 mA at 100 V

Calculate the reverse/forward resistance ratio for each of these diodes.

Troubleshooting

2-19. Suppose the voltage across the diode of Fig. 2-27*a* is 5 V. Is the diode open or shorted?

Fig. 2-27

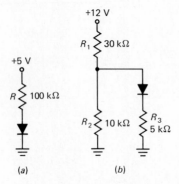

(a)   (b)

2-20. Something causes R to short in Fig. 2-27*a*. What will the diode voltage be? What will happen to the diode?

2-21. You measure 0 V across the diode of Fig. 2-27*a*. Next you check the source voltage, and it reads $+5$ V with respect to ground. What is wrong with the circuit?

2-22. In Fig. 2-27b, you measure a potential of +3 V at the junction of R_1 and R_2. (Remember, potentials are always with respect to ground.) Next you measure 0 V at the junction of the diode and the 5-kΩ resistor. Name some possible troubles.

2-23. You measure 0 V at the junction of R_1 and R_2 in Fig. 2-27b. What are some of the things that can be wrong with this circuit?

Design

2-24. In Fig. 2-27a, what value should R be to get a diode current of 10 mA?

2-25. What value should R_2 be in Fig. 2-27b to set up a diode current of 0.25 mA?

2-26. Redesign the circuit of Fig. 2-27b to meet the following specifications: $V_{TH} = 4$ V, $R_3 = 100$ kΩ, and a stiff voltage divider. (See Sec. 1-2 if you don't know what "stiff" means.)

2-27. Design a circuit like Fig. 2-27b that meets these specifications: $V_S = 25$ V, R_3 can vary from 50 to 250 kΩ, $V_{TH} = 5$ V, and a stiff voltage divider.

Challenging

2-28. A silicon diode has a forward current of 50 mA at 1 V. Use the third approximation to calculate its bulk resistance.

2-29. Given a silicon diode with a reverse current of 5 μA at 25°C and 100 μA at 100°C, calculate the saturation current and the surface leakage current.

2-30. The power is turned off and the upper end of R_1 is grounded in Fig. 2-27b. Next you use an ohmmeter to read the forward and reverse resistance of the diode. Both readings are identical. What does the ohmmeter read?

2-31. Some systems, like burglar alarms, computers, etc., use battery backup just in case the main source of power should fail. Describe how the circuit of Fig. 2-28 works.

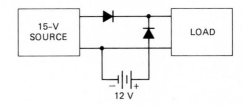

Fig. 2-28

Computer

2-32. The LET statement allows you to set the value of a variable. For instance, if you run this program

```
10 LET Y = 108
20 LET X = 25
```

then the computer sets Y equal to 108 and $X = 25$. In fact, you can omit the word LET in most computers. The program

```
10 Y = 108
20 X = 25
```

still sets Y equal to 108 and $X = 25$.

What does the computer screen display after this program is run:

10 R1 = 4700
20 R2 = 6800
30 PRINT R1
40 PRINT R2

2-33. The arithmetic operators are +, −, *, and /. They stand for plus, minus, times, and divide. What does this program do:

10 R1 = 4700
20 R2 = 6800
30 RT = R1 + R2
40 PRINT RT

2-34. Write a program that works out the parallel resistance of 4.7 kΩ and 6.8 kΩ. (Hint: The product of R_1 and R_2 appears as R1 * R2.)

Diode Circuits

Most electronic circuits need a dc voltage in order to work properly. Since line voltage is alternating, the first thing that has to be done in any electronic equipment is to convert ac voltage to dc voltage. In this chapter we discuss rectifying circuits that perform the required ac-to-dc conversion. We also include capacitor-input filters, voltage multipliers, diode limiters, clampers, and peak-to-peak detectors.

3-1 THE SINE WAVE

The sine wave is the most basic of electric signals. It is often used, for example, to test electronic circuits. Furthermore, complicated signals can be expressed as the super-position of several sine waves. This section is a brief review of sine-wave quantities needed for our discussion of diode circuits.

PEAK VALUE

Look at the sine wave shown in Fig. 3-1. This is a graph of

$$v = V_P \sin \theta \qquad (3\text{-}1)$$

where v = instantaneous voltage
V_P = peak voltage
θ = angle in degrees or radians

Notice how the voltage increases from zero to a positive maximum at 90°, decreases to zero at 180°, reaches a negative maximum at 270°, and returns to zero at 360°.

Table 3-1 lists some of the instantaneous values you should know. Because of the symmetry of a sine wave, you can easily figure out the values at 120°, 150°, 180°, 210°, etc. You should memorize Table 3-1. If you know these few values, you can estimate almost any intermediate value along the sine wave.

As indicated, V_P is the peak value of a sine wave, the maximum value it reaches. The sine wave has a positive peak at 90° and a negative peak at 270°.

Fig. 3-1
Sine wave.

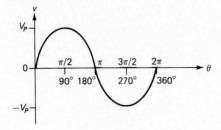

**Table 3-1. Values
of a Sine Wave**

θ	v
0°	0
30°	$0.5V_p$
45°	$0.707V_p$
60°	$0.866V_p$
90°	V_p

PEAK-TO-PEAK VALUE

The peak-to-peak value of any signal is the difference between its algebraic maximum and minimum:

$$V_{pp} = V_{max} - V_{min} \tag{3-2}$$

For the sine wave of Fig. 3-1, the peak-to-peak value is

$$V_{pp} = V_P - (-V_P) = 2V_P$$

In other words, the peak-to-peak value of a sine wave is twice the peak value. Given a sine wave with a peak value of 18 V, the peak-to-peak value is 36 V.

RMS VALUE

If a *sinusoidal* (same as sine-wave) voltage is across a resistor, it produces an in-phase sinusoidal current through the resistor. The product of instantaneous voltage and current gives instantaneous power, which averaged over the cycle results in an average power dissipation. In other words, the resistor gives off a steady amount of heat, as though a dc voltage were across it.

The *rms* (root-mean-square) value of a sine wave, also called the effective or heating value, is defined as the dc voltage that produces the same amount of heat as the sine wave. Basic courses show that

$$V_{rms} = 0.707V_P \tag{3-3}$$

We can prove this relation experimentally by building two circuits: one with a dc source driving a resistor and another with a sinusoidal source driving a resistor of the same value. If the dc source is adjusted to produce the same amount of heat as the sine wave, we will measure a dc voltage equal to 0.707 times the peak value of the sine wave. (Another way to prove $V_{rms} = 0.707V_P$ is with advanced mathematics.)

LINE VOLTAGE

Power companies in the United States typically supply a line voltage of 115 V rms with a tolerance of ±10 percent and a frequency of 60 Hz. (On schematic diagrams 115 V rms is usually written as 115 V ac, where V ac stands for volts ac.) With Eq. (3-3), we can calculate the peak value as follows:

$$115 \text{ V} = 0.707V_P$$

or

$$V_P = \frac{115 \text{ V}}{0.707} = 163 \text{ V}$$

(In this book, all answers are rounded off to three places unless otherwise indicated.)

A peak value of 163 V implies a peak-to-peak value of 326 V, definitely a dangerous level of voltage. If you ever accidentally come in contact with both sides of the power line, you will never forget the experience, assuming you live through it.

AVERAGE VALUE

The *average* value of a sine wave over a cycle is zero. This is because the sine wave is symmetrical: Each positive value in the first half cycle is offset by an equal negative value in the second half cycle. Therefore, if you add up all the sine-wave values between 0° and 360°, you get zero, which implies an average value of zero.

Another way of saying the same thing is saying that a dc voltmeter will read zero if used to measure a sinusoidal wave. Why? Because the needle of a dc voltmeter tries to fluctuate positively and negatively by equal amounts, but the inertia of the moving parts prevents it from moving, and so it indicates an average value of zero. (This assumes a frequency greater than approximately 10 Hz, so that the needle cannot track the rapid changes.)

3-2 THE TRANSFORMER

The reason the power line is so dangerous is because its Thevenin resistance approaches zero. This means it can supply hundreds of amperes. Even with a circuit breaker, it can still deliver tens of amperes, depending on the size of the circuit breaker.

VOLTAGE STEPDOWN

Some electronics equipment includes a transformer like Fig. 3-2 to step the line voltage up or down, as required in the application. When connected to a wall outlet, the middle pin of the power plug grounds the chassis of your equipment; this ensures that all equipment with three-wire plugs is at the same ground potential.

Fig. 3-2
Transformer steps voltage up or down.

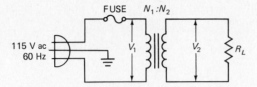

As discussed in basic ac theory, the primary and secondary voltages of an ideal transformer are related as follows:

$$\frac{V_2}{V_1} = \frac{N_2}{N_1} \tag{3-4}$$

where V_1 = primary voltage, either rms or peak
V_2 = secondary voltage, either rms or peak
N_1 = number of turns in primary winding
N_2 = number of turns in second winding

For instance, if the turns ratio is $9:1$ in Fig. 3-2, then

$$\frac{V_2}{115\text{ V}} = \frac{1}{9}$$

or

$$V_2 = \frac{115\text{ V}}{9} = 12.8\text{ V rms}$$

This lower voltage is a lot safer to work with than 115 V rms and is typically what semiconductor circuits require. Furthermore, the transformer isolates the load (all the circuits you will be measuring) from the line. This means that the only connection with the power line is through the magnetic field linking the primary and secondary windings. This further reduces the hazard of electric shock because there no longer is a direct electrical contact with either side of the line.

FUSE

In an ideal transformer, the currents are given by

$$\frac{I_1}{I_2} = \frac{N_2}{N_1} \tag{3-5}$$

You can use this equation to work out the size of the fuse in Fig. 3-2. For instance, if the load current is 1.5 A rms and the turns ratio is $9:1$, then

$$\frac{I_1}{1.5\text{ A}} = \frac{1}{9}$$

or

$$I_1 = \frac{1.5\text{ A}}{9} = 0.167\text{ A rms}$$

This means that the fuse must have a value greater than 0.167 A, plus 10 percent in case the line voltage is high, plus approximately another 10 percent for transformer losses (these produce extra primary current). The next higher standard fuse size, 0.25 A (slow-blow in case of line surges), would probably be satisfactory. The purpose of the fuse is to prevent excessive damage in case the load resistance is accidentally shorted.

REAL TRANSFORMERS

The transformers you buy from a parts supplier are not ideal because the windings have resistance that produces power losses. Furthermore, the laminated core has eddy currents, which produce additional power losses. Because of these unwanted power losses, a real transformer is a difficult device to specify fully. The data sheets for transformers rarely list the turns ratio, the resistances of windings, and other transformer quantities. Usually, all you get is the secondary voltage at a rated current. For instance, the F25X is an industrial transformer whose data sheet gives only the following specifications: for a primary voltage of 115 V ac, the secondary voltage is 12.6 V ac when the secondary current is 1.5 A. If the secondary current is less than 1.5 A, the secondary voltage will rise slightly because of lower IR drops across the winding resistance.

Therefore, the discussions that follow will no longer specify the turns ratio. Instead, only the secondary voltage will be given. When it's necessary to know the primary current, you can estimate the turns ratio of a real transformer by using Eq. (3-4) and find the approximate primary current with Eq. (3-5).

3-3 THE HALF-WAVE RECTIFIER

Figure 3-3a shows a circuit known as a *half-wave rectifier.* On the positive half cycle of secondary voltage, the diode is forward-biased for all instantaneous voltages greater than the offset voltage (approximately 0.7 V for silicon diodes and 0.3 V for germanium diodes). This produces approximately a half sine wave of voltage across the load resistor. To simplify our discussions, we will use the ideal-diode approximation because the peak source voltage is usually much larger than the offset voltage of the diode. With this in mind, the peak of rectified voltage is equal to the peak secondary voltage, as shown in Fig. 3-3b. On the negative half cycle, the diode is

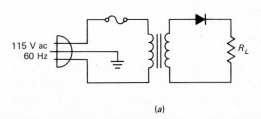

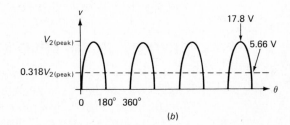

Fig. 3-3 (a) *Half-wave rectifier.* (b) *Rectified output.*

reverse-biased. Ignoring leakage current (the same as reverse current), the load current drops to zero; this is why the load voltage drops to zero between 180° and 360°.

RECTIFICATION

The important thing to notice about the half-wave rectifier is this: It has converted the ac input voltage to a pulsating dc voltage. In other words, the load voltage is always positive or zero, depending on which half cycle it's in. Stated another way, the load current is always in the same direction. This conversion from ac to dc is known as *rectification.*

AVERAGE VOLTAGE

Ignoring the diode drop, the average or dc value of the half-wave signal in Fig. 3-3b is

$$V_{dc} = 0.318 V_{2(peak)} \tag{3-6}$$

This is sometimes written as

$$V_{dc} = \frac{V_{2(peak)}}{\pi}$$

For example, suppose the secondary voltage is 12.6 V ac. Ideally, the peak secondary voltage is

$$V_{2(peak)} = \frac{12.6 \text{ V}}{0.707} = 17.8 \text{ V}$$

and the average value is

$$V_{dc} = 0.318 \, (17.8 \text{ V}) = 5.66 \text{ V}$$

Figure 3-3b indicates the peak and average values for a secondary voltage of 12.6 V rms.

The average voltage is called the dc voltage because it is what a dc voltmeter connected across the load resistor would read. Given a peak load voltage of 17.8 V out of a half-wave rectifier, a dc voltmeter would read 5.66 V.

Equation (3-6) is easy to prove. If you set up a half-wave rectifier in the laboratory, you will discover that the average voltage equals 0.318 times the peak voltage. Alternatively, you can derive the same formula mathematically by averaging the values of a rectified sine wave.

CURRENT RATING OF DIODE

With Eq. (3-6), we can calculate the average or dc load voltage. If the load resistance is known, we then can calculate the average load current I_{dc}. Because the half-wave rectifier is a single-loop circuit, the dc diode current is equal to the dc load current. On data sheets, I_{dc} is usually listed as I_O. Therefore, one of the things a designer must

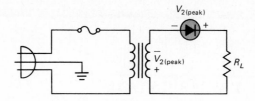

Fig. 3-4
*Peak inverse
voltage.*

look for is the I_O rating of the diode. This rating tells him how much direct current the diode can handle.

For example, the data sheet of a 1N4001 gives an I_O rating of 1 A. If the dc voltage is 5.66 V and the load resistance is 10 Ω, then the dc load current is 0.566 A. The 1N4001 would be all right to use in a half-wave rectifier because its I_O rating (1 A) is greater than the average rectified current (0.566 A).

PEAK INVERSE VOLTAGE

Figure 3-4 shows the half-wave rectifier at the instant the secondary voltage reaches its maximum negative peak. The diode is shaded or dark to indicate that it is off. Since the diode is reverse-biased, there is no load voltage. For Kirchhoff's voltage law to be satisfied, all the secondary voltage must appear across the diode as shown. This maximum reverse voltage is called the *peak inverse voltage* (PIV). To avoid breakdown, the peak inverse voltage must be less than the PIV rating of the diode. For instance, if the peak inverse voltage is 75 V, the diode needs a PIV rating greater than 75 V.

3-4 THE FULL-WAVE RECTIFIER

Figure 3-5*a* shows a *full-wave rectifier.* During the positive half cycle of secondary voltage, the upper diode is forward-biased and the lower diode is reverse-biased; therefore, the current is through the upper diode, the load resistor, and the upper half winding (Fig. 3-5*c*). During the negative half cycle, current is through the lower diode, the load resistor, and the lower half winding (Fig. 3-5*d*). Notice that the load voltage has the same polarity in Fig. 3-5*c* and *d* because the current through the load resistor is in the same direction no matter which diode is conducting. This is why the load voltage is the full-wave rectified signal shown in Fig. 3-5*b*.

EFFECT OF CENTER-TAPPED SECONDARY

A full-wave rectifier is like two back-to-back half-wave rectifiers with one rectifier handling the first half cycle and the other handling the alternate half cycle. Because of the center-tapped secondary winding, each diode circuit receives only half the secondary voltage. Assuming ideal diodes, this means the peak rectified output voltage is

$$V_{out(peak)} = 0.5 V_{2(peak)} \qquad (3\text{-}7)$$

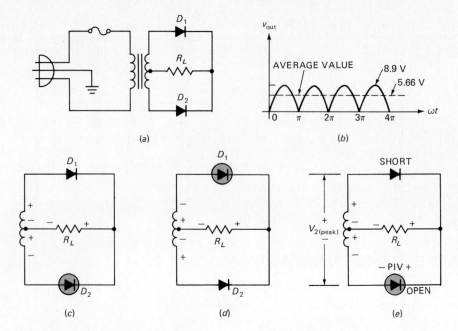

Fig. 3-5 (a) *Full-wave rectifier.* (b) *Rectified output.* (c) *Positive half cycle.* (d) *Negative half cycle.* (e) *Peak inverse voltage.*

AVERAGE VOLTAGE

The average or dc value of a full-wave rectified output is twice the output of a half-wave rectified output driven by the same secondary voltage:

$$V_{dc} = 0.636 V_{out(peak)} \qquad (3\text{-}8)$$

An alternative formula is

$$V_{dc} = \frac{2V_{out(peak)}}{\pi}$$

As an example, suppose the secondary voltage is 12.6 V ac. As found earlier, this implies a peak secondary voltage of 17.8 V. Because of the center tap, however, the peak voltage driving each diode circuit is only half of this, 8.9 V. Ignoring diode drop, the full-wave output has a peak value of 8.9 V and an average value of

$$V_{dc} = 0.636 V_{out(peak)} = 0.636(8.9 \text{ V}) = 5.66 \text{ V}$$

Figure 3-5b indicates the peak and average values for a secondary voltage of 12.6 V ac.

As before, you can prove Eq. (3-8) experimentally or mathematically. (See Prob. 3-33.)

CURRENT RATING OF DIODES

Given a dc load voltage of 5.66 V and a load resistance of 10 Ω, the dc load current is

$$I_{dc} = \frac{5.66 \text{ V}}{10 \ \Omega} = 0.566 \text{ A}$$

Here is an interesting point. The I_O rating of each diode need only be greater than half the dc load current, or 0.283 A. Why? Look carefully at Fig. 3-5*a* and realize that each diode conducts for only half a cycle. This means that the current through a diode is a half-wave rectified current. Therefore, the dc current through each diode is half the dc load current.

Here is another way to understand what's going on. Assume that dc ammeters are in series with each of the diodes and the load resistor. Then each of the diode ammeters would read 0.283 A and the load ammeter would read 0.566 A. This makes sense; it means that Kirchhoff's current law is being satisfied.

FREQUENCY

The *period T* of a repetitive waveform is the time between corresponding or equivalent points on the waveform. Frequency *f* is the reciprocal of period *T*. In a half-wave rectifier, the period of the output equals the period of the input, which means that the output frequency is the same as the input frequency. In other words, you get one output cycle for each input cycle. For this reason, the frequency out of a half-wave rectifier is 60 Hz, the same as the line frequency.

A full-wave rectifier is different. Look carefully at Fig. 3-5*b* and notice that the period is half the input period. Stated another way, two full cycles of output occur for each input cycle. This is because the full-wave rectifier has inverted the negative half cycle of the input voltage. As a result, the output of a full-wave rectifier has a frequency of 120 Hz, exactly double the line frequency.

PEAK INVERSE VOLTAGE

Figure 3-5*e* shows the full-wave rectifier at the instant the secondary voltage reaches its maximum positive value. If we apply Kirchhoff's voltage law around the outside loop, we get

$$V_{2(\text{peak})} - \text{PIV} + 0 = 0$$

where the 0 on the left side of this equation represents the ideal voltage of the upper diode. Solving gives the peak inverse voltage across the lower diode:

$$\text{PIV} = V_{2(\text{peak})}$$

Therefore, it follows that each diode in a full-wave rectifier must have a PIV rating greater than $V_{2(\text{peak})}$.

Fig. 3-6
Grounded center tap.

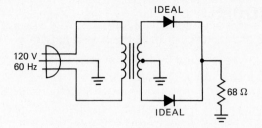

EXAMPLE 3-1

In Fig. 3-6, the secondary voltage is 40 V ac. Using ideal diodes, calculate the dc load voltage. Also, work out the I_O rating and PIV rating for the diodes.

SOLUTION

First, notice that the center tap has been grounded. Since the lower end of the load resistor is also grounded, the circuit is equivalent to the full-wave rectifier analyzed earlier. Given a secondary voltage of 40 V ac, the peak secondary voltage is

$$V_{2(peak)} = \frac{40 \text{ V}}{0.707} = 56.6 \text{ V}$$

Ignoring diode drop, the peak rectified voltage is half of this:

$$V_{out(peak)} = 0.5(56.6 \text{ V}) = 28.3 \text{ V}$$

and the dc output is

$$V_{dc} = 0.636(28.3 \text{ V}) = 18 \text{ V}$$

The peak inverse voltage equals the peak secondary voltage, which is 56.6 V.
Next, calculate the dc load current:

$$I_{dc} = \frac{18 \text{ V}}{68 \text{ }\Omega} = 265 \text{ mA}$$

The direct current through each diode is half of this:

$$I_O = \frac{265 \text{ mA}}{2} = 132 \text{ mA}$$

Therefore, each diode must have an I_O rating greater than 132 mA.
Finally, the peak inverse voltage across the diodes of a full-wave rectifier is always equal to the peak secondary voltage. In this case,

$$PIV = 56.6 \text{ V}$$

This means that the diode must have a PIV rating greater than 56.6 V.

3-5 THE BRIDGE RECTIFIER

Now we come to the *bridge rectifier,* the most popular way to rectify because it features the full peak voltage of a half-wave rectifier and the higher average value of a full-wave rectifier. Figure 3-7*a* shows a bridge rectifier. During the positive half cycle of secondary voltage (Fig. 3-7*c*), diodes D_2 and D_3 are forward-biased; therefore, the load voltage has the polarity shown: minus on the left and plus on the right. During the negative half cycle (Fig. 3-7*d*), diodes D_1 and D_4 are forward-biased; again, the load voltage has the minus-plus polarity shown. On either half cycle, the load voltage has the same polarity because the load current is in the same direction no matter which diodes are conducting. This is why the load voltage is the full-wave rectified signal of Fig. 3-7*b*.

AVERAGE VOLTAGE

Ignoring diode drop in Fig. 3-7*c*, the peak load voltage is

$$V_{\text{out(peak)}} = V_{2\text{(peak)}} \tag{3-9}$$

Notice that the full secondary voltage appears across the load resistor; this is one of the things that make the bridge rectifier better than the full-wave rectifier discussed earlier, where only half the secondary voltage reached the output. Furthermore, a center-tapped transformer that produces equal voltages on each half of the secondary winding is difficult and expensive to manufacture. By using a bridge rectifier, a designer eliminates the need for an accurate center tap; the savings more than compensate for the cost of the two additional diodes.

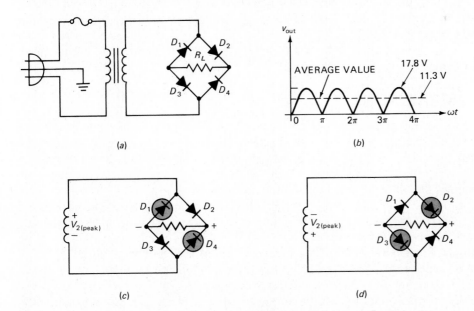

Fig. 3-7 (a) *Full-wave rectifier.* (b) *Rectified output.* (c) *Positive half cycle.* (d) *Negative half cycle.*

Because the bridge output is a full-wave signal, the average or dc value is

$$V_{dc} = 0.636 V_{out(peak)} \qquad\qquad (3\text{-}10)$$

For instance, if the secondary voltage is 12.6 V ac, the peak secondary voltage is 17.8 V (found earlier). Ideally,

$$V_{out(peak)} = 17.8 \text{ V}$$

and

$$V_{dc} = 0.636(17.8 \text{ V}) = 11.3 \text{ V}$$

Figure 3-7*b* indicates these ideal voltages for a secondary voltage of 12.6 V ac.

RATINGS AND FREQUENCY

Given a dc load voltage of 11.3 V and a load resistance of 10 Ω, the dc load current is 1.13 A. Since each diode conducts for only half a cycle, the I_O rating of the diodes must be at least half the dc load current, or 0.565 A.

In Fig. 3-7*c*, diode D_2 is ideally shorted and diode D_4 is ideally open. Summing voltages around the outside loop gives

$$V_{2(peak)} - \text{PIV} + 0 = 0$$

where the 0 on the left side of the equation is the ideal voltage across D_2. Therefore, the peak inverse voltage across D_4 is

$$\text{PIV} = V_{2(peak)}$$

By a similar argument, each of the remaining diodes must withstand a peak inverse voltage equal to the peak secondary voltage. Therefore, the PIV rating of the diodes must be greater than $V_{2(peak)}$.

Because the output is a full-wave signal, the output frequency is double the input frequency, or 120 Hz.

COMPARISON

For historical reasons, a rectifier with a center-tapped transformer is called a full-wave rectifier. In practice, the bridge rectifier is sometimes called a full-wave bridge rectifier, but this is unnecessary because a bridge rectifier always produces a full-wave output. In this book, we refer to the three circuits simply as the half-wave rectifier, the full-wave rectifier, and the bridge rectifier.

Table 3-2 summarizes the rectifiers discussed so far. These circuits are called *average rectifiers* because their dc output equals the average value of a rectified sine wave. Notice the final entry of Table 3-2. This has been included for troubleshooting purposes. Normally, you will measure the rms value of secondary voltage on one of the ac ranges of a floating VOM (one that does not plug into a wall socket). Then you will measure the dc voltage on one of the dc ranges. As you can see in Table 3-2, the bridge rectifier produces a dc load voltage that is ideally 90 percent of the rms secondary voltage; the other rectifiers produce a dc load voltage of only 45 percent.

Everything considered, the bridge rectifier is the best compromise for most applications, and you see it used in industry more than the others.

Table 3-2. Ideal Average Rectifiers

	Half-wave	Full-wave	Bridge
Number of diodes	1	2	4
Peak output voltage	$V_{2(peak)}$	$0.5V_{2(peak)}$	$V_{2(peak}$
Dc output voltage	$0.318V_{out(peak)}$	$0.636V_{out(peak)}$	$0.636V_{out(peak)}$
Dc diode current	I_{dc}	$0.5I_{dc}$	$0.5I_{dc}$
Peak inverse voltage	$V_{2(peak)}$	$V_{2(peak)}$	$V_{2(peak)}$
Ripple frequency	f_{in}	$2f_{in}$	$2f_{in}$
Dc output voltage	$0.45V_{2(rms)}$	$0.45V_{2(rms)}$	$0.9V_{2(rms)}$

ENCAPSULATED BRIDGE RECTIFIERS

Bridge rectifiers are so common that manufacturers are packaging them as a module. For instance, the MDA920-3 is a commercially available bridge-rectifier assembly. It consists of four hermetically sealed diodes interconnected and encapsulated in plastic to provide a single rugged package. It has two input pins for secondary voltage and two output pins for the load resistance.

SECOND APPROXIMATION

For most troubleshooting and design, the ideal diode is a practical approximation for rectifier circuits. The reason it is acceptable is because all voltages and currents already have a built-in tolerance of more than ± 10 percent (line-voltage variations and transformer effects).

Nevertheless, you may occasionally require more accurate answers. If so, you can subtract 0.6 to 0.7 V from the ideal peak load voltage of the half-wave and full-wave rectifiers, or subtract 1.2 to 1.4 V from the peak load voltage of bridge rectifiers. For instance, if the secondary voltage is 12.6 V ac, then the ideal peak load voltage of a bridge rectifier is 17.8 V. Including diode drops, this means that the output peak is reduced by about 1.2 to 1.4 V, so that $V_{out(peak)}$ is between 16.4 and 16.6 V.

The diode voltage drop is significant only when the output voltage is low. For example, if the peak secondary voltage to a bridge rectifier is only 5 V, then the peak rectified voltage is only 3.6 to 3.8 V. On the other hand, if the peak secondary voltage is 50 V, then the peak rectified voltage is 48.6 to 48.8 V, which is quite close to 50 V.

In the problems at the end of the chapter, we will keep things simple by ignoring diode drops. In other words, always use the ideal-diode approximation when analyzing rectifier circuits unless otherwise indicated.

3-6 THE CAPACITOR-INPUT FILTER

The output of an average rectifier is a pulsating dc voltage. The uses for this kind of output are limited to charging batteries, running dc motors, and a few other applications. What we really need for most electronic circuits is a dc voltage that is constant in value, similar to the voltage from a battery. To convert half-wave and full-wave signals into a constant dc voltage, we need to use a filter.

HALF-WAVE FILTERING

Figure 3-8*a* shows a *capacitor-input filter.* An ac source generates a sinusoidal voltage with a peak value of V_P. During the first quarter cycle of source voltage, the diode is forward-biased. Ideally, it looks like a closed switch (see Fig. 3-8*b*). Since the diode connects the source directly across the capacitor, the capacitor charges to the peak voltage V_P.

Just past the positive peak, the diode stops conducting, which means that the switch opens, as shown in Fig. 3-8*c*. Why? Because the capacitor has V_P volts across it with the polarity shown. Since the source voltage is slightly less than V_P, the diode goes into reverse bias.

With the diode now open, the capacitor discharges through the load resistance. But here is the key idea behind the capacitor-input filter: by deliberate design, the discharging time constant (the product of R_L and C) is much greater than the period T of the input signal. Because of this, the capacitor will lose only a small part of its charge during the off time of the diode, as shown in Fig. 3-8*d*.

When the source voltage again reaches its peak, the diode conducts briefly and recharges the capacitor to the peak voltage. In other words, after the capacitor is initially charged during the first quarter cycle, its voltage is approximately equal to the peak source voltage (see Fig. 3-8*d*).

The load voltage is now almost a perfect dc voltage. The only deviation from a pure dc voltage is the small *ripple* caused by charging and discharging the capacitor. The smaller the ripple is, the better. One way to reduce this ripple is by increasing the discharging time constant, which equals $R_L C$.

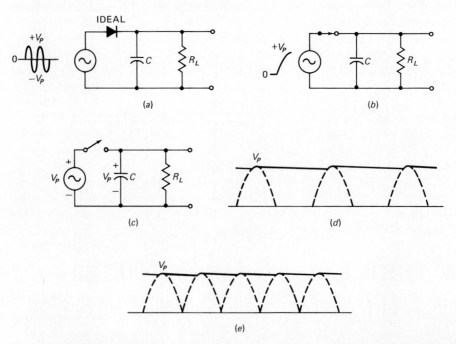

Fig. 3-8 (a) *Peak rectifier.* (b) *At positive peak.* (c) *Slightly past positive peak.* (d) *Half-wave output.* (e) *Full-wave output.*

FULL-WAVE FILTERING

Another way to reduce the ripple is to use a full-wave rectifier or bridge rectifier; then the ripple frequency is 120 Hz instead of 60 Hz. In this case, the capacitor is charged twice as often and has only half the discharge time (see Fig. 3-8*e*). As a result, the ripple is smaller and the dc output voltage more closely approaches the peak voltage. From now on, our analysis will emphasize the bridge rectifier driving a capacitor-input filter because this is the most commonly used configuration.

CONDUCTION ANGLE OF DIODE

In the average rectifiers discussed earlier, each diode had a conduction angle of 180°, meaning that each diode was turned on for approximately 180° of the cycle. In the peak rectifiers we are now discussing, each diode has a conduction angle of only a few degrees; this is because the diodes driving a capacitor-input filter turn on briefly near the peak and are off during the rest of the cycle.

MATHEMATICAL DERIVATION OF RIPPLE

In basic ac courses, capacitance is defined as

$$C = \frac{Q}{V}$$

An alternative form is

$$V = \frac{Q}{C}$$

Suppose the capacitor discharge begins when $t = T_1$. Then the initial voltage can be written as

$$V_1 = \frac{Q_1}{C}$$

If the capacitor discharge ends at $t = T_2$, then the final voltage is

$$V_2 = \frac{Q_2}{C}$$

The peak-to-peak ripple equals the difference of the foregoing voltages:

$$V_1 - V_2 = \frac{Q_1 - Q_2}{C}$$

To get something practical out of this, let us divide both sides by the time of discharge:

$$\frac{V_1 - V_2}{T_1 - T_2} = \frac{Q_1 - Q_2}{C(T_1 - T_2)}$$

When the time constant is much longer than the period of the ripple, the discharge time $T_1 - T_2$ approximately equals T, the period of ripple; therefore,

$$\frac{V_1 - V_2}{T} = \frac{Q_1 - Q_2}{CT}$$

Because the load voltage is almost constant, the load current is approximately constant, and the foregoing equation reduces to

$$\frac{V_1 - V_2}{T} = \frac{I}{C}$$

To get the final formula for ripple voltage, we will let V_{rip} represent $V_1 - V_2$, the peak-to-peak ripple. Furthermore, the ripple frequency f equals the reciprocal of the ripple period T. With the foregoing in mind, we can write

$$V_{rip} = \frac{I}{fC} \qquad\qquad (3\text{-}11)$$

where V_{rip} = peak-to-peak ripple voltage
   I = dc load current
   f = ripple frequency
   C = capacitance

Now you have arrived. You will use this formula many times when you are troubleshooting, analyzing, or designing a capacitor-input filter (the most common type nowadays). Memorize this key formula.

AN EXAMPLE

Suppose the dc load current is approximately 10 mA and the capacitance is 470 μF. Assuming a bridge rectifier and a line frequency of 60 Hz, the peak-to-peak ripple out of a capacitor-input filter is

$$V_{rip} = \frac{10 \text{ mA}}{120 \text{ Hz} \times 470 \ \mu\text{F}} = 0.177 \text{ V}$$

Given the same conditions in a half-wave rectifier, the ripple would be twice as much because the ripple frequency is only 60 Hz.

DESIGN GUIDELINE

If you are designing a capacitor-input filter, you need to choose a capacitor that is large enough to keep the ripple small. How small is small? This depends on how bulky a capacitor you're willing to use. As the ripple decreases, the capacitor becomes larger and more expensive.

As a compromise between small ripple and large capacitance, many designers use the 10 percent rule, which says to select a capacitor that will hold the peak-to-peak ripple at approximately 10 percent of the peak voltage. For instance, if the peak voltage is 15 V, then select a capacitor that makes the peak-to-peak ripple around 1.5 V. This may sound like too much ripple, but it is not. As you will see later, additional filtering is usually done with electronic circuits called *voltage regulators.*

Incidentally, the derivation of Eq. (3-11) assumed small ripple, but we will use it to estimate large ripple as well. Even though the answers we get will contain some error, they still are useful in practice. Why? Because the filter capacitor is typically an electrolytic capacitor with a tolerance of ±20 percent or more, so exact answers are unnecessary.

SAWTOOTH RATHER THAN EXPONENTIAL

The output of a capacitor-input filter is often connected to a voltage regulator (discussed in Chap. 19). This type of load does not act like a resistance. It acts like a constant-current sink, a load whose current is fixed even though its voltage varies. Because the filter capacitor supplies a constant current during its discharge, the ripple is a sawtooth wave rather than an exponential wave. Furthermore, this means that Eq. (3-11) is more accurate than an exponential equation when the filter capacitor drives a voltage regulator.

DC VOLTAGE

Ideally, the dc load voltage equals the peak voltage. Since we are going to allow up to 10 percent ripple, we can use a slightly more accurate formula:

$$V_{dc} = V_{2(peak)} - \frac{V_{rip}}{2} \tag{3-12}$$

As an example, if $V_{2(peak)} = 15$ V and $V_{rip} = 1.5$ V, then

$$V_{dc} = 15 \text{ V} - \frac{1.5 \text{ V}}{2} = 14.25 \text{ V}$$

With the 10 percent design rule, the dc load voltage is 95 percent of the peak voltage. As you see, the load voltage is still quite close to the peak value.

DIODE RATINGS

Figure 3-9*a* shows a half-wave rectifier driving a capacitor-input filter. Ideally, the dc load voltage equals $V_{2(peak)}$, producing a dc load current of I_{dc}. Since the average or dc current through a capacitor is zero, Kirchhoff's current law tells us that the dc diode current equals I_{dc}. In other words, the I_O rating of the diode in a half-wave rectifier must be greater than the dc load current.

At the negative peak of secondary voltage, the diode is reverse-biased and a peak inverse voltage appears across it. Summing voltages around the loop gives

$$\text{PIV} - V_{2(peak)} - V_{2(peak)} = 0$$

or

$$\text{PIV} = 2V_{2(peak)}$$

This means that the PIV rating of the diode must be greater than two times the peak secondary voltage.

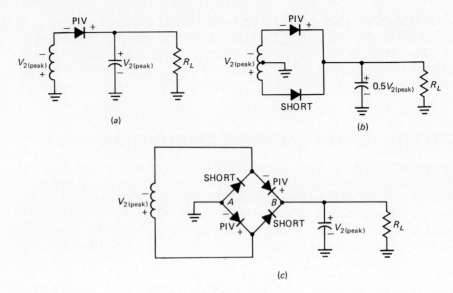

Fig. 3-9 *Peak inverse voltage.* (a) *Half-wave.* (b) *Full-wave.* (c) *Bridge.*

With the full-wave rectifier and filter shown in Fig. 3-9*b*, each diode has a dc current of half the dc load current. The argument is the same as before. To satisfy Kirchhoff's current law, the dc diode currents add to produce the dc load current. Furthermore, summing voltages around the left loop gives

$$\text{PIV} - V_{2(\text{peak})} + 0 = 0$$

where the 0 on the left side of this equation is the ideal voltage of the conducting diode. By solving, we get

$$\text{PIV} = V_{2(\text{peak})}$$

A similar proof exists for the other diodes. This means that the PIV rating of the diodes must be greater than the peak secondary voltage.

With the bridge circuit of Fig. 3-9*c*, the sum of the dc diode currents at node *B* equals the dc load current, so each of these diodes has an I_O of half the dc load current. Similarly in the ground return path, the sum of the dc diode currents at node *A* equals the dc load current, implying an I_O of half the load current for each diode. Therefore, each diode in the bridge must have an I_O rating greater than $I_{\text{dc}}/2$, half the dc load current.

Finally, we can sum voltages for the left loop (the one that includes node *A*) to get

$$\text{PIV} + 0 - V_{2(\text{peak})} = 0$$

or

$$\text{PIV} = V_{2(\text{peak})}$$

So the PIV rating of each diode in the bridge must be greater than the peak secondary voltage.

Table 3-3. Ideal Rectifiers with Capacitor-input Filters

	Half-wave	Full-wave	Bridge
Number of diodes	1	2	4
Dc output voltage	$V_{2(peak)}$	$0.5V_{2(peak)}$	$V_{2(peak)}$
Dc diode current	I_{dc}	$0.5I_{dc}$	$0.5I_{dc}$
Peak inverse voltage	$2V_{2(peak)}$	$V_{2(peak)}$	$V_{2(peak)}$
Ripple frequency	f_{in}	$2f_{in}$	$2f_{in}$
Dc output voltage	$1.41V_{2(rms)}$	$0.707V_{2(rms)}$	$1.41V_{2(rms)}$

COMPARISON

Table 3-3 summarizes the key ideas we have discussed. These values are for ideal diodes with negligible ripple and will serve as a starting point in troubleshooting and design. The entries in this table are ideal because they ignore the effects of ripple and diode drops. Both of these reduce the dc load voltage slightly. When you are dealing with low-voltage rectifiers, you may want to improve the answers as follows: Subtract 0.6 to 0.7 V from the dc load voltage of the half-wave and full-wave rectifiers; subtract 1.2 to 1.4 V from the dc load voltage of bridge rectifiers.

SURGE CURRENT

Before the power is turned on, the filter capacitor is uncharged. At the instant the circuit is energized, the capacitor looks like a short; therefore, the initial charging current may be quite large. This sudden gush of current is called the *surge current.*

In the worst case, the circuit may be energized at the instant when line voltage is at its maximum. This means that $V_{2(peak)}$ is across the secondary winding, and the capacitor is uncharged. The only thing that impedes current is the resistance of the windings and the bulk resistance of the diodes. We can symbolize this resistance as R_{TH}, the Thevenin resistance looking from the capacitor back toward the rectifier. Therefore, in the worst case,

$$I_{surge} = \frac{V_{2(peak)}}{R_{TH}} \tag{3-13}$$

For instance, suppose that the secondary voltage is 12.6 V ac and the Thevenin resistance facing the capacitor is 1.5 Ω. As found earlier, $V_{2(peak)} = 17.8$ V, which means a maximum surge current of

$$I_{surge} = \frac{17.8 \text{ V}}{1.5 \text{ Ω}} = 11.9 \text{ A}$$

This current starts to decrease as soon as the capacitor charges. If the capacitor is extremely large, however, the surge current can remain at a high level for a while and may cause diode damage.

Here is something to give you more insight into the problem. The secondary voltage has a period of

$$T = \frac{1}{f} = \frac{1}{60 \text{ Hz}} = 16.7 \text{ ms}$$

For a Thevenin resistance of 1 Ω, a capacitor of 1000 μF produces a time constant of 1 ms. This means that the capacitor can charge within a few milliseconds, a fraction of a cycle. This usually is not long enough to damage the diode.

When the capacitance is much larger than 1000 μF, the time constant becomes very long, and it may take many cycles to get the capacitor fully charged. If the surge current is too high, diode damage can easily occur, as can capacitor damage from heating and gas formation in the electrolyte.

DATA SHEETS

Data sheets list the surge-current rating as I_{surge}, $I_{FM(surge)}$, I_{FSM}, etc. You have to read the fine print here because this rating depends on the number of cycles needed to charge the filter capacitor. For instance, the surge-current rating of a 1N4001 is 30 A for one cycle, 24 A for two cycles, 18 A for four cycles, and so on. Most of the designs in this book will charge the filter capacitor within a fraction of a cycle. In fact, whenever the filter capacitor is less than 1000 μF, the capacitor usually charges in less than one cycle.

DESIGN HINTS

Suppose you are designing a rectifier circuit with a capacitor-input filter. What do you do about surge current? As before, you select a capacitance that will produce a ripple of about 10 percent of the dc load voltage. If the capacitance is less than 1000 μF, you usually can ignore the surge current, because it's unlikely to damage the rectifier diodes of a typical circuit.

On the other hand, if the filter capacitance is greater than 1000 μF, you may need to use winding resistance and bulk resistance to calculate the surge current, using Eq. (3-13). You can measure the winding resistances with an ohmmeter. And here is how to estimate the bulk resistance:

$$r_B = \frac{V_F - 0.7}{I_F} \qquad (3\text{-}14)$$

where r_B = bulk resistance
  V_F = forward voltage
  I_F = forward current

The quantities V_F and I_F are listed on data sheets. After you calculate the surge current, you select a diode whose surge-current rating is larger than your estimated surge current. Another solution is to include a surge resistor (see Example 3-4).

TROUBLESHOOTING

Let's wrap up our discussion of the capacitor-input filter by talking about troubleshooting. Every piece of electronic equipment has a *power supply*, typically a rectifier driving a capacitor-input filter followed by a voltage regulator (discussed later). This power supply provides dc voltages needed by transistors and other devices. If the

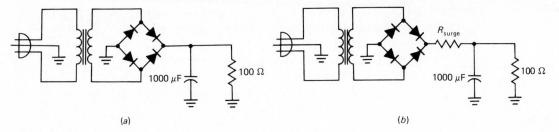

Fig. 3-10 *Surge current.*

equipment is not working properly, the first thing to test is the dc voltage out of the power supply.

You can check out the rectifier and capacitor-input filter as follows. Use a floating VOM to measure the secondary voltage (ac range). The same VOM (dc range) can be used to measure the dc load voltage. The ratio of dc to ac should agree with Table 3-2, at least approximately. If it doesn't, then look at the dc load voltage with an oscilloscope, checking the ripple. A peak-to-peak ripple of around 10 percent of the ideal load voltage is reasonable. (Note: The ripple may be somewhat more or less than this, depending on the design.) Furthermore, the ripple frequency should be 120 Hz for a full-wave rectifier or a bridge rectifier.

Here are some common troubles that arise and the symptoms they produce in bridge rectifiers with capacitor-input filters. If a diode is defective, the dc load voltage will be lower than it should be, and the ripple frequency will equal 60 Hz instead of 120 Hz. If the filter capacitor is open, the dc load voltage will be low, equal to the average instead of the peak because the output will be a full-wave signal. On the other hand, if the filter capacitor is shorted, one or more diodes may be ruined and the transformer may be damaged. Sometimes the filter capacitor becomes leaky with age, and this reduces the dc load voltage. Occasionally, shorted windings in the transformer reduce the dc output voltage. Besides these troubles, you can also have solder bridges, cold-solder joints, etc.

EXAMPLE 3-2

What is the dc load voltage and ripple in Fig. 3-10*a* for a secondary voltage of 17.7 V ac?

SOLUTION

First, calculate the peak secondary voltage:

$$V_{2(\text{peak})} = \frac{17.7 \text{ V}}{0.707} = 25 \text{ V}$$

Estimate the dc load voltage as

$$V_{\text{dc}} = V_{2(\text{peak})} = 25 \text{ V}$$

For this estimated dc load voltage, the dc load current is

$$I_{dc} = \frac{25 \text{ V}}{100 \text{ }\Omega} = 0.25 \text{ A}$$

Now calculate the peak-to-peak ripple:

$$V_{rip} = \frac{I}{fC} = \frac{0.25 \text{ A}}{120(1000 \text{ }\mu\text{F})} = 2.08 \text{ V}$$

Next, we can refine the answer for dc load voltage, originally estimated as 25 V:

$$V_{dc} = V_{2(peak)} - \frac{V_{rip}}{2} = 25 \text{ V} - \frac{2.08 \text{ V}}{2} = 24 \text{ V}$$

As a further refinement, we can include the effect of diode drops:

$$V_{dc} = 24 \text{ V} - 1.4 \text{ V} = 22.6 \text{ V}$$

and

$$I_{dc} = \frac{22.6 \text{ V}}{100 \text{ }\Omega} = 0.226 \text{ A}$$

EXAMPLE 3-3

The diodes used in Fig. 3-10*a* are 1N4001s. The data sheet lists a typical forward current of 1.5 A for a forward voltage of 1 V. The winding resistance is 0.8 Ω. Calculate the surge current.

SOLUTION

With Eq. (3-14), the bulk resistance is approximately

$$r_B = \frac{1 \text{ V} - 0.7 \text{ V}}{1.5 \text{ A}} = 0.2 \text{ }\Omega$$

The total bulk resistance (two diodes) is 0.4 Ω. Now calculate the surge current:

$$I_{surge} = \frac{V_{2(peak)}}{R_{TH}} = \frac{25 \text{ V}}{0.8 \text{ }\Omega + 0.4 \text{ }\Omega} = 20.8 \text{ A}$$

The data sheet of a 1N4001 gives a surge-current rating of 30 A for one cycle; therefore, there is no problem with surge current.

EXAMPLE 3-4

Sometimes a designer includes a surge resistor, as shown in Fig. 3-10*b*. This limits the surge current because it is added to the winding and bulk resistances. Calculate the maximum possible surge current for a secondary voltage of 17.7 V ac and a surge resistance of 2 Ω.

SOLUTION

As found earlier, the peak secondary voltage is 25 V. Suppose that we don't know the value of winding or bulk resistance. Because of the surge resistor, we are guaranteed a minimum Thevenin resistance of 2 Ω. Therefore, the surge current must be less than

$$I_{surge} = \frac{25 \text{ V}}{2 \text{ Ω}} = 12.5 \text{ A}$$

Using a surge resistor is an effective way to limit the surge current. You don't have to worry about winding resistance (which is seldom known) or bulk resistance (which varies from diode to diode). By including a surge resistor, the designer immediately nails down the maximum possible surge current.

Don't use surge resistors indiscriminately; they lower the dc load voltage. As previously mentioned, filter capacitors of less than 1000 μF usually do not produce surge current for a long enough time to damage the diodes of a typical rectifier circuit. Well above 1000 μF, the surge current will become a problem if the Thevenin resistance is too low. In this case, you may want to include a surge resistor instead of selecting a bigger diode.

EXAMPLE 3-5

Redesign the capacitor-input filter of Fig. 3-10*a* using the 10 percent rule.

SOLUTION

We already know that the peak secondary voltage is 25 V and the ideal load current is 0.25 A (found in Example 3-2). The 10 percent rule says to select a capacitor that produces a peak-to-peak ripple equal to 10 percent of the peak secondary voltage. This means a ripple of 2.5 V. Use Eq. (3-11), the ripple equation you memorized:

$$V_{rip} = \frac{I}{fC}$$

Substituting known values gives

$$2.5 \text{ V} = \frac{0.25 \text{ A}}{(120 \text{ Hz})C}$$

Solve for *C* to get

$$C = \frac{0.25 \text{ A}}{(120 \text{ Hz})(2.5 \text{ V})} = 833 \ \mu\text{F}$$

Actually, the circuit is well designed as it stands. The next standard size above 833 μF is 1000 μF. Remember that electrolytic capacitors usually have at least $\pm$20 percent tolerance. So, 1000 μF is a solid design choice because it will vary from 800 μF to 1200 μF in mass production.

3-7 *RC* AND *LC* FILTERS

With the 10 percent rule we get a dc load voltage with a peak-to-peak ripple of around 10 percent. Before the 1970s, passive filters were connected between the filter capacitor and the load to reduce the ripple to less than 1 percent. The whole idea was to get an almost perfect dc voltage, similar to what you get from a battery.

RC FILTER

Figure 3-11*a* shows two *RC filters* between the input capacitor and the load resistor. By deliberate design, R is much greater than X_C at the ripple frequency. Therefore, the ripple is dropped across the series resistors instead of across the load resistor. Typically, R is at least 10 times greater than X_C; this means that each section attenuates (reduces) the ripple by a factor of at least 10. The main disadvantage of the *RC* filter is the loss of dc voltage across each R. This means that the *RC* filter is suitable only for light loads (small load current or large load resistance).

LC FILTER

When the load current is large, the *LC filters* of Fig. 3-11*b* are an improvement over *RC* filters. Again, the idea is to drop the ripple across the series components, in this case, the inductors. This is accomplished by making X_L much greater than X_C at the ripple frequency. In this way, the ripple can be reduced to extremely low levels.

Fig. 3-11
(a) *RC filter.*
(b) *LC filter.*

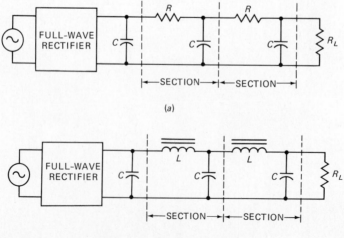

(a)

(b)

Furthermore, the dc voltage drop across the inductors is much smaller because only the winding resistance is involved.

The *LC* filter was quite popular at one time. Now, it's becoming obsolete in typical power supplies because of the size and cost of inductors. For low-voltage power supplies, the *LC* filter has been replaced by IC voltage regulators, active filters that reduce ripple and hold the final dc voltage constant. Chapter 19 tells you more about voltage regulators.

3-8 VOLTAGE MULTIPLIERS

A *voltage multiplier* is two or more peak rectifiers that produce a dc voltage equal to a multiple of the peak input voltage ($2V_P$, $3V_P$, $4V_P$, and so on). These power supplies are used for high-voltage/low-current devices like cathode-ray tubes (the picture tubes in TV receivers, oscilloscopes, and computer displays).

HALF-WAVE VOLTAGE DOUBLER

Figure 3-12a is a *voltage doubler*. At the peak of the negative half cycle, D_1 is forward-biased and D_2 is reverse-biased. Ideally, this charges C_1 to the peak voltage V_P with the polarity shown in Fig. 3-12b. At the peak of the positive half cycle, D_1 is reverse-biased and D_2 is forward-biased. Because the source and C_1 are in series, C_2 will try to charge toward $2V_P$. After several cycles, the voltage across C_2 will equal $2V_P$, as shown in Fig. 3-12c.

By redrawing the circuit and connecting a load resistance, we get Fig. 3-12d. Now it's clear that the final capacitor discharges through the load resistor. As long as R_L is large, the output voltage equals $2V_P$ (ideally). That is, provided the load is light (long time constant), the output voltage is double the peak input voltage. This input voltage normally comes from the secondary winding of a transformer.

For a given transformer, you can get twice as much output voltage as you get from a standard peak rectifier. This is useful when you are trying to produce high voltages

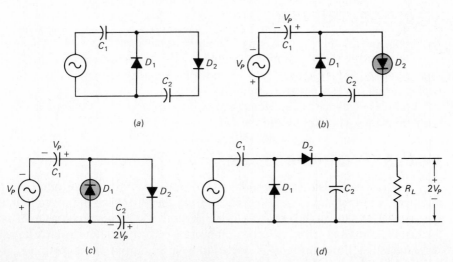

Fig. 3-12
Half-wave voltage doubler.

Fig. 3-13
*Full-wave voltage
doubler.*

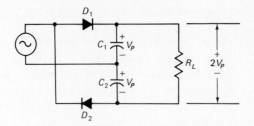

(several hundred volts or more). Why? Because higher secondary voltages result in bulkier transformers. At some point, a designer may prefer to use voltage doublers instead of bigger transformers.

The circuit is called a *half-wave doubler* because the output capacitor C_2 is charged only once during each cycle. As a result, the ripple frequency is 60 Hz. Sometimes you will see a surge resistor in series with C_1.

FULL-WAVE VOLTAGE DOUBLER

Figure 3-13 shows a *full-wave* voltage doubler. On the positive half cycle of the source, the upper capacitor charges to the peak voltage with the polarity shown. On the next half cycle, the lower capacitor charges to the peak voltage with the indicated polarity. For a light load, the final output voltage is approximately $2V_P$.

The circuit is called a full-wave voltage doubler because one of the output capacitors is being charged during each half cycle. Stated another way, the output ripple is 120 Hz. This ripple frequency is an advantage because it is easier to filter. Another advantage of the full-wave doubler is that the PIV rating of the diodes need only be greater than V_P.

The disadvantage of a full-wave doubler is the lack of a common ground between input and output. In other words, if we ground the lower end of the load resistor in Fig. 3-13, the source is floating. In the half-wave doubler of Fig. 3-12*d*, grounding the load resistor also grounds the source, an advantage in some applications.

VOLTAGE TRIPLER

By connecting another section, we get the *voltage tripler* of Fig. 3-14*a*. The first two peak rectifiers act like a doubler. At the peak of the negative half cycle, D_3 is forward-biased. This charges C_3 to $2V_P$ with the polarity shown in Fig. 3-14*b*. The tripler output appears across C_1 and C_3.

The load resistance is connected across the tripler output. As long as the time constant is long, the output approximately equals $3V_P$.

VOLTAGE QUADRUPLER

Figure 3-14*b* is a *voltage quadrupler*, four peak rectifiers in cascade (one after another). The first three are a tripler, and the fourth makes the overall circuit a quadrupler. As shown, the first capacitor charges to V_P; all others charge to $2V_P$. The quadrupler output is across the series connection of C_2 and C_4. As usual, a large load resistance (long time constant) is needed to have an output of approximately $4V_P$.

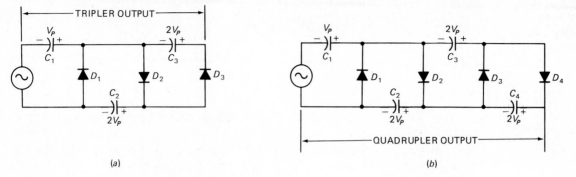

Fig. 3-14 (a) *Voltage tripler.* (b) *Voltage quadrupler.*

Theoretically, we can add sections indefinitely; however, the ripple gets worse as additional sections are added. This is why voltage multipliers are not used in low-voltage supplies, which are the supplies you encounter the most. As stated earlier, voltage multipliers are almost always used to produce high voltages, well into the hundreds or thousands of volts.

3-9 THE LIMITER

The diodes used in power supplies are rectifier diodes, those with a power rating greater than 0.5 W and optimized for use at 60 Hz. In the remainder of the chapter, we will be using small-signal diodes; these have power ratings of less than 0.5 W (with current in milliamperes rather than amperes) and are typically used at frequencies much greater than 60 Hz.

The first small-signal circuit to be discussed is the *limiter;* it removes signal voltages above or below a specified level. This is useful not only for signal shaping, but also for protecting circuits that receive the signal.

POSITIVE LIMITER

Figure 3-15 shows a positive limiter (sometimes called a clipper), a circuit that removes positive parts of the signal. As shown, the output voltage has all positive half cycles clipped off. The circuit works as follows: During the positive half of input voltage, the diode turns on. Ideally, the output voltage is zero; to a second approximation, it is approximately +0.7 V.

During the negative half cycle, the diode is reverse-biased and looks open. In many limiters, the load resistor R_L is at least 100 times greater than series resistor R. For this reason, the source is stiff and the negative half cycle appears at the output. (See Sec. 1-2 for a discussion of stiff voltage sources.)

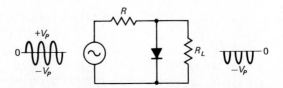

Fig. 3-15
Positive limiter.

Fig. 3-16
*Biased positive
limiter.*

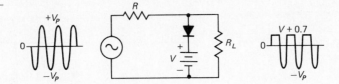

Figure 3-15 shows the output waveform. The positive half cycle has been clipped off. The clipping is not perfect. To a second approximation, a conducting silicon diode drops approximately 0.7 V. Because the first 0.7 V is used to overcome the barrier potential, the output signal is clipped near +0.7 V rather than 0 V.

If you reverse the polarity of the diode in Fig. 3-15, you get a negative limiter that removes the negative half cycle. In this case, the clipping level is near −0.7 V.

BIASED LIMITER

With the biased limiter of Fig. 3-16, you can move the clipping level to $V + 0.7$. When the input voltage is greater than $V + 0.7$, the diode conducts and the output is held at $V + 0.7$. When the input voltage is less than $V + 0.7$, the diode opens and the circuit becomes a voltage divider. As before, load resistance should be much greater than the series resistance; then the source is stiff and all the input voltage reaches the output.

A COMBINATION LIMITER

You can combine biased positive and negative limiters, as shown in Fig. 3-17. Diode D_1 turns on when the input voltage exceeds $V_1 + 0.7$; this is the positive clipping level. Similarly, diode D_2 conducts when the input is more negative than $-V_2 - 0.7$; this is the negative clipping level. When the input signal is large, that is, when V_P is much greater than the clipping levels, the output signal resembles a square wave like that of Fig. 3-17.

VARIATIONS

Using batteries to set the clipping level is impractical. One approach is to add more silicon diodes because each produces an offset of 0.7 V. For instance, Fig. 3-18*a* shows two diodes in a positive limiter. Since each diode has an offset of around 0.7 V, the pair of diodes produces a clipping level of approximately +1.4 V. Figure 3-18*b* extends this idea to four diodes; this results in a clipping level of approximately +2.8 V. There is no limit on the number of diodes used, and this is practical because diodes are inexpensive.

Fig. 3-17
*Combination
limiter.*

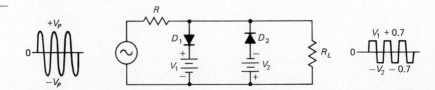

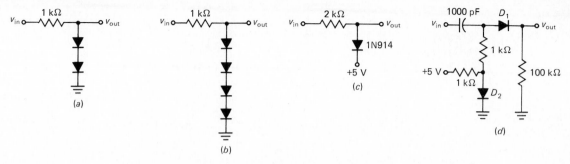

Fig. 3-18 *Limiters.* (a) *Two-diode offset.* (b) *Four-diode offset.* (c) *Diode clamp.* (d) *Biased near zero.*

Limiters are sometimes used to protect the load against excessive voltage. For instance, Fig. 3-18c shows a 1N914 protecting the load (not shown) against excessively large input voltages. The 1N914 conducts when the input exceeds +5.7 V. In this way, a destructively large input voltage like +100 V never reaches the load because the diode clips at +5.7 V, the maximum voltage to reach the load.

Incidentally, a circuit like Fig. 3-18c is often called a *diode clamp* because it holds the signal at a fixed level. It literally clamps the output voltage at +5.7 V when the input voltage exceeds this level. A typical use for a diode clamp is to protect the load.

Sometimes a variation like Fig. 3-18d is used to remove the offset of limiting diode D_1. Here is the idea: Diode D_2 is biased slightly into forward conduction, so that it has approximately 0.7 V across it. This 0.7 V is applied to 1 kΩ in series with D_1 and 100 kΩ. This means that diode D_1 is on the verge of conduction. Therefore, when a signal comes in, diode D_1 conducts near 0 V.

3-10 THE DC CLAMPER

The diode clamp is a variation of the limiter discussed in the preceding section. The *dc clamper* is different, so don't confuse the similar-sounding words. A dc clamper adds a dc voltage to the signal. For instance, if the incoming signal swings (varies) from −10 V to +10 V, a positive dc clamper will produce an output that ideally swings from 0 to +20 V. (A negative dc clamper would produce an output between 0 and −20 V.)

POSITIVE CLAMPER

Figure 3-19a shows a positive dc clamper. Ideally, here is how it works. On the first negative half cycle of input voltage, the diode turns on, as shown in Fig. 3-19b. At the negative peak, the capacitor must charge to V_P with the polarity shown.

Slightly beyond the negative peak, the diode shuts off, as shown in Fig. 3-19c. The R_LC time constant is deliberately made much greater than the period T of the incoming signal. For this reason, the capacitor remains almost fully charged during the off time of the diode. To a first approximation, the capacitor acts like a battery of V_P volts. This is why the output voltage in Fig. 3-19a is a positively clamped signal.

Fig. 3-19
*Positive dc
clamper.*

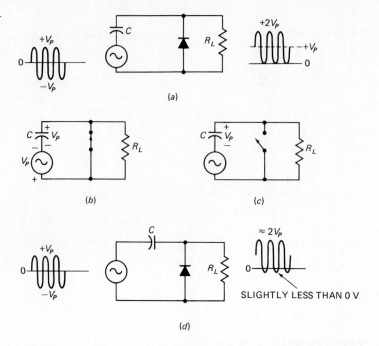

Figure 3-19*d* shows the circuit as it is usually drawn. Since the diode drops 0.7 V when conducting, the capacitor voltage does not quite reach V_P. For this reason, the dc clamping is not perfect, and the negative peaks are at -0.7 V.

NEGATIVE CLAMPER

What happens if we turn the diode in Fig. 3-19*d* around? The polarity of capacitor voltage reverses, and the circuit becomes a negative clamper. Both positive and negative clampers are widely used. Television receivers, for instance, use a dc clamper to add a dc voltage to the video signal. In television work, the dc clamper is usually called a *dc restorer.*

MEMORY AID

To remember which way the dc level of a signal moves, look at Fig. 3-19*d*. Notice that the diode arrow points upward, the same direction as the dc shift. In other words, when the diode points upward, you have a positive dc clamper. When the diode points downward, the circuit is a negative dc clamper.

3-11 THE PEAK-TO-PEAK DETECTOR

If you cascade a dc clamper and a peak detector (same as a peak rectifier), you get a *peak-to-peak detector* (see Fig. 3-20). The input sine wave is positively clamped; therefore, the input to the peak detector has a peak value of $2V_P$. This is why the output of the peak detector is a dc voltage equal to $2V_P$.

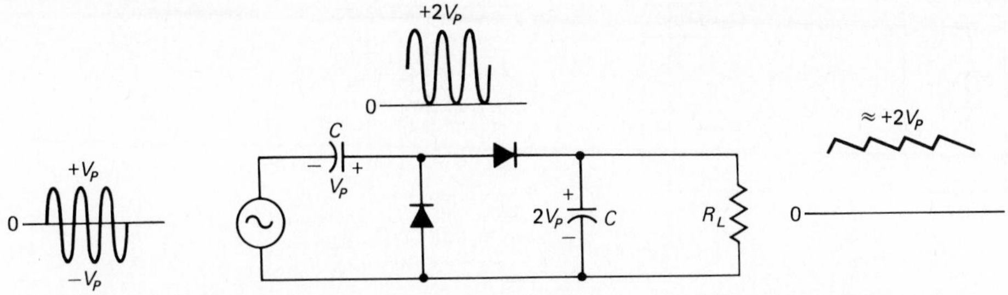

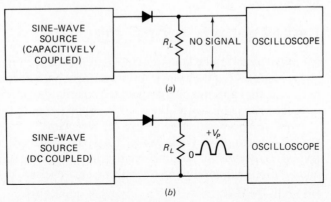

Fig. 3-20
*Peak-to-peak
detector.*

As usual, the discharge time constant R_LC must be much greater than the period of the incoming signal. By satisfying this condition, you get good clamping action and good peak detection. The output ripple will therefore be small.

Where are peak-to-peak detectors used? Sometimes the output of a peak-to-peak detector is applied to a dc voltmeter. The combination acts like a peak-to-peak ac voltmeter. For instance, suppose that a signal swings from -20 V to $+50$ V. If you try to measure this with an ordinary ac voltmeter, you will get an incorrect reading. If you use a peak-to-peak detector in front of a dc voltmeter, you will read 70 V for the peak-to-peak value of the signal.

3-12 THE DC RETURN

One of the most baffling things that may happen in the laboratory is this: You connect a signal source to a circuit and for some reason the circuit will not work, yet nothing is defective in the circuit or the signal source. As a concrete example, Fig. 3-21a shows a sine-wave source driving a half-wave rectifier. When you look at the output with an oscilloscope, you see no signal at all; the rectifier refuses to work. To add to the confusion, you may try another sine-wave source and find a normal half-wave signal across the load (Fig. 3-21b).

The phenomenon just described is a classic in electronics; it occurs again and again in practice. It may happen with diode circuits, transistor circuits, integrated circuits, etc. Unless you understand why one kind of source works and another doesn't, you will be confused and possibly discouraged every time the problem arises.

Fig. 3-21
*The dc-return
problem.*

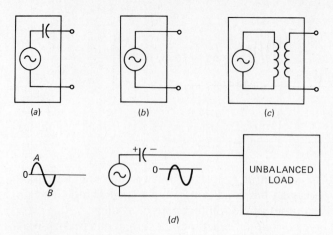

Fig. 3-22   (a) *Capacitively coupled source.* (b) *Direct-coupled source.* (c) *Transformer-coupled source.* (d) *Unbalanced load causes unequal charging currents.*

TYPES OF COUPLING

The signal source of Fig. 3-22*a* is *capacitively coupled;* this means that it has a capacitor in the signal path. Many commercial signal generators use a capacitor to dc-isolate the source from the load. The idea of a capacitively coupled source is to let only ac signal pass from source to load.

The *dc-coupled* source of Fig. 3-22*b* is different. It has no capacitor; therefore, it provides a path for both alternating and direct currents. When you connect this kind of source to the load, it is possible for the load to force a direct current through the source. As long as this direct current is not too large, no damage to the source occurs. Many commercial signal generators are dc-coupled like this.

Sometimes a signal source is *transformer-coupled,* as in Fig. 3-22*c*. The advantage is that it passes the ac signal and at the same time provides a dc path through the secondary winding.

All circuits discussed earlier in this chapter work with dc-coupled and transformer-coupled sources. It is only with capacitively coupled sources that the trouble may arise.

UNBALANCED DIODE CIRCUITS

An *unbalanced load* is one that has more resistance on one half cycle than on the other. Figure 3-22*d* shows an unbalanced load. If the current is greater on the positive half cycle, the capacitor charges with the polarity shown. As we saw with dc clampers, a charged capacitor causes the dc level of the signal to shift.

Now we know why a half-wave rectifier won't work when it is connected to a capacitively coupled source. In Fig. 3-23*a*, the capacitor charges to V_P during the first few cycles. Because of this, the signal coming from the source is negatively clamped, and the diode cannot turn on after the first few cycles. This is why we see no signal on the oscilloscope.

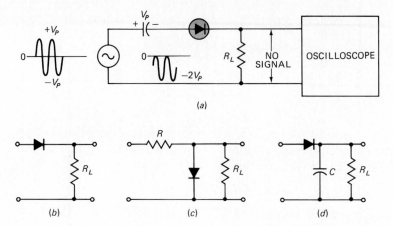

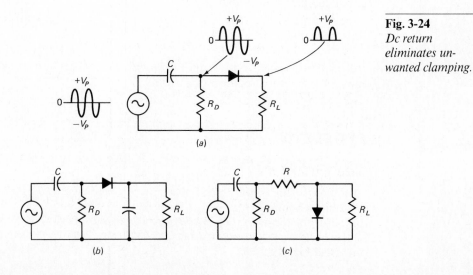

Fig. 3-23 *Capacitively coupled source produces unwanted clamping.*

Among the diode circuits discussed earlier, the following are unbalanced loads: the half-wave rectifier, the limiter, the peak detector, the dc clamper, and the peak-to-peak detector. The last two are supposed to dc-clamp the signal; therefore, they work fine with a capacitively coupled source. But the half-wave rectifier, the limiter, and the peak detector of Fig. 3-23b, c, and d will not work with a capacitively coupled source because of unwanted dc clamping.

DC RETURN

Is there a remedy for unwanted dc clamping? Yes. You can add a *dc-return* resistor across the input to the unbalanced circuit (see Fig. 3-24a). This resistor R_D allows the capacitor to discharge during the off time of the diode. In other words, any charge deposited on the capacitor plates is removed during the alternate half cycle.

Fig. 3-24
Dc return eliminates unwanted clamping.

Fig. 3-25
*Bridge is a
balanced load.*

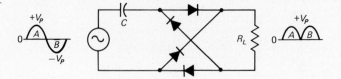

The size of R_D is not critical. The main idea in preventing unwanted dc clamping is to keep the discharging resistance R_D less than or equal to the charging resistance in series with the diode. In Fig. 3-24a, this means

$$R_D \leq R_L$$

When this condition is satisfied, only a slight shift occurs in the dc level of the signal. The same rule applies to Fig. 3-24b. (For best results, make R_D less than one-tenth of R_L. This gives a highly balanced load with negligible dc shift.)

The limiter of Fig. 3-24c is slightly different. When the diode is conducting, the charging resistance in series with the diode equals R instead of R_L. Therefore, in Fig. 3-24c, the rule is

$$R_D \leq R$$

When possible, R_D should be less than one-tenth of R.

BALANCED DIODE CIRCUITS

Some diode circuits are *balanced loads.* Examples are the full-wave rectifier and the bridge rectifier. These circuits work fine with a capacitively coupled source. In other words, no dc return is needed because the equal and opposite half-cycle currents produce an average capacitor voltage of zero. For instance, Fig. 3-25 shows a capacitively coupled source. Assuming identical diodes, the currents during each half cycle are equal and opposite, so that the capacitor voltage equals zero and the bridge rectifier works normally.

In summary, you may get clamping action with a capacitively coupled source. This unwanted clamping can occur in diode circuits, transistor circuits, integrated circuits, etc. In general, whenever a capacitor drives a device that conducts for only part of the ac cycle, you will get unwanted clamping action. When it happens, you can eliminate it by adding a dc return.

PROBLEMS

Straightforward

3-1. Line voltage typically is 115 V ac ±10 percent. Calculate the peak value for a low line and a high line.

3-2. The transformer of Fig. 3-26a has a secondary voltage of 30 V ac. What is the peak voltage across the load resistance? The average voltage? The average current through the load resistance?

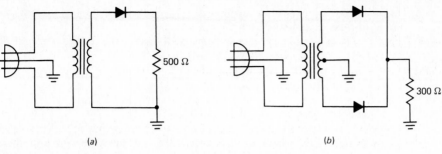

Fig. 3-26

(a)   (b)

3-3. Here are some diodes and their I_O ratings:
  a. 1N914: $I_O = 50$ mA
  b. 1N3070: $I_O = 100$ mA
  c. 1N4002: $I_O = 1$ A
  d. 1N1183: $I_O = 35$ A
  If the secondary voltage is 115 V ac in Fig. 3-26a, which of the foregoing diode types can be used?

3-4. Here are some diodes and their PIV ratings:
  a. 1N914: 20 V
  b. 1N1183: 50 V
  c. 1N4002: 100 V
  d. 1N3070: 175 V
  Given a secondary voltage of 60 V ac in Fig. 3-26a, what is the PIV across the diode? Which of the foregoing diodes can be used?

3-5. In Fig. 3-26b, the secondary voltage is 40 V ac. What is the peak load voltage? The dc load voltage? The dc load current?

3-6. If the secondary voltage is 60 V ac in Fig. 3-26b, which of the diodes listed in Probs. 3-3 and 3-4 has sufficient I_O and PIV ratings to be used?

3-7. Given a secondary voltage of 40 V ac in Fig. 3-26b, calculate the dc load current and the PIV across each diode. What is the average rectified current through each diode?

3-8. If the secondary voltage in Fig. 3-27a is 30 V ac, what is the dc load voltage? The dc load current? The PIV across each diode?

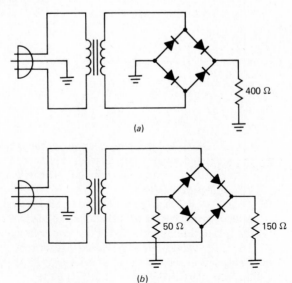

Fig. 3-27

(a)

(b)

3-9. The secondary voltage in Fig. 3-27a is 60 V ac. What is the dc load current? The dc current through each diode? The PIV across each diode?

3-10. In Fig. 3-27b, the secondary voltage is 40 V ac. What are the dc load voltages? The dc load currents? The dc current through each diode? The PIV across each diode?

3-11. The diodes of Fig. 3-27a have an I_O rating of 150 mA and a PIV rating of 75 V. Will these diodes be adequate if the secondary voltage is 40 V ac?

3-12. If the diodes of Fig. 3-27b have an I_O rating of 0.5 A and a PIV rating of 50 V, will they be adequate with a secondary voltage of 60 V ac?

3-13. A bridge rectifier with a capacitor-input filter has a peak output voltage of 25 V. If the load resistance is 220 Ω and the capacitance is 500 μF, what is the peak-to-peak ripple?

3-14. The secondary voltage is 21.2 V ac in Fig. 3-28a. What is the dc load voltage if $C = 220$ μF? The peak-to-peak ripple? The minimum I_O and PIV ratings of the diodes?

Fig. 3-28

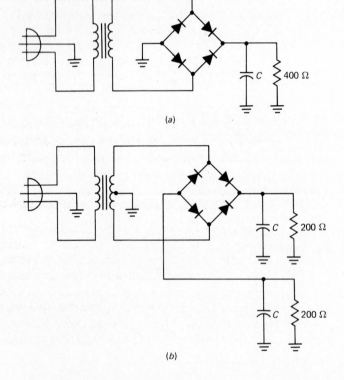

(a)

(b)

3-15. Figure 3-28b is a split supply. Because of the grounded center tap, the output voltages are equal and opposite in polarity. What are the dc output voltages for a secondary voltage of 17.7 V ac and $C = 500$ μF? The peak-to-peak ripple? The minimum I_O and PIV ratings of the diodes?

3-16. A surge resistor of 4.7 Ω is added to Fig. 3-28a. If the secondary voltage is 25 V ac, what is the maximum possible value of surge current?

3-17. Given a surge resistor of 3.3 Ω between the secondary winding and the bridge of Fig. 3-28b, what is the largest possible surge current if the secondary voltage is 25 V ac?

3-18. In Fig. 3-29a, what is the ideal load voltage? The PIV across each diode?

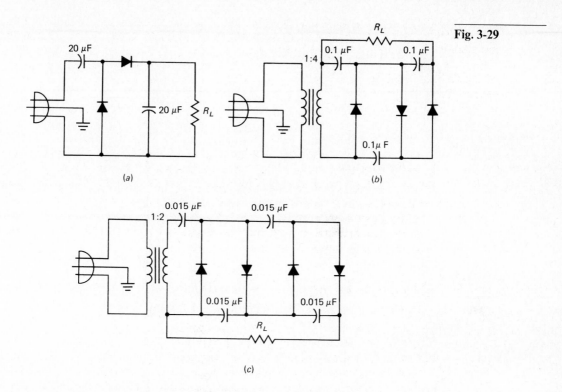

Fig. 3-29

(a)

(b)

(c)

3-19. In Fig. 3-29*b*, what is the ideal load voltage? The PIV across each diode?

3-20. In Fig. 3-29*c*, what is the ideal load voltage? The PIV across each diode? The voltage across each capacitor?

3-21. In Fig. 3-30*a*, the secondary voltage is 900 V ac. What is the ideal dc load voltage? The dc load current? Because the two capacitors are in series, the net filter capacitance is 1 μF. Calculate the peak-to-peak ripple.

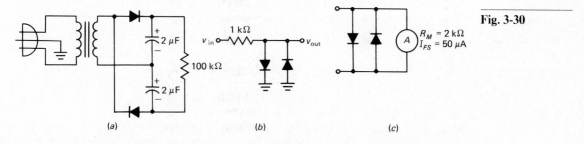

Fig. 3-30

(a)

(b)

(c)

3-22. A sine wave with a peak of 50 V drives the circuit shown in Fig. 3-30*b*. Describe the output voltage.

3-23. The ammeter of Fig. 3-30*c* has a meter resistance of 2 kΩ and a full-scale current of 50 μA. What is the voltage across this ammeter when it reads full scale? Diodes are sometimes shunted across an ammeter, as shown in Fig. 3-30*c*. If the ammeter is connected in series with a circuit whose Thevenin resistance is 1 kΩ, the diodes across the meter may serve a very useful purpose. What do you think this purpose is?

3-24. Figure 3-31 shows a power supply sometimes used in inexpensive consumer products. What is the peak-to-peak ripple? The dc load voltage? What advantage do you think the circuit has? What disadvantage?

Fig. 3-31

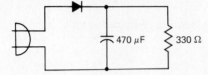

Troubleshooting

3-25. You measure 24 V ac across the secondary of Fig. 3-28*a*. Next you measure 21.6 V dc across the load resistor. Suggest some possible troubles.

3-26. The dc load voltage of Fig. 3-28*a* measures somewhat lower than it should. Looking at the ripple with a scope, you discover that it has a frequency of 60 Hz. What are some of the possible causes of this?

3-27. There is no voltage out of the circuit of Fig. 3-28*a*. What are some of the possible causes?

3-28. Checking with an ohmmeter, you find all diodes in Fig. 3-28*a* open. You replace the diodes. What else should you check before you power up?

Design

3-29. You are designing a bridge rectifier with a capacitor-input filter. The specifications are a dc load voltage of 15 V and a ripple of 1 V for a load resistance of 680 Ω. How much rms voltage should the secondary winding produce for a line voltage of 115 V ac? What size should the filter capacitor be? What are the minimum I_O and PIV ratings for your diodes?

3-30. Design a full-wave rectifier using a 48-V ac center-tapped transformer that produces a 10 percent ripple across a capacitor-input filter with a load resistance of 330 Ω. What are the minimum I_O and PIV ratings of the diodes?

3-31. The split supply of Fig. 3-28*b* has a secondary voltage of 25 V ac. Select filter capacitors, using the 10 percent rule for ripple.

3-32. Design a power supply to meet the following specifications: The secondary voltage is 12.6 V ac, one dc output is approximately 17.8 V at 120 mA, and a second dc output is around 35.6 V at 75 mA. What are the minimum I_O and PIV ratings of the diodes?

Challenging

3-33. The full-wave signal of Fig. 3-5*b* has a dc value of 0.636 times the peak value. With your calculator or a table of sine values you can derive the average value of 0.636. Describe how you would do it.

3-34. The secondary voltage in Fig. 3-32 is 25 V ac. With the switch in the position shown, what is the ideal output voltage? With the switch in the lower position, what is the ideal output voltage?

Fig. 3-32

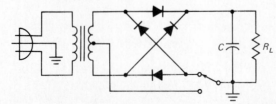

3-35. A rectifier diode has a forward voltage of 1.2 V at 2 A. The winding resistance is 0.3 Ω. If the secondary voltage is 25 V ac, what is the surge current?

Computer

3-36. The INPUT statement allows you to enter data during a computer run. For instance, here is a program:

10 PRINT "ENTER R"
20 INPUT R
30 PRINT "THE VALUE OF R IS"
40 PRINT R

When this program is run, the computer will stop at line 20 and wait until you enter the value of *R*. After you key in a value, the computer will display the value of *R* on the screen.

Here is a program:

10 PRINT "ENTER R1"
20 INPUT R1
30 PRINT "ENTER R2"
40 INPUT R2
50 RT = R1 + R2
60 PRINT "THE TOTAL RESISTANCE IS"
70 PRINT RT

Describe what the program does.

3-37. Write a program that works out the parallel resistance of two resistances that will be entered from the keyboard.

3-38. Write a program that computes the peak-to-peak ripple. Use three INPUT statements to get the current, frequency, and capacitance into the computer.

Special-Purpose Diodes

Rectifier and small-signal diodes are optimized for rectification. But this is not all a diode can do. Now we will discuss diodes used in nonrectifier applications. The chapter begins with the zener diode, which is optimized for its breakdown properties. Zener diodes are very important because they are the key to voltage regulation. The chapter also includes optoelectronic diodes, Schottky diodes, varactors, and other diodes.

4-1 THE ZENER DIODE

Small-signal and rectifier diodes are never intentionally operated in the breakdown region because this may damage them. A *zener diode* is different; it is a silicon diode that the manufacturer has optimized for operation in the breakdown region. In other words, unlike ordinary diodes that never work in the breakdown region, zener diodes work best in the breakdown region. Sometimes called a breakdown diode, the zener diode is the backbone of voltage regulators, circuits that hold the load voltage almost constant despite large changes in line voltage and load resistance.

I-V GRAPH

Figure 4-1*a* shows the schematic symbol of a zener diode; Fig. 4-1*b* is an alternative symbol. In either symbol, the lines resemble a "z," which stands for zener. By varying the doping level of silicon diodes, a manufacturer can produce zener diodes with breakdown voltages from about 2 to 200 V. These diodes can operate in any of three regions: forward, leakage, or breakdown.

Figure 4-1*c* shows the *I-V* graph of a zener diode. In the forward region, it starts conducting around 0.7 V, just like an ordinary silicon diode. In the leakage region (between zero and breakdown), it has only a small leakage or reverse current. In a zener diode, the breakdown has a very sharp knee, followed by an almost vertical increase in current. Note that the voltage is almost constant, approximately equal to V_Z over most of the breakdown region. Data sheets usually specify the value of V_Z at a particular test current I_{ZT}.

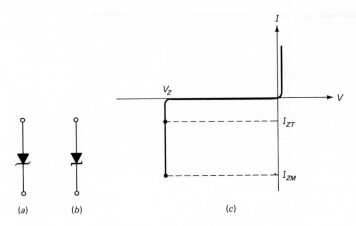

Fig. 4-1 *Zener diode. (a) Symbol. (b) Alternative symbol. (c) I-V curve.*

MAXIMUM RATINGS

The power dissipation of a zener diode equals the product of its voltage and current. In symbols,

$$P_Z = V_Z I_Z \qquad (4\text{-}1)$$

For instance, if $V_Z = 12$ V and $I_Z = 10$ mA, then

$$P_Z = 12 \text{ V} \times 10 \text{ mA} = 120 \text{ mW}$$

As long as P_Z is less than the power rating, the zener diode can operate in the breakdown region without being destroyed. Commercially available zener diodes have power ratings from $\frac{1}{4}$ W to more than 50 W.

Data sheets sometimes include the *maximum current* a zener diode can handle without exceeding its power rating. This maximum current is related to the power rating as follows:

$$I_{ZM} = \frac{P_{ZM}}{V_Z} \qquad (4\text{-}2)$$

where I_{ZM} = maximum rated zener current
P_{ZM} = power rating
V_Z = zener voltage

As an example, a 12-V zener diode with a power rating of 400 mW has a current rating of

$$I_{ZM} = \frac{400 \text{ mW}}{12 \text{ V}} = 33.3 \text{ mA}$$

In other words, if there's enough current-limiting resistance to keep the zener current less than 33.3 mA, then the zener diode can operate in the breakdown region without burning out.

ZENER RESISTANCE

When a zener diode is operating in the breakdown region, an increase in current produces a slight increase in voltage. This implies that a zener diode has a small ac resistance. Data sheets specify *zener resistance* (often called zener impedance) at the same test current I_{ZT} used to measure V_Z. The zener resistance at this test current is designated R_{ZT} (or Z_{ZT}). For instance, the data sheet of a 1N3020 lists the following: $V_{ZT} = 10$ V, $I_{ZT} = 25$ mA, and $Z_{ZT} = 7\ \Omega$. This tells us that the zener diode has a voltage of 10 V and a zener resistance of 7 Ω when the current is 25 mA.

VOLTAGE REGULATION

A zener diode is sometimes called a *voltage-regulator* diode because it maintains a constant output voltage even though the current through it changes. For normal operation, you have to reverse-bias the zener diode, as shown in Fig. 4-2a. Furthermore, to produce breakdown, the source voltage V_S must be greater than the zener breakdown voltage V_Z. A series resistor R_S is always used to limit the zener current to less than its current rating; otherwise, the zener diode will burn out like any device with too much power dissipation.

The voltage across the series resistor equals the difference of the source voltage and the zener voltage, $V_S - V_Z$. Therefore, the current through the resistor is

$$I_S = \frac{V_S - V_Z}{R_S} \tag{4-3}$$

Because this is a one-loop circuit, the zener current I_Z equals I_S.

Equation (4-3) can be used to construct the load line as previously discussed. For instance, suppose $V_S = 20$ V and $R_S = 1$ kΩ. Then the foregoing equation reduces to

$$I_Z = \frac{20 - V_Z}{1000}$$

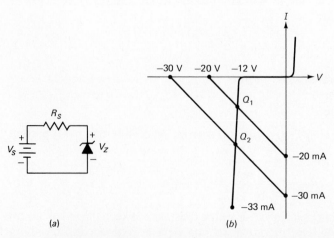

Fig. 4-2   (a) *Zener-diode circuit.* (b) *Two Q points have approximately the same voltage.*

As before, we get the saturation point (vertical intercept) by setting V_Z equal to zero and solving for I_Z to get 20 mA. Similarly, to get the cutoff point (horizontal intercept), we set I_Z equal to zero and solve for V_Z to get 20 V.

Alternatively, you can get the ends of the load line as follows. Visualize Fig. 4-2a with $V_S = 20$ V and $R_S = 1$ kΩ. With the zener diode shorted, the maximum diode current is 20 mA. With the diode open, the maximum diode voltage is 20 V.

Suppose the zener diode has a breakdown voltage of 12 V. Then its *I-V* graph appears as shown in Fig. 4-2b. When we plot the load line for $V_S = 20$ V and $R_S = 1$ kΩ, we get the upper load line with an intersection point of Q_1. The voltage across the zener diode will be slightly more than the knee voltage at breakdown because the *I-V* curve slopes slightly.

To understand how voltage regulation works, assume that the source voltage changes to 30 V. Then the zener current changes to

$$I_Z = \frac{30 - V_Z}{1000}$$

This implies that the ends of the load line are 30 mA and 30 V, as shown in Fig. 4-2b. The new intersection is at Q_2. Compare Q_2 with Q_1 and you can see that there is more current through the zener diode but approximately the same zener voltage. Therefore, even though the source voltage has changed from 20 to 30 V, the zener voltage is still approximately equal to 12 V. This is the basic idea of voltage regulation; the output voltage has remained almost constant even though the input voltage has changed by a large amount.

IDEAL ZENER DIODE

For troubleshooting and preliminary design, we can approximate the breakdown region as vertical. This means that the voltage is constant even though the current changes, which is equivalent to ignoring the zener resistance. Figure 4-3a shows the ideal approximation of a zener diode. This means that a zener diode operating in the breakdown region ideally acts like a battery. In a circuit, it means that you can mentally replace a zener diode by a voltage source V_Z, provided the zener diode is operating in the breakdown region.

SECOND APPROXIMATION

In the *I-V* graph the breakdown region is not quite vertical, implying a zener resistance. Sometimes in design problems it is necessary to take this zener resistance

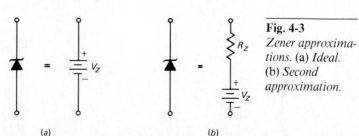

(a)   (b)

Fig. 4-3
Zener approxima-tions. (a) *Ideal.*
(b) *Second approximation.*

into account. Even though R_Z is small, it causes changes of a few tenths of a volt when the current changes significantly.

Figure 4-3*b* shows how to visualize a zener diode when using the second approximation. Here you see a zener resistance (relatively small) in series with an ideal battery. This resistance produces more *IR* drop as the current increases. For instance, the voltage at Q_1 is

$$V_1 = I_1 R_Z + V_Z$$

and the voltage at Q_2 is

$$V_2 = I_2 R_Z + V_Z$$

The change in voltage is

$$V_2 - V_1 = (I_2 - I_1)R_Z$$

This is usually written as

$$\Delta V_Z = \Delta I_Z R_Z \tag{4-4}$$

where ΔV_Z = change in zener voltage
ΔI_Z = change in zener current
R_Z = zener resistance

This tells us that the change in zener voltage equals the change in zener current times the zener resistance. Usually R_Z is small, and so the voltage change is slight.

EXAMPLE 4-1

The zener diode of Fig. 4-4*a* has $V_Z = 10$ V. Use the ideal zener approximation to work out the minimum and maximum zener current.

SOLUTION

The applied voltage, 20 to 40 V, is greater than the breakdown voltage of the zener diode. Therefore, we can visualize the zener diode like the battery shown in Fig. 4-4*b*. This means that the output voltage is a constant 10 V for any source voltage between 20 and 40 V.

The current through the series resistor is

$$I_S = \frac{V_S - 10 \text{ V}}{820 \ \Omega}$$

Because of the one-loop circuit, the zener current equals this resistor current. Furthermore, the minimum zener current is

$$I_{Z(\text{min})} = \frac{20 \text{ V} - 10 \text{ V}}{820 \ \Omega} = 12.2 \text{ mA}$$

and the maximum zener current is

$$I_{Z(\text{max})} = \frac{40 \text{ V} - 10 \text{ V}}{820 \ \Omega} = 36.6 \text{ mA}$$

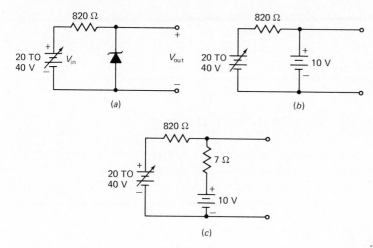

Fig. 4-4 *Voltage regulation.* (a) *Circuit.* (b) *Ideal zener approximation.* (c) *Second zener approximation.*

Here is the point: In a voltage regulator like Fig. 4-4a, the output voltage is held constant at 10 V, despite the change in source voltage from 20 to 40 V. The greater source voltage produces more zener current, but ideally the output voltage holds rock-solid at 10 V.

EXAMPLE 4-2

Suppose the zener diode of the preceding example has a zener resistance of 7 Ω. Use the second approximation to calculate the change in zener voltage when the source voltage changes from 20 to 40 V.

SOLUTION

Figure 4-4c shows the second approximation of the zener diode. Because the zener resistance is only 7 Ω, it is swamped out by the much larger series resistance of 820 Ω. In other words, R_Z has practically no effect on the zener current, which still varies from approximately 12.2 to 36.6 mA when the source voltage increases from 20 to 40 V.

There is a slight change in output voltage as the current varies. With Eq. (4-4),

$$\Delta V_Z = (36.6 \text{ mA} - 12.2 \text{ mA}) \times 7 \text{ } \Omega = 0.171 \text{ V}$$

This means that the zener voltage, nominally 10 V, increases 0.171 V when the source changes from 20 to 40 V. Again, the example illustrates voltage regulation; there is only a small output voltage change even though there is a large input voltage change.

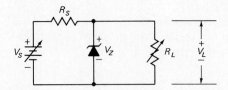

4-2 THE ZENER REGULATOR

Figure 4-5 shows a zener diode used to regulate the voltage across a load resistance. This is slightly more complicated than the zener circuits previously analyzed because the circuit has two loops. Nevertheless, the basic idea remains the same; the zener diode operates in the breakdown region and holds the load voltage almost constant.

THEVENIN VOLTAGE

Whether you are troubleshooting or designing, you will have to know certain basic relationships between the currents and voltages. To begin with, the first thing to check for is breakdown operation of the zener diode. Because of the load resistor, the Thevenin voltage driving the zener diode is less than the source voltage.

What is the Thevenin voltage driving the zener diode? Visualize the zener diode removed from the circuit. A voltage divider remains, formed by R_S and R_L. If you are visualizing this correctly, you will see a Thevenin voltage of

$$V_{TH} = \frac{R_L}{R_S + R_L} V_S \tag{4-5}$$

For breakdown operation of the zener diode, V_{TH} must be greater than V_Z. This is the first relationship that must be satisfied in any zener regulator.

SERIES CURRENT

Assuming that the zener diode is operating in the breakdown region, we proceed as follows. The current through the series resistor is given by

$$I_S = \frac{V_S - V_Z}{R_S} \tag{4-6}$$

This is Ohm's law applied to the current-limiting resistor.

LOAD CURRENT

Since the zener resistance typically has a very small effect, we can closely approximate the load voltage by

$$V_L \cong V_Z \tag{4-7}$$

(The symbol $\cong$ means "approximately equals.") This allows us to use Ohm's law to calculate the load current:

$$I_L = \frac{V_L}{R_L} \tag{4-8}$$

ZENER CURRENT

Because of the two-loop circuit, the series current splits at the junction of the zener diode and the load resistor. With Kirchhoff's current law,

$$I_S = I_Z + I_L$$

We can rearrange this to get a relationship for zener current:

$$I_Z = I_S - I_L \tag{4-9}$$

AN EXAMPLE

Equations (4-5) through (4-9) are adequate for all preliminary analysis of zener regulators. For instance, in Fig. 4-6 the first thing to work out is the Thevenin voltage driving the zener diode:

$$V_{TH} = \frac{2 \text{ k}\Omega}{2.82 \text{ k}\Omega} \, 40 \text{ V} = 28.4 \text{ V}$$

This is enough to produce breakdown operation because it is greater than the zener voltage.

Next, calculate the series current:

$$I_S = \frac{40 \text{ V} - 10 \text{ V}}{820 \text{ }\Omega} = 36.6 \text{ mA}$$

We know that the load voltage is approximately 10 V, and so the load current is

$$I_L = \frac{10 \text{ V}}{2 \text{ k}\Omega} = 5 \text{ mA}$$

The zener current is the difference between the series current and the load current:

$$I_Z = 36.6 \text{ mA} - 5 \text{ mA} = 31.6 \text{ mA}$$

RIPPLE ACROSS THE LOAD RESISTOR

In Fig. 4-7*a*, notice that a rectifier with a capacitor-input filter drives a zener regulator. This has two effects. First, the load voltage is held approximately constant despite changes in peak rectified voltage caused by power-line changes. Second, the zener regulator reduces the peak-to-peak ripple. This should make sense; after all, ripple is another way of saying that the source voltage changes.

How much reduction is there in ripple? Visualize the zener diode replaced by its

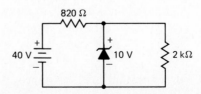

Fig. 4-6

Fig. 4-7
Effect on ripple.
(a) *Zener regula-*
tor. (b) *Second*
approximation.

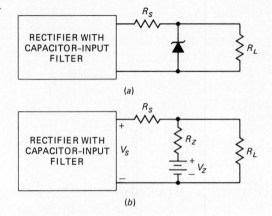

(a)

(b)

second approximation, as shown in Fig. 4-7*b*. At beginning of the discharge, the current through the series resistor is

$$I_{S(\text{max})} \cong \frac{V_{S(\text{max})} - V_Z}{R_S}$$

At the end of the discharge,

$$I_{S(\text{min})} \cong \frac{V_{S(\text{min})} - V_Z}{R_S}$$

Subtracting these equations gives

$$I_{S(\text{max})} - I_{S(\text{min})} \cong \frac{V_{S(\text{max})} - V_{S(\text{min})}}{R_S}$$

which is usually written as

$$\Delta I_S \cong \frac{\Delta V_S}{R_S}$$

Rearranging gives

$$\Delta V_S \cong \Delta I_S R_S$$

This says that the peak-to-peak input ripple equals the change in series current times the series resistance.

Earlier, we derived the change in zener voltage:

$$\Delta V_Z \cong \Delta I_Z R_Z$$

If these are the maximum changes in Fig. 4-7*b*, then the peak-to-peak ripple across the zener diode equals the change in zener current times the zener resistance. Taking the ratio of output ripple to input ripple gives

$$\frac{\Delta V_Z}{\Delta V_S} = \frac{\Delta I_Z R_Z}{\Delta I_S R_S}$$

For a constant load resistance, the change in zener current equals the change in source current, and so the foregoing ratio reduces to

$$\frac{\Delta V_Z}{\Delta V_S} \cong \frac{R_Z}{R_S} \tag{4-10}$$

where ΔV_Z = output ripple
ΔV_S = input ripple
R_Z = zener resistance
R_S = series resistance

This is a useful equation because it tells you at a glance how the output and input ripple are related. The equation says that the ratio of output to input ripple equals the ratio of zener to series resistance. For instance, if zener resistance is 7 Ω and the series resistance is 700 Ω, then the output ripple is $\frac{1}{100}$ of the input ripple.

ZENER DROPOUT POINT

For a zener regulator to hold the output voltage constant, the zener diode must remain in the breakdown region under all operating conditions; this is equivalent to saying that there must be zener current for all source voltages and load currents. The worst case occurs for minimum source voltage and maximum load current because the zener current drops to a minimum. For this case,

$$I_{S(\text{min})} = \frac{V_{S(\text{min})} - V_Z}{R_{S(\text{max})}}$$

which can be rearranged as

$$R_{S(\text{max})} = \frac{V_{S(\text{min})} - V_Z}{I_{S(\text{min})}} \tag{4-11}$$

As shown earlier,

$$I_Z = I_S - I_L$$

In the worst case, this is written as

$$I_{Z(\text{min})} = I_{S(\text{min})} - I_{L(\text{max})}$$

The critical point occurs when the maximum load current equals the minimum series current:

$$I_{L(\text{max})} = I_{S(\text{min})}$$

At this point the zener current drops to zero and regulation is lost.

By substituting $I_{L(\text{max})}$ for $I_{S(\text{min})}$ in Eq. (4-11), we get this useful design relationship:

$$R_{S(\text{max})} = \frac{V_{S(\text{min})} - V_Z}{I_{L(\text{max})}} \tag{4-12}$$

where $R_{S(\text{max})}$ = critical value of series resistance
$V_{S(\text{min})}$ = minimum source voltage
V_Z = zener voltage
$I_{L(\text{max})}$ = maximum load current

The *critical* resistance $R_{S(\text{max})}$ is the maximum allowable series resistance. The series resistance R_S must always be less than the critical value; otherwise, breakdown operation is lost and the regulator stops working. In this case, we lose the constant load voltage, and the ripple becomes almost as large as the input ripple.

STIFF ZENER REGULATOR

In this book, a zener regulator is *stiff* when it satisfies these two conditions:

$$R_Z \leq 0.01R_S$$
$$R_Z \leq 0.01R_L$$

By meeting the first condition, the zener regulator reduces source-voltage changes, including ripple, by a factor of at least 100.

By meeting the second condition, the zener regulator appears as a stiff voltage source to the load. Why? Because in Fig. 4-7b the load resistor looks back at a Thevenin voltage of approximately V_Z and a Thevenin resistance of approximately R_Z. In other words, thevenizing the circuit gives the equivalent circuit shown in Fig. 4-8. If R_Z is equal to or less than $0.01R_L$, then the zener regulator appears as a stiff voltage source to the load.

When you design zener regulators, try to satisfy the 100:1 rule for series and load resistances; this will guarantee that the zener diode is regulating very well against source and load changes. Sometimes it is impossible to satisfy the 100:1 rule using a zener regulator. Then you have two choices. First, you can settle for a less-than-stiff zener regulator. Second, you can include a transistor in the design to get a stiff voltage regulator; this is discussed in Chap. 8.

TEMPERATURE COEFFICIENT

One final point: Raising the *ambient* (surrounding) temperature changes the zener voltage slightly. On data sheets the effect of temperature is listed under *temperature coefficient,* which is the percent change per degree change. When designing, you may need to calculate the change in zener voltage at the highest ambient temperature.

For zener diodes with breakdown voltages of less than 5 V, the temperature coefficient is negative. For zener diodes with breakdown voltages of more than 6 V, the temperature coefficient is positive. Between 5 and 6 V, the temperature coefficient changes from negative to positive; this means that you can find an operating point for a zener diode at which the temperature coefficient is zero. This is important in some applications where a solid zener voltage is needed over a large temperature range.

Fig. 4-8
Equivalent circuit seen from load.

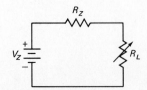

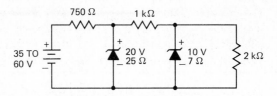

Fig. 4-9
*Preregulator
drives output reg-
ulator.*

EXAMPLE 4-3

A zener regulator has an input voltage from 15 to 20 V and a load current from 5 to 20 mA. If the zener voltage is 6.8 V, what value should the series resistor have?

SOLUTION

The worst case occurs for minimum source voltage and maximum load current. With Eq. (4-12), the critical series resistance is

$$R_{S(\text{max})} = \frac{15 \text{ V} - 6.8 \text{ V}}{20 \text{ mA}} = 410 \text{ } \Omega$$

Therefore, the series resistor must be less than 410 Ω for the zener diode to operate in the breakdown region under the worst-case conditions.

EXAMPLE 4-4

Figure 4-9 shows an input regulator (called a *preregulator*) driving an output regulator. The source has a large ripple that varies from 35 to 60 V. What is the final output voltage and ripple?

SOLUTION

First, notice that the preregulator has an output voltage of 20 V. This is the input to the second zener regulator, whose output is 10 V. The basic idea is to provide the second regulator with a well-regulated input, so that the final output is extremely well regulated.

The first zener diode has a zener resistance of 25 Ω. For the preregulator, the ratio of zener to series resistance is

$$\frac{R_Z}{R_S} = \frac{25 \text{ } \Omega}{750 \text{ } \Omega} = 0.0333$$

This means that the ripple coming out of the preregulator is

$$\Delta V = 0.0333(25 \text{ V}) = 0.833 \text{ V}$$

The second zener diode has a zener resistance of 7 Ω, which gives a ratio of

$$\frac{R_Z}{R_S} = \frac{7 \text{ } \Omega}{1000 \text{ } \Omega} = 0.007$$

So, the final output ripple is

$$\Delta V_{\text{out}} = 0.007(0.833 \text{ V}) = 0.00583 \text{ V}$$

Fig. 4-10
Zener diode used in combination limiter.

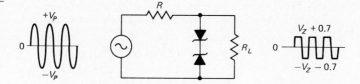

The use of a preregulator is common whenever you want very stiff regulation against changes in source voltage. In a cascade connection of zener regulators (cascade means that the output of one circuit is the input of another), the overall ripple reduction is given by the product of the resistance ratios. In symbols,

$$\frac{\Delta V_{\text{out}}}{\Delta V_{\text{in}}} = \frac{R_{Z1}}{R_{S1}} \times \frac{R_{Z2}}{R_{S2}}$$

EXAMPLE 4-5

What does the circuit of Fig. 4-10 do?

SOLUTION

In most applications, zener diodes are used in voltage regulators, where they remain in the breakdown region. But there are exceptions. Sometimes zener diodes are used in waveshaping circuits like limiters. Figure 4-10 is an example.

Notice the back-to-back connection of two zener diodes. On the positive half cycle, the upper diode conducts at approximately 0.7 V and the lower diode breaks down at V_Z. Therefore, when the input voltage exceeds $V_Z + 0.7$, the output is clipped as shown.

On the negative half cycle, the action is reversed. The lower diode conducts and the upper diode breaks down. In this way, the output is almost a square wave. The larger the input sine wave, the better looking the output square wave.

The circuit of Fig. 4-10 is an alternative way to build the combination clipper discussed in Chap. 3. By selecting different zener voltages, we can move the clipping level to whatever level we want.

4-3 OPTOELECTRONIC DEVICES

Optoelectronics is the technology that combines optics and electronics. This exciting field includes many devices based on the action of a *pn* junction. Examples of optoelectronic devices are light-emitting diodes, photodiodes, optocouplers, etc.

LIGHT-EMITTING DIODE

In a forward-biased diode, free electrons cross the junction and fall into holes. As these electrons fall from a higher to a lower energy level, they radiate energy. In ordinary diodes this energy goes off in the form of heat. But in the *light-emitting diode* (LED), the energy radiates as light. LEDs have replaced incandescent lamps in many applications because of their low voltage, long life, and fast on-off switching.

Ordinary diodes are made of silicon, an opaque material that blocks the passage of light. LEDs are different. By using elements like gallium, arsenic, and phosphorus, a manufacturer can produce LEDs that radiate red, green, yellow, blue, orange, or infrared (invisible). LEDs that produce visible radiation are useful with instruments, calculators, etc. The infrared LED finds applications in burglar-alarm systems and other areas requiring invisible radiation.

LED VOLTAGE AND CURRENT

LEDs have a typical voltage drop from 1.5 to 2.5 V for currents between 10 and 50 mA. The exact voltage drop depends on the LED current, color, tolerance, etc. Unless otherwise specified, use a nominal drop of 2 V when troubleshooting or analyzing the LED circuits in this book. If you get into design, you will have to consult data sheets, because LED voltages have a large tolerance.

Figure 4-11*a* shows the schematic symbol for a LED; the outward arrows symbolize the radiated light. Allowing 2 V for the LED drop, we can calculate a LED current of

$$I = \frac{10\text{ V} - 2\text{ V}}{680\ \Omega} = 11.8\text{ mA}$$

Typically, LED current is between 10 and 50 mA because this range provides adequate light for most applications.

DESIGN GUIDELINE

The brightness of a LED depends on the current. Ideally, the best way to control brightness is by driving the LED with a current source. The next best thing to a current source is a large supply voltage and a large series resistance. In this case, the LED current is given by

$$I = \frac{V_S - V_{\text{LED}}}{R_S}$$

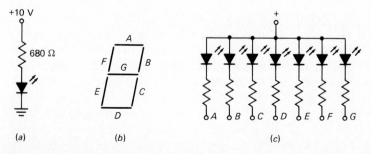

Fig. 4-11 (a) *LED circuit.* (b) *Seven-segment indicator.* (c) *Schematic diagram.*

Fig. 4-12
Photodiode.

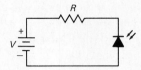

The larger the source voltage, the less effect V_{LED} has. In other words, a large V_S swamps out the variation in LED voltage.

For example, a TIL222 is a green LED with a minimum drop of 1.8 V and a maximum drop of 3 V for a current of approximately 25 mA. If you drive a TIL222 with a 20-V source and a 750-Ω resistor, the current varies from 22.7 to 24.3 mA. This implies a brightness that is essentially the same for all TIL222s. On the other hand, suppose your design uses a 5-V source and a 120-Ω resistor. Then the current varies from about 16.7 to 26.7 mA; this results in a noticeable change in brightness. Therefore, to get approximately constant brightness with LEDs, use as large a supply voltage and series resistor as possible.

SEVEN-SEGMENT DISPLAY

Figure 4-11*b* shows a *seven-segment* display; it contains seven rectangular LEDs (A through G). Each LED is called a segment because it forms part of the character being displayed. Figure 4-11*c* is the schematic diagram of the seven-segment display; external series resistors are included to limit the currents to safe levels. By grounding one or more resistors, we can form any digit from 0 through 9. For instance, by grounding A, B, and C, we get a 7. Grounding A, B, C, D, and G produces a 3.

A seven-segment display can also display the capital letters A, C, E, and F, plus lower-case letters b and d. Microprocessor trainers often use seven-segment displays that show all digits from 0 through 9, plus A, b, C, d, E, and F.

PHOTODIODE

As previously discussed, one component of reverse current in a diode is the flow of minority carriers. These carriers exist because thermal energy keeps dislodging valence electrons from their orbits, producing free electrons and holes in the process. The lifetime of the minority carriers is short, but while they exist they can contribute to the reverse current.

When light energy bombards a *pn* junction, it too can dislodge valence electrons. Stated another way, the amount of light striking the junction can control the reverse current in a diode. A *photodiode* is one that has been optimized for its sensitivity to light. In this diode, a window lets light pass through the package to the junction. The incoming light produces free electrons and holes. The stronger the light, the greater the number of minority carriers and the larger the reverse current.

Figure 4-12 shows the schematic symbol for a photodiode. The inward arrows represent the incoming light. Especially important, the source and the series resistor reverse-bias the photodiode. As the light becomes brighter, the reverse current increases. With typical photodiodes, the reverse current is in the tens of microamperes.

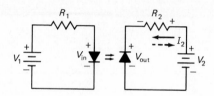

Fig. 4-13
Optocoupler.

The photodiode is one example of a *photodetector,* an optoelectronic device that can convert incoming light into an electrical quantity. Later chapters will introduce you to other photodetectors.

OPTOCOUPLER

An *optocoupler* (also called an optoisolator or an optically coupled isolator) combines a LED and a photodetector in a single package. Figure 4-13 shows one example of an optocoupler; it has a LED on the input side and a photodiode on the output side. Source voltage V_1 and series resistor R_1 produce a current through the LED. In turn, the light from the LED hits the photodiode, and this sets up a reverse current I_2. Summing voltages around the output loop gives

$$V_{out} - V_2 + I_2R_2 = 0$$

or

$$V_{out} = V_2 - I_2R_2$$

Notice that the output voltage depends on the reverse current I_2. If input voltage V_1 is varying, the amount of light is fluctuating. This means that the output voltage is varying in step with the input voltage. This is why the combination of a LED and a photodiode is called an optocoupler; the device can couple an input signal to the output circuit.

The key advantage of an optocoupler is the electrical isolation between the input and output circuits. With an optocoupler, the only contact between the input and the output is a beam of light. Because of this, it is possible to have an insulation resistance between the two circuits in the thousands of megohms. Isolation like this comes in handy in high-voltage applications, where the potentials of the two circuits may differ by several thousand volts.

4-4 THE SCHOTTKY DIODE

At lower frequencies an ordinary diode can easily turn off when the bias changes from forward to reverse. But as the frequency increases, the diode reaches a point where it cannot turn off fast enough to prevent noticeable current during part of the reverse half cycle. This section discusses the cause of the problem and one solution to it.

CHARGE STORAGE

Figure 4-14*a* shows a forward-biased diode, and Fig. 4-14*b* illustrates the energy bands. As you can see, conduction-band electrons have diffused across the junction

Fig. 4-14
(a) *Forward-biased
diode.* (b) *Charge
storage in energy
bands.*

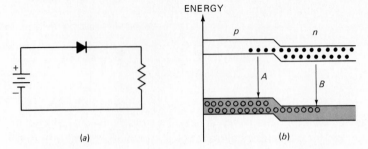

and traveled into the *p* region before recombining (path *A*). Similarly, holes have crossed the junction and traveled into the *n* region before recombination occurs (path *B*). If the lifetime equals 1 μs, free electrons and holes exist for an average of 1 μs before recombination takes place.

Because of the lifetime of minority carriers, charges in a forward-biased diode are temporarily stored in different energy bands near the junction. The greater the forward current, the larger the number of stored charges. This effect is referred to as *charge storage*.

REVERSE RECOVERY TIME

Charge storage is important when you try to switch a diode from on to off. Why? Because if you suddenly reverse-bias a diode, the stored charges can flow in the reverse direction for a while. The greater the lifetime, the longer these charges can contribute to reverse current.

For example, suppose a forward-biased diode is suddenly reverse-biased, as shown in Fig. 4-15*a*. A large reverse current can exist for a while because of the stored charges shown in Fig. 4-15*b*. Until the stored charges either cross the junction or recombine, the reverse current can remain large.

The time it takes to turn off a forward-biased diode is called the *reverse recovery time* t_{rr}. The conditions for measuring t_{rr} vary from one manufacturer to the next. As a guide, t_{rr} is the time it takes for reverse current to drop to 10 percent of forward current. For instance, the 1N4148 has a t_{rr} of 4 ns. If this diode has a forward current

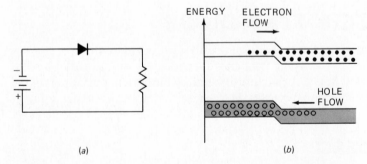

Fig. 4-15   (a) *Reverse-biased diode.* (b) *Stored charges can flow for a while in reverse direction.*

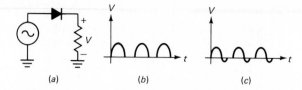

(a) (b) (c)

Fig. 4-16 (a) *Half-wave rectifier.* (b) *Normal output.* (c) *Distorted output with charge storage.*

of 10 mA and it is suddenly reverse-biased, it will take approximately 4 ns for the reverse current to decrease to 1 mA.

Reverse recovery time is so short in small-signal diodes that you don't even notice its effect at frequencies below 10 MHz or so. It's only when you get well above 10 MHz that you have to take t_{rr} into account.

EFFECT ON RECTIFICATION

What effect does reverse recovery time have on rectification? Take a look at the half-wave rectifier shown in Fig. 4-16a. At low frequencies the output is well behaved because it is the classic half-wave rectified signal, shown in Fig. 4-16b.

As the frequency increases well into megahertz, however, the output signal begins to deviate from its normal shape, as shown in Fig. 4-16c. As you can see, some conduction is noticeable near the beginning of the reverse half cycle. The reverse recovery time is now becoming a significant part of the period. For instance, if $t_{rr} = 4$ ns and the period is 50 ns, then the early part of the reverse half cycle will have a wiggle in it similar to Fig. 4-16c.

ELIMINATING CHARGE STORAGE

What is the solution to reverse recovery time? A *Schottky* diode. This special-purpose diode uses a metal like gold, silver, or platinum on one side of the junction and doped silicon (typically *n*-type) on the other side (see Fig. 4-17a). When a Schottky diode (also called a hot-carrier diode) is unbiased, free electrons on the *n* side are in smaller orbits than free electrons on the metal side. This difference in orbit size is called the Schottky barrier.

When the diode is forward-biased, free electrons on the *n* side can gain enough energy to travel in larger orbits. Because of this, free electrons can cross the junction and enter the metal, producing a large forward current. Since the metal has no holes, there is no charge storage and no reverse recovery time.

Figure 4-17b shows the schematic symbol of a Schottky diode. As a memory aid,

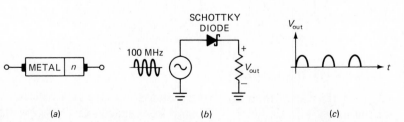

(a) (b) (c)

Fig. 4-17
Schottky diode.
(a) *Structure.* (b)
Circuit. (c) *Output at 100 MHz.*

notice that the lines look almost like a rectangular S. The lack of charge storage means that the Schottky diode can switch off faster than an ordinary diode. In fact, a Schottky diode can easily rectify frequencies above 300 MHz. As an example, the signal source of Fig. 4-17b has a frequency of 100 MHz. Despite this, the rectified output appears like the half-wave signal of Fig. 4-17c.

One important application of Schottky diodes is in digital computers. The speed of computers depends on how fast their diodes and transistors can turn on and off. This is where the Schottky diode comes in. Because it has no charge storage, the Schottky diode has become the backbone of low-power Schottky TTL, a group of widely used digital devices.

FORWARD VOLTAGE DROP

One final point: In the forward direction, a Schottky diode has an offset voltage of approximately 0.25 V. Therefore, another important application of Schottky diodes is in low-voltage rectifiers because you have to subtract only 0.25 V instead of 0.7 V for each diode when using the second approximation. You often will see Schottky diodes used instead of silicon diodes in low-voltage power supplies.

4-5 THE VARACTOR

The *varactor* (also called a voltage-variable capacitance, varicap, epicap, and tuning diode) is widely used in television receivers, FM receivers, and other communications equipment. Let's find out more about this special-purpose silicon diode.

DIODE CAPACITANCE

When reverse-biased, a small-signal diode has a reverse resistance well into the megohms. At low frequencies, the diode is approximated as an open circuit. But at high frequencies there is another path for current that has to be taken into account.

In Fig. 4-18a, the depletion layer is between the p region and the n region. When reverse-biased, a silicon diode resembles a capacitor; the p and n regions are like the plates of a capacitor, and the depletion layer is like the dielectric. The external circuit can charge this capacitance by removing valence electrons from the p side and adding free electrons to the n side. The action is the same as removing electrons from one capacitor plate and depositing them on the other plate.

The foregoing diode capacitance is called the *transition* capacitance, designated C_T. The word "transition" refers to the transition from p-type to n-type material. Transition capacitance is also known as depletion-layer capacitance, barrier capacitance, and junction capacitance. The thing that makes transition capacitance useful is this: Since the depletion layer gets wider with more reverse voltage, the transition capacitance becomes smaller. It's as though you moved the plates of a capacitor apart. The key idea is that capacitance is controlled by voltage.

EQUIVALENT CIRCUIT

Figure 4-18b shows the equivalent circuit for a reverse-biased diode. A large reverse resistance R_R is in parallel with the transition capacitance C_T. At low frequencies, the

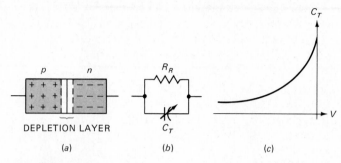

Fig. 4-18   *Varactor. (a) Structure. (b) Equivalent circuit. (c) Transition capacitance versus reverse voltage.*

capacitance is negligible, and the diode appears essentially open because R_R approaches infinity. At higher frequencies, however, the voltage-controlled capacitance becomes the dominant effect. Figure 4-18c indicates how the capacitance varies with reverse voltage.

When a varactor is connected in parallel with an inductor, you get a resonant circuit whose tuned frequency is

$$f_r = \frac{1}{2\pi\sqrt{LC}}$$

This is the principle behind tuning in a radio station, a TV channel, etc. More will be said about this in later chapters.

REFERENCE CAPACITANCE, TUNING RANGE, VOLTAGE RANGE

Varactors are silicon diodes optimized for their variable capacitance. Because the capacitance is voltage-controlled, they have replaced mechanically tuned capacitors in many applications, such as television receivers and automobile radios. Data sheets for varactors list a reference value of capacitance measured at a specific reverse voltage, typically −4 V. For instance, the data sheet of a 1N5142 lists a reference capacitance of 15 pF at −4 V.

Besides the reference value of capacitance, data sheets give a tuning range and a voltage range. For example, along with the reference value of 15 pF, the data sheet of a 1N5142 shows a tuning range of 3:1 for a voltage range of −4 to −60 V. This means that the capacitance decreases from 15 pF to 5 pF when the voltage varies from −4 to −60 V.

ABRUPT AND HYPERABRUPT JUNCTIONS

The tuning range of a varactor depends on the doping level. For instance, Fig. 4-19a shows the doping profile for an *abrupt-junction* diode (the ordinary type of diode). Notice that the doping is uniform on both sides of the junction; this means that the number of holes and free electrons are equally distributed. The tuning range of an abrupt-junction diode is between 3:1 and 4:1.

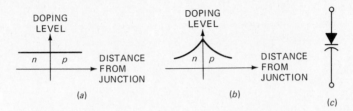

Fig. 4-19 (a) *Abrupt doping profile.* (b) *Hyperabrupt doping profile.* (c) *Symbol.*

To get larger tuning ranges, some varactors have a *hyperabrupt* junction, one whose doping profile looks like Fig. 4-19*b*. This profile tells us that the density of charges increases as we approach the junction. The heavier concentration leads to a narrower depletion layer and a larger capacitance. Furthermore, changes in reverse voltage have more pronounced effects on capacitance. A hyperabrupt varactor has a tuning range of about 10:1, enough to tune an AM radio through its frequency range (535 to 1605 kHz).

Whether a varactor is abrupt or hyperabrupt, we will use the schematic symbol shown in Fig. 4-19*c*.

4-6 OTHER DIODES

Besides the special-purpose diodes discussed so far, there are a few others you should know about. What follows is a brief description. Later chapters will supplement this information as needed.

CONSTANT-CURRENT DIODES

There are diodes that work in a way exactly opposite to zener diodes. Instead of holding the voltage constant, these diodes hold the current constant. Known as *constant-current* diodes (and also as current-regulator diodes), these devices keep the current through them fixed when the voltage changes. For example, the 1N5305 is a constant-current diode with a typical current of 2 mA over a voltage range of 2 to 100 V.

Incidentally, the *voltage compliance* (or simply compliance) of a current source is the range of voltage over which it can operate. The 1N5305 has a compliance of 98 V because it can operate normally between 2 and 100 V.

STEP-RECOVERY DIODES

The *step-recovery* diode has the unusual doping profile shown in Fig. 4-20*a*. This profile indicates that the density of carriers decreases near the junction. This unusual distribution of carriers causes a phenomenon called reverse snapoff. Figure 4-20*b* shows the schematic symbol for a step-recovery diode. During the positive half cycle, the diode conducts like any silicon diode. But during the negative half cycle, reverse current exists for a while because of the stored charges, then suddenly the reverse current drops to zero. Figure 4-20*c* shows the resulting output voltage. It's as though

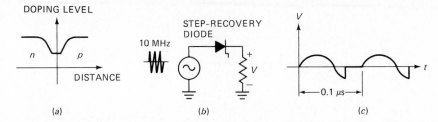

Fig. 4-20 *Step-recovery diode.* (a) *Doping profile.* (b) *Circuit.* (c) *Output contains sharp spike.*

the diode suddenly snapped open; this is why a step-recovery diode is often called a snap diode.

The *Fourier theorem* (discussed in basic courses) says that any nonsinusoidal periodic waveform is equivalent to the superposition of sinusoidal components called harmonics; these harmonics have frequencies of f, $2f$, $3f$, . . . , nf. Frequency f is the fundamental frequency of the waveform, calculated by taking the reciprocal of the period ($f = 1/T$).

The output waveform of Fig. 4-20c is rich in harmonics and can be filtered to produce a sine wave of a higher frequency. Because of this, step-recovery diodes are useful in frequency multipliers, circuits whose output frequency is a multiple of the input frequency. As an example, the period in Fig. 4-20c is 0.1 μs, which implies a fundamental frequency of 10 MHz (the input frequency). With resonant circuits, we can filter out the second harmonic (20 MHz), the third harmonic (30 MHz), or whatever harmonic we want.

BACK DIODES

Zener diodes normally have breakdown voltages greater than 2 V. By increasing the doping level, we can get the zener effect to occur near zero, as shown in Fig. 4-21a. Forward conduction still occurs around $+0.7$ V, but now reverse conduction (breakdown) starts at approximately -0.1 V.

A diode with an *I-V* curve like Fig. 4-21a is called a *back* diode (or backward diode) because it conducts better in the reverse than in the forward direction. As an example, Fig. 4-21b shows a sine wave with a peak of 0.5 V driving a back diode. (Notice that the zener symbol is used for a back diode.) The 0.5 V is not enough to turn on the diode in the forward direction, but it is enough to break down the diode in the reverse direction. For this reason, the output is a half-wave signal with a peak of 0.4 V.

Back diodes are occasionally used to rectify weak signals whose peak amplitudes are between 0.1 and 0.7 V.

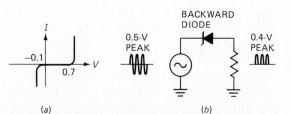

Fig. 4-21
Back diode. (a)
I-V curve. (b)
Rectifier circuit.

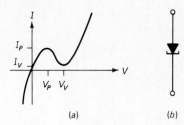

Fig. 4-22
Tunnel diode. (a)
I-V curve. (b)
Symbol.

(a)   (b)

TUNNEL DIODES

By increasing the doping level of a back diode, we can get breakdown to occur at 0 V. Furthermore, the heavier doping distorts the forward curve, as shown in Fig. 4-22*a*. A diode like this is known as a *tunnel* diode or Esaki diode (see Fig. 4-22*b* for the schematic symbol). In Fig. 4-22*a*, forward bias produces immediate conduction. The current reaches a maximum value I_P (peak current) when the diode voltage equals V_P. Then the current decreases to a minimum value I_V (valley current) at voltage V_V.

The region between the peak and valley points is called a *negative-resistance* region. Why? Because in this region an increase in voltage produces a decrease in current. The negative resistance of tunnel diodes is useful in high-frequency circuits called oscillators. These circuits are able to convert dc power to ac power; that is, they create a sinusoidal signal where none previously existed. Later chapters will tell you more about oscillators.

VARISTORS

Lightning, power-line faults, reactive-load switching, etc., can pollute the line voltage by superimposing dips, spikes, and other transients on the normal 115 V ac. Dips are severe voltage drops lasting microseconds or less. Spikes are brief overvoltages of 500 to more than 2000 V. In some equipment, filters are used between the power line and the primary of the transformer to eliminate the problems caused by line transients.

One of the devices used for line filtering is the *varistor* (also called a transient suppressor). This semiconductor device is like two back-to-back zener diodes with a high breakdown voltage in either direction. For instance, the V130LA2 has a breakdown voltage of 184 V (equivalent to 130 V ac) and a peak current rating of 400 A. Connect one of these across the primary winding and you don't have to worry about spikes. The varistor will clip all spikes at the 184-V level and protect your equipment.

PROBLEMS

Straightforward

4-1. A zener diode has 15 V across it and 20 mA through it. What is the power dissipation?

4-2. If a zener diode has a power rating of 5 W and a zener voltage of 20 V, what does its I_{ZM} equal?

4-3. A zener diode has a zener resistance of 5 Ω. If the current changes from 10 to 20 mA, what is the voltage change across the zener diode?

4-4. A current change of 2 mA through a zener diode produces a voltage change of 15 mV. What does the zener resistance equal?

4-5. The zener diode of Fig. 4-23*a* has a zener voltage of 15 V and a power rating of 0.5 W. If $V_S = 40$ V, what is the minimum value of R_S that prevents the zener diode from being destroyed?

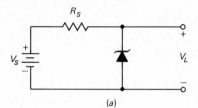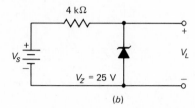

(a)   (b)

Fig. 4-23

4-6. Use the same data as in Prob. 4-5, except that $R_S = 2$ kΩ. What does the zener current equal? The power dissipation of the zener diode?

4-7. In Fig. 4-23*a*, $V_Z = 18$ V, $Z_{ZT} = 2$ Ω, $R_S = 68$ Ω, and $V_S = 27$ V. What does the zener current equal? If V_S increases to 40 V, what does the change in load voltage equal?

4-8. In Fig. 4-23*b*, what is the minimum value of source voltage that keeps the zener diode operating in the breakdown region?

4-9. The source voltage changes from 40 to 60 V in Fig. 4-23*b*. If the zener diode has a zener resistance of 10 Ω, what is the change in load voltage?

4-10. In Fig. 4-24*a*, what is the approximate value of zener current for each of these load resistances:

    a. $R_L = 100$ kΩ
    b. $R_L = 10$ kΩ
    c. $R_L = 1$ kΩ

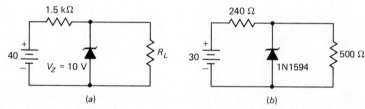

(a)   (b)

Fig. 4-24

4-11. Suppose the source of Fig. 4-24*a* has a peak-to-peak ripple of 4 V. If the zener resistance is 10 Ω, what is the output ripple?

4-12. For what value of load resistance does the zener regulator of Fig. 4-24*a* stop working?

4-13. What is the critical value of series resistance in Fig. 4-24*a* if the load resistance is 1 kΩ?

4-14. R_L can vary from infinity to 1 kΩ in Fig. 4-24*a*. If $Z_{ZT} = 10$ Ω, what is the change in load voltage over this range?

4-15. The 1N1594 of Fig. 4-24*b* has $V_Z = 12$ V and $R_Z = 1.4$ Ω. What is the load voltage? The zener current? The output ripple if the input ripple is 5 V p-p?

4-16. Use the same data as in Prob. 4-15, except that the load resistance may vary from infinity to 500 Ω. What are the minimum and maximum zener current? The change in load voltage?

4-17. Use the same data as in Prob. 4-15, except that the load resistance can decrease to 50 Ω. For what value of R_L does the zener regulator stop working?

4-18. A zener regulator has $V_Z = 15$ V. V_S may vary from 22 to 40 V. R_L may vary from 1 kΩ to 50 kΩ. What is the critical value of series resistance?

4-19. Draw the *I-V* curve of a zener diode and explain all the key ideas.

4-20. Two zener regulators are cascaded. The first has a series resistance of 680 Ω and a zener resistance of 10 Ω. The second has a series resistance of 1.2 kΩ and a zener resistance of 6 Ω. If the source ripple is 9 V p-p, what is the output ripple?

4-21. Assume that the dc load voltage is 25 V in Fig. 4-25. How much LED current is there if $R_1 = 1$ kΩ?

Fig. 4-25

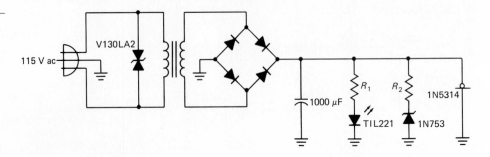

4-22. The secondary voltage is 12.6 V ac in Fig. 4-25. Ignore rectifier-diode drops. What is the LED current if $R_1 = 1$ kΩ?

4-23. The LED of Fig. 4-25 has a minimum drop of 1.5 V and a maximum drop of 2.3 V. If the load voltage is 10 V and $R_1 = 470$ Ω, what are the minimum and maximum values of LED current?

4-24. In the seven-segment display of Fig. 4-11*b* and *c*, which segments have to be lit to display each of the following:
  a. Numeral 4
  b. Numeral 9
  c. Letter A
  d. Letter d

4-25. In the optocoupler circuit of Fig. 4-13, $V_2 = 20$ V and $R_2 = 47$ kΩ. If I_2 changes from 2 to 10 μA, what is the voltage change across the photodiode?

4-26. The secondary voltage in Fig. 4-25 is only 3.6 V ac. If the diodes are replaced by Schottky diodes, what is the ideal peak load voltage? The peak load voltage after subtracting the voltage drop of the Schottky diodes? In this circuit, what advantage do the Schottky diodes have?

4-27. An inductor has an inductance of 20 μH. A varactor has a reference capacitance of 30 pF with a tuning range of 3:1. If the inductor and the varactor are in parallel, what is the minimum resonant frequency? The maximum resonant frequency?

4-28. A constant-current diode has a current of 1.5 mA over a voltage range of 2 to 150 V. What is its compliance?

4-29. A step-recovery diode is used in a frequency multiplier. The period of the rectified output signal is 0.04 μs. What is the frequency of the fifth harmonic?

Troubleshooting

4-30. In Fig. 4-24*a*, what is the load voltage for each of these conditions:
  a. Zener diode shorted   **c.** Series resistor open
  b. Zener diode open   **d.** Load resistor shorted
Also, what happens to the load voltage and the zener diode if the series resistor is shorted?

4-31. In Fig. 4-24*b*, the 1N1594 has a zener voltage of 12 V and a zener resistance of 1.4 Ω. If you measure approximately 20 V for the load voltage, what component do you think is defective? Explain why.

4-32. Use the same data as in Prob. 4-31, except that you measure 30 V across the load and an ohmmeter indicates that the zener diode is burned out. Before replacing the zener diode, what should you check for?

4-33. In Fig. 4-25, the LED does not light up. Which of the following are possible troubles:
   a. V130LA2 is open
   b. Ground between the two left bridge diodes is open
   c. Filter capacitor is open
   d. Filter capacitor is shorted
   e. Load resistor is open
   f. Load resistor is shorted

Design

4-34. Select a value of series resistance for a zener regulator to meet the following specifications: source voltage varies from 30 to 50 V, load current varies from 10 to 25 mA, and load voltage is 12 V.

4-35. Design a zener regulator to meet these specifications: load voltage is 6.8 V, source voltage is 20 V ± 20 percent, and load current is 30 mA ± 50 percent.

4-36. The input ripple to a zener regulator is 4 V p-p. The source voltage is 30 V, the source resistance is 1 kΩ, and the load resistance is 820 Ω. The load voltage is to be 12 V. Specify a zener diode that will produce a stiff zener regulator.

4-37. A TIL312 is a seven-segment indicator similar to Fig. 4-11*c*. Each segment has a voltage drop between 1.5 and 2 V at 20 mA. You have your choice of using a 5-V supply or a 12-V supply. Design a seven-segment display circuit controlled by on-off switches that has a maximum current drain of 140 mA.

Challenging

4-38. The secondary voltage of Fig. 4-25 is 12.6 V ac when the line voltage is 115 V ac; during the day the power line varies ±10 percent. The silicon diodes have an offset voltage of 0.7 V, which you should take into account in this problem. The resistors have tolerances of ±5 percent. The 1N753 has a zener voltage of 6.2 ± 10 percent and a zener resistance of 7 Ω. If R_2 equals 560 Ω, what is the maximum possible value of zener current at any instant during the day?

4-39. In Fig. 4-25, the secondary voltage is 12.6 V ac and diode drops are 0.7 V each. The 1N5314 is a constant-current diode with a current of 4.7 mA. The LED current is 15.6 mA, and the zener current is 21.7 mA. The filter capacitor has a tolerance of ±20 percent. What is the maximum possible ripple?

4-40. Figure 4-26 shows part of a bicycle lighting system. Use the second approximation to calculate the voltage across the filter capacitor.

Fig. 4-26

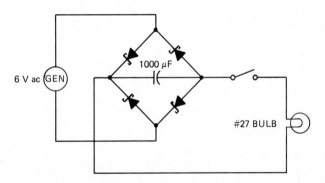

Computer

4-41. The GOTO statement (pronounced go to) allows you to repeat part of a program. For instance, here is a program with a GOTO statement:

```
10 PRINT "ENTER POWER RATING"
20 INPUT PZM
30 PRINT "THE POWER RATING IS"
40 PRINT PZM
50 GOTO 10
```

Here is what happens. The computer asks you to input the value of P_{ZM}. After you enter it, the computer prints out the value of P_{ZM}. The GOTO statement forces the computer to return to line 10, and the program repeats. Each time you enter a new value of P_{ZM}, the computer prints it out on the screen and then jumps back to line 10.

What does the following program do?

```
10 PRINT "ENTER POWER RATING"
20 INPUT PZM
30 PRINT "ENTER ZENER VOLTAGE"
40 INPUT VZ
50 IZM = PZM/VZ
60 PRINT "THE CURRENT RATING IS"
70 PRINT IZM
80 GOTO 10
```

4-42. Write a program that prints out the value of zener current after you input the values of V_S, V_Z, R_S, and R_L. Include a GOTO statement to repeat the program.

4-43. Write a program that prints out the critical value of series resistance in a zener regulator after you have input the values of $V_{S(min)}$, V_Z, and $I_{L(max)}$. The last instruction should be a GOTO that forces the program to run again.

Bipolar Transistors

Before 1950 all electronics equipment used vacuum tubes, those soft-glow bulbs that once dominated our industry. The heater of a typical vacuum tube consumed a couple of watts of power. Because of this, vacuum-tube equipment required a bulky power supply and created a lot of heat that the designer had to worry about. The result was the heavy and antiquated equipment that was once so widespread.

Then it happened. In 1951 Shockley invented the first junction transistor. It was one of those big breakthroughs that changes all the rules. Everyone was excited at the time and predicted great things to come. As it turned out, the wildest predictions weren't even close to the new world that was coming.

The transistor's impact on electronics has been enormous. Besides starting the multi-billion-dollar semiconductor industry, the transistor has led to all kinds of related inventions like integrated circuits, optoelectronic devices, and microprocessors. Almost all electronic equipment now being designed uses semiconductor devices.

The changes have been most noticeable in computers. The transistor did not revise the computer industry, it created it. Before 1950 a computer filled an entire room and cost millions of dollars. Today, a better computer sits on a desk and costs a few thousand dollars (inflated dollars at that).

5-1 SOME BASIC IDEAS

Figure 5-1*a* shows an *npn* crystal. The *emitter* is heavily doped; its job is to emit, or inject, electrons into the base. The *base* is lightly doped and very thin; it passes most of the emitter-injected electrons on to the *collector*. The doping level of the collector is intermediate, between the heavy doping of the emitter and the light doping of the base. The collector is so named because it collects or gathers electrons from the base. The collector is the largest of the three regions; it must dissipate more heat than the emitter or the base.

EMITTER AND COLLECTOR DIODES

The transistor of Fig. 5-1*a* has two junctions, one between the emitter and the base, and another between the base and the collector. Because of this, a transistor is similar

113

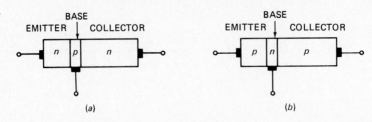

Fig. 5-1
The three transistor regions.
(a) npn transistor.
(b) pnp transistor.

to two diodes. We call the diode on the left the emitter-base diode or simply the *emitter diode.* The diode on the right is the collector-base diode or the *collector diode.*

Figure 5-1*b* shows the other possibility, a *pnp* transistor. The *pnp* transistor is the complement of the *npn* transistor; the majority carriers in the emitter are holes instead of free electrons. This means that opposite currents and voltages are involved in the action of a *pnp* transistor. To avoid confusion, we will concentrate on the *npn* transistor during our early discussions.

UNBIASED TRANSISTOR

The diffusion of free electrons across the junction produces two depletion layers (Fig. 5-2*a*). For each of these depletion layers, the barrier potential is approximately equal to 0.7 V at 25°C for a silicon transistor (0.3 V for a germanium transistor). As with diodes, we emphasize silicon transistors because they are more widely used than germanium transistors. Silicon transistors provide higher voltage ratings, greater current ratings, and less temperature sensitivity. In the discussions that follow, the transistors are silicon unless otherwise indicated.

Because the three regions have different doping levels, the depletion layers do not have the same width. The more heavily doped a region is, the greater the concentration of ions near the junction. This means that the depletion layer penetrates only slightly into the emitter region (heavily doped), but deeply into the base (lightly doped). The other depletion layer extends well into the base, and penetrates the collector region to a lesser amount. Figure 5-2*b* summarizes this idea. The emitter depletion layer is small and the collector depletion layer is large. Notice how the depletion layers are shaded dark to indicate the lack of majority carriers.

BIASING THE TRANSISTOR

In Fig. 5-3*a* the left battery forward-biases the emitter diode, and the right battery forward-biases the collector diode. Free electrons (majority carriers) enter the emitter and collector of the transistor, join at the base, and flow down the common wire

Fig. 5-2
Depletion layers.

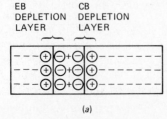

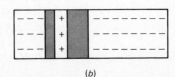

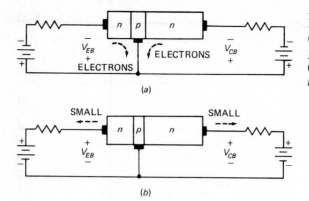

Fig. 5-3
(a) *Both junctions forward-biased.*
(b) *Both junctions reverse-biased.*

as shown. Because both diodes are forward-biased, the emitter and collector currents are large.

Figure 5-3*b* shows another way to bias the transistor. Now both diodes are reverse-biased. For this condition, the flow is small, consisting only of two types of minority carriers: thermally produced and surface-leakage. The thermally produced component is temperature-dependent; it approximately doubles for every 10°C rise in ambient temperature. The surface-leakage component, on the other hand, increases with voltage. These reverse currents are usually negligible.

In Fig. 5-3*a* and *b*, nothing unusual happens. We get either a lot of current (when both diodes are forward-biased) or practically none (when both are reverse-biased). Transistors are rarely biased like this in linear circuits.

5-2 FORWARD-REVERSE BIAS

Forward-bias the emitter diode and reverse-bias the collector diode, and the unexpected happens. In Fig. 5-4*a*, we expect a large emitter current because the emitter diode is forward-biased. But we do not expect a large collector current because the collector diode is reverse-biased. Nevertheless, the collector current is almost as large as the emitter current.

PRELIMINARY EXPLANATION

Here is a brief explanation of why we get a large collector current in Fig. 5-4*a*. At the instant the forward bias is applied to the emitter diode, electrons in the emitter have not yet entered the base region (see Fig. 5-4*b*). If V_{EB} is greater than the barrier potential (0.6 to 0.7 V for silicon transistors), many emitter electrons enter the base region, as shown in Fig. 5-4*c*. These electrons in the base can flow in either of two directions: down the thin base into the external base lead, or across the collector junction into the collector region. The downward component of base current is called *recombination current*. It is small because the base is lightly doped with only a few holes.

A second crucial idea in transistor action is that the base is very thin. In Fig. 5-4*c*, the base is teeming with injected conduction-band electrons, causing diffusion into

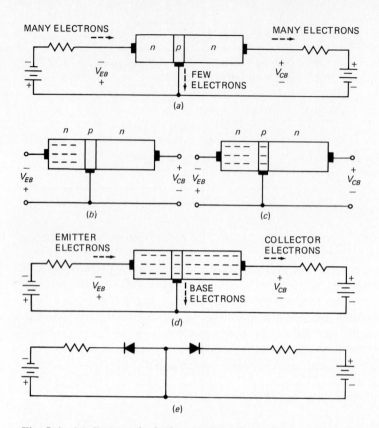

Fig. 5-4 (a) *Emitter diode forward-biased and collector diode reverse-biased.* (b) *Emitter has many free electrons.* (c) *Free electrons injected into base.* (d) *Free electrons pass through base to collector.* (e) *Two discrete back-to-back diodes are not a transistor.*

the collector depletion layer. Once inside this layer, the free electrons are pushed by the depletion-layer field into the collector region (see Fig. 5-4d). These collector electrons can then flow into the external collector lead as shown.

Here's our final picture of what's going on. In Fig. 5-4d we visualize a steady stream of electrons leaving the negative source terminal and entering the emitter region. The V_{EB} forward bias forces these emitter electrons to enter the base region. The thin and lightly doped base gives almost all these electrons enough lifetime to diffuse into the collector depletion layer. The depletion-layer field then pushes a steady stream of electrons into the collector region. These electrons leave the collector, enter the external collector lead, and flow into the positive terminal of the voltage source. In most transistors, more than 95 percent of the emitter-injected electrons flow to the collector; less than 5 percent fall into base holes and flow out the external base lead.

Incidentally, don't get the idea that you can connect two discrete diodes back to back, as shown in Fig. 5-4e, to get a transistor. Each diode has two doped regions, so that the overall circuit has four doped regions. This won't work because the base region is not the same as that of a transistor. The key to transistor action is the lightly doped base between the heavily doped emitter and the intermediate-doped collector.

Free electrons passing through the base to the collector have a short lifetime. As long as the base is thin, the free electrons can reach the collector. With two discrete back-to-back diodes, you have four doped regions instead of three, and there is nothing that resembles a thin base region between an emitter and a collector.

THE ENERGY VIEWPOINT

An energy diagram is the next step to a deeper understanding of the transistor. Forward-biasing the emitter diode of Fig. 5-5 allows some of the free electrons to move from the emitter into the base. In other words, the orbits of some emitter electrons are now large enough to match some of the available base orbits. Because of this, emitter electrons can diffuse from the emitter conduction band to the base conduction band.

Upon entering the base conduction band, the electrons become minority carriers because they are inside a *p* region. Now the base has a higher density of minority carriers. In almost any transistor, more than 95 percent of these minority carriers have a long enough lifetime to diffuse into the collector depletion layer and fall down the collector energy hill. As they fall, they give up energy, mostly in the form of heat. The collector must be able to dissipate this heat, and for this reason, it is usually the largest of the three doped regions. Less than 5 percent of the emitter-injected electrons fall along the recombination path shown in Fig. 5-5; those that do recombine become valence electrons and flow through base holes into the external base lead.

Here's a summary:

1. The forward bias on the emitter diode controls the number of free electrons injected into the base. The larger V_{BE} is, the greater the number of injected electrons.
2. The reverse bias on the collector diode has little influence on the number of electrons that enter the collector. Increasing V_{CB} steepens the collector energy hill, but this does not significantly change the number of free electrons arriving at the collector depletion layer.

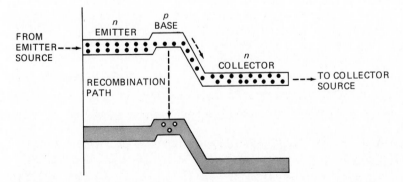

Fig. 5-5 *Energy bands when the emitter diode is forward-biased and the collector diode is reverse-biased.*

DC ALPHA

Saying that more than 95 percent of injected electrons reach the collector is the same as saying that collector current almost equals emitter current. The *dc alpha* of a transistor indicates how close in value the two currents are; it is defined as

$$\alpha_{dc} = \frac{I_C}{I_E} \tag{5-1}$$

For instance, if we measure an I_C of 4.9 mA and an I_E of 5 mA, then

$$\alpha_{dc} = \frac{4.9 \text{ mA}}{5 \text{ mA}} = 0.98$$

The thinner and more lightly doped the base is, the higher the α_{dc}. Ideally, if all injected electrons went on to the collector, α_{dc} would equal unity. Many transistors have α_{dc} greater than 0.99, and almost all have α_{dc} greater than 0.95. Because of this, we can approximate α_{dc} as 1 in most analysis.

BASE-SPREADING RESISTANCE

With two depletion layers penetrating the base, the base holes are confined to the thin channel of *p*-type semiconductor shown in Fig. 5-6. The resistance of this thin channel is called the *base spreading resistance r_b'*. Increasing the V_{CB} reverse bias on the collector diode decreases the width of the *p* channel, which is equivalent to an increase in r_b'.

Recombination current in the base must flow down through r_b'. When it does, it produces a voltage. We discuss the importance of this voltage later. For now, just be aware that r_b' exists and that it depends on the width of the *p* channel in Fig. 5-6 as well as on the doping of the base. In rare cases, r_b' may be as high as 1000 Ω. Typically, it is in the range of 50 to 150 Ω. The effects of r_b' are important in high-frequency circuits. At low frequencies, r_b' usually has little effect. For this reason, we will ignore the effects of r_b' until later chapters.

BREAKDOWN VOLTAGES

Since the two halves of a transistor are diodes, too much reverse voltage on either diode can cause breakdown. This breakdown voltage depends on the width of the depletion layer and the doping levels. Because of the heavy doping level, the emitter

Fig. 5-6
Base-spreading resistance.

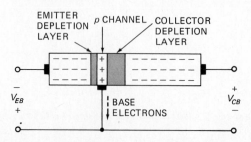

diode has a low breakdown voltage BV_{BE}, approximately 5 V to 30 V. The collector diode, on the other hand, is less heavily doped. So its breakdown voltage BV_{CE} is higher, roughly 20 to 300 V.

For normal transistor action, the collector diode is reverse-biased. When V_{CB} is too large, the collector diode breaks down and may be damaged by excessive power dissipation. In most designs, therefore, we must keep the collector voltage less than the maximum rating for BV_{CE} specified on a manufacturer's data sheet. Likewise, in some transistor circuits the emitter diode may be temporarily reverse-biased; at no time should the reverse voltage exceed the maximum rating for BV_{BE}.

5-3 THE CE CONNECTION

In Fig. 5-7a, the emitter and the two voltage sources connect to the common point shown. Because of this, we call the circuit a *common-emitter* (CE) connection. The circuit first discussed (Fig. 5-4a) is a *common-base* (CB) connection because the base and the two voltage sources are connected to a common point.

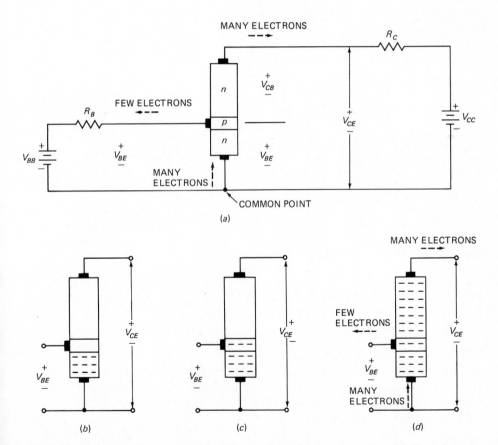

Fig. 5-7 *CE connection.* (a) *Actual circuit.* (b) *Free electrons in emitter.* (c) *Free electrons enter base.* (d) *Free electrons diffuse into collector.*

CE ACTION

Just because you change from a CB connection (Fig. 5-4a) to a CE connection (Fig. 5-7a), you do not change the way a transistor operates. Free electrons move exactly the same as before. That is, the emitter is full of free electrons (Fig. 5-7b). When V_{BE} is greater than 0.7 V or so, the emitter injects these electrons into the base (Fig. 5-7c). As before, the thin, lightly doped base gives almost all these electrons enough lifetime to diffuse into the collector depletion layer. With a reverse-biased collector diode, the depletion-layer field pushes the electrons into the collector region, where they flow out to the external voltage source (Fig. 5-7d).

DC BETA

We have related the collector current to the emitter current by using α_{dc}. We can also relate collector current to base current by defining the *dc beta* of a transistor as

$$\beta_{dc} = \frac{I_C}{I_B} \tag{5-2}$$

For instance, if we measure a collector current of 5 mA and a base current of 0.05 mA, the transistor has a β_{dc} of

$$\beta_{dc} = \frac{5 \text{ mA}}{0.05 \text{ mA}} = 100$$

For almost any transistor, less than 5 percent of the emitter-injected electrons recombine with base holes to produce I_B; therefore, β_{dc} is almost always greater than 20. Usually, it is between 50 and 300. And some transistors have β_{dc} as high as 1000.

In another system of analysis, called h parameters, h_{FE} rather than β_{dc} is used for the dc current gain. In symbols,

$$\beta_{dc} = h_{FE} \tag{5-2a}$$

Remember this relation, because data sheets use the symbol h_{FE} for the dc current gain. For example, the data sheet of a 2N3904 lists a minimum h_{FE} of 100 and a maximum h_{FE} of 300. This means that β_{dc} varies from 100 to 300. Chapter 9 discusses h parameters in detail.

RELATION BETWEEN α_{dc} AND β_{dc}

Kirchhoff's current law gives

$$I_E = I_C + I_B \tag{5-3}$$

This tells you that emitter current is the sum of collector current and base current. Always remember the following: Emitter current is the largest of the three currents, collector current is almost as large, and base current is much smaller.

Dividing both sides of Eq. (5-3) by I_C gives

$$\frac{I_E}{I_C} = 1 + \frac{I_B}{I_C}$$

or

$$\frac{1}{\alpha_{dc}} = 1 + \frac{1}{\beta_{dc}}$$

With algebra, we can rearrange to get

$$\beta_{dc} = \frac{\alpha_{dc}}{1 - \alpha_{dc}} \tag{5-4}$$

As an example, if $\alpha_{dc} = 0.98$, the value of β_{dc} is

$$\beta_{dc} = \frac{0.98}{1 - 0.98} = \frac{0.98}{0.02} = 49$$

Occasionally we need a formula for α_{dc} in terms of β_{dc}. With algebra we can rearrange Eq. (5-4) to get

$$\alpha_{dc} = \frac{\beta_{dc}}{\beta_{dc} + 1} \tag{5-5}$$

For instance, for a β_{dc} of 100,

$$\alpha_{dc} = \frac{100}{100 + 1} = \frac{100}{101} = 0.99$$

ACTIVE REGION

As a summary, here are the conditions needed to operate a transistor in a linear circuit:

1. The emitter diode must be forward-biased.
2. The collector diode must be reverse-biased.
3. The voltage across the collector diode must be less than the breakdown voltage.

When these conditions are satisfied, the transistor is an active device because it can amplify an input signal to produce a larger output signal. Much more will be said about this in later chapters.

TWO EQUIVALENT CIRCUITS

To remember the main ideas of transistor action, look at the equivalent circuit of Fig. 5-8a, shown for conventional current. The voltage V_{BE} is the voltage across the emitter depletion layer. When this voltage is greater than approximately 0.7 V, the emitter injects electrons into the base. As mentioned earlier, the current in the emitter diode controls the collector current. For this reason, the collector current source forces a current of $\alpha_{dc}I_E$ to flow in the collector circuit. The equivalent circuit of Fig. 5-8a assumes that V_{CE} is greater than a volt or so but less than breakdown. In other words, the equivalent circuit assumes that the transistor is operating in the active region.

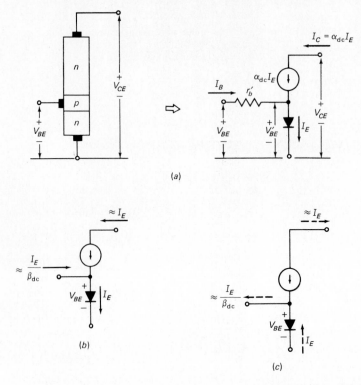

Fig. 5-8   (a) *Equivalent circuit for transistor.* (b) *Ebers-Moll model.* (c) *Electron-flow model.*

The internal voltage V'_{BE} differs from the applied voltage V_{BE} by the drop across r'_b:

$$V_{BE} = V'_{BE} + I_B r'_b$$

When the $I_B r'_b$ drop is small, V_{BE} is approximately equal to V'_{BE}.

PRACTICAL CIRCUITS

Figure 5-8b shows a common way to draw the equivalent circuit of a transistor. This is for conventional-current users. If you prefer electron flow, use the equivalent circuit of Fig. 5-8c. This equivalent circuit of a transistor, an emitter diode in series with a collector current source, is called the *Ebers-Moll* model.

In using the Ebers-Moll model, we usually approximate as follows:

1. Let V_{BE} equal 0.7 V for silicon transistors (0.3 V for germanium).
2. Disregard the $I_B r'_b$ voltage (this is equivalent to treating the product of I_B and r'_b as negligibly small).
3. Treat I_C as equal to I_E because α_{dc} approaches unity.
4. Approximate I_B as I_E/β_{dc} because I_C almost equals I_E.

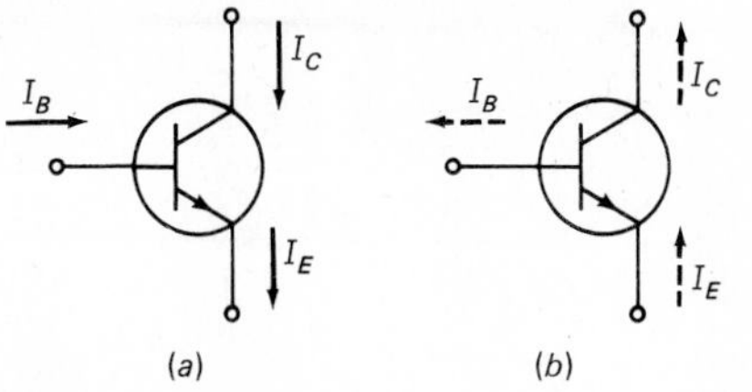

Fig. 5-9
Transistor currents. (a) *Conventional flow.* (b) *Electron flow.*

(a) (b)

CONTROLLED CURRENT SOURCE

The CE connection is used much more than the CB connection because a small input current (base) controls a large output current (collector). For instance, if $\beta_{dc} = 200$, a small base current produces a collector current that is 200 times larger. Furthermore, the output current appears to come out of a current source. This means that the transistor is a controlled current source. The base current controls the current source.

SCHEMATIC SYMBOL

Figure 5-9*a* shows the schematic symbol for an *npn* transistor with conventional flow; Fig. 5-9*b* is the same symbol with electron flow. The emitter has an arrowhead, but the collector does not. We will be using this symbol extensively in the chapters that follow. Since industrial schematics do not show current direction, future diagrams will not show current direction. You must remember these current directions for future discussions, so memorize the schematic symbol and your preferred current direction. Furthermore, you should remember the following:

1. Emitter current is the sum of the collector current and the base current:

$$I_E = I_C + I_B$$

2. Collector current approximately equals emitter current:

$$I_C \cong I_E$$

3. Base current is much smaller than the other two:

$$I_B \ll I_C$$
$$I_B \ll I_E$$

5-4 TRANSISTOR CHARACTERISTICS

One way to visualize how a transistor operates is with graphs that relate transistor currents and voltages. These *I-V* curves will be more complicated than those of a diode because we have to include the effect of base current.

COLLECTOR CURVES

You can get data for CE collector curves by setting up a circuit like Fig. 5-10*a* or by using a transistor curve tracer. Either way, the idea is to vary the V_{BB} and V_{CC} sup-

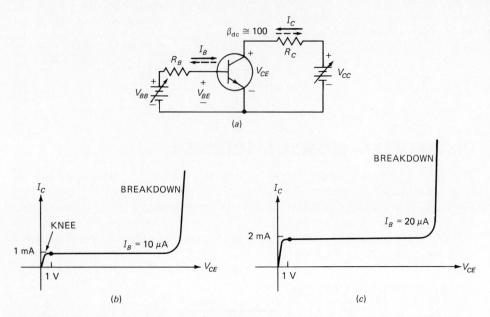

Fig. 5-10   (a) *Circuit for measuring collector current and voltage of transistor with* β_{dc} *of 100.* (b) *Curve for* $I_B = 1\ \mu A$. (c) *Curve for* $I_B = 2\ \mu A$.

plies to set up different transistor voltages and currents. To keep things orderly, the usual procedure is to set a value of I_B and hold it fixed while you vary V_{CC}. By measuring I_C and V_{CE}, you can get data for graphing I_C versus V_{CE}. For instance, suppose you set I_B to 10 μA in Fig. 5-10a. Next, you can vary V_{CC} and measure the resulting I_C and V_{CE}. Then, plotting the data gives Fig. 5-10b.

There is nothing mysterious about Fig. 5-10b. It echoes our explanation of transistor action. When V_{CE} is 0, the collector diode is not reverse-biased; therefore, the collector current is negligibly small. For V_{CE} between 0 and approximately 1 V, the collector current rises sharply and then becomes almost constant. This is tied in with the idea of reverse-biasing the collector diode. It takes approximately 0.7 V to reverse-bias the collector diode; once you reach this level, the collector is gathering all electrons that reach its depletion layer.

Above the knee, the exact value of V_{CE} is not too important because making the collector hill steeper cannot appreciably increase the collector current. The small increase in collector current with increasing V_{CE} is caused by the collector depletion layer getting wider and capturing a few more base electrons before they fall into holes (less recombination). Since the transistor has a β_{dc} of approximately 100, the collector current is approximately 100 times the base current.

If we increase V_{CE} too much, the collector diode breaks down and normal transistor action is lost. Then, the transistor no longer acts like a current source. The *voltage compliance* (or simply compliance) of a transistor is the range of collector-emitter voltage over which the transistor acts like a current source. In Fig. 5-10b, the compliance is the V_{CE} voltage range between 1 V and breakdown. Keep the transistor operating in this active region, and it will act like a controlled current source. Get outside of this range, and you lose normal transistor action.

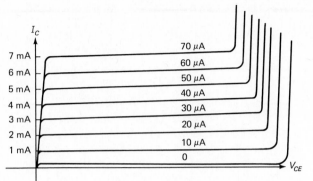

Fig. 5-11
*Transistor curves
for β_{dc} of 100.*

If we measure I_C and V_{CE} for $I_B = 20\ \mu A$, we can plot the graph of Fig. 5-10c. The curve is similar, except that above the knee the collector current approximately equals 2 mA. Again, an increase in V_{CE} produces a small increase in collector current because the widening depletion layer captures a few additional base electrons. Also, too much collector voltage produces collector breakdown.

If several curves for different I_B are drawn on the same graph, we get collector curves like Fig. 5-11. Since we used a transistor with a β_{dc} of approximately 100, the collector current is approximately 100 times greater than the base current for any point in the active region. These curves are sometimes called *static collector curves* because dc currents and voltages are being plotted.

Notice the bottom curve, where base current is zero. A small collector current exists here because of the leakage current of the collector diode. For silicon transistors this leakage current is usually small enough to ignore in most applications. For instance, a 2N3904 has a leakage current of only 50 nA, which is so small that you would not even see the bottom curve of Fig. 5-11.

Also notice the breakdown voltages; they get lower at higher currents. This means that the voltage compliance of a transistor decreases for larger collector currents. All that's necessary is to avoid breakdown under all conditions. This guarantees that the transistor will operate in the active region.

Most transistor data sheets do not show collector curves. If you want to see the collector curves of a particular transistor, use a curve tracer. This instrument displays collector curves similar to those in Fig. 5-11. If you try different transistors, you will notice changes in knee voltage, β_{dc}, breakdown voltage, etc. As a rule, transistor curves show a wide variation from one transistor to another of the same type number.

BASE CURVES

Figure 5-12a shows a graph of base current versus base-emitter voltage. Since the base-emitter section of a transistor is a diode, we expect to see a graph that resembles a diode curve. And that is what we get, almost. Remember, there are more variables in a transistor than in a diode.

At higher collector voltages, the collector gathers in a few more electrons. This reduces the base current. Figure 5-12b illustrates the idea. The curve with the higher

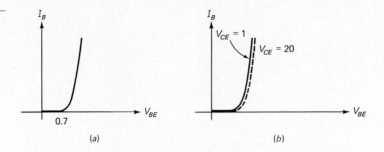

Fig. 5-12
Base curves. (a)
Ideal. (b) *Early
effect.*

V_{CE} has slightly less base current for a given V_{BE}. This phenomenon, called the *Early
effect,* results from internal transistor feedback from the collector diode to the emitter
diode. The gap between the curves of Fig. 5-12b is quite small, not even noticeable on
an oscilloscope. For this reason, we ignore the Early effect in all preliminary analysis.
(Chapter 9 discusses h parameters, a high-level method of analysis that includes the
Early effect.)

CURRENT-GAIN CURVES

The β_{dc} of a transistor, also called the *current gain,* varies enormously. Figure 5-13
shows the typical variation in β_{dc}. At a fixed temperature, β_{dc} increases to a maximum
value when collector current increases. For further increases in I_C, the β_{dc} drops off.
The variation in β_{dc} may be as much as 3:1 over the useful current range of the
transistor; it depends on the type of transistor.

A change in ambient temperature also has an effect on β_{dc}. As shown in Fig. 5-13,
increasing the temperature will increase β_{dc} at a given collector current. Over a large
temperature range, the variations in β_{dc} may approach 3:1, depending on the
transistor type.

In the worst case, where both collector current and temperature vary significantly,
β_{dc} may vary over a 9:1 range. Keep this in mind, because any design that requires an
exact value of β_{dc} is doomed from the start. Good design involves coming up with
circuits that do not depend too much on the exact value of β_{dc}. Later sections will tell
you how to design circuits that are relatively immune to variations in β_{dc}.

CUTOFF AND BREAKDOWN

In the collector curves of Fig. 5-11, the lowest curve is for zero base current. The
condition $I_B = 0$ is equivalent to having an *open base lead* (see Fig. 5-14a). The

Fig. 5-13
*Variation in β_{dc}
with collector
current and
temperature.*

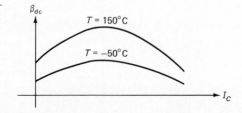

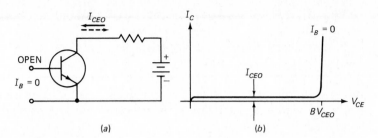

Fig. 5-14
*Cutoff current and
breakdown
voltage.*

collector current with open base lead is designated I_{CEO}, where the subscript CEO stands for collector to emitter with open base. I_{CEO} is caused partly by thermally produced carriers and partly by surface-leakage current.

Figure 5-14b shows the $I_B = 0$ curve. With a large enough collector voltage, we reach a breakdown voltage, labeled BV_{CEO}, where the subscript again stands for collector to emitter with open base. For normal transistor operation, we must keep V_{CE} less than BV_{CEO}. Most transistor data sheets list the value of BV_{CEO} among the maximum ratings. This breakdown voltage may be less than 20 or over 200 V, depending on the transistor type.

As a rule, a good design includes a safety factor to keep V_{CE} well below BV_{CEO}. Transistor lifetime may be shortened by a design that pushes the absolute maximum ratings of a transistor. A safety factor of 2 (V_{CE} less than half BV_{CEO}) is common. Some conservative designs use a safety factor as large as10 (V_{CE} less than one-tenth of BV_{CEO}).

COLLECTOR SATURATION VOLTAGE

Figure 5-15 shows one of the collector curves. The comments that follow apply to any collector curve. The initial part of the curve is called the *saturation* region; this is all the curve between the origin and the knee. The flat part of the curve is the *active* region; this is where the transistor should operate if you want it to act like a controlled

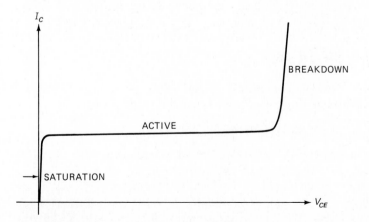

Fig. 5-15 *One of the collector curves showing the three regions: saturation, active, and breakdown.*

current source. The final part of the curve is the *breakdown* region, which you should avoid at all costs.

In the saturation region, the collector diode goes into forward bias. Because of this, normal transistor action is lost and the transistor acts like a small ohmic resistance rather than a current source. Any further increase in base current cannot produce a further increase in collector current. The collector-emitter voltage in the saturation region is usually only a few tenths of a volt, depending on how much collector current there is.

For the transistor to operate in the active region, the collector diode must be reverse-biased; this requires a V_{CE} greater than a volt or so. As a guide, many data sheets list $V_{CE(sat)}$, the value of V_{CE} at some point in the saturation region. Typically, $V_{CE(sat)}$ is only a few tenths of a volt. Therefore, normal operation requires a V_{CE} that is greater than $V_{CE(sat)}$.

TRANSISTOR RATINGS

Small-signal transistors can dissipate half a watt or less; power transistors can dissipate more than half a watt. When you look at a data sheet for either type of transistor, you should start with the maximum ratings because these set the limits on the transistor currents, voltages, and other such quantities.

As an example, the maximum ratings for a 2N3904 are listed as follows:

V_{CEO}	40 V
V_{CBO}	60 V
V_{EBO}	6 V
I_C	200 mA dc
P_D	310 mW

All voltage ratings are reverse breakdown voltages. The first rating is V_{CEO}, which stands for the voltage from collector to emitter with the base open. V_{CBO} is the voltage between the collector and the base with the emitter open. V_{EBO} is the voltage from the emitter to the base with the collector open. I_C is the maximum dc collector current rating; this means that a 2N3904 can handle up to 200 mA of steady current.

The last rating is P_D, the maximum power rating of the device. You calculate the power dissipation of a transistor by

$$P_D = V_{CE}I_C \tag{5-6}$$

For instance, if a 2N3904 has $V_{CE} = 20$ V and $I_C = 10$ mA, then

$$P_D = 20 \text{ V} \times 10 \text{ mA} = 200 \text{ mW}$$

This is less than the power rating, 310 mW. As with breakdown voltage, a good design should include a safety factor to ensure a longer operating life for the transistor. Safety factors of 2 or more are common. For example, if 310 mW is the maximum power rating, a safety factor of 2 requires a power dissipation less than 155 mW. (Some transistors need a heat sink, a mass of metal attached to the transistor to keep it from getting too hot. Chapter 10 tells you more about this.)

5-5 DC LOAD LINES

A load line can be drawn on the collector curves to give more insight into how a transistor works and what region it operates in. The approach is similar to that used with a diode. In the circuit of Fig. 5-16a, the supply voltage V_{CC} reverse-biases the collector diode through R_C. The voltage across this resistor is $V_{CC} - V_{CE}$. Therefore, the current through it equals

$$I_C = \frac{V_{CC} - V_{CE}}{R_C} \qquad (5\text{-}7)$$

This is the equation of the dc load line.

AN EXAMPLE

Suppose the supply voltage is 10 V and the collector resistance is 5 kΩ. Then the equation of the dc load line is

$$I_C = \frac{10 - V_{CE}}{5000} \qquad (5\text{-}8)$$

We can calculate the upper end of the load line by setting V_{CE} equal to 0 and solving for I_C:

$$I_C = \frac{10 - 0}{5000} = 2 \text{ mA}$$

Next we calculate the lower end of the load line by setting I_C equal to 0 in Eq. (5-8) and solving for V_{CE}:

$$V_{CE} = 10 \text{ V}$$

An alternative way to find the upper end of the load line is to visualize the collector-emitter terminals shorted in Fig. 5-16a and calculate the resulting collector current, which is V_{CC}/R_C. To get the lower end of the load line, visualize the terminals open and calculate the resulting collector-emitter voltage, which equals V_{CC}.

ANY LOAD LINE

We can solve Eq. (5-7) for the ends of the load line to get these two formulas:

$$I_C = \frac{V_{CC}}{R_C} \qquad \text{(upper)} \qquad (5\text{-}9)$$

and

$$V_{CE} = V_{CC} \qquad \text{(lower)} \qquad (5\text{-}10)$$

Figure 5-16b shows the dc load line superimposed on the collector curves. The vertical intercept is V_{CC}/R_C and the horizontal intercept is V_{CC}. The intersection of the dc load line with the calculated base current is the *Q point* of the transistor (also called the operating point or the quiescent point).

Fig. 5-16
(a) *Base bias.* (b)
Dc load line.

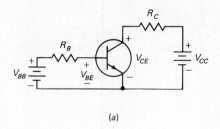

(a)

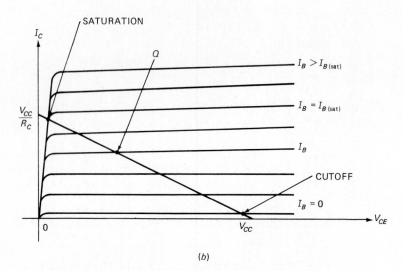

(b)

CUTOFF AND SATURATION

The point at which the load line intersects the $I_B = 0$ curve is known as *cutoff.* At this point, base current is zero and collector current is negligibly small (only leakage current I_{CEO} exists). At cutoff, the emitter diode comes out of forward bias, and normal transistor action is lost. To a close approximation, the collector-emitter voltage equals the lower end of the load line:

$$V_{CE(\text{cutoff})} \cong V_{CC}$$

The intersection of the load line and the $I_B = I_{B(\text{sat})}$ curve is called *saturation.* At this point, base current equals $I_{B(\text{sat})}$ and the collector current is maximum. At saturation, the collector diode comes out of reverse bias, and normal transistor action is again lost. To a close approximation, the collector current at saturation equals the upper end of the load line:

$$I_{C(\text{sat})} \cong \frac{V_{CC}}{R_C}$$

In Fig. 5-16*b*, $I_{B(\text{sat})}$ represents the amount of base current that just produces saturation. If the base current is less than $I_{B(\text{sat})}$, the transistor operates in the active region somewhere between saturation and cutoff. In other words, the operating point is somewhere along the dc load line. On the other hand, if the base current is greater than $I_{B(sat)}$, the collector current approximately equals V_{CC}/R_C, the maximum possible value. Graphically, this means that the intersection of the load line with any base current greater than $I_{B(sat)}$ produces the saturation point in Fig. 5-16b.

COMPLIANCE

The dc load line tells us at a glance what the compliance (active V_{CE} voltage range) of a transistor is. In Fig. 5-16b, the transistor has a compliance from approximately 0 to V_{CC}. In other words, the transistor acts like a current source anywhere along the dc load line, excluding saturation or cutoff, where the current-source action is lost.

5-6 THE TRANSISTOR SWITCH

The simplest way to use a transistor is as a *switch,* meaning that we operate it at either saturation or cutoff but nowhere else along the load line. When a transistor is saturated, it is like a closed switch from the collector to the emitter. When a transistor is cut off, it's like an open switch.

BASE CURRENT

Figure 5-17a shows the circuit we have been analyzing up to now; Fig. 5-17b is the way you will usually see the circuit drawn. Summing voltages around the input loop gives

$$I_B R_B + V_{BE} - V_{BB} = 0$$

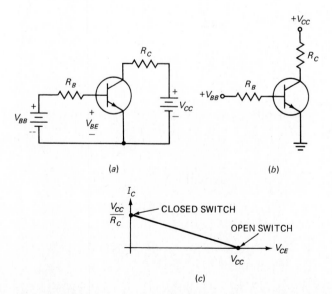

(a)   (b)

(c)

Fig. 5-17 (a) *Transistor switching circuit.* (b) *Usual way circuit is drawn.* (c) *Dc load line.*

Solving for I_B, we get

$$I_B = \frac{V_{BB} - V_{BE}}{R_B} \tag{5-11}$$

This is Ohm's law for the base resistor. For instance, suppose $V_{BB} = 5$ V and $R_B = 1$ MΩ. Then,

$$I_B = \frac{5 \text{ V} - 0.7 \text{ V}}{1 \text{ M}\Omega} = \frac{4.3 \text{ V}}{1 \text{ M}\Omega} = 4.3 \ \mu\text{A}$$

If the base current is greater than or equal to $I_{B(\text{sat})}$, the operating point Q is at the upper end of the load line (Fig. 5-17c). In this case, the transistor appears like a closed switch. On the other hand, if the base current is zero, the transistor operates at the lower end of the load line, and the transistor appears like an open switch.

DESIGN RULE

Soft saturation means that we barely saturate the transistor; that is, the base current is just enough to operate the transistor at the upper end of the load line. Soft saturation is not reliable in mass production because of the variation in β_{dc} and $I_{B(\text{sat})}$. Don't try to use soft saturation in a transistor switching circuit.

Hard saturation means having sufficient base current to saturate the transistor at all the values of β_{dc} encountered in mass production. For the worst case of temperature and current, almost all small-signal silicon transistors have a β_{dc} greater than 10. Therefore, a design guideline for hard saturation is to have a base current that is approximately *one-tenth* of the saturated value of collector current; this guarantees hard saturation under all operating conditions. For instance, if the upper end of the load line has a collector current of 10 mA, then we will set up a base current of 1 mA. This guarantees saturation for all transistors, currents, temperatures, etc.

Unless otherwise indicated, we will use the 10:1 rule when designing transistor switching circuits. Remember, this is only a guideline. If standard resistance values produce an I_C/I_B ratio slightly greater than 10, almost any small-signal transistor will still go into hard saturation.

AN EXAMPLE

Figure 5-18a shows a transistor switching circuit driven by a voltage step. When the input voltage is zero, the transistor is cut off. In this case, it appears like an open switch. With no current through the collector resistor, the output voltage equals +15 V.

When the input voltage is +5 V, the base current is

$$I_B = \frac{5 \text{ V} - 0.7 \text{ V}}{3 \text{ k}\Omega} = 1.43 \text{ mA}$$

Visualize the transistor shorted between the collector and the emitter. Then the output voltage ideally drops to zero and the saturation current is

$$I_{C(\text{sat})} = \frac{15 \text{ V}}{1 \text{ k}\Omega} = 15 \text{ mA}$$

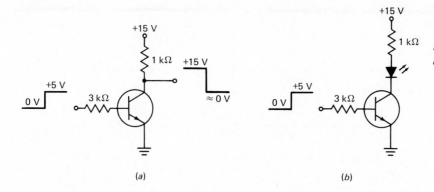

Fig. 5-18
*Examples of
transistor used as
a switch.*

(a)

(b)

This is approximately 10 times the base current, easily enough to produce hard saturation in almost any small-signal transistor. This means that the transistor acts like a closed switch and V_{out} is approximately zero.

Figure 5-18*b* shows a slight variation of the design. The circuit is called an *LED driver* because the transistor controls the LED. When input voltage is low, the transistor is cut off and the LED is dark. When the input voltage is high, the transistor is saturated and the LED is lit up. Assuming an LED voltage drop of 2 V, the LED current is

$$I_{C(sat)} \cong \frac{15\ V - 2\ V}{1\ k\Omega} = 13\ mA$$

5-7 TRANSISTOR CURRENT SOURCE

The LED driver of Fig. 5-18*b* is all right as long as the collector supply voltage is large. But at lower supply voltages, the tolerance of the LED voltage drop affects the brightness. For instance, if $V_{CC} = 5$ V, an LED variation from 1.5 to 2.5 V produces a noticeable change in LED brightness. The larger V_{CC} is compared with V_{LED}, the less sensitive the LED current is to changes in V_{LED}. In fact, the best way to drive an LED is with a current source; this sets up a fixed LED current, regardless of the LED voltage drop. This section discusses a transistor current source, another fundamental way to use a transistor.

EMITTER RESISTOR

Figure 5-19*a* has a resistor R_E between the emitter and the common point. Emitter current flows through this resistor, producing a voltage drop $I_E R_E$. Summing voltages around the output loop gives

$$V_{CE} + I_E R_E - V_{CC} + I_C R_C = 0$$

Because collector current approximately equals emitter current, we can rewrite the foregoing equation as

$$I_C \cong \frac{V_{CC} - V_{CE}}{R_C + R_E} \tag{5-12}$$

Fig. 5-19
*Transistor used as
a current source.*
(a) *Circuit.* (b) *Dc
load line.*

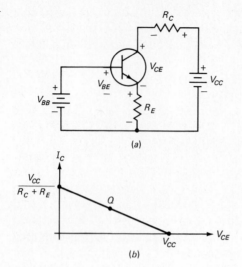

(a)

(b)

This is the equation of the dc load line. It is almost the same as that of the load line discussed earlier, except for the presence of R_E in the denominator.

Figure 5-19*b* shows the dc load line. You can get the upper end of the load line as follows: Visualize the collector-emitter terminals shorted in Fig. 5-19*a*. Then R_C and R_E are in series; this results in a collector current of $V_{CC}/(R_C + R_E)$. Similarly, visualize the collector-emitter terminals open; then all the supply voltage appears across the collector-emitter terminals, and $V_{CE} = V_{CC}$.

Unlike the transistor switch, the circuit of Fig. 5-19*a* is intended to operate in the active region at some Q point along the load line of Fig. 5-19*b*. With this new circuit it is possible to set up a rock-solid collector current that is virtually immune to changes in β_{dc}.

SETTING UP EMITTER CURRENT

To understand how we can set up a fixed value of collector current, let us sum voltages around the input loop of Fig. 5-19*a*:

$$V_{BE} + I_E R_E - V_{BB} = 0$$

Solving for I_E, we get

$$I_E = \frac{V_{BB} - V_{BE}}{R_E} \qquad (5\text{-}13)$$

$V_{BB} - V_{BE}$ is the voltage across the emitter resistor. Therefore, Eq. (5-13) is nothing more than Ohm's law applied to the emitter resistor.

Since the collector current is almost equal to the emitter current, we can use Eq. (5-13) to calculate the approximate value of collector current. For instance, if $V_{BB} = 2$ V and $R_E = 100$ Ω, then

$$I_E = \frac{2 \text{ V} - 0.7 \text{ V}}{100 \text{ Ω}} = 13 \text{ mA}$$

Therefore, the collector current approximately equals 13 mA. Especially important, this value of collector current does not depend on the value of β_{dc}, because β_{dc} does not appear in Eq. (5-13). You can change transistors all day and you will still have close to 13 mA of collector current.

EMITTER CURRENT IS FIXED

Figure 5-20a is the usual way you will see the transistor current source. Given a base voltage V_{BB}, you can set up a fixed emitter current by selecting a value of R_E. This is useful in many applications because the circuit is relatively immune to changes in β_{dc}. Here's why. If β_{dc} changes, the base current changes, but the collector current remains essentially the same. This is because the circuit of Fig. 5-20a produces a fixed value of emitter current. The use of an *emitter resistor* is the key to rock-solid values of collector current. The larger R_E is, the more stable the collector current.

The action is quite different from that of the transistor switch of Fig. 5-20b, where the base current is fixed by V_{BB} and R_B. In a transistor switch, we set up a fixed base current that is large enough to drive the transistor into hard saturation. No attempt is made to operate in the active region because the variations in β_{dc} would cause the operating point to drift all over the load line. In a transistor switch *the emitter is grounded;* this is how you can recognize a transistor switch.

BOOTSTRAP CONCEPT

The voltage across the emitter resistor of Fig. 5-20a is sometimes written as V_E. This voltage equals the base source voltage minus the drop across the base-emitter terminals. In symbols,

$$V_E = V_{BB} - V_{BE}$$

Because V_{BE} is fixed at approximately 0.7 V, V_E will follow the changes in V_{BB}. For instance, if V_{BB} rises from 2 V to 10 V, V_E will rise from 1.3 V to 9.3 V. This kind of follow-the-leader action is called *bootstrapping.* In a transistor current source the emitter is bootstrapped to the input voltage, which means that it is always within 0.7 V of it.

Again, notice how different this is from the transistor switch of Fig. 5-20b. Since the emitter in a transistor switch is grounded, the emitter is not bootstrapped to the

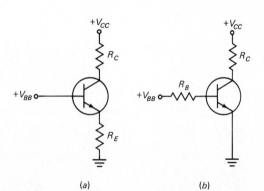

(a)　　　　(b)

Fig. 5-20
(a) *Transistor current source.* (b) *Transistor switch.*

input voltage. Instead, the emitter remains at ground potential no matter what the input voltage does.

VOLTAGE SOURCE VERSUS CURRENT SOURCE

Another way to distinguish a transistor current source from a transistor switch is by the type of source driving the base. In Fig. 5-20a, the source voltage is applied directly to the base; no base resistor is used. Therefore, a voltage source drives the base. Because of the small V_{BE} drop, most of this source voltage appears across the emitter resistor. This produces a fixed emitter current and a solid Q point in the active region.

On the other hand, the base resistor in the transistor switch of Fig. 5-20b makes the base supply act more like a current source because most of V_{BB} is dropped across the base resistor. Current-sourcing the base is all right in switching circuits because the variations in β_{dc} are swamped out by the hard saturation.

USEFUL VOLTAGES

When troubleshooting a circuit like 5-20a, you should measure transistor voltages with respect to ground because most electronic voltmeters have the common lead grounded. The emitter is bootstrapped to within 0.7 V of the base voltage, so you should read an emitter voltage of

$$V_E = V_{BB} - 0.7 \tag{5-14}$$

Collector current flows through the collector resistor, producing a drop of $I_C R_C$. Therefore, you should read a collector-to-ground voltage of

$$V_C = V_{CC} - I_C R_C \tag{5-15}$$

Incidentally, single subscripts (V_C, V_E, V_B) refer to the voltage of a transistor terminal with respect to ground. Double subscripts (V_{BE}, V_{CE}, V_{CB}) refer to the voltage between two transistor terminals. You can calculate a double-subscript voltage by subtracting the corresponding single-subscript voltages. For instance, to get V_{CE}, subtract V_E from V_C:

$$V_{CE} = V_C - V_E \tag{5-16}$$

To get V_{CB}, subtract V_B from V_C:

$$V_{CB} = V_C - V_B \tag{5-17a}$$

To get V_{BE}, subtract V_E from V_B:

$$V_{BE} = V_B - V_E \tag{5-17b}$$

EXAMPLE 5-1

Calculate the LED current in Fig. 5-21a.

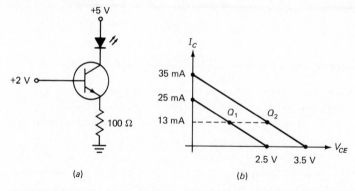

Fig. 5-21 (a) *Transistor current source driving LED.* (b) *LED current remains constant for two different LED voltages.*

SOLUTION

Since the emitter is bootstrapped to within 0.7 V of the input, the voltage across the emitter resistor is 1.3 V and the emitter current is

$$I_E = \frac{1.3 \text{ V}}{100 \text{ }\Omega} = 13 \text{ mA}$$

The LED current is approximately 13 mA.

Notice that we do not include LED voltage drop in these calculations because the transistor is acting like a current source rather than a switch.

EXAMPLE 5-2

The circuit of Fig. 5-21*a* is to be mass produced. The LED can have a voltage drop anywhere from 1.5 to 2.5 V. Draw the load lines and *Q* points for low and high LED voltage.

SOLUTION

Sum the voltages around the collector loop to get

$$V_{CE} + I_E R_E - V_{CC} + V_{LED} = 0$$

Since collector current approximately equals emitter current, we can rearrange the equation to get

$$I_C \cong \frac{V_{CC} - V_{LED} - V_{CE}}{R_E}$$

or

$$I_C \cong \frac{5 - V_{LED} - V_{CE}}{100}$$

For the smaller LED voltage, $V_{LED} = 1.5$ V, and the load-line equation simplifies to

$$I_C \cong \frac{3.5 - V_{CE}}{100}$$

At the other extreme, $V_{LED} = 2.5$ V, and the load-line equation is

$$I_C \cong \frac{2.5 - V_{CE}}{100}$$

Figure 5-21*b* shows the load lines for low and high LED voltage. Regardless of the actual LED drop, the circuit acts like a current source because the operating point is in the active region for low or high LED voltage. This tells us that all LEDs will glow with equal brightness, no matter what their voltage drop, because the current is constant.

5-8 MORE OPTOELECTRONIC DEVICES

As mentioned earlier, a transistor with an open base has a small collector current consisting of thermally produced minority carriers and surface leakage. By exposing the collector junction to light, a manufacturer can produce a *phototransistor*, a transistor that has more sensitivity to light than a photodiode.

BASIC IDEA OF PHOTOTRANSISTOR

Figure 5-22*a* shows a transistor with an open base. As mentioned earlier, a small collector current exists in this circuit. Forget about the surface-leakage component and concentrate on the thermally produced carriers in the collector diode. Visualize the reverse current produced by these carriers as an ideal current source in parallel with the collector-base junction of an ideal transistor (Fig. 5-22*b*).

Because the base lead is open, all the reverse current is forced into the base of the transistor. The resulting collector current is

$$I_{CEO} = \beta_{dc} I_R$$

This says that the collector current is higher than the original reverse current by a factor of β_{dc}. The collector diode is sensitive to light as well as heat. In a phototransistor, light passes through a window and strikes the collector-base junction. As the light increases, I_R increases, and so does I_{CEO}.

PHOTOTRANSISTOR VERSUS PHOTODIODE

The main difference between a phototransistor and a photodiode is the current gain β_{dc}. The same amount of light striking both devices produces β_{dc} times more current in a phototransistor than in a photodiode. The increased sensitivity of a phototransistor is the big advantage it has over a photodiode.

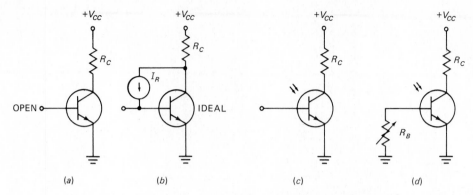

Fig. 5-22 (a) *Transistor with open base.* (b) *Reverse current in parallel with collector diode.* (c) *Phototransistor.* (d) *Controlling sensitivity to light.*

Figure 5-22c shows the schematic symbol of a phototransistor. Notice the open base. This is the usual way to operate a phototransistor. You can control the sensitivity with a variable base return resistor (Fig. 5-22d), but the base is usually left open to get maximum sensitivity to light.

The price paid for increased sensitivity is reduced speed. A phototransistor is more sensitive than a photodiode, but it cannot turn on and off as fast. For instance, a photodiode has typical output currents in microamperes and can switch on or off in nanoseconds. On the other hand, the phototransistor has typical output currents in milliamperes but switches on or off in microseconds.

OPTOCOUPLER

Figure 5-23 shows an LED driving a phototransistor. This is a much more sensitive optocoupler than the LED-photodiode example given earlier. The idea is straightforward. Any changes in V_S produce changes in LED current, which changes the current through the phototransistor. In turn, this produces a changing voltage across the collector-emitter terminals. Therefore, a signal voltage is coupled from the input circuit to the output circuit.

Again, the big advantage of an optocoupler is the *electrical isolation* between the input and output circuits. Stated another way, the common for the input circuit is different from the common for the output circuit. Because of this, no conductive path exists between the two circuits. This means that you can ground one of the circuits and float the other. For instance, the input circuit can be grounded to the chassis of the equipment, while the common of the output side is ungrounded.

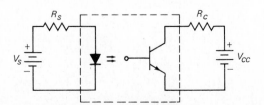

Fig. 5-23
Optocoupler with LED and phototransistor.

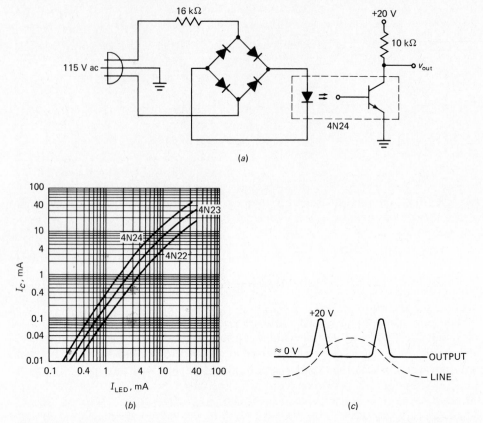

Fig. 5-24 (a) *Zero-crossing detector.* (b) *Static curves for optocouplers.* (c) *Output of detector.*

EXAMPLE 5-3

The 4N24 optocoupler of Fig. 5-24*a* provides isolation from the line and detects *zero crossings* of line voltage. Use the transfer characteristic of Fig. 5-24*b* to calculate the peak output voltage from the optocoupler.

SOLUTION

The bridge rectifier produces a full-wave current through the LED. Ignoring diode drops, the peak current through the LED is

$$I_{\text{LED}} \cong \frac{1.414(115 \text{ V})}{16 \text{ k}\Omega} = 10.2 \text{ mA}$$

The saturated value of phototransistor current is

$$I_{C(\text{sat})} \cong \frac{20 \text{ V}}{10 \text{ k}\Omega} = 2 \text{ mA}$$

Figure 5-24*b* shows the static curves of phototransistor current versus LED current for three different optocouplers. With a 4N24 (top curve), an I_{LED} of 10.2 mA produces an I_C of approximately 15 mA when the load resistance is zero. In Fig. 5-24*a*, the phototransistor current never reaches 15 mA because the phototransistor saturates at 2 mA. In other words, there is more than enough LED current to produce saturation. Since the peak LED current is 10.2 mA, the transistor is saturated during most of the cycle. At this time the output voltage is approximately zero, as shown in Fig. 5-24*c*.

The zero crossings occur when line voltage is changing polarity, from positive to negative, or vice versa. At a zero crossing, the LED current drops to zero. At this instant, the phototransistor becomes an open circuit and the output voltage increases to approximately 20 V, as indicated in Fig. 5-24*c*. As you see, the output voltage is near zero most of the cycle. At the zero crossings, it increases rapidly to 20 V and then decreases back to the baseline.

A circuit like Fig. 5-24*a* is useful because it does not require a transformer to provide isolation from the line. The photocoupler takes care of this. Furthermore, the circuit detects zero crossings; this is desirable in applications where you want to synchronize some other circuit to the frequency of line voltage.

5-9 TROUBLESHOOTING

Many things can go wrong with a transistor. Since it contains two diodes, exceeding any of the breakdown voltages, maximum currents, or power ratings can damage either or both diodes. The troubles may include shorts, opens, high leakage currents, reduced β_{dc}, and other troubles.

OUT-OF-CIRCUIT TESTS

One way to test transistors is with an ohmmeter. You can begin by measuring the resistance between the collector and the emitter. This should be very high in both directions because the collector and emitter diodes are back-to-back in series. One of the most common troubles is a collector-emitter short, produced by exceeding the power rating. If you read zero to a few thousand ohms in either direction, the transistor is shorted and should be thrown away.

Assuming that the collector-emitter resistance is very high in both directions (in megohms), you can read the reverse and forward resistances of the collector diode (collector-base terminals) and the emitter diode (base-emitter terminals). You should get a high reverse/forward ratio for both diodes, typically more than 1000:1 (silicon). If you do not, the transistor is defective.

Even if the transistor passes the ohmmeter tests, it still may have some faults. After all, the ohmmeter only tests each transistor junction under dc conditions. You can use a curve tracer to look for more subtle faults, like too much leakage current, low β_{dc}, or insufficient breakdown voltage. Commercial transistor testers are also available; these check the leakage current, β_{dc}, and other such quantities.

IN-CIRCUIT TESTS

The simplest in-circuit tests are to measure transistor voltages with respect to ground. For instance, measuring the collector voltage V_C and the emitter voltage V_E is a good start. The difference $V_C - V_E$ should be more than 1 V but less than V_{CC}. If the reading is less than 1 V, the transistor may be shorted. If the reading equals V_{CC}, the transistor may be open.

The foregoing test usually pins down a dc trouble if one exists. Many people include a test of V_{BE}, done as follows. Measure the base voltage V_B and the emitter voltage V_E. The difference of these readings is V_{BE}, which should be 0.6 to 0.7 V for small-signal transistors operating in the active region. For power transistors, V_{BE} may be 1 V or more because of the additional $I_B r'_b$ drop. If the V_{BE} reading is less than 0.6 V, the emitter diode is not being forward-biased. The trouble could be in the transistor or in the biasing components.

Some people include a cutoff test, performed as follows. Short the base-emitter terminals with a jumper wire. This removes the forward bias on the emitter diode and should force the transistor into cutoff. The collector-to-ground voltage should equal the collector supply voltage. If it does not, something is wrong with the transistor or the circuitry.

The foregoing tests apply to transistor switches and current sources. Later chapters discuss transistor amplifiers and the troubles you encounter with them.

5-10 DISCRETE AND INTEGRATED CIRCUITS

In the coming chapters we will be analyzing all kinds of transistor circuits. At first the emphasis will be on *discrete circuits,* the kind you build when you solder individual components, like resistors, diodes, and transistors.

Integrated circuits are different. Figure 5-25 shows a *chip,* a small piece of semiconductor material that serves as a chassis for the integrated circuit. The dimensions are representative; chips are often smaller than this, sometimes larger. Using advanced photographic techniques, a manufacturer can produce circuits on the surface of this chip, circuits with many resistors, diodes, transistors, and other components. The finished network is so small that you need a microscope to see the connections. We call a circuit like this an integrated circuit (IC).

Fig. 5-25
Semiconductor chip.

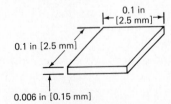

The word "chip" has a second meaning. It stands for the entire integrated circuit, including the package and its external pins. For instance, an LM741 is a chip with eight pins that are externally connected to supply voltages, input signals, etc. The 741 will be discussed in a later chapter. For now, keep in mind that discrete circuits are those in which the components are individually soldered together, whereas integrated circuits are those in which the components are atomically connected during the manufacturing process.

PROBLEMS

Straightforward

5-1. Suppose that only 2 percent of the electrons injected into the base recombine with base holes. If 1 million electrons enter the emitter in 1 μs, how many electrons come out the base lead in this period? How many come out the collector lead during this time?

5-2. If the emitter current is 6 mA and the collector current is 5.75 mA, what does the base current equal? What is the value of α_{dc}?

5-3. A transistor has an I_C of 100 mA and an I_B of 0.5 mA. What do α_{dc} and β_{dc} equal?

5-4. A transistor has a β_{dc} of 150. If the collector current equals 45 mA, what does the base current equal?

5-5. A 2N5067 is a power transistor with an r'_b of 10 Ω. How much $I_B r'_b$ drop is there when $I_B = 1$ mA? When $I_B = 10$ mA? When $I_B = 50$ mA?

5-6. A 2N3298 has a typical β_{dc} of 90. Calculate the approximate collector and base currents for an emitter current of 10 mA.

5-7. A transistor has a β_{dc} of 400. What does the base current equal when the collector current equals 50 mA?

5-8. Figure 5-26a shows one of the collector curves. Calculate β_{dc} at point A and point B.

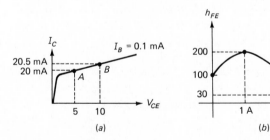

Fig. 5-26

5-9. Sketch collector curves for a transistor with the following specifications: $V_{CE(sat)}$ less than 1 V, $\beta_{dc} = 200$, $V_{CEO} = 40$ V, and $I_{CEO} = 50$ nA. Show six curves distributed between $I_B = 0$ and the I_B needed to produce 50 mA of collector current. Indicate the saturation region, the active region, the breakdown region, and the cutoff region.

5-10. A 2N5346 has the β_{dc} variations shown in Fig. 5-26b. What does β_{dc} equal if I_C is 1 mA? How much base current is there when $I_C = 1$ A? When $I_C = 7$ A?

5-11. Figure 5-27a shows a transistor circuit with an open base lead. If we measure a V_{CE} of 9 V, what value does I_{CEO} have? If we change the collector resistor from 10 MΩ to 10 kΩ, as shown in Fig. 5-27b, what is the new value of V_{CE}? (Assume that I_{CEO} remains the same.)

Fig. 5-27

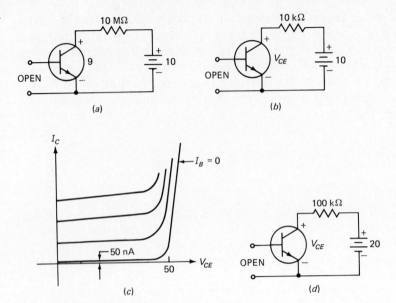

(a)   (b)

(c)   (d)

5-12. A transistor has the collector curves of Fig. 5-27c. If this transistor is used in the circuit of Fig. 5-27d, what will V_{CE} equal? What is the value of BV_{CEO}? Is the transistor of Fig. 5-27d in danger of breakdown?

5-13. A transistor has a collector current of 10 mA and a collector-emitter voltage of 12 V. What is the power dissipation?

5-14. A 2N3904 has a power rating of 310 mW at room temperature (25°C). If the collector-emitter voltage is 10 V, what is the maximum current that the transistor can handle without exceeding its power rating?

5-15. Draw the load line for Fig. 5-28a. What is the saturation current? The cutoff voltage?

Fig. 5-28

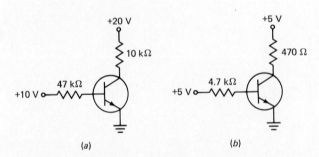

(a)   (b)

5-16. In the load line of Fig. 5-28b, what is the maximum possible collector current? If the base voltage is removed, what does V_{CE} equal?

5-17. What is the base current in Fig. 5-28a? The collector-emitter voltage? Is the transistor in hard saturation?

5-18. Suppose we connect an LED in series with the 10-kΩ resistor of Fig. 5-28a. What does the LED current equal? Comment on the brightness of the LED.

5-19. What does the base current equal in Fig. 5-28b? The collector current? The collector-emitter voltage?

5-20. Draw the load line for Fig. 5-29*a*. What is the saturated value of collector current? The cutoff voltage?

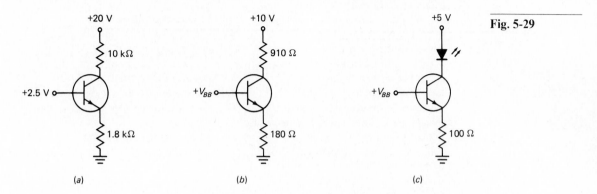

(a) (b) (c)

Fig. 5-29

5-21. How much collector current is there in Fig. 5-29*a*? What is the voltage between the collector and ground? The collector-emitter voltage?

5-22. What is the maximum possible collector current in Fig. 5-29*b*? If V_{BB} is 2 V, what is the collector-to-ground voltage?

5-23. In Fig. 5-29*b*, $V_{BB} = 10$ V. What is the collector-emitter voltage?

5-24. If $V_{BB} = 5$ V in Fig. 5-29*c*, what is the LED current? The collector-to-ground voltage?

5-25. Figure 5-30*a* shows a 4N33 optocoupler isolating a low-voltage circuit (the input) from a high-voltage circuit (common at +1000 V). Figure 5-30*b* is the transfer characteristic of a 4N33 for an unsaturated phototransistor. Answer the following questions:

   a. What is the maximum possible phototransistor current?

   b. How much LED current is there when V_{BB} is +5 V? What is the collector-emitter voltage of the phototransistor for this condition?

   c. If V_{BB} is zero, what does the collector-emitter voltage of the phototransistor equal?

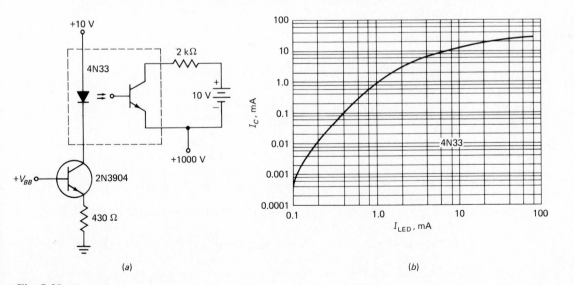

(a) (b)

Fig. 5-30

Troubleshooting

5-26. In Fig. 5-28a, the collector-to-ground voltage is +20 V. Which of the following is the probable cause of the trouble?
  a. Collector-emitter terminals shorted   **c.** 47-kΩ resistor open
  b. 10-kΩ resistor open   **d.** Collector-base terminals shorted
5-27. The collector-to-ground voltage of Fig. 5-29a reads approximately 3 V. Which of these is the root of the trouble?
  a. 10-kΩ resistor is shorted   **c.** Base-emitter terminals are shorted
  b. 1.8-kΩ resistor is open   **d.** Collector-emitter terminals are shorted.
5-28. With the base voltage removed in Fig. 5-28b, the collector-emitter voltage is approximately zero. Name some possible causes.
5-29. Is the LED of Fig. 5-29c on or off for each of the following conditions:
  a. Collector-emitter terminals shorted
  b. 100-Ω resistor open
  c. Collector-emitter terminals open
  d. Cold-solder joint on grounded end of 100-Ω resistor

Design

5-30. Redesign the transistor switch of Fig. 5-28a to get a saturated collector current of 5 mA.
5-31. Redesign the current source of Fig. 5-29c to set up a LED current of approximately 35 mA for a V_{BB} of 5 V.
5-32. Design a transistor switch similar to Fig. 5-28a to meet these specifications: $V_{CC} = 15$ V, $V_{BB} = 0$ or 15 V and $I_{C(\text{sat})} = 5$ mA.
5-33. Design a LED driver like Fig. 5-29c to meet these specifications: $V_{CC} = 10$ V, $V_{BB} = 0$ or 10 V, and $I_{\text{LED}} = 20$ mA.

Challenging

5-34. If $V_{BB} = V_{CC}$ in a transistor switch, a quick design rule for hard saturation is to satisfy this condition:

$$\frac{R_B}{R_C} = 10$$

Use mathematics to prove that this is approximately correct.

Fig. 5-31

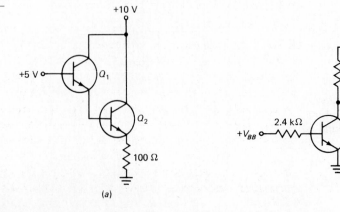

(a)   (b)

5-35. Figure 5-31*a* shows a *Darlington* connection of two transistors. Answer these questions:
   a. What is the voltage across the 100-Ω resistor?
   b. What is the approximate value of collector current in the first transistor if the second transistor has a β_{dc} of 150?
   c. If the first transistor has a β_{dc} of 100 and the second transistor has a β_{dc} of 150, what is the base current in the first transistor?

5-36. Answer these questions for the circuit of Fig. 5-31*b*:
   a. How much LED current is there when V_{BB} is zero?
   b. What is the LED current if V_{BB} is 10 V?

Computer

5-37. A multiple-statement line has two or more statements separated by a colon. For instance,

```
10 PRINT "ENTER VCC": INPUT VCC
20 END
```

After printing "ENTER VCC" on the screen, the computer will wait until you enter a value for V_{CC}.

Here is another program with multiple-statement lines:

```
10 PRINT "ENTER VCC": INPUT VCC
20 PRINT "ENTER RC": INPUT RC
30 IC = VCC/RC: IB = 0.1 * IC
40 PRINT "ENTER VBB": INPUT VBB
50 RB = (VBB − 0.7)/IB
60 PRINT "BASE RESISTOR SHOULD BE": PRINT RB
70 GOTO 10
```

What does the program calculate in line 30? What is the program supposed to do?

5-38. Write a program for a transistor current source to do the following: ask that the values of V_{BB} and the desired I_C be input, then print out the correct value of R_E.

5-39. If the PRINT statement is used without anything following it, the computer will print a blank space on the screen. This is one way of inserting spaces between printed lines. For instance, here is a sample program:

```
10 PRINT "1. TRANSISTOR SWITCH"
20 PRINT "2. CURRENT SOURCE"
30 PRINT
40 PRINT "ENTER CHOICE"
50 INPUT C
```

If this program is run, the screen will display

```
1. TRANSISTOR SWITCH
2. CURRENT SOURCE

ENTER CHOICE
?
```

The computer will then wait until you enter a value for *C*.

The foregoing program shows how you create a menu, a list of choices on the computer screen. Write a program that creates a menu with these choices: 1. TRANSISTOR SWITCH, 2. CURRENT SOURCE, 3. DC BETA, 4. POWER DISSIPATION, 5. SATURATION CURRENT.

CHAPTER 6

Transistor Biasing Circuits

Digital circuits are those that use the transistor as a switch. Linear circuits are those that use it as a current source. Driving an LED with a transistor current source is one example of a linear circuit. Another example is an *amplifier,* a circuit that increases the amplitude of a signal. The idea is to put a small ac signal into a transistor and get out a larger ac signal of the same frequency. Amplifiers are essential for radio, television, and other communication circuits.

Before the ac signal can be coupled into a transistor, however, we have to set up a *quiescent* (Q) point of operation, typically near the middle of the dc load line. Then the incoming ac signal can produce fluctuations above and below this Q point. For the circuit to remain linear, the emitter diode must stay in forward bias and the collector diode must stay in reverse bias. Stated another way, the fluctuations in current and voltage must not drive the transistor into either saturation or cutoff.

This chapter discusses ways to bias a transistor for linear operation, which means setting up a Q point near the middle of the dc load line. The next chapter will discuss what happens when a small ac signal drives the circuits.

6-1 BASE BIAS

Figure 6-1a is an example of *base bias* (also called fixed bias). Usually, the base power supply is the same as the collector power supply; that is, $V_{BB} = V_{CC}$. In this case, the base and collector resistors are both returned to the positive side of the collector supply, and the circuit is drawn as shown in Fig. 6-1b.

In either case, this is the worst possible way to bias a transistor for linear operation because the Q point is unstable. As discussed in the preceding chapter, β_{dc} may have a 9:1 variation with current and temperature. This means that it is impossible to set up a stable Q point, one that we can rely on in mass production. Therefore, we never use base bias in linear circuits.

The primary use for base bias is in digital circuits in which the transistor is used as a switch between cutoff and saturation. In this case, we use hard saturation to overcome the variations in β_{dc}.

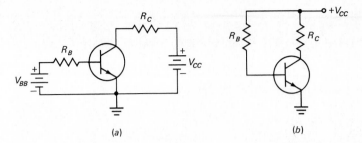

Fig. 6-1 *Base bias.* (a) *Complete circuit.* (b) *Simplified drawing.*

6-2 EMITTER-FEEDBACK BIAS

Figure 6-2a shows an early attempt at compensating for the variations in β_{dc}. Usually, the base and collector supplies are equal and the circuit is drawn as shown in Fig. 6-2b. In either case, the idea is to try to use the voltage across the emitter resistor to offset the changes in β_{dc}. For instance, if β_{dc} increases, the collector current increases. This increases the emitter voltage, which decreases the voltage across the base resistor and reduces the base current. The reduced base current results in less collector current, which partially offsets the original increase in β_{dc}.

PRACTICAL OBSERVATION

Emitter-feedback bias relies on the increased collector current producing more voltage across the emitter resistor, which reduces the base current and consequently the collector current. The basic idea sounds good, but the circuit does not work very well for practical values of resistance. To be effective, the circuit needs an emitter resistor that is as large as possible. This is where the snag is. You have to keep the emitter resistor relatively small to avoid collector saturation. The mathematical analysis that follows will show you why.

Incidentally, the word "feedback" is used here to indicate an output quantity (collector current) that produces a change in an input quantity (base current). The emitter resistor is the feedback element because it is common to the output and input circuits.

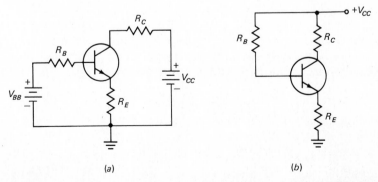

Fig. 6-2 *Emitter-feedback bias.* (a) *Complete circuit.* (b) *Simplified drawing.*

DC LOAD LINE

If we sum the voltages around the collector loop of Fig. 6-2*b*, we get

$$V_{CE} + I_E R_E - V_{CC} + I_C R_C = 0$$

Since I_E approximately equals I_C, this equation can be rearranged as

$$I_C \cong \frac{V_{CC} - V_{CE}}{R_C + R_E} \tag{6-1}$$

By now, you should be able to see at a glance that the upper end of the load line has a saturation current of $V_{CC}/(R_C + R_E)$ and the lower end of the load line has a cutoff voltage of V_{CC}.

EFFECT OF β_{dc}

Next, we can sum voltages around the base loop to get

$$V_{BE} + I_E R_E - V_{CC} + I_B R_B = 0$$

Since $I_E \cong I_C$ and $I_B = I_C/\beta_{dc}$, we can rewrite the equation as

$$I_C \cong \frac{V_{CC} - V_{BE}}{R_E + R_B/\beta_{dc}} \tag{6-2}$$

The intent of emitter-feedback bias is to swamp out the variations in β_{dc}; this is equivalent to R_E being much larger than R_B/β_{dc}. In practical circuits, however, you cannot make R_E large enough to swamp out the effects of β_{dc} without saturating the transistor. For typical designs, it turns out that emitter-feedback bias is almost as sensitive to changes in β_{dc} as is base bias. Therefore, emitter-feedback bias is not a preferred form of bias, and we will avoid using it. Study Example 6-1 to see how ineffective the circuit is against variations in β_{dc}.

SATURATION

If $R_B = \beta_{dc} R_C$, then Eq. (6-2) produces

$$I_C = \frac{V_{CC} - V_{BE}}{R_E + R_C}$$

This is slightly less than the saturation current found earlier, $V_{CC}/(R_E + R_C)$. Therefore, we arrive at the following conclusion: A base resistance of slightly less than $\beta_{dc} R_C$ produces saturation in an emitter-feedback-biased circuit.

EXAMPLE 6-1

Calculate the saturation value of collector current in Fig. 6-3*a*. Then calculate the collector current for these two values of β_{dc}: 100 and 300.

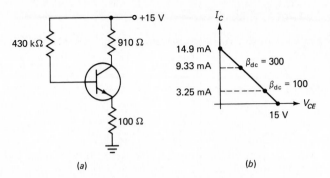

SOLUTION

The saturated collector current is

$$I_{C(\text{sat})} = \frac{15 \text{ V}}{910 \ \Omega + 100 \ \Omega} = 14.9 \text{ mA}$$

When $\beta_{\text{dc}} = 100$, Eq. (6-2) gives

$$I_C = \frac{15 \text{ V} - 0.7 \text{ V}}{100 \ \Omega + 430 \text{ k}\Omega/100} = 3.25 \text{ mA}$$

When $\beta_{\text{dc}} = 300$,

$$I_C = \frac{15 \text{ V} - 0.7 \text{ V}}{100 \ \Omega + 430 \text{ k}\Omega/300} = 9.33 \text{ mA}$$

Figure 6-3*b* summarizes the calculations by showing the dc load line and the two *Q* points. As you see, a 3:1 change in β_{dc} produces almost a 3:1 change in collector current. This change is unacceptable. If you try to select other values for the circuit, you will find that emitter-feedback bias remains too sensitive to variations in β_{dc} to be a preferred biasing circuit.

6-3 COLLECTOR-FEEDBACK BIAS

Figure 6-4*a* shows *collector-feedback bias* (also called self-bias). The base resistor is returned to the collector rather than to the power supply. This is what distinguishes collector-feedback bias from base bias.

FEEDBACK ACTION

Here's how the feedback works. Suppose the temperature increases, causing β_{dc} in Fig. 6-4*a* to increase. This produces more collector current. As soon as the collector current increases, the collector-emitter voltage decreases (there's a greater drop across R_C). This means that there is less voltage across the base resistor, and this causes a decrease in base current. The smaller base current offsets the original

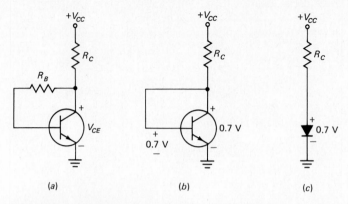

Fig. 6-4 *Collector-feedback bias. (a) Circuit. (b) Base shorted to collector. (c) Equivalent circuit when $R_B = 0$.*

increase in collector current. As you will see, the collector-feedback bias has certain advantages over emitter-feedback bias.

DC LOAD LINE

Summing voltages around the collector loop gives

$$V_{CE} - V_{CC} + (I_C + I_B)R_C = 0$$

Since I_B is much smaller than I_C in the active region, we can ignore I_B and rearrange the equation as

$$I_C \cong \frac{V_{CC} - V_{CE}}{R_C} \tag{6-3}$$

At a glance, you can see that the upper end of the load line has a saturation current of V_{CC}/R_C and the lower end has a cutoff voltage of V_{CC}.

EFFECT OF β_{dc}

If we sum the voltages around the base loop,

$$V_{BE} - V_{CC} + (I_C + I_B)R_C + I_B R_B = 0$$

or

$$V_{BE} - V_{CC} + I_C R_C + I_B R_B \cong 0$$

Since $I_B = I_C/\beta_{dc}$, the foregoing equation can be solved for I_C:

$$I_C \cong \frac{V_{CC} - V_{BE}}{R_C + R_B/\beta_{dc}} \tag{6-4}$$

Collector-feedback bias is somewhat more effective than emitter-feedback bias. Although the circuit is still sensitive to changes in β_{dc}, it is used in practice. It has the advantage of simplicity (there are only two resistors) and improved frequency response (discussed later). Example 6-3 gives you a concrete idea of how effective the circuit is in overcoming changes in β_{dc}.

SPECIAL CASE

Collector-feedback bias has another advantage over emitter bias: you cannot saturate the transistor. As you decrease the base resistance, the operating point moves toward the saturation point on the dc load line. But it can never reach saturation, no matter how low the base resistance is.

Figure 6-4b shows collector-feedback bias with a shorted base resistor. Notice that the V_{CE} can be no less than 0.7 V because this is the drop across the base-emitter terminals. The collector current is approximately

$$I_C = \frac{V_{CC} - 0.7}{R_C} \tag{6-5}$$

This is a little less than V_{CC}/R_C, the upper end of the dc load line; therefore, the transistor cannot be saturated.

Figure 6-4c shows the equivalent circuit for a shorted base resistor. A transistor with the base shorted to the collector acts the same as a diode. This idea is important to integrated circuits. Later, you will see why.

DESIGN GUIDELINE

In this book, we usually set the Q point near the middle of the dc load line. With collector-feedback bias, this requires

$$R_B = \beta_{dc} R_C \tag{6-6}$$

The easiest way to see this is by substituting this value into Eq. (6-4):

$$I_C = \frac{V_{CC} - V_{BE}}{R_C + \beta_{dc} R_C / \beta_{dc}} = \frac{V_{CC} - V_{BE}}{2R_C}$$

This is approximately half of V_{CC}/R_C, the saturation current. Therefore, satisfying Eq. (6-6) produces a Q point near the middle of the load line. Unless otherwise indicated, we will design collector-feedback bias circuits by satisfying the rule $R_B = \beta_{dc} R_C$.

EXAMPLE 6-2

Design a midpoint collector-feedback-biased circuit to meet these specifications: $V_{CC} = 15$ V, $R_C = 1$ kΩ, and $\beta_{dc} = 200$.

SOLUTION

The base resistance should be

$$R_B = 200 \times 1 \text{ k}\Omega = 200 \text{ k}\Omega$$

Figure 6-5a shows the circuit.

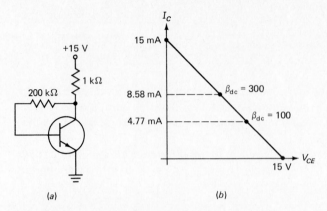

Fig. 6-5 (a) *Collector-feedback-biased circuit.* (b) *Dc load line with two Q points.*

EXAMPLE 6-3

In Fig. 6-5a, calculate the collector current for these two values of β_{dc}: 100 and 300.

SOLUTION

When $\beta_{dc} = 100$, Eq. (6-4) gives

$$I_C = \frac{15\text{ V} - 0.7\text{ V}}{1\text{ k}\Omega + 200\text{ k}\Omega/100} = 4.77\text{ mA}$$

When $\beta_{dc} = 300$,

$$I_C = \frac{15\text{ V} - 0.7\text{ V}}{1\text{ k}\Omega + 200\text{ k}\Omega/300} = 8.58\text{ mA}$$

Figure 6-5b shows the dc load line and the operating points. As you can see, a 3:1 change in β_{dc} produces less than a 2:1 change in collector current. The Q point is not rock solid, but at least there is an improvement over emitter-feedback bias. Furthermore, the circuit cannot saturate no matter how large β_{dc} becomes. For this reason, you sometimes see collector-feedback bias used in small-signal amplifiers.

6-4 VOLTAGE-DIVIDER BIAS

Figure 6-6a shows *voltage-divider bias* (also called universal bias). This is the bias most widely used in linear circuits. The name "voltage divider" comes from the voltage divider formed by R_1 and R_2. The voltage across R_2 forward-biases the emitter diode.

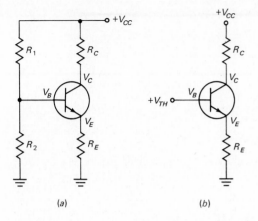

Fig. 6-6 *Voltage-divider bias.* (a) *Complete circuit.* (b) *Simplified drawing.*

EMITTER CURRENT

Voltage-divider bias works like this: Mentally open the base lead in Fig. 6-6*a*. Then you are looking back at an unloaded voltage divider whose Thevenin voltage is

$$V_{TH} = \frac{R_2}{R_1 + R_2} \, V_{CC} \qquad (6\text{-}7)$$

Now mentally reconnect the base lead. If the voltage divider is stiff, more than 99 percent of the Thevenin voltage drives the base. In other words, the circuit simplifies to Fig. 6-6*b* and the transistor acts like the controlled current source discussed in Chap. 5. Because the emitter is bootstrapped to the base,

$$I_E = \frac{V_{TH} - V_{BE}}{R_E} \qquad (6\text{-}8)$$

The collector current approximately equals this value.

Notice that β_{dc} does not appear in the formula for emitter current. This means that the circuit is immune to variations in β_{dc}, which implies a fixed Q point. Because of this, voltage-divider bias is the preferred form of bias in linear transistor circuits. You see it used almost universally.

STIFF VOLTAGE DIVIDER

The key to a well-designed circuit is the *stiffness* of the voltage divider. Here is how to get a design guideline for stiffness. If we thevenize the circuit of Fig. 6-6*a*, we get the equivalent circuit of Fig. 6-7, in which

$$R_{TH} = \frac{R_1 R_2}{R_1 + R_2} \qquad (6\text{-}9)$$

For simplicity, this is often written as

$$R_{TH} = R_1 \parallel R_2 \qquad (6\text{-}10)$$

Fig. 6-7
Equivalent circuit for voltage-divider bias.

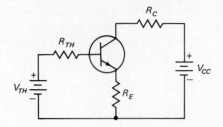

where the vertical bars stand for "in parallel with." You read Eq. (6-10) as "R_{TH} equals R_1 in parallel with R_2." Summing voltages around the base loop of Fig. 6-7 gives

$$V_{BE} + I_E R_E - V_{TH} + I_B R_{TH} = 0$$

Since $I_B \cong I_E/\beta_{dc}$, the foregoing equation reduces to

$$I_E \cong \frac{V_{TH} - V_{BE}}{R_E + R_{TH}/\beta_{dc}} \tag{6-11}$$

If R_E is 100 times greater than R_{TH}/β_{dc}, then the second term is swamped out and the equation simplifies to

$$I_E = \frac{V_{TH} - V_{BE}}{R_E} $$

In this book, a *stiff* voltage-divider-biased circuit is one that satisfies this condition:

$$R_{TH} \leqslant 0.01\beta_{dc}R_E \tag{6-12}$$

This 100:1 rule must be satisfied for the minimum β_{dc} encountered under all conditions. For instance, if a transistor has β_{dc} that varies from 80 to 400, use the lower value (80).

Usually, R_2 is smaller than R_1 and Eq. (6-12) is simplified to

$$R_2 \leqslant 0.01\beta_{dc}R_E \tag{6-13}$$

This is conservative because satisfying Eq. (6-13) automatically satisfies Eq. (6-12). For convenience, we will use Eq. (6-13) when designing stiff voltage dividers.

FIRM VOLTAGE DIVIDER

Sometimes a stiff design results in such small values of R_1 and R_2 that other problems arise (discussed later). In this case, many designers compromise by using this rule:

$$R_{TH} \leqslant 0.1\beta_{dc}R_E \tag{6-14}$$

Once more, it will be convenient to work with this design rule:

$$R_2 \leqslant 0.1\beta_{dc}R_E \tag{6-15}$$

In the worst case, satisfying this rule means that the collector current will be approximately 10 percent lower than the ideal value given by Eq. (6-8).

From now on, we will refer to a voltage divider as "firm" when it satisfies Eq. (6-15). As a guideline, we usually try to make the voltage divider stiff. For reasons given later (input impedance), we sometimes compromise by using a firm voltage divider because this may give us a better all-around circuit design.

DC LOAD LINE

If you sum voltages around the collector loop of Fig. 6-7 and solve for I_C, you get

$$I_C = \frac{V_{CC} - V_{CE}}{R_C + R_E} \qquad (6\text{-}16)$$

From this it is clear that the upper end of the load line has a saturation current of $V_{CC}/(R_C + R_E)$ and the lower end has a cutoff voltage of V_{CC}.

TRANSISTOR VOLTAGES

For troubleshooting you will find it convenient to measure the transistor voltages with respect to ground. The collector-to-ground voltage V_C equals the supply voltage minus the drop across the collector resistor:

$$V_C = V_{CC} - I_C R_C \qquad (6\text{-}17)$$

The emitter-to-ground voltage is

$$V_E = I_E R_E \qquad (6\text{-}18)$$

which is also given by

$$V_E = V_{TH} - V_{BE}$$

because the emitter is bootstrapped to within one V_{BE} drop of the base. In a stiff design, the base-to-ground voltage is

$$V_B = V_{TH} \qquad (6\text{-}19)$$

DESIGN GUIDELINES

Figure 6-8 shows an amplifier. The capacitors couple the ac signal into and out of the amplifier. As far as the dc is concerned, the capacitors appear like open circuits. A small ac input voltage drives the base, and an amplified ac output voltage appears at the collector. Chapter 8 analyzes this amplifier in depth. All we are doing now is learning how to design these circuits with a stable Q point.

In this book, unless otherwise indicated, use the one-tenth rule, which makes the emitter voltage approximately one-tenth of the supply voltage:

$$V_E = 0.1 V_{CC} \qquad (6\text{-}20)$$

This design rule is suitable for most circuits, but remember, it is only a guideline. Not everyone uses this rule, so don't be surprised to find emitter voltages at values different from one-tenth of supply voltage.

Fig. 6-8
An amplifier.

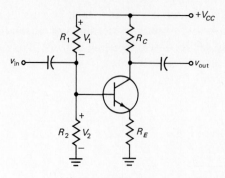

Start by calculating R_E needed to set up the specified collector current:

$$R_E = \frac{V_E}{I_E} \tag{6-21}$$

Also, locate the Q point at approximately the middle of the dc load line. This means that about $0.5V_{CC}$ appears across the collector-emitter terminals. The remaining $0.4V_{CC}$ appears across the collector resistor; therefore,

$$R_C = 4R_E \tag{6-22}$$

Next, you can design a stiff voltage divider using the 100:1 rule:

$$R_2 \leqslant 0.01\beta_{dc}R_E$$

If you prefer a firm voltage divider, then apply the 10:1 rule:

$$R_2 \leqslant 0.1\beta_{dc}R_E$$

Finally, calculate R_1 by using proportion:

$$R_1 = \frac{V_1}{V_2}R_2 \tag{6-23}$$

EXAMPLE 6-4

The circuit of Fig. 6-9a has a stiff voltage divider. Draw the dc load line and show the Q point.

SOLUTION

Mentally open the transistor from collector to emitter. Then all the supply voltage appears across the collector-emitter terminals. This means that the lower end of the load line has a cutoff voltage of 30 V.

Next, mentally short the transistor from collector to emitter. Then you can see that R_C is in series with R_E with a saturation current of

$$I_{C(sat)} = \frac{30 \text{ V}}{3000 \text{ }\Omega + 750 \text{ }\Omega} = 8 \text{ mA}$$

This represents the upper end of the dc load line.

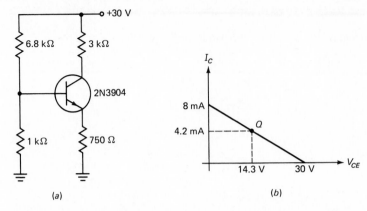

Fig. 6-9   (a) *Voltage-divider-biased circuit.* (b) *Dc load line with Q point.*

The stiff voltage divider produces a Thevenin voltage of

$$V_{TH} = \frac{1000}{6800 + 1000} \, 30 \text{ V} = 3.85 \text{ V}$$

The emitter current is

$$I_E \cong \frac{3.85 \text{ V} - 0.7 \text{ V}}{750 \, \Omega} = 4.2 \text{ mA} \cong I_C$$

The collector voltage is

$$V_C = 30 \text{ V} - (4.2 \text{ mA}) \, (3 \text{ k}\Omega) = 17.4 \text{ V}$$

The emitter voltage is

$$V_E = 3.85 \text{ V} - 0.7 \text{ V} = 3.15 \text{ V}$$

Therefore, the collector-emitter voltage is

$$V_{CE} = V_C - V_E = 17.4 \text{ V} - 3.15 \text{ V} = 14.3 \text{ V}$$

Figure 6-9*b* shows the dc load line and the *Q* point. As you can see, the *Q* point is near the middle of the dc load line.

EXAMPLE 6-5

Figure 6-10 shows a two-stage amplifier. (A stage is each transistor with its biasing resistors, including R_C and R_E.) What are the dc emitter voltages for each stage? The dc collector voltages?

SOLUTION

The capacitors are open to dc; therefore, we can analyze each stage separately because the dc voltages and currents do not interact. The stages are identical because they use the same values of resistance. In either stage, the voltage

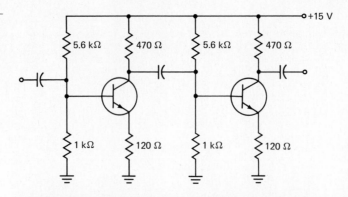

Fig. 6-10
*A two-stage
amplifier.*

across the 1-kΩ resistor of the voltage divider is 2.27 V. The emitter voltage is 0.7 V less, or

$$V_E = 1.57 \text{ V}$$

The emitter current is

$$I_E \cong \frac{1.57 \text{ V}}{120 \text{ }\Omega} = 13.1 \text{ mA}$$

To a close approximation, $I_C = 13.1$ mA and

$$V_C \cong 15 \text{ V} - (13.1 \text{ mA})(470 \text{ }\Omega) = 8.84 \text{ V}$$

EXAMPLE 6-6

Design a voltage-divider-biased circuit with these specifications: $V_{CC} = 20$ V, $I_C = 5$ mA, and β_{dc} varies from 80 to 400.

SOLUTION

The emitter voltage should be approximately one-tenth of the supply voltage, so $V_E = 2$ V. The quiescent collector current is specified as 5 mA; therefore, the required emitter resistance is

$$R_E = \frac{2 \text{ V}}{5 \text{ mA}} = 400 \text{ }\Omega$$

The nearest standard value is 390 Ω. To operate near the middle of the dc load line, the collector resistance must be approximately four times the emitter resistance:

$$R_C = 4(390 \text{ }\Omega) = 1560 \text{ }\Omega$$

The nearest standard value is 1.6 kΩ.

The base voltage is 0.7 V higher than the emitter voltage, so $V_B = 2.7$ V. This is the voltage across R_2. The voltage across R_1 is

$$V_1 = V_{CC} - V_2 = 20 \text{ V} - 2.7 \text{ V} = 17.3 \text{ V}$$

To get a stiff voltage divider,

$$R_2 \leqslant 0.01(80)(390) = 312 \ \Omega$$

The nearest standard value is 300, so

$$R_2 = 300 \ \Omega$$

and

$$R_1 = \frac{V_1}{V_2} R_2 = \frac{17.3 \ \text{V}}{2.7 \ \text{V}} (300 \ \Omega) = 1922 \ \Omega$$

The nearest standard value is 2 kΩ. Therefore, the final design values are

$$R_E = 390 \ \Omega$$
$$R_C = 1.6 \ \text{k}\Omega$$
$$R_1 = 2 \ \text{k}\Omega$$
$$R_2 = 300 \ \Omega$$

6-5 EMITTER BIAS

Figure 6-11*a* shows *emitter bias,* which is sometimes used when a split supply is available (positive and negative voltages). Figure 6-11*b* is a simplified way to draw the circuit.

Here is how to analyze an emitter-biased circuit. If R_B is small enough, the base voltage is approximately zero. The emitter voltage is one V_{BE} drop below this. Therefore, the voltage across the emitter resistor is $V_{EE} - V_{BE}$ and the emitter current is

$$I_E \cong \frac{V_{EE} - V_{BE}}{R_E} \tag{6-24}$$

Since β_{dc} does not appear in this formula, the Q point is fixed. Whenever a split supply is available, emitter bias may be used, because it provides a rock-solid Q point, just as voltage-divider bias does.

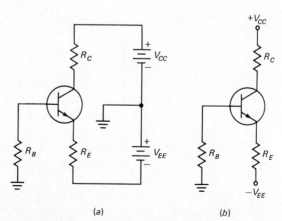

Fig. 6-11
Emitter bias. (a)
Complete circuit.
(b) *Simplified drawing.*

(a)   (b)

The key to a well-designed emitter-biased circuit is the size of R_B. It must be small. But how small? By a derivation similar to that given for voltage-divider bias, the exact formula for emitter current is

$$I_E = \frac{V_{EE} - V_{BE}}{R_E + R_B/\beta_{dc}}$$

Note the similarity to Eq. (6-11). In a stiff design, R_E is at least 100 times greater than R_B/β_{dc}. This is equivalent to

$$R_B \le 0.01\beta_{dc} R_E \tag{6-25}$$

When troubleshooting an emitter-biased circuit, you need to estimate the transistor voltages with respect to ground. The collector voltage is

$$V_C = V_{CC} - I_C R_C \tag{6-26}$$

In a stiff design, the base voltage is approximately 0 V and the emitter voltage is approximately −0.7 V.

EXAMPLE 6-7

In Fig. 6-11b, $R_C = 5.1$ kΩ, $R_E = 10$ kΩ, and $R_B = 6.8$ kΩ. If the split supply is +15 V and −15V, what is the collector voltage to ground?

SOLUTION

Assume a stiff design. Then the base is essentially grounded. The emitter is one V_{BE} drop below ground, or −0.7 V. The emitter current is

$$I_E = \frac{15 \text{ V} - 0.7 \text{ V}}{10 \text{ k}\Omega} = 1.43 \text{ mA}$$

The collector voltage is

$$V_C = 15 - (1.43 \text{ mA})(5.1 \text{ k}\Omega) = 7.71 \text{ V}$$

6-6 MOVING GROUND AROUND

Ground is a reference point that you can move around as you please. For instance, Fig. 6-12a shows collector-feedback bias. Figure 6-12b shows the same circuit where we do not rely on ground to conduct current. In a circuit like this, we can remove the ground to get the floating circuit of Fig. 6-12c. In this isolated circuit, all transistor voltages and currents have the same values as before; the same emitter current is flowing, and the same collector-emitter voltage is set up.

Following this line of thinking, we can ground the positive terminal of the supply, as shown in Fig. 6-12d. And finally, we can draw the circuit as shown in Fig. 6-12e. The voltages with respect to ground differ in Fig. 6-12a through e; $V_E = 0$ in Fig. 6-12a, but $V_E = -20$ V in Fig. 6-12e.

What is important is that the three transistor currents (I_E, I_C, and I_B) and the three

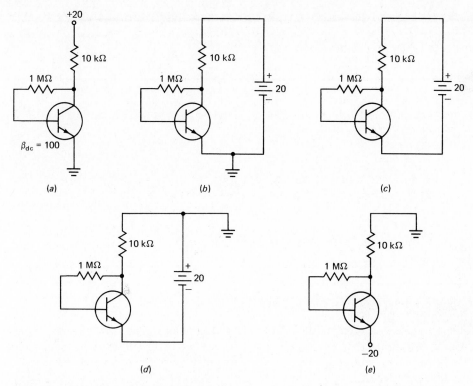

Fig. 6-12 *Relocating the ground point.* (a) *Original circuit.* (b) *First modification.* (c) *Floating circuit.* (d) *Positive end of supply grounded.* (e) *Final circuit.*

transistor voltages (V_{BE}, V_{CE}, and V_{CB}) have the same values in Fig. 6-12*a* through *e*. The key idea to remember is that a circuit can work with the negative end of the supply grounded (Fig. 6-12*a*), with neither end grounded (Fig. 6-12*c*), or with the positive end grounded (Fig. 6-12*e*).

6-7 *PNP* CIRCUITS

Figure 6-13*a* shows a *pnp* transistor. Since the emitter and collector diodes point in opposite directions, all currents and voltages are reversed. In other words, to forward-bias the emitter diode of a *pnp* transistor, V_{BE} has a minus-plus polarity, as

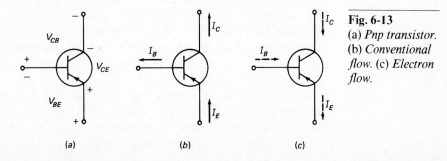

Fig. 6-13
(a) *Pnp transistor.*
(b) *Conventional flow.* (c) *Electron flow.*

Fig. 6-14
*Complementary
circuits.*

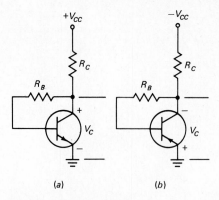

(a)   (b)

shown in Fig. 6-13*a*. To reverse-bias the collector diode, V_{CB} has the minus-plus polarity indicated. Figure 6-13*b* shows the directions for conventional flow, and Fig. 6-13*c* shows them for electron flow.

COMPLEMENTARY CIRCUITS

The *pnp* transistor is called the *complement* of the *npn* transistor. The word "complement" signifies that all voltages and currents are opposite to those of the *npn* transistor. Every *npn* circuit has a complementary *pnp* circuit. To find the complementary *pnp* circuit,

1. Replace the *npn* transistor by a *pnp* transistor.
2. Reverse all voltages and currents.

As an example, Fig. 6-14*a* shows collector-feedback bias using an *npn* transistor. The collector voltage is positive with respect to ground. Figure 6-14*b* shows the complementary *pnp* transistor circuit; all we have done is complement the voltages and replace the *npn* by a *pnp* transistor.

THE UPSIDE-DOWN CONVENTION
FOR *PNP* TRANSISTORS

When the power supply is positive, *pnp* transistors are usually drawn upside down. For instance, Fig. 6-15*a* shows a *pnp* amplifier using collector-feedback bias. If we float the supply, we get the circuit of Fig. 6-15*b*. If we ground the negative terminal of the supply, we get Fig. 6-15*c*. It does not matter how a transistor is oriented in space; therefore, it works just as well upside down, as shown in Fig. 6-15*d*. The collector current in this circuit has exactly the same value as the collector current of Fig. 6-15*a, b,* and *c*.

Upside-down *pnp* transistors look a little strange at first, but you will get used to the convention quickly. This practice leads to neater drawings when there are *npn* and *pnp* transistors in the same circuit. You may as well start using this convention, because the practice is widespread in industry.

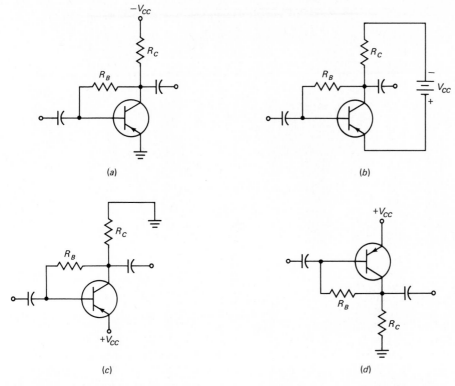

Fig. 6-15 *Drawing pnp circuit upside down.* (a) *Original circuit.* (b) *Floating circuit.* (c) *Moving ground.* (d) *Upside-down version.*

EXAMPLE 6-8

In Fig. 6-16, GND is an abbreviation for ground. What are the emitter and collector voltages in each stage?

SOLUTION

We analyzed this two-stage amplifier earlier (Example 6-5). The only difference is that upside-down *pnp* transistors are used in place of *npn* transistors. In either stage, the voltage across the 1-kΩ resistor is 2.27 V. The voltage across the emitter resistor is still 0.7 V less, or 1.57 V. Therefore, the emitter current is still

$$I_E = \frac{1.57 \text{ V}}{120 \ \Omega} = 13.1 \text{ mA}$$

Each collector resistor has 13.1 mA through it. As a result, the collector-to-ground voltage in either stage is

$$V_C = (13.1 \text{ mA})(470 \ \Omega) = 6.16 \text{ V}$$

Fig. 6-16
Two-stage amplifier with pnp transistors.

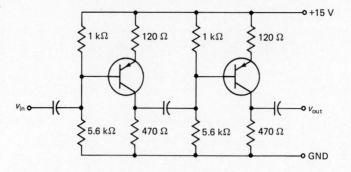

The emitter-to-ground voltage in each stage is

$$V_E = 15 \text{ V} - 1.57 \text{ V} = 13.4 \text{ V}$$

Incidentally, when a circuit is drawn like Fig. 6-16, the supply line is often called the *supply rail,* and the ground line is called the *ground rail.* The 1-kΩ resistors and the 120-Ω resistors are returned to the supply rail, while the 5.6-kΩ resistors and the 470-Ω resistors are returned to the ground rail.

EXAMPLE 6-9

What is the LED current in Fig. 6-17a?

SOLUTION

Assuming a stiff voltage divider, the voltage across the 680-Ω resistor is 6.28 V. So, the voltage across the emitter resistor is 0.7 V less, or 5.58 V. This means

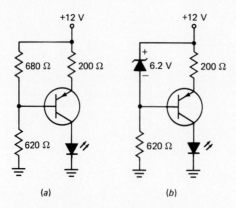

Fig. 6-17 (a) *Pnp current source drives LED.* (b) *Zener diode produces stiff source.*

that the emitter current is

$$I_E = \frac{5.58 \text{ V}}{200 \ \Omega} = 27.9 \text{ mA}$$

Therefore, the LED current is approximately 27.9 mA. This is an excellent way of sourcing current through a LED. The upside-down *pnp* circuit has the advantage of allowing us to ground one side of the LED.

Figure 6-17*b* shows a similar circuit with a zener diode. It sets up roughly the same LED current. Because the zener resistance is so small, this circuit represents a very stiff current source.

PROBLEMS

Straightforward

6-1. The transistor of Fig. 6-18 has an h_{FE} of 80. What is the voltage between the collector and ground?

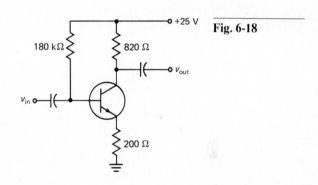

Fig. 6-18

6-2. Draw the dc load line for Fig. 6-18.

6-3. For what approximate value of β_{dc} does the circuit of Fig. 6-18 saturate?

6-4. If $\beta_{dc} = 125$ in Fig. 6-18, calculate the base voltage, the emitter voltage, and the collector voltage. (All voltages are with respect to ground.)

6-5. If $V_{CC} = 10$ V in Fig. 6-19, what is the collector voltage for each stage?

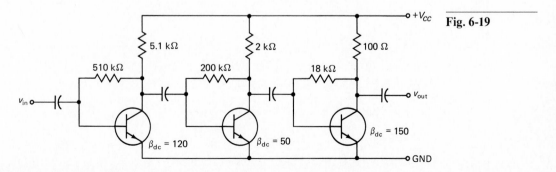

Fig. 6-19

6-6. If $V_{CC} = 15$ V in Fig. 6-19, what is the power dissipation of each transistor?

6-7. What is the emitter voltage to ground for each stage of Fig. 6-20 if the supply voltage is 10 V?

Fig. 6-20

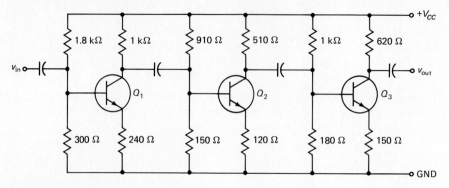

6-8. Calculate the collector saturation current for each stage of Fig. 6-20 for a V_{CC} of 15 V.

6-9. Do a complete analysis of Fig. 6-20 for $V_{CC} = 20$ V by calculating the following for each stage: V_B, V_E, V_C, I_C, and P_D (power dissipation).

6-10. What is the collector current in each stage of Fig. 6-21? The collector-to-ground voltage?

Fig. 6-21

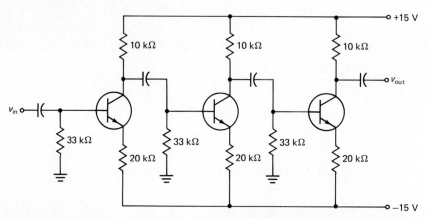

6-11. What is the power dissipation of each transistor in Fig. 6-21?

6-12. What is the collector current in Fig. 6-22a? The following voltages to ground: V_B, V_E, V_C?

Fig. 6-22

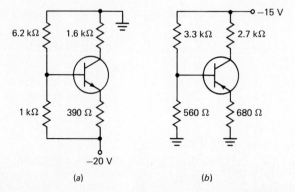

6-13. In Fig. 6-22*b*, calculate V_B, V_E, and V_C.

6-14. Work out the following voltages for each stage of Fig. 6-23: V_B, V_E, and V_C. Also calculate the power dissipation in each transistor.

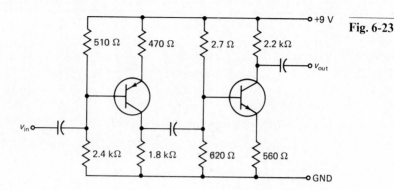

Fig. 6-23

6-15. The zener diode of Fig. 6-17*b* is replaced by one with $V_Z = 4.7$ V. What is the LED current?

Troubleshooting

6-16. In Fig. 6-20 assume $V_{CC} = 20$ V. Indicate whether the collector voltage of Q_1 increases, decreases, or remains the same for each of these troubles:
  a. 1.8 kΩ is open
  b. Q_1 collector-emitter short
  c. 240 Ω is open
  d. 240 Ω is shorted
  e. 300 Ω is shorted
  f. 1 kΩ is open
  g. 910 Ω is open

6-17. Examine the Q_3 stage of Fig. 6-20 for a V_{CC} of 20 V. Is the collector-to-ground voltage higher, lower, or normal for each of these conditions:
  a. 1 kΩ is open
  b. 1 kΩ is shorted
  c. 180 Ω is open
  d. 180 Ω is shorted
  e. 620 Ω is open
  f. 620 Ω is shorted
  g. Q_3 collector-emitter short
  h. Q_3 collector-emitter open
  i. 150 Ω is open
  j. 150 Ω is shorted

6-18. The supply voltage is 15 V in Fig. 6-20. Here are the voltmeter readings:
  First stage: $V_E = 1.5$ V and $V_C = 8.9$ V
  Second stage: $V_E = 0$ V and $V_C = 15$ V
  Third stage: $V_E = 1.6$ V and $V_C = 8.6$ V
  Name three possible troubles.

6-19. The collector voltage of the first stage is zero in Fig. 6-23. What are some of the possible troubles?

Design

6-20. Design a collector-feedback-biased stage to meet these specifications: $V_{CC} = 20$ V, $I_C = 5$ mA, and $\beta_{dc} = 150$.

6-21. Design a two-stage amplifier using stiff voltage-divider bias. The supply voltage is 15 V, and the quiescent collector current should be 1.5 mA for each stage. Assume an h_{FE} of 125.

6-22. Redesign the two-stage amplifier of Prob. 6-21 to have firm voltage dividers.

6-23. Design a *pnp* current source like Fig. 6-17*b* to meet these specifications: $V_{CC} = 15$ V, $V_Z = 7.5$ V, $I_C = 40$ mA.

Challenging

6-24. Calculate the following quantities in Fig. 6-24*a*: V_E, I_C, and V_C. (Assume silicon diodes.)

Fig. 6-24

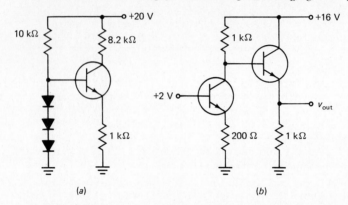

(a)   (b)

6-25. Figure 6-24*b* is an example of direct coupling between stages. What is the value of v_{out}? If v_{in} changes from 2 to 3 V, what is the new value of v_{out}?

6-26. In Fig. 6-24*b*, the voltage compliance of the output is the maximum voltage range the output can swing through when the input is varied. What is the output voltage compliance of the circuit?

Computer

6-27. The ON . . . GOTO . . . statement allows you to branch to different points in a program. For instance, here is part of a program:

```
10 PRINT "ENTER CHOICE: 1, 2, OR 3"
20 INPUT C
30 ON C   GOTO 500, 1000, 1500
```

If you enter a 1 for C, the program goes to line 500. Enter a 2 and it goes to line 1000. A 3 sends it to line 1500.

Here is a program:

```
10 PRINT "1. EMITTER-FEEDBACK BIAS"
20 PRINT "2. COLLECTOR-FEEDBACK BIAS"
30 PRINT "3. VOLTAGE-DIVIDER BIAS"
40 PRINT "4. EMITTER BIAS"
50 PRINT
60 PRINT "ENTER CHOICE"
70 INPUT C
80 ON C   GOTO 5000, 6000, 7000, 8000
```

Answer the following questions:

a. Where does the program go if you enter a 2?

b. What line does the computer branch to if you select voltage-divider bias?

6-28. Here is a program:

```
10 PRINT "ENTER VCC": INPUT VCC
20 PRINT "ENTER RC": INPUT RC
30 PRINT "ENTER RB": INPUT RB
40 PRINT "ENTER DC BETA": INPUT BDC
50 NUM = VCC − 0.7: DEN = RC + RB/BDC
60 IC = NUM/DEN
70 PRINT "THE COLLECTOR CURRENT IS": PRINT IC
```

What kind of transistor bias does this program analyze? What does the computer print out?

6-29. Write a program that prints out V_B, V_E, V_C, I_C, and P_D for a voltage-divider biased stage.

6-30. Write a program that calculates the collector current for emitter-feedback bias, collector-feedback bias, voltage-divider bias, and emitter bias. The program should include a menu like the one discussed in Prob. 6-28.

7

CE Amplifiers

After a transistor has been biased with a Q point near the middle of the dc load line, we can couple a small ac signal into the base. This produces fluctuations in the collector current of the same shape and frequency. For instance, if the input is a sine wave with a frequency of 1 kHz, the output will be an enlarged sine wave with a frequency of 1 kHz. The amplifier is called a linear (or high-fidelity) amplifier if it does not change the shape of the signal. As long as the amplitude of the signal is small, the transistor will use only a small part of the load line and the operation will be linear.

On the other hand, if the input signal is too large, the fluctuations along the load line will drive the transistor into saturation and cutoff. This clips the peaks off a sine wave, and the amplifier is no longer linear. If you listen to such an output with a loudspeaker, it sounds terrible because the signal is grossly distorted.

This chapter introduces some ideas needed to analyze small-signal amplifiers. We will begin with coupling capacitors, the devices that allow us to couple ac signals into and out of a transistor stage without changing the dc bias voltages. After you understand coupling capacitors, we will examine the superposition theorem as it applies to small-signal amplifiers. Then the chapter will conclude with an ac model of the transistor that allows rapid troubleshooting and design of basic amplifiers.

7-1 COUPLING AND BYPASS CAPACITORS

A *coupling capacitor* passes an ac signal from one point to another. For instance, in Fig. 7-1a the ac voltage at point A is transmitted to point B. For this to happen, the capacitive reactance X_C must be very small compared with the series resistances.

In Fig. 7-1a the circuit to the left of point A may be a single source and resistor, or it may be the Thevenin equivalent circuit of something more complicated. Likewise, resistance R_L may be a single load resistor, or it may be the equivalent resistance of a more complex network. It doesn't matter what the actual circuits are on either side of the capacitor; as long as we can reduce the circuit to a single loop as shown, the alternating current flows through a total resistance of $R_{TH} + R_L$.

As you will recall from basic circuit theory, the magnitude of the alternating

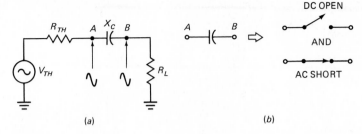

Fig. 7-1 (a) *Coupling capacitor between source and load.* (b) *Equivalent circuit for coupling capacitor.*

current in a one-loop *RC* circuit equals

$$I = \frac{V}{\sqrt{R^2 + X_C^2}} \tag{7-1}$$

where R is the total resistance of the loop. In Fig. 7-1a, $R = R_{TH} + R_L$. As the frequency increases, X_C decreases until it becomes much smaller than R. In this case, the current reaches a maximum value of V/R. In other words, the capacitor couples the signal properly from A to B when $X_C \ll R$.

STIFF COUPLING

The size of the coupling capacitor depends on the lowest frequency you are trying to couple. We will use this rule:

$$X_C \leq 0.1R \tag{7-2}$$

at the lowest input frequency to the amplifier. This says that the capacitive reactance of the coupling capacitor must be less than or equal to one-tenth of the total series resistance. Satisfying this 10:1 rule means that the alternating current is down less than 1 percent at the lowest frequency. To see this, substitute the worst case, $X_C = 0.1R$, into Eq. (7-1) to get

$$I = \frac{V}{\sqrt{R^2 + (0.1R)^2}} = \frac{V}{1.005R} = 0.995\frac{V}{R}$$

This shows that the alternating current is down only half a percent at the lowest frequency. We will refer to any coupling capacitor that satisfies the 10:1 rule as a *stiff coupling capacitor.*

Here is an example. Suppose we are designing a transistor stage for the audio range, 20 Hz to 20 kHz. If the input coupling capacitor sees a total series resistance of 10 kΩ, then X_C must be equal to or less than 1 kΩ at the lowest frequency, 20 Hz. We proceed as follows:

$$X_C = \frac{1}{2\pi f C} = \frac{1}{2\pi(20 \text{ Hz})C} = 1 \text{ k}\Omega$$

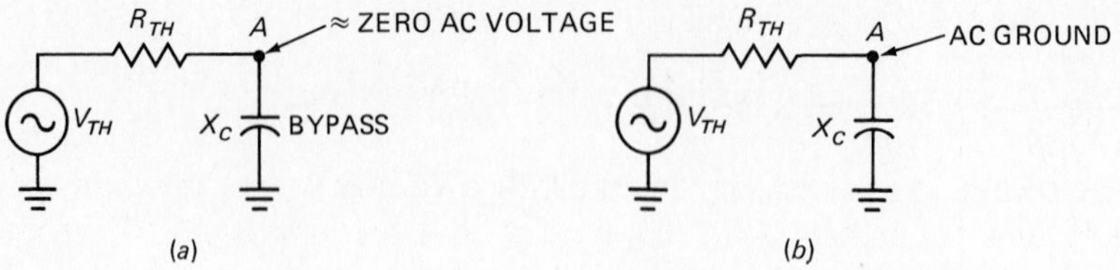

Fig. 7-2
Bypass capacitor.

(a)   (b)

Solving for C gives

$$C = \frac{1}{2\pi(20 \text{ Hz})(1 \text{ k}\Omega)} = 7.96 \ \mu\text{F}$$

This is the minimum capacitance needed for stiff coupling. In practice, we would use 10 μF, the next higher standard value (see Appendix 3). This provides stiff coupling for all frequencies above 20 Hz.

A coupling capacitor has the equivalent circuit shown in Fig. 7-1b. It acts like a switch that is open to direct current but shorted to alternating current. Because of this, a coupling capacitor blocks direct current but passes alternating current. This action allows us to get an ac signal from one stage to another without disturbing the dc biasing of each stage.

AC GROUND

A *bypass capacitor* is similar to a coupling capacitor, except that it couples an ungrounded point to a grounded point, as shown in Fig. 7-2a. Again, V_{TH} and R_{TH} may be a single source and resistor, as shown, or may be a Thevenin circuit. To the capacitor it makes no difference. It sees a total resistance of R_{TH}. Equations (7-1) and (7-2) still apply because we have a one-loop RC circuit. The only difference is $R = R_{TH}$.

In Fig. 7-2b, the capacitor ideally looks like a short to an ac signal. Because of this, point A is shorted to ground as far as the ac signal is concerned. This is why we have labeled point A as ac ground. A bypass capacitor will not disturb the dc voltage at point A because it looks open to dc current. However, a bypass capacitor makes point A an ac ground point.

Unless otherwise indicated, all coupling and bypass capacitors are considered stiff, which means that they act approximately like open circuits to direct current and short circuits to alternating current. Remember this because we use the concept of *dc open* and *ac short* over and over in subsequent chapters.

7-2 THE SUPERPOSITION THEOREM FOR AMPLIFIERS

In a transistor amplifier, the dc source sets up quiescent currents and voltages. The ac source then produces fluctuations in these currents and voltages. The simplest way to analyze the circuit is to split the analysis into two parts: a dc analysis and an ac analysis. In other words, we can use the *superposition theorem* when analyzing transistor amplifiers.

AC AND DC EQUIVALENT CIRCUITS

Here are the steps in applying superposition to transistor circuits:

1. Reduce the ac source to zero; this means shorting a voltage source or opening a current source. Open all capacitors. The circuit that remains is called the *dc equivalent circuit*. With this circuit, we can calculate any dc currents and voltages that we are interested in.
2. Reduce the dc source to zero; this is equivalent to shorting a voltage source or opening a current source. Short all coupling and bypass capacitors. The circuit that remains is called the *ac equivalent circuit*. This is the circuit to use in calculating ac currents and voltages.
3. The total current in any branch of the circuit is the sum of the dc current and the ac current through that branch. The total voltage across any branch is the sum of the dc voltage and the ac voltage across that branch.

Here is how to apply the superposition theorem to the transistor amplifier of Fig. 7-3*a*. First, short the ac voltage source and open all capacitors. All that remains is the circuit of Fig. 7-3*b*; this is the dc equivalent circuit. This is all that matters as far as

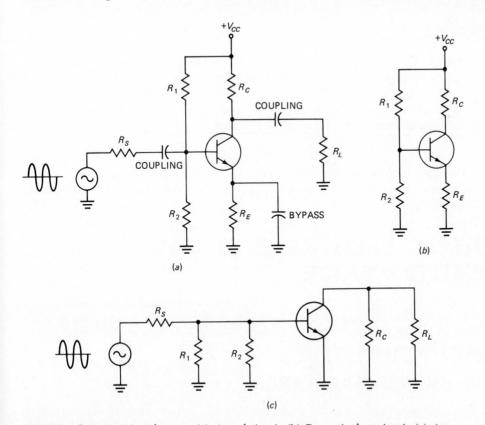

Fig. 7-3 *Superposition theorem.* (a) *Actual circuit.* (b) *Dc equivalent circuit.* (c) *Ac equivalent circuit.*

dc currents and voltages are concerned. With this circuit, we can calculate the quiescent current and voltage.

Second, short the dc voltage source; short all coupling and bypass capacitors. What remains is the ac equivalent circuit shown in Fig. 7-3c. Notice how the emitter is at ac ground because of the bypass capacitor across R_E. Also, when the dc supply is shorted, it grounds one side of R_1 and R_C. Stated another way, the dc supply point is an ac ground because it has an internal impedance that approaches zero. With the ac equivalent circuit of Fig. 7-3c, we can calculate any ac currents and voltages that we are interested in.

NOTATION

To keep dc distinct from ac, it is standard practice to use capital letters and subscripts for dc quantities. For instance, we use

I_E, I_C, I_B for the dc currents

V_E, V_C, V_B for the dc voltages to ground

V_{BE}, V_{CE}, V_{CB} for the dc voltages between terminals

Likewise, we will use lowercase letters and subscripts for ac currents and voltages:

i_e, i_c, i_b for the ac currents

v_e, v_c, v_b for the ac voltages to ground

v_{be}, v_{ce}, v_{cb} for the ac voltages between terminals

It is also standard practice to use a minus sign to indicate two sinusoidal voltages that are 180° out of phase. For instance, the equation

$$v_{\text{out}} = -v_{\text{in}}$$

means that the output voltage is 180° out of phase with the input voltage.

Become familiar with these notations because we will use them in the remainder of this book.

7-3 AC RESISTANCE OF THE EMITTER DIODE

In setting up the Q point, we have been visualizing the transistor of Fig. 7-4a replaced by the equivalent circuit of Fig. 7-4b (the Ebers-Moll model). Up to now, we have approximated V_{BE} as 0.7 V. In this section, we discuss what the Ebers-Moll model implies for an ac signal.

AC EMITTER RESISTANCE

Figure 7-4d shows the diode curve relating I_E and V_{BE}. In the absence of an ac signal, the transistor operates at the Q point, usually set near the middle of the dc load line. When an ac signal drives a transistor, however, the emitter current and voltage

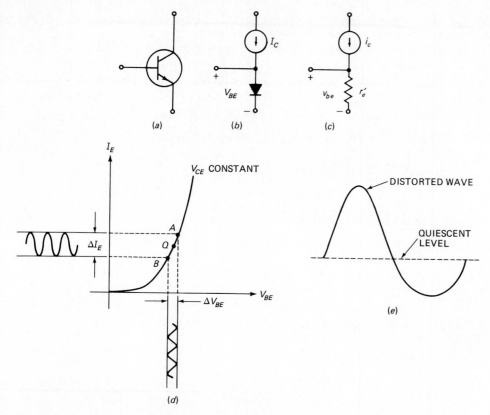

Fig. 7-4 (a) *Transistor.* (b) *Ebers-Moll dc model.* (c) *Ebers-Moll ac model.* (d) *Changes in base-emitter voltage produce changes in emitter current.* (e) *Large-signal distortion.*

change. If the signal is small, the operating point swings sinusoidally from Q to a positive current peak at A, then to a negative current peak at B, and back to Q, where the cycle repeats. This implies sinusoidal variations in I_E and V_{BE} as shown.

If the signal is small, peaks A and B are close to Q, and the operation is approximately linear. In other words, the arc from A to B is almost a straight line. Because of this, the changes in voltage and current are approximately proportional. Put another way, as far as the ac signal is concerned, the diode appears to be a resistance given by

$$r'_e = \frac{\Delta V_{BE}}{\Delta I_E} \qquad (7\text{-}3)$$

where r'_e = ac emitter resistance
ΔV_{BE} = small change in base-emitter voltage
ΔI_E = corresponding change in emitter current

Because the changes in V_{BE} and I_E are equivalent to ac voltage and current, Eq. (7-3) is often written as

$$r'_e = \frac{v_{be}}{i_e} \qquad (7\text{-}4)$$

where r'_e = ac emitter resistance

v_{be} = ac voltage across base-emitter terminals

i_e = ac current through emitter

For instance, if $v_{be} = 10$ mV and $i_e = 0.4$ mA, then

$$r'_e = \frac{10 \text{ mV}}{0.4 \text{ mA}} = 25 \text{ } \Omega$$

Figure 7-4*c* shows the Ebers-Moll *ac model.* This is what we will use when we analyze the ac equivalent circuit of an amplifier. In this model, the base-emitter diode is replaced by the ac emitter resistance.

HOW SMALL IS SMALL

Suppose the base-emitter variation is sinusoidal in Fig. 7-4*d.* If the signal is small, the variations in emitter current will also be sinusoidal. But when the input signal is large, the emitter current is no longer sinusoidal because of the nonlinearity of the diode curve. When the signal is too large, the emitter current is elongated on the positive half cycle and compressed on the negative half cycle, as shown in Fig. 7-4*e.* A distorted wave like this does not sound like the input signal when it drives a loudspeaker.

A later chapter gives a mathematical discussion of nonlinear distortion. Until then, we need some guidelines for how small a signal has to be in a small-signal amplifier. From now on, we will consider a signal as small if the peak-to-peak swing in emitter current is less than 10 percent of the quiescent emitter current. For instance, if $I_E = 10$ mA, then a peak-to-peak swing of less than 1 mA means that we have small-signal operation. This 10:1 rule keeps the distortion down to low levels for most applications.

THE FORMULA FOR r'_e

Since r'_e is the ratio of the change in V_{BE} to the change in I_E, its value depends on the location of the Q point. The higher up the curve Q is, the smaller r'_e becomes because the same change in base-emitter voltage produces a larger change in emitter current. The slope of the diode curve at point Q determines the value of r'_e. Appendix 1 shows how to use calculus to find this slope and proves that

$$r'_e = \frac{25 \text{ mV}}{I_E} \tag{7-5}$$

For instance, if the Q point has an I_E of 1 mA, then

$$r'_e = \frac{25 \text{ mV}}{1 \text{ mA}} = 25 \text{ } \Omega$$

Or, for a higher Q point with $I_E = 5$ mA, r'_e decreases to

$$r'_e = \frac{25 \text{ mV}}{5 \text{ mA}} = 5 \text{ } \Omega$$

We already know how to calculate I_E using the biasing formulas of Chap. 6. Therefore, once we have I_E, we can find the corresponding value of r'_e. Try to remember that r'_e is an ac quantity whose value depends on a dc quantity (I_E). This means that the Q point determines the value of r'_e.

Equation (7-5) is valid at room temperature, around 25°C. Resistance r'_e increases approximately 1 percent for each 3°C rise. Furthermore, Eq. (7-5) assumes a rectangular *pn* junction. Because the shape of a diode curve changes with a nonrectangular junction, the value of r'_e may differ somewhat from Eq. (7-5). Nevertheless, for preliminary troubleshooting and design of small-signal amplifiers, $r'_e = 25$ mV/I_E is an excellent starting point for any transistor. For more accurate answers, we will have to use *h* parameters (Chap. 9).

7-4 AC BETA

Figure 7-5 shows a typical graph of I_C versus I_B. β_{dc} is the ratio of dc collector current I_C to dc base current I_B. Since the graph is nonlinear, β_{dc} depends on where the Q point is located. This is why data sheets specify β_{dc} for a particular value of I_C.

The *ac beta* (designated β_{ac} or simply β) is a small-signal quantity that depends on the location of point Q. In Fig. 7-5, β is defined as

$$\beta = \frac{\Delta I_C}{\Delta I_B} \tag{7-6}$$

or, since alternating currents are the same as the changes in total currents,

$$\beta = \frac{i_c}{i_b} \tag{7-7}$$

Graphically, β is the slope of the curve at point Q. For this reason, it has different values at different Q locations.

On data sheets β is listed as h_{fe}. Especially note that the subscripts on h_{fe} are lowercase letters, whereas the subscripts on h_{FE} are capital letters. Therefore, when reading data sheets, do not confuse the direct and alternating current gains. The quantity h_{FE} is the ratio I_C/I_B, also known as β_{dc}. Quantity h_{fe} is the ratio i_c/i_b, the same as β.

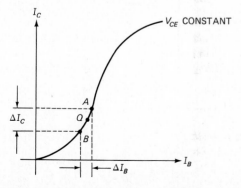

Fig. 7-5 *Curve of dc collector current versus dc base current is nonlinear.*

7-5 THE GROUNDED-EMITTER AMPLIFIER

Figure 7-6*a* shows a CE amplifier. Since the emitter is bypassed to ground, this amplifier is sometimes called a *grounded-emitter* amplifier; this means that the emitter is at ac ground, but not dc ground. A small sine wave is coupled into the base. This produces variations in the base current. Because of β, the collector current is an amplified sine wave of the same frequency. This sinusoidal collector current flows through the collector resistance and produces an amplified output voltage.

PHASE INVERSION

Because of the ac fluctuations in collector current, the output voltage of Fig. 7-6*a* swings sinusoidally above and below the quiescent voltage. Notice that the ac output voltage is inverted with respect to the ac input voltage, meaning that it is 180° out of phase with the input. During the positive half cycle of input voltage, the base current increases, causing the collector current to increase. This produces a larger voltage drop across the collector resistor; therefore, the collector voltage decreases, and we get the first negative half cycle of output voltage. Conversely, on the negative half cycle of input voltage, less collector current flows and the voltage drop across the collector resistor decreases. For this reason, the collector-to-ground voltage rises and we get the positive half cycle of output voltage.

LOAD-LINE VIEWPOINT

Figure 7-6*b* shows the ac load line and the *Q* point. The ac input voltage produces ac variations in base current. This results in sinusoidal variations about the *Q* point, as shown. For small-signal operation the peak-to-peak swing in collector current should be less than 10 percent of the quiescent collector current. (This keeps the distortion down to acceptable levels for most applications.)

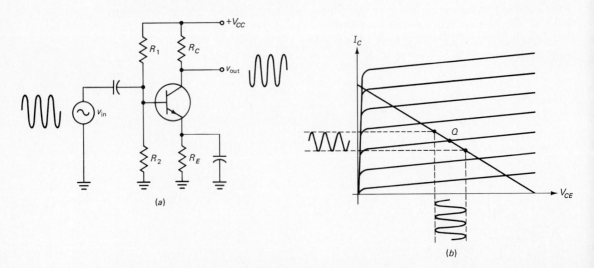

Fig. 7-6 (a) *A grounded-emitter amplifier.* (b) *Ac load line.*

For large signals the operating point swings further along the load line. If the signal is too large, the transistor goes into saturation and cutoff. This will clip the positive and negative peaks off the signal. In some applications we may want clipping, but with linear amplifiers the transistor should operate in the active region at all times. This means never going into saturation or cutoff during the cycle. Furthermore, with a simple amplifier like Fig. 7-6a, the peak-to-peak collector current must be less than 10 percent of the quiescent current to avoid the nonlinear distortion described earlier.

VOLTAGE GAIN

The voltage gain of an amplifier is the ratio of ac output voltage to ac input voltage. In symbols,

$$A = \frac{v_{\text{out}}}{v_{\text{in}}} \tag{7-8}$$

If we measure an ac output voltage of 250 mV and an input voltage of 2.5 mV, then the voltage gain is

$$A = \frac{250 \text{ mV}}{2.5 \text{ mV}} = 100$$

When you are troubleshooting, it helps to have some idea of what the voltage gain should be. Here is how to derive a simple formula for the voltage gain of Fig. 7-6a. Replace the circuit by its ac equivalent circuit. This means shorting the supply voltage to ground and shorting all capacitors because they look like ac shorts in a well-designed amplifier.

Figure 7-7a shows the ac equivalent circuit. Collector resistance R_C is grounded because the supply voltage point appears like an ac short. Likewise, resistor R_1 is now grounded, so it appears in parallel with R_2 and the emitter diode. Because of the parallel circuit on the input side, v_{in} appears directly across the emitter diode. Therefore, we can visualize the ac equivalent circuit as shown in Fig. 7-7b. Compare this simplified circuit with the original amplifier of Fig. 7-6a. In either circuit, the voltage across r'_e is equal to v_{in}. Let this idea sink in before you read on.

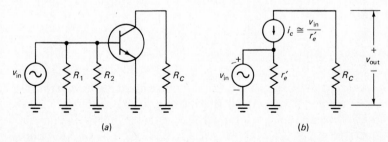

(a)   (b)

Fig. 7-7 (a) *Ac equivalent circuit of grounded-emitter amplifier.* (b) *Ebers-Moll ac model used for transistor.*

The input voltage of Fig. 7-7*b* is shown with a plus-minus polarity to indicate the positive half cycle of input voltage. Ohm's law tells us that the ac emitter current is

$$i_e = \frac{v_{in}}{r'_e}$$

Because collector current approximately equals emitter current,

$$i_c \cong i_e$$

This ac collector current flows through the collector resistor, producing an output voltage of

$$v_{out} \cong - i_e R_C$$

The minus sign is used here to indicate the phase inversion. In other words, on the positive half cycle of input voltage, collector current increases, producing the negative half cycle of output voltage. Since $i_e = v_{in}/r'_e$, the foregoing equation becomes

$$v_{out} \cong \frac{-v_{in}R_C}{r'_e}$$

Now rearrange this to get the voltage gain:

$$A = \frac{v_{out}}{v_{in}} \cong - \frac{R_C}{r'_e} \tag{7-9}$$

What a neat and simple result! Here you see that the voltage gain is the ratio of the collector resistance to the ac emitter resistance. This means that you can rapidly calculate the approximate voltage gain of a grounded-emitter amplifier. Then you can measure the input and output voltages of a built-up amplifier to see if the measured voltage gain agrees with the theoretical voltage gain. As an example, if $R_C = 4.7$ kΩ and $r'_e = 25$ Ω, then the voltage gain is

$$A = - \frac{4700}{25} = -188$$

With an amplifier like this, a base voltage of 1 mV produces an output voltage of 188 mV. (Voltages may be rms, peak, peak-to-peak, etc., as long as the input and output are consistent.)

RATIO OF RESISTANCES

Equation (7-9) makes sense. With the same current (approximately) flowing through R_C and r'_e, the ratio of voltages has to equal the ratio of resistances. In other words, since $i_c \cong i_e$, essentially the same current flows through R_C and r'_e. Output voltage v_{out} appears across R_C, and input voltage v_{in} appears across r'_e. Because of this, the voltage ratio v_{out}/v_{in} must equal the resistance ratio R_C/r'_e.

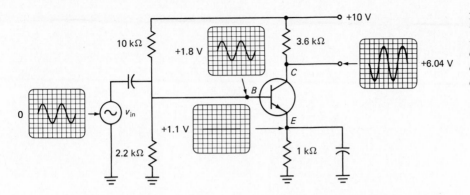

Fig. 7-8
*Dc and ac
components seen
on a dc-coupled
oscilloscope.*

EXAMPLE 7-1

A dc-coupled oscilloscope displays the total signal with its dc and ac components. Figure 7-8 shows the voltages at the base, emitter, and collector of a grounded-emitter amplifier. Explain what these voltages represent.

SOLUTION

To begin with, the ac input signal is a small sine wave of voltage. This passes through the input coupling capacitor and appears at the base. The quiescent level of base voltage is 1.8 V because this is the dc voltage out of the voltage divider:

$$V_B = \frac{2.2 \text{ k}\Omega}{10 \text{ k}\Omega + 2.2 \text{ k}\Omega} 10 \text{ V} = 1.8 \text{ V}$$

So, the total input voltage to the base is 1.8 V dc, plus the small ac source signal. Assuming a stiff coupling capacitor, all the ac source voltage appears at the base.

Notice the emitter voltage. It is 1.1 V dc because the emitter is bootstrapped to within 0.7 V of the input dc voltage. No ac signal appears here because the emitter is bypassed to ground through the emitter bypass capacitor. Stated another way, the stiff bypass capacitor shorts the ac signal to ground. This is equivalent to saying that there is no alternating current through R_E because all the alternating current is shunted through the bypass capacitor. This is why we see no ac signal at the emitter. The dc emitter current equals

$$I_E = \frac{1.1 \text{ V}}{1 \text{ k}\Omega} = 1.1 \text{ mA}$$

An amplified and inverted ac voltage appears at the collector. It has a quiescent level of 6.04 V, calculated as follows:

$$V_C = 10 \text{ V} - (1.1 \text{ mA})(3.6 \text{ k}\Omega) = 6.04 \text{ V}$$

EXAMPLE 7-2

The ac input signal has a peak value of 1 mV in Fig. 7-8. What is the peak value of the ac output voltage?

SOLUTION

In the preceding example, we calculated a dc emitter current of 1.1 mA. Therefore, the ac emitter resistance is

$$r_e' = \frac{25 \text{ mV}}{1.1 \text{ mA}} = 22.7 \text{ }\Omega$$

The voltage gain is

$$A = -\frac{3.6 \text{ k}\Omega}{22.7 \text{ }\Omega} = -159$$

This means that the ac output voltage is

$$v_{\text{out}} = A v_{\text{in}} = -159(1 \text{ mV}) = -159 \text{ mV}$$

The minus sign reminds us of the phase inversion.

Here is the conclusion. An ac input signal with a peak of 1 mV results in an ac output signal with a peak of 159 mV. The output signal is 180° out of phase with the input signal.

7-6 THE AC MODEL OF A CE STAGE

The challenge begins when we cascade amplifier stages together. Then we can get overall gains in the thousands. But before we are ready to analyze multistage amplifiers, we need a simple ac equivalent circuit for a single stage. To get such a model, we must discuss *input* and *output impedance.*

INPUT IMPEDANCE

The ac source driving an amplifier has to supply alternating current to the amplifier. Usually, the less current the amplifier draws from the source, the better. The input impedance of an amplifier determines how much current the amplifier takes from the ac source.

In the normal frequency range of an amplifier, where coupling and bypass capacitors look like ac shorts and all other reactances are negligible, the ac input impedance is defined as

$$z_{\text{in}} = \frac{v_{\text{in}}}{i_{\text{in}}} \qquad (7\text{-}10)$$

where v_{in} and i_{in} are peak values, peak-to-peak values, or rms values.

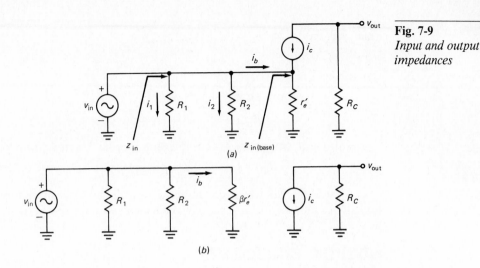

Fig. 7-9
*Input and output
impedances*

Looking into a grounded-emitter amplifier, the ac source sees the biasing resistors in parallel with the emitter diode, as shown in Fig. 7-9a. Conventional current i_1 flows into R_1, i_2 into R_2, and i_b into the base. (If you use electron flow, reverse all arrows.) The impedance looking directly into the base is symbolized $z_{in(base)}$, given by

$$z_{in(base)} = \frac{v_{in}}{i_b}$$

Ohm's law tells us that

$$v_{in} = i_e r'_e$$

Since $i_e \cong i_c = \beta i_b$, this becomes

$$v_{in} \cong \beta i_b r'_e$$

Then $z_{in(base)}$ neatly simplifies to

$$z_{in(base)} \cong \frac{\beta i_b r'_e}{i_b} = \beta r'_e$$

Remember, this is only the input impedance looking into the base of the transistor. It does not include the effect of external biasing resistors connected to the base. In other words, the input impedance looking into the base of a grounded-emitter amplifier equals the ac current gain times the ac emitter resistance.

Why isn't the input impedance of the base equal to r'_e? The ac source looks into the base and only has to supply the base current. Inside the transistor, the collector current adds to the base current to produce the emitter current through r'_e. Because base current is β times smaller than emitter current, the input impedance of the base is β times larger than r'_e.

To emphasize the point, some people use the equivalent circuit of Fig. 7-9b. This produces the same effect as Fig. 7-9a. Now the ac source sees the parallel circuit consisting of R_1, R_2, and $\beta r'_e$. The base current through $\beta r'_e$ is amplified to get a collector current of βi_b.

Fig. 7-10
*Ac model of a
grounded-emitter
amplifier.*

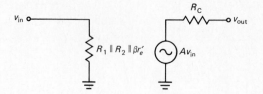

In summary, the grounded-emitter amplifier has an input impedance of

$$z_{in} \cong R_1 \parallel R_2 \parallel \beta r'_e \qquad (7\text{-}11)$$

This is the total input impedance because it includes the biasing resistors and the impedance looking into the base of the transistor.

OUTPUT IMPEDANCE

Now let's do something interesting to the output side of the amplifier. Let's thevenize it. In Fig. 7-9*b*, the Thevenin voltage appearing at the output is

$$v_{out} = Av_{in}$$

The Thevenin impedance is the parallel combination of R_C and the internal impedance of the collector current source. In the Ebers-Moll model the collector current source is ideal; therefore, it has an infinite internal impedance. This is an approximation, but it's satisfactory for preliminary work. Therefore, the Thevenin impedance is

$$z_{out} \cong R_C$$

SIMPLE AC MODEL

Figure 7-10 summarizes what we have said about input and output impedances of a grounded-emitter amplifier. Here you see an input impedance of $R_1 \parallel R_2 \parallel \beta r'_e$. This is what the ac input source also sees. On the output side, you see an ac voltage source Av_{in} in series with an output impedance R_C. Once you get used to it, this simple model of a grounded-emitter amplifier allows you to analyze cascaded stages quickly.

EXAMPLE 7-3

Figure 7-11*a* shows the grounded-emitter amplifier analyzed in Example 7-2. If the transistor has a β of 150, what is the ac output voltage?

SOLUTION

Two new things appear in Fig. 7-11*a*. First, the ac source now has a source impedance of 1 kΩ. Therefore, some of the source signal will be dropped across

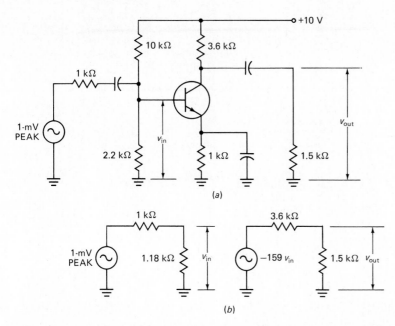

Fig. 7-11 (a) *Source and load resistors connected to a grounded-emitter amplifier.* (b) *Ac equivalent circuit.*

this resistance before reaching the base. On the output side, the capacitor couples the ac signal into a load resistance of 1.5 kΩ. This should produce some loading effect. As a result, we expect the output signal to be lower than before.

To see the effects of source and load resistance, replace the grounded-emitter amplifier by its ac model. First, calculate the input impedance of the base:

$$z_{in(base)} = \beta r'_e = 150(22.7 \ \Omega) = 3.4 \ k\Omega$$

Next, work out the input impedance of the amplifier:

$$z_{in} = R_1 \parallel R_2 \parallel \beta r'_e = 10 \ k\Omega \parallel 2.2 \ k\Omega \parallel 3.4 \ k\Omega$$
$$= 1.18 \ k\Omega$$

The unloaded voltage gain, found earlier, is −159. The output impedance is equal to R_C. Therefore, we can visualize the circuit as shown in Fig. 7-11b.

We are looking at two voltage dividers. The input voltage divider reduces the signal at the base:

$$v_{in} = \frac{1.18 \ k\Omega}{2.18 \ k\Omega} 1 \ mV = 0.541 \ mV$$

The Thevenin output voltage is

$$A v_{in} = -159(0.541 \ mV) = -86 \ mV$$

This is the unloaded output. The actual output is what appears across the 1.5 kΩ:

$$v_{out} = \frac{1.5\ k\Omega}{5.1\ k\Omega}(-86\ mV) \cong -25\ mV$$

This means that the output has a peak voltage of 25 mV.

7-7 THE SWAMPED AMPLIFIER

The quantity r'_e ideally equals 25 mV/I_E. As was pointed out earlier, the actual r'_e depends on the temperature and the type of junction. Because of this, the r'_e of a transistor may vary over a 2:1 range for different temperatures and transistors. Any change in the value of r'_e will change the voltage gain in a grounded-emitter amplifier. In some applications, a change in voltage gain is acceptable. For instance, in a transistor radio you can offset changes in the voltage gain by adjusting the volume control. But there are many applications in which you need as stable a voltage gain as possible.

WHAT SWAMPING MEANS

Many people insert a resistor r_E in series with the emitter, as shown in Fig. 7-12a. Now the emitter is no longer at ac ground. Because of this, the ac emitter current flows through r_E and produces an ac voltage at the emitter. If r_E is much greater than r'_e, almost all of the ac input signal appears at the emitter. Stated another way, when r_E is much greater than r'_e, the emitter is *bootstrapped* to the base for ac as well as dc.

Figure 7-12b shows the ac equivalent circuit. Because r_E is in series with r'_e, the total resistance is $r_E + r'_e$. The ac input appears across this resistance and produces an ac emitter current of

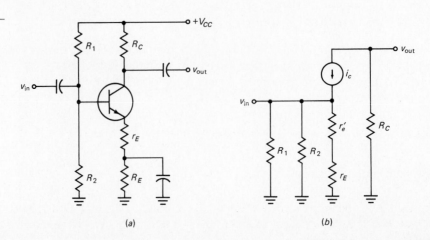

Fig. 7-12
(a) *Swamped amplifier.* (b) *Ac equivalent circuit.*

(a)

(b)

$$i_e = \frac{v_{in}}{r_E + r'_e}$$

Swamping the emitter diode means making r_E much greater than r'_e.

EFFECT OF SWAMPING ON VOLTAGE GAIN

In Fig. 7-12b, the ac output voltage is

$$v_{out} = -i_c R_C$$

The ac voltage across $r_E + r'_e$ is

$$v_{in} = i_e(r_E + r'_e)$$

Therefore, the ratio of v_{out} to v_{in} is

$$\frac{v_{out}}{v_{in}} = \frac{-i_c R_C}{i_e(rE + r'_e)}$$

Since $i_c \cong i_e$, the equation can be rewritten as

$$A \cong -\frac{R_C}{r_E + r'_e} \qquad (7\text{-}12)$$

This equation tells us two things: First, the swamping resistor reduces the voltage gain, and second, changes in r'_e have less effect on voltage gain. As an example, suppose r'_e increases from 25 to 50 Ω over a large temperature range. If $R_C = 10$ kΩ and $r_E = 510$ Ω, the maximum voltage gain is

$$A = -\frac{10,000}{510 + 25} = -18.7$$

and the minimum voltage gain is

$$A = -\frac{10,000}{510 + 50} = -17.9$$

The decrease in voltage gain is less than 5% even though r'_e increases by 100%.

What price do we pay for stable gain? We have to be satisfied with less voltage gain. If no swamping resistor is used, for example, the voltage gain varies from

$$A = -\frac{10,000}{25} = -400$$

to

$$A = -\frac{10,000}{50} = -200$$

The voltage gain is much larger, but look at the variation: a decrease from −400 to −200.

EFFECT OF SWAMPING ON DISTORTION

Another advantage of swamping the emitter diode is reduced distortion. It is the nonlinearity of the emitter diode that produces distortion in large-signal amplifiers.

Since $r_e' = 25$ mV/I_E, any significant change in I_E produces a noticeable change in r_e'. In a large-signal unswamped amplifier, this means that the voltage gain changes throughout the cycle because r_e' is changing. The changing voltage gain results in a distorted output signal. This is why an unswamped amplifier uses only a small part of the load line to avoid excessive distortion of the signal. We can use more of the load line and still have acceptable distortion levels if we swamp the emitter diode. In a heavily swamped amplifier, almost all the sinusoidal input signal appears across the swamping resistor. This means that the collector current will be sinusoidal, and it remains sinusoidal even though most of the load line is used.

The mathematical details of nonlinear distortion are covered in a later chapter. The important point to grasp now is this: Swamping the emitter diode greatly reduces the distortion produced by the emitter diode because the bulk of the ac signal appears across the swamping resistor. Since this resistor is a linear device, we can use almost the entire load line and still have only a small amount of distortion.

HOW MUCH SWAMPING?

If you swamp too lightly, you get a voltage gain that shows wide variation with temperature and transistor replacement. Furthermore, you are restricted to small-signal operation. Light swamping or none at all may be perfectly all right in some applications. But in other applications, the designer must swamp the emitter diode. If you swamp too heavily, you get almost no distortion, but the voltage gain may be too small to be useful. The amount of swamping to be used finally comes down to a design decision that takes everything into account. Remember, you can always cascade two swamped stages to get the same voltage gain as a single unswamped stage. In short, it depends on what you are trying to do in the particular application.

EMITTER FOLLOWS BASE

In a swamped amplifier, the emitter is bootstrapped to the base for alternating current as well as direct current. Because of this, the dc emitter voltage is within 0.7 V of the dc base voltage, and the ac emitter voltage is approximately equal to the ac base voltage. Remember this for troubleshooting. Also, remember that the bottom of r_E should be at ac ground. One common trouble is an open bypass capacitor; this decreases the voltage gain because R_E adds to r_E, producing an overswamped amplifier with very little gain.

AC MODEL OF SWAMPED AMPLIFIER

The ac input voltage of Fig. 7-12b appears across r_E in series with r_e'. Ohm's law tells us that

$$v_{\text{in}} = i_e(r_E + r_e')$$

Since $i_e \cong i_c = \beta i_b$, we can rearrange the equation as

$$\frac{v_{\text{in}}}{i_b} \cong \beta(r_E + r_e')$$

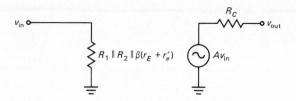

Fig. 7-13
Ac model of swamped amplifier.

or

$$z_{\text{in(base)}} \cong \beta(r_E + r_e') \qquad (7\text{-}13)$$

Here is another bonus from swamping. The input impedance of the base is much higher in a swamped amplifier. For instance, if $\beta = 150$, $r_E = 510\ \Omega$, and $r_e' = 25\ \Omega$, then

$$z_{\text{in(base)}} = 150(510\ \Omega + 25\ \Omega) = 80.3\ \text{k}\Omega$$

(If $r_E = 0$, no swamping, the input impedance of the base would be only 3.75 kΩ.) When the ac source has a source resistance, more of the signal will reach the input of a swamped amplifier than of an unswamped amplifier.

Figure 7-13 shows the ac model of a swamped amplifier. The input impedance is the parallel combination of R_1, R_2, and $\beta(r_E + r_e')$. The voltage gain equals $-R_C/(r_E + r_e')$. The output impedance is R_C. With this ac model, we can analyze swamped amplifiers inside a multistage amplifier.

EXAMPLE 7-4

Figure 7-14 shows a swamped amplifier. Explain the voltages.

SOLUTION

As far as direct current is concerned, the circuit is identical to the grounded-emitter amplifier discussed in Example 7-1. For this reason, the dc base voltage is still 1.8 V, the dc emitter voltage is 1.1 V, and the dc collector voltage is 6.04 V. Notice that the total dc emitter resistance is still 1 kΩ; therefore, r_e' is still 22.7 Ω.

Now for the differences. The ac source voltage is larger, 100 mV instead of 1 mV. This signal is coupled into the base. Because of the bootstrapping effect, almost all the 100 mV appears across the swamping resistor. The lower end of the swamping resistor is at ac ground, and this is why an oscilloscope would show a horizontal line (no ac signal) at +0.902 V. The ac collector voltage is amplified and inverted as before.

EXAMPLE 7-5

Calculate the ac output voltage in Fig. 7-14.

Fig. 7-14
*Dc and ac
components seen
on a dc-coupled
oscilloscope.*

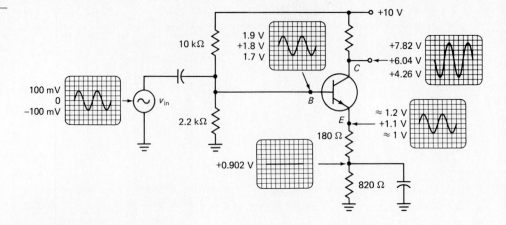

SOLUTION

The voltage gain is

$$A = -\frac{3600}{180 + 22.7} = -17.8$$

Therefore, the ac output voltage is

$$v_{\text{out}} = -17.8(100 \text{ mV}) = -1.78 \text{ V}$$

This means that the peak output voltage is 1.78 V. So, the peak-to-peak swing along the load line is twice as much, 3.56 V. The operation is large-signal, made possible by the reduced distortion that accompanies swamping (also called local feedback or negative feedback).

EXAMPLE 7-6

Suppose that the bypass capacitor of Fig. 7-14 is open. What happens to the voltage gain?

SOLUTION

An open bypass capacitor means that

$$r_E = 180 \text{ }\Omega + 820 \text{ }\Omega = 1000 \text{ }\Omega$$

because the bottom of the 180-Ω resistor is no longer at ac ground. In a case like this, the swamping becomes enormously heavy, and the voltage gain drops to

$$A = -\frac{3600}{1000 + 22.7} = -3.52$$

The ac output voltage is only

$$v_{\text{out}} = -3.52(100 \text{ mV}) = -352 \text{ mV}$$

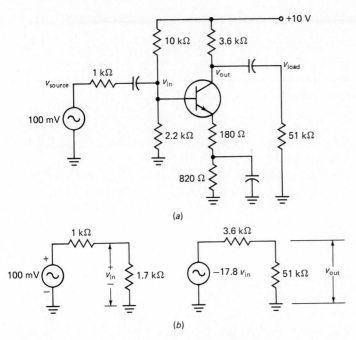

Fig. 7-15 (a) *Source and load resistors connected to swamped amplifier.* (b) *Ac equivalent circuit.*

When you are troubleshooting an amplifier whose output is much lower than it should be, the place to start is the bypass capacitor. A dc-coupled oscilloscope connected across the bypass capacitor should display a horizontal line at the correct dc level (+0.902 V in Fig. 7-14). If you see an ac signal across this capacitor with a peak-to-peak value almost as large as the ac emitter voltage, you will know that the bypass capacitor is open.

EXAMPLE 7-7

If $\beta = 150$, what is the ac output voltage in Fig. 7-15?

SOLUTION

From Eq. (7-13), the input impedance of the base is

$$z_{\text{in(base)}} = 150(180 + 22.7) = 30.4 \text{ k}\Omega$$

The input impedance of the amplifier is

$$z_{\text{in}} = 10 \text{ k}\Omega \parallel 2.2 \text{ k}\Omega \parallel 30.4 \text{ k}\Omega = 1.7 \text{ k}\Omega$$

The unloaded voltage gain from base to collector is -17.8 (found earlier). So, the ac model of the amplifier appears as shown in Fig. 7-15*b*.

The input voltage reaching the amplifier is

$$v_{in} = \frac{1700}{2700} (100 \text{ mV}) = 63 \text{ mV}$$

The Thevenin output voltage is

$$Av_{in} = -17.8(63 \text{ mV}) = -1.12 \text{ V}$$

This is the unloaded output voltage. The actual ac output voltage appearing at the collector and across the load resistor is

$$v_{out} = \frac{51,000}{54,600} (-1.12 \text{ V}) = -1.05 \text{ V}$$

This means that the peak output voltage is 1.05 V.

7-8 CASCADED STAGES

If you have understood everything up to this point, you will have no problem with cascaded CE stages. The idea here is to use the amplified output from one stage as the input to another stage. In this way, we can build a multistage amplifier with a very large overall voltage gain.

Figure 7-16a shows a two-stage amplifier using cascaded CE circuits. An ac source with a source resistance of R_S drives the input of the amplifier. The grounded-emitter stage amplifies the signal, which is then coupled into the next CE stage. Then the

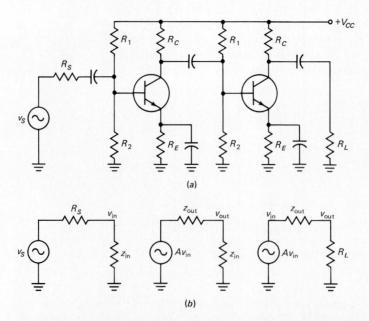

Fig. 7-16   (a) *Two-stage amplifier with grounded-emitter stages.* (b) *Ac equivalent circuit.*

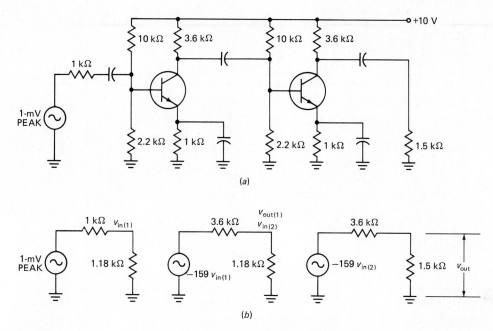

Fig. 7-17 (a) *Source and load resistors connected to two-stage amplifier.* (b) *Ac equivalent circuit.*

signal is amplified once more to get a final output that is considerably larger than the source signal.

Figure 7-16*b* shows the ac model for the two-stage amplifier. Each stage has an input impedance, given by the parallel combination of R_1, R_2, and $\beta r'_e$. Also, each stage has an unloaded voltage gain of $-R_C/r'_e$ and an output impedance of R_C. Therefore, the analysis is straightforward and boils down to calculating z_{in}, A, and z_{out} for each stage; then, by analyzing the source and loading effects, you can work out the final output voltage. The same approach can be applied to any number of stages, some of which may be swamped amplifiers.

EXAMPLE 7-8

The transistors of Fig. 7-17*a* each have $\beta = 150$. What is the ac output voltage?

SOLUTION

Each stage is identical. In Example 7-3, we analyzed this CE stage, with the following results: $r'_e = 22.7 \; \Omega$, $z_{in} = 1.18 \; k\Omega$, $A = -159$, and $z_{out} = 3.6 \; k\Omega$. Figure 7-17*b* shows the ac model of the two-stage amplifier. The ac input voltage to the first stage is

$$v_{in} = \frac{1.18}{1 + 1.18} \; 1 \; mV = 0.541 \; mV$$

The Thevenin ac voltage out of the first stage is

$$Av_{in} = -159(0.541 \text{ mV}) = -86 \text{ mV}$$

The ac input to the second stage is

$$v_{in} = \frac{1.18}{3.6 + 1.18}(-86 \text{ mV}) = -21.2 \text{ mV}$$

The Thevenin voltage out of the second stage is

$$Av_{in} = -159(-21.2 \text{ mV}) = 3.37 \text{ V}$$

And the final ac output voltage is

$$v_{out} = \frac{1.5}{3.6 + 1.5}3.37 \text{ V} = 0.991 \text{ V}$$

Incidentally, since there are two stages of inverted gain, the final output signal is in phase with the input signal. Therefore, if we couple in a sine wave with a peak of 1 mV, we get out a sine wave of the same frequency and phase but with a peak of 991 mV. The two-stage amplifier has a voltage gain of 991 from the source to the final output.

PROBLEMS
Straightforward

7-1. The ac source of Fig. 7-18a can have a frequency between 100 Hz and 200 kHz. To have stiff coupling over this range, what size should the coupling capacitor be?

Fig. 7-18

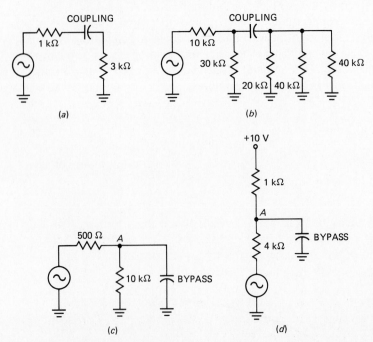

7-2. In Fig. 7-18*b*, we want a stiff coupling capacitor for all frequencies between 500 Hz and 1 MHz. What size should it be?

7-3. To bypass point *A* to ground in Fig. 7-18*c* for all frequencies from 20 Hz up, what size should the bypass capacitor have?

7-4. We want point *A* in Fig. 7-18*d* to look like ac ground from 10 Hz to 200 kHz. What size should the bypass capacitor be?

7-5. Draw the dc equivalent circuit for the amplifier of Fig. 7-19*a*; label the three currents with standard dc notation. Next, draw the ac equivalent circuit.

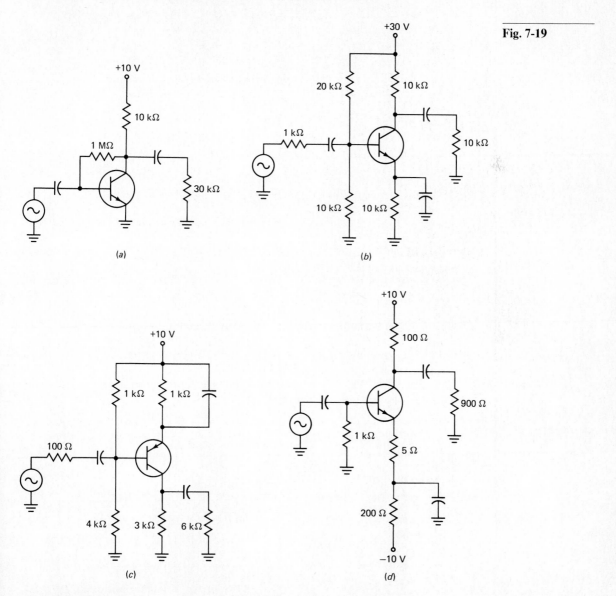

Fig. 7-19

(a)

(b)

(c)

(d)

7-6. Draw the dc and ac equivalent circuits for Fig. 7-19*b*.

7-7. Draw the dc and ac equivalent circuits for the upside-down *pnp* amplifier of Fig. 7-19*c*.

7-8. Draw the dc and ac equivalent circuits for Fig. 7-19*d*.

7-9. Using Eq. (7-5), calculate the value of r'_e for each of these dc emitter currents: 0.01 mA, 0.05 mA, 0.1 mA, 0.5 mA, 1 mA, and 10 mA.

7-10. What is the value of r'_e in the amplifier of Fig. 7-19*b*?

7-11. In the *pnp* amplifier of Fig. 7-19*c*, what value does r'_e have?

7-12. If the transistor of Fig. 7-19*a* has a β_{dc} of 100, what value does r'_e have?

7-13. If $v_{in}=1$ mV in Fig. 7-20*a*, what does v_{out} equal?

Fig. 7-20

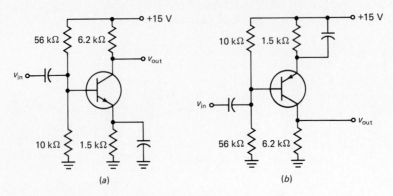

(a) (b)

7-14. In Fig. 7-20*b*, $v_{in}=2$ mV. What does v_{out} equal?

7-15. The resistors of Fig. 7-20*a* have a tolerance of ±5 percent. What is the minimum voltage gain? The maximum voltage gain?

7-16. In Fig. 7-19*b*, $\beta=125$ and the ac source voltage has an rms value of 5 mV. What does the ac output voltage equal?

7-17. The ac source voltage has a peak value of 2.5 mV in Fig. 7-19*c*. If $\beta=200$, what does the ac output voltage equal?

7-18. Suppose we insert a 100-Ω swamping resistor in Fig. 7-19*b*. What is the input impedance of the stage if $\beta=175$? The ac output voltage for an ac source voltage of 1 mV rms?

7-19. In Fig. 7-19*d*, what does the ac output voltage equal if the ac input voltage is 10 mV peak?

7-20. A variable swamping resistor is used in Fig. 7-20*a*. If r_E can be adjusted from 0 to 100 Ω, what are the minimum and maximum voltage gains?

7-21. Each transistor of Fig. 7-21 has a β of 100. If the ac source voltage equals 10 μV peak, what does the ac output voltage equal?

Fig. 7-21

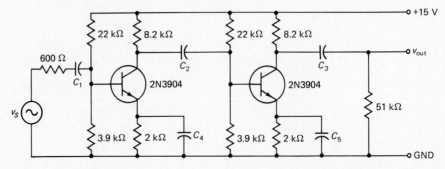

7-22. In Fig. 7-21, the first transistor has a β of 125 and the second transistor has a β of 90. What does the ac output voltage equal if the ac source voltage is 15 μV rms?

7-23. A 200-Ω swamping resistor is added to each stage of Fig. 7-21. If $\beta=170$ for each transistor, what does the ac output voltage equal for an ac source voltage of 1 mV peak?

Troubleshooting

7-24. Does the ac output voltage of Fig. 7-21 increase, decrease, or remain the same for each of these troubles:

    a. Capacitor C_3 opens

    b. Capacitor C_3 shorts

    c. Capacitor C_4 opens

    d. 12-kΩ resistor mistakenly used instead of 8.2-kΩ resistor in first stage

7-25. Capacitor C_5 in Fig. 7-21 opens. If the transistors each have a β of 80, what does the ac output voltage equal when the ac input voltage is 1 mV peak?

7-26. In Fig. 7-21, you measure an ac output voltage of 200 mV for an ac source voltage of 1 mV. Next, you measure 57 mV for the ac input voltage to the second stage. What is the most likely trouble? How can you verify this with an oscilloscope?

7-27. Describe the symptoms in Fig. 7-21 for each of these troubles:

    a. Capacitor C_1 open

    b. Capacitor C_2 shorted

    c. Capacitor C_3 open

    d. Capacitor C_4 open

    e. Capacitor C_5 shorted

    f. 8.2-kΩ resistor of first stage open

    g. 8.2-kΩ resistor of second stage shorted

Design

7-28. The transistors of Fig. 7-21 each have a β of 125. Select values for C_1, C_2, and C_3 to get stiff coupling for all frequencies greater than 20 Hz.

7-29. Design a swamped amplifier stage like Fig. 7-12a to meet these specifications: $V_{CC} = 15$ V, $I_C = 10$ mA, $h_{FE} = 300$, $h_{fe} = 250$, $z_{in} \geqslant 500$ Ω, and $A = 27.5$.

7-30. Design a two-stage amplifier with swamping resistors to meet these specifications: $V_{CC} = 10$ V, $R_S = 1$ kΩ, $R_L = 10$ kΩ, $\beta = 125$, 1 V rms output for 1 mV rms input.

Challenging

7-31. Prove that the Thevenin voltage gain of a grounded-emitter stage is approximately $16V_{CC}$. Ignore the dimension of volts. In other words, a grounded-emitter stage has an unloaded voltage gain of 160 for a 10-V supply, 320 for a 20-V supply, 480 for a 30-V supply, and so on.

7-32. Each transistor in Fig. 7-22 has an h_{FE} of 100 and an h_{fe} of 100. Calculate each of the following:

    a. The input impedance looking into the base of Q_2.

    b. The input impedance looking into the base of Q_1.

    c. The input impedance of the amplifier.

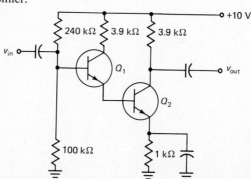

Fig. 7-22

Computer

7-33. The REM statement (REM stands for reminder) is used by a programmer as a memory aid. The computer ignores all words on any line following a REM. For instance, here is a program:

```
10 REM CALCULATION OF THEVENIN VOLTAGE GAIN
20 PRINT "ENTER IE": INPUT IE
30 PRINT "ENTER RC": INPUT RC
40 RPE = 0.025/IE
50 A = RC/RPE
60 PRINT "THE VOLTAGE GAIN IS": PRINT A
```

In this program, the computer ignores the words "CALCULATION OF THEVENIN VOLTAGE GAIN." The only reason we include the REM statement is to tell us what the program does. In complicated programs, well-placed REM statements can jog your memory and help you understand programs written days or months earlier.

What does the foregoing program do?

7-34. Here is a program:

```
10 REM CALCULATE INPUT IMPEDANCE
20 PRINT "ENTER R1": INPUT R1
30 PRINT "ENTER R2": INPUT R2
40 PRINT "ENTER AC BETA": INPUT BETA
50 PRINT "ENTER R PRIME E": INPUT RPE
60 REM CALCULATE ZIN(BASE)
70 ZB = BETA * RPE
80 REM CALCULATE ADMITTANCE
90 YIN = 1/R1 + 1/R2 + 1/ZB
100 ZIN = 1/YIN
110 PRINT "THE INPUT IMPEDANCE IS": PRINT ZIN
```

What does the program do in line 70? In line 90? In line 100?

7-35. Given R_1, R_2, β, r'_e, and r_E, write a program that prints out the value of input impedance for a swamped amplifier stage.

7-36. Write a program that calculates the ac output voltage in the amplifier of Fig. 7-21. Assume that both stages are identical. Use INPUT statements to enter V_{CC}, R_1, R_2, R_E, R_C, v_S, R_S, R_L, and β. The program should calculate the dc emitter current, r'_e, etc., with a final PRINT statement to display v_{out}.

CC and CB Amplifiers

If you connect a high-impedance source to a low-impedance load, most of the ac signal is dropped across the internal impedance of the source. One way to get around this problem is to use an emitter follower between the high-impedance source and the low-impedance load. The emitter follower steps up the impedance level and reduces the signal loss. In addition to emitter followers, this chapter also discusses Darlington amplifiers, types of interstage coupling, improved voltage regulators, and CB amplifiers.

8-1 THE CC AMPLIFIER

Figure 8-1a shows a circuit called a *common-collector* (CC) amplifier. Since R_C is zero, the collector is at ac ground. This is why the circuit is also known as a *grounded-collector* amplifier. When a dc voltage V_{in} drives the base, a dc voltage V_{out} appears across the emitter resistor.

BASIC IDEA

A CC amplifier is like a heavily swamped CE amplifier with the collector resistor shorted and the output taken from the emitter rather than the collector. Since the emitter is bootstrapped to the base, the dc output voltage is

$$V_{out} = V_{in} - V_{BE} \qquad (8\text{-}1)$$

The circuit is also called an *emitter follower* because the dc emitter voltage follows the dc base voltage. If V_{in} is 2 V, as shown in Fig. 8-1b, V_{out} is 1.3 V. If V_{in} increases to 3 V, V_{out} increases to 2.3 V (Fig. 8-1c). This means that changes in V_{out} are in phase with changes in V_{in}.

PHASE RELATIONS

In Fig. 8-1a, V_{CE} is given by

$$V_{CE} = V_{CC} - V_{out}$$

201

Fig. 8-1
CC amplifier.

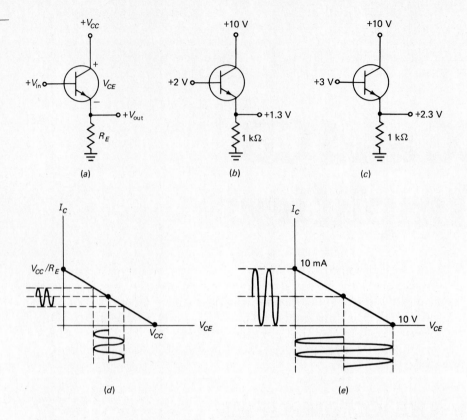

(a)

(b)

(c)

(d)

(e)

If V_{out} increases, V_{CE} decreases. Therefore, V_{CE} is out of phase with both V_{out} and V_{in}. You can see this out-of-phase relation by referring to Fig. 8-1*b* and *c*. When V_{in} increases 1 V, V_{out} increases 1 V but V_{CE} decreases 1 V.

DC LOAD LINE

Summing dc voltages around the collector loop of Fig. 8-1*a* gives

$$V_{CE} + I_E R_E - V_{CC} = 0$$

Since the collector current approximately equals the emitter current, we can solve for I_C to get

$$I_C = \frac{V_{CC} - V_{CE}}{R_E} \tag{8-2}$$

This is the equation of the dc load line, shown in Fig. 8-1*d*.

When the input voltage contains an ac component as well as a dc component, the ac load line is the same as the dc load line because I_C and V_{CE} have the sinusoidal fluctuations shown in Fig. 8-1*d*. If the input signal is large enough to use the entire ac load line, the transistor goes into saturation and cutoff at the peaks. This limits the output voltage swing to a peak-to-peak value of V_{CC} as shown in Fig. 8-1*e*.

VOLTAGE GAIN

Figure 8-2a shows an emitter follower driven by a small ac voltage. Figure 8-2b shows the ac equivalent circuit. The ac output voltage equals

$$v_{\text{out}} = i_e R_E$$

Since the ac input voltage is

$$v_{\text{in}} = i_e(R_E + r'_e)$$

the ratio of v_{out} to v_{in} is

$$A = \frac{v_{\text{out}}}{v_{\text{in}}} = \frac{R_E}{R_E + r'_e} \tag{8-3}$$

In most emitter followers, R_E swamps out r'_e and the voltage gain approaches unity:

$$A \cong 1$$

For example, if $R_E = 5.1 \text{ k}\Omega$ and $r'_e = 25 \text{ }\Omega$, Eq. (8-3) gives a voltage gain of

$$A = \frac{5100 \text{ }\Omega}{5125 \text{ }\Omega} = 0.995$$

Notice that the output voltage is within half a percent of the input voltage.

SMALL DISTORTION

The emitter follower is inherently a low-distortion amplifier. Since the emitter resistor is unbypassed, the swamping is extremely heavy and the nonlinearity of the emitter diode is almost eliminated.

Since the voltage gain is approximately 1, the output voltage is a replica of the input voltage. Given a perfect input sine wave, we get an almost perfect output sine wave.

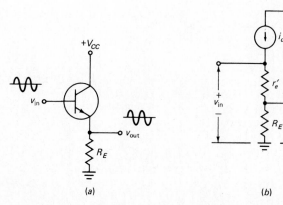

Fig. 8-2
(a) *Emitter follower.* (b) *Ac equivalent circuit.*

(a)

(b)

Fig. 8-3
*Emitter follower
with dc and ac
components.*

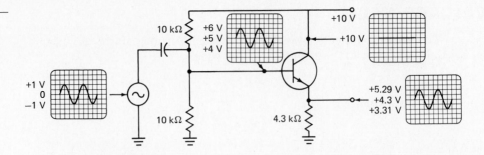

EXAMPLE 8-1

Figure 8-3 shows an emitter follower using voltage-divider bias. Calculate the dc voltages and explain the waveforms.

SOLUTION

The Thevenin dc voltage out of the voltage divider is

$$V_{TH} = \frac{10 \text{ k}\Omega}{20 \text{ k}\Omega} 10 \text{ V} = 5 \text{ V}$$

This sets up a quiescent emitter voltage of

$$V_E = 5 \text{ V} - 0.7 \text{ V} = 4.3 \text{ V}$$

The collector is tied directly to the supply voltage; therefore,

$$V_C = 10 \text{ V}$$

The source signal is a sine wave with a dc level of 0 V. This is coupled into the base of the emitter follower. The quiescent or dc level at the base is +5 V. So, the input voltage to the base consists of +5 V dc plus an ac signal from the source. Because the emitter is bootstrapped to the base, it has a dc voltage of +4.3 V with an ac output signal approximately equal to the ac input signal.

Notice that the collector voltage has no ac signal because the supply voltage is an ac ground point. This is because the positive supply consists of a dc source and a small supply impedance. Ideally, the supply impedance is zero, which means that the voltage at the supply point is constant, equivalent to having no ac component.

EXAMPLE 8-2

Calculate the voltage gain in Fig. 8-3. If the ac input voltage has a peak of 1 V, what is the output voltage?

SOLUTION

The dc emitter current is

$$I_E = \frac{4.3 \text{ V}}{4.3 \text{ k}\Omega} = 1 \text{ mA}$$

and the ac resistance of the emitter diode is

$$r'_e = \frac{25 \text{ mV}}{1 \text{ mA}} = 25 \text{ }\Omega$$

The voltage gain is

$$A = \frac{4300}{4300 + 25} = 0.994$$

The output voltage has a peak of

$$v_{\text{out}} = 0.994(1 \text{ V}) = 0.994 \text{ V}$$

This is so close to the input voltage that you cannot distinguish v_{in} and v_{out} with an oscilloscope. In almost any emitter follower, you will find the ac output voltage to be almost equal to the ac input voltage.

8-2 THE AC MODEL OF AN EMITTER FOLLOWER

We need a practical ac equivalent circuit for an emitter follower, something that allows us to calculate the effects of source and load resistances. In this section, we derive a simple ac model for an emitter follower.

INPUT IMPEDANCE

An ac source with a resistance of R_S drives the emitter follower of Fig. 8-4a. Because of the biasing resistors and the input impedance of the base, some of the ac signal is lost across the source resistance. The question is how much.

The first thing to do is to get the input impedance of the base. Since the emitter follower is a heavily swamped amplifier, it has an input impedance of

$$z_{\text{in(base)}} = \beta(R_E + r'_e)$$

In many emitter followers, R_E swamps out r'_e, and the foregoing simplifies to

$$z_{\text{in(base)}} = \beta R_E \qquad (8\text{-}4)$$

Notice how the emitter follower steps up the impedance by a factor of β. If $\beta = 200$ and $R_E = 4.3 \text{ k}\Omega$, then

$$z_{\text{in(base)}} = 200(4.3 \text{ k}\Omega) = 860 \text{ k}\Omega$$

As you can see, the input impedance of the base is very high.

The total input impedance of an emitter follower includes the biasing resistors in parallel with the input impedance of the base:

$$z_{\text{in}} = R_1 \parallel R_2 \parallel \beta R_E \qquad (8\text{-}5a)$$

Fig. 8-4
*Deriving an ac
model for the
emitter follower.*

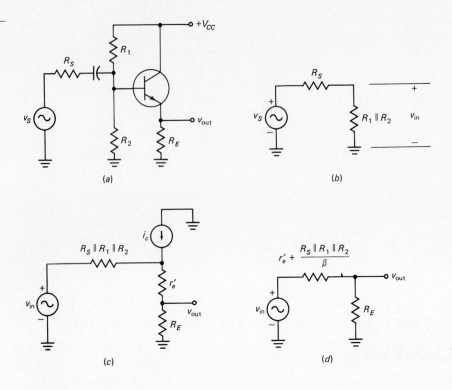

(a)

(b)

(c)

(d)

When the voltage-divider bias is stiff, βR_E is at least 100 times larger than $R_1 \parallel R_2$, and the foregoing equation simplifies to

$$z_{\text{in}} = R_1 \parallel R_2 \tag{8-5b}$$

In other words, the input impedance of a typical emitter follower consists mainly of the parallel resistance of the biasing resistors, as shown in Fig. 8-4*b*. Because some of the source signal is dropped across R_S, the input voltage driving the base is less than the source voltage.

OUTPUT IMPEDANCE

To get an ac model for the emitter follower, we need to apply Thevenin's theorem to Fig. 8-4*b*. The Thevenin voltage is v_{in}. The Thevenin resistance is $R_S \parallel R_1 \parallel R_2$. Figure 8-4*c* shows the ac equivalent circuit of Fig. 8-4*a*. Summing ac voltages around the loop gives

$$i_e r'_e + i_e R_E - v_{\text{in}} + i_b(R_S \parallel R_1 \parallel R_2) = 0$$

Since $i_b = i_c/\beta \cong i_e/\beta$, we can solve for i_e, to get

$$i_e \cong \frac{v_{\text{in}}}{R_E + r'_e + (R_S \parallel R_1 \parallel R_2)/\beta}$$

Figure 8-4*d* shows the equivalent circuit for this output current. The emitter resistor

R_E is driven by an ac source with an ac output impedance of

$$z_{out(emitter)} = r'_e + \frac{R_S \parallel R_1 \parallel R_2}{\beta} \tag{8-6}$$

Notice how the ac Thevenin resistance of the base has been reduced by the β factor. This implies a low output impedance. For instance, if $r'_e = 25\ \Omega$, $R_S = 10\ k\Omega$, $R_1 = 10\ k\Omega$, $R_2 = 10\ k\Omega$, and $\beta = 200$, then

$$R_S \parallel R_1 \parallel R_2 = 10\ k\Omega \parallel 10\ k\Omega \parallel 10\ k\Omega = 3.33\ k\Omega$$

and

$$z_{out(emitter)} = 25\ \Omega + \frac{3.33\ k\Omega}{200} = 41.7\ \Omega$$

As you can see, the output impedance of the emitter is very low.

Usually, the ac signal out of an emitter follower is coupled into a load resistance. For this reason, it will be convenient to have the Thevenin equivalent circuit of the output side. In Fig. 8-4d, the unloaded output voltage is

$$v_{out} = Av_{in}$$

where

$$A = \frac{R_E}{R_E + r'_e + (R_S \parallel R_1 \parallel R_2)/\beta}$$

Resistance R_E usually swamps out the other terms in the denominator, so that

$$A \cong 1$$

Therefore, the Thevenin voltage is approximately equal to v_{in}.

The Thevenin resistance is

$$z_{out} = R_E \parallel \left(r'_e + \frac{R_S \parallel R_1 \parallel R_2}{\beta} \right)$$

Since R_E is usually large enough to ignore, the output impedance of an emitter follower is approximately

$$z_{out} = r'_e + \frac{R_S \parallel R_1 \parallel R_2}{\beta}$$

AC MODEL

Figure 8-5a shows an ac model for an emitter follower. It has an input impedance z_{in}, an output source Av_{in}, and an output impedance z_{out}. Using the approximations discussed earlier, we can simplify this equivalent circuit to Fig. 8-5b. This simplified ac model is accurate enough for most troubleshooting and design. In this model, we approximate the input impedance by $R_1 \parallel R_2$. Furthermore, we approximate the Thevenin or unloaded voltage gain by 1. This means that the Thevenin output voltage equals the input voltage.

Fig. 8-5
Emitter follower.
(a) Ac model. (b)
Simplified model.

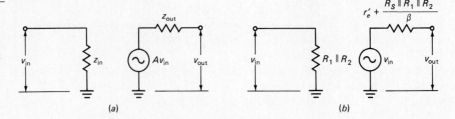

(a)   (b)

EXAMPLE 8-3

Figure 8-6*a* shows an ac source of 100 mV rms with an output impedance of 3.6 kΩ driving an emitter follower. If $\beta = 100$, what is the ac output voltage?

SOLUTION

The input impedance of the stage is approximately

$$z_{in} = R_1 \parallel R_2 = 10 \text{ k}\Omega \parallel 10 \text{ k}\Omega = 5 \text{ k}\Omega$$

Next, we can calculate the output impedance as follows:

$$R_S \parallel R_1 \parallel R_2 = 3.6 \text{ k}\Omega \parallel 10 \text{ k}\Omega \parallel 10 \text{ k}\Omega = 2.09 \text{ k}\Omega$$

Earlier, we calculated an r'_e of 25 Ω, and so

$$z_{out(emitter)} = 25 \text{ }\Omega + \frac{2.09 \text{ k}\Omega}{100} = 45.9 \text{ }\Omega$$

Fig. 8-6

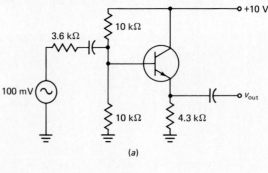

(a)

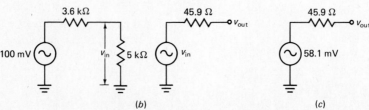

(b)   (c)

The exact output impedance is 45.9 Ω in parallel with 4.3 kΩ, the value of R_E. Since R_E is much larger, let's ignore it and approximate the output impedance as

$$z_{out} \cong 45.9 \ \Omega$$

Figure 8-6b shows the emitter follower replaced by its ac model. The input voltage to the emitter follower is

$$v_{in} = \frac{5 \text{ k}\Omega}{8.6 \text{ k}\Omega} 100 \text{ mV} = 58.1 \text{ mV}$$

Since the voltage gain is approximately 1, the final output voltage is approximately 58.1 mV.

Figure 8-6c shows the Thevenin equivalent circuit for the output side of the emitter follower. It is a Thevenin voltage source of 58.1 mV and a Thevenin impedance of 45.9 Ω.

EXAMPLE 8-4

What is the ac output voltage in the two-stage amplifier of Fig. 8-7a?

SOLUTION

The first stage should look familiar. It is the CE amplifier analyzed throughout Chap. 7. Earlier, we calculated the following quantities:

$$A = -159$$
$$z_{in} = 1.18 \text{ k}\Omega$$
$$z_{out} = 3.6 \text{ k}\Omega$$

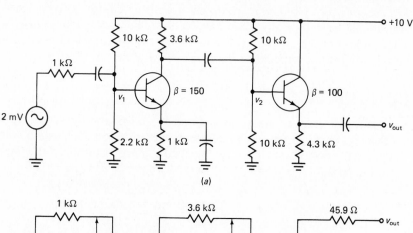

Fig. 8-7
Two-stage amplifier with CE and CC stages.

The second stage is the emitter follower analyzed in the preceding example.

Figure 8-7*b* shows the ac equivalent circuit of the two-stage amplifier. The ac input voltage to the first stage is

$$v_1 = \frac{1.18 \text{ k}\Omega}{2.18 \text{ k}\Omega} \, 2 \text{ mV} = 1.08 \text{ mV}$$

The Thevenin ac voltage out of the first stage is

$$Av_1 = -159(1.08 \text{ mV}) = -172 \text{ mV}$$

The input voltage to the emitter follower is

$$v_2 = \frac{5 \text{ k}\Omega}{8.6 \text{ k}\Omega} \, (-172 \text{ mV}) = -100 \text{ mV}$$

Since the voltage gain of the emitter follower is approximately unity, the final output voltage is

$$v_{\text{out}} = -100 \text{ mV}$$

Notice the improvement in voltage and stiffness. The source driving the two-stage amplifier has a Thevenin voltage of 2 mV and a Thevenin resistance of 1 kΩ. The output stage of the amplifier has a Thevenin voltage of −100 mV and a Thevenin resistance of 45.9 Ω. The first stage increases the voltage, while the second stage decreases the impedance.

8-3 THE DARLINGTON AMPLIFIER

The *Darlington* amplifier is a popular transistor circuit. It consists of cascaded emitter followers, typically a pair like Fig. 8-8. The overall voltage gain is close to 1. The main result is a very large increase in input impedance and an equally dramatic decrease in output impedance.

DC ANALYSIS

To begin with, the first transistor has one V_{BE} drop, and the second transistor has another V_{BE} drop. As usual, the voltage divider produces a Thevenin voltage that is applied to the input base. Because of the two V_{BE} drops, the dc emitter current in the second stage is

$$I_{E2} = \frac{V_{TH} - 2V_{BE}}{R_E} \tag{8-7}$$

The dc emitter current in the first stage equals the dc base current in the second stage; therefore,

$$I_{E1} \cong \frac{I_{E2}}{\beta_{\text{dc}}} \tag{8-8}$$

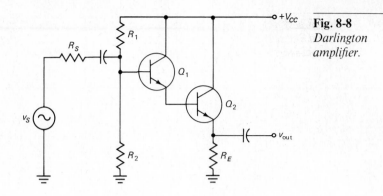

Fig. 8-8
Darlington amplifier.

AC ANALYSIS

If we ignore r'_{e2}, the input impedance of the second stage is

$$z_{\text{in}(2)} \cong \beta_2 R_E$$

where β_2 is the ac beta of the second transistor. This is the impedance seen by the emitter of the first transistor. If we ignore r'_{e1}, the input impedance looking into the base of the first transistor is

$$z_{\text{in}(1)} = \beta_1 \beta_2 R_E$$

This input impedance is extremely high because of the product of the ac betas, and so the approximate input impedance of the Darlington amplifier is

$$z_{\text{in}} = R_1 \parallel R_2$$

Although this is the same formula as before, the difference is that we can use much higher values of R_1 and R_2.

Next, let us find the output impedance of the Darlington amplifier. As before, the ac Thevenin impedance at the input is

$$r_{th} = R_S \parallel R_1 \parallel R_2$$

The output impedance of the first stage is

$$z_{\text{out}(1)} = r'_{e1} + \frac{r_{th}}{\beta_1}$$

The second stage has an output impedance of

$$z_{\text{out}(2)} = r'_{e2} + \frac{r'_{e1} + r_{th}/\beta_1}{\beta_2} \qquad (8\text{-}9)$$

where $z_{\text{out}(2)}$ = output impedance of Darlington
r'_{e1} = ac resistance of first emitter diode
r'_{e2} = ac resistance of second emitter diode
$r_{th} = R_S \parallel R_1 \parallel R_2$
β_1 = ac beta of first transistor
β_2 = ac beta of second transistor

This output impedance can be very small.

Fig. 8-9

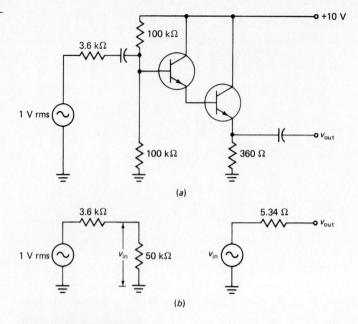

(a)

(b)

AN EXAMPLE

An ac source of 1 V rms with an impedance of 3.6 kΩ drives the Darlington amplifier of Fig. 8-9a. To keep the analysis simple, let us use $\beta_{dc} = 100$ and $\beta = 100$ for both transistors. The dc emitter current in the second stage is

$$I_{E2} = \frac{5 \text{ V} - 1.4 \text{ V}}{360 \ \Omega} = 10 \text{ mA}$$

Therefore, the ac resistance of the second emitter diode is

$$r'_{e2} = \frac{25 \text{ mV}}{10 \text{ mA}} = 2.5 \ \Omega$$

The dc emitter current in the first stage is

$$I_{E1} = \frac{10 \text{ mA}}{100} = 0.1 \text{ mA}$$

So, the ac resistance of the first emitter diode is

$$r'_{e1} = \frac{25 \text{ mV}}{0.1 \text{ mA}} = 250 \ \Omega$$

The approximate input impedance looking into the base of the first transistor is

$$z_{in(1)} \cong 100(100)(360 \ \Omega) = 3.6 \text{ M}\Omega$$

Since this is much larger than the biasing resistors, the approximate input impedance of the Darlington amplifier is

$$z_{in} = 100 \text{ k}\Omega \parallel 100 \text{ k}\Omega = 50 \text{ k}\Omega$$

The ac Thevenin impedance at the input is

$$r_{th} = 3.6 \text{ k}\Omega \parallel 100 \text{ k}\Omega \parallel 100 \text{ k}\Omega = 3.36 \text{ k}\Omega$$

So, the output impedance of the first emitter is

$$z_{out(1)} = 250 \ \Omega + \frac{3.36 \text{ k}\Omega}{100} = 284 \ \Omega$$

If we ignore the 360 Ω, the output impedance of the second emitter is

$$z_{out(2)} = 2.5 \ \Omega + \frac{284 \ \Omega}{100} = 5.34 \ \Omega$$

Figure 8-9*b* summarizes the calculations.

Because the input impedance is much higher than before, almost all the ac source voltage reaches the input of the amplifier:

$$v_{in} = \frac{50 \text{ k}\Omega}{53.6 \text{ k}\Omega} 1 \text{ V} = 0.933 \text{ V}$$

The voltage gain is approximately equal to 1. Therefore, the final output is a source voltage of 0.933 V and a source impedance of 5.34 Ω. A source like this will look stiff to all loads greater than 534 Ω and firm to all loads greater than 53.4 Ω.

ISOLATION

As you have seen, input impedance is increased and output impedance is decreased. Because of this, emitter followers and Darlington amplifiers are used to isolate high-impedance sources from low-impedance loads. If you try to run a signal directly from a high-impedance source into a low-impedance load, most of the ac voltage will be lost across the source impedance. By inserting an emitter follower or Darlington amplifier between the source and the load, you prevent excessive signal loss and wind up with a source stiff enough to drive most load resistances.

Remember this idea of isolating a high-impedance source from a low-impedance load. If you don't, you will be constantly losing signal voltage when you cascade high-impedance circuits to low-impedance loads.

DARLINGTON PAIR

Transistor manufacturers sometimes put two transistors connected as a Darlington pair inside a single transistor housing. This three-terminal device acts like a single transistor with an extremely high β. For instance, the 2N2785 is an *npn* Darlington transistor connected like Fig. 8-10*a*. The data sheet of this device indicates a minimum β of 2000 and a maximum of 20,000.

Pnp transistors can also be connected like the Darlington pair of Fig. 8-10*b*. This connection acts the same as a single *pnp* transistor with an equivalent β of

$$\beta = \beta_1 \beta_2$$

Fig. 8-10
(a) *Npn Darling-*
ton. (b) *Pnp*
Darlington. (c)
Photo-Darlington.

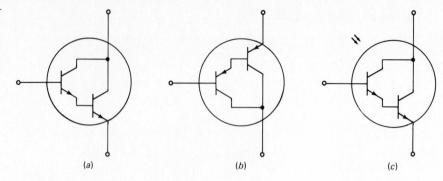

(a)   *(b)*   *(c)*

PHOTO-DARLINGTON

If the base of the first transistor of a Darlington pair is left open, a small collector current will exist because of thermally produced carriers. By exposing the first collector junction to light, a manufacturer can produce a *photo-Darlington,* a device that is β times more sensitive to light than a phototransistor. Figure 8-10c shows the schematic symbol of a photo-Darlington.

8-4 TYPES OF COUPLING

There are several ways to couple a signal from the output of one stage to the input of the next. We can use capacitors, transformers, direct coupling, and other methods. This section discusses the types of coupling.

RC COUPLING

Figure 8-11a illustrates *resistance-capacitance* (*RC*) coupling, the most common method of coupling the signal from one stage to the next. In this approach, the signal developed across the collector resistor of each stage is coupled into the base of the next stage. The cascaded stages amplify the signal, and the overall gain equals the product of the individual gains.

The coupling capacitors pass alternating current but block direct current. Because of this, the stages are isolated as far as direct current is concerned. This is necessary to prevent dc interaction and shifting of Q points. The drawback of this approach is the lower frequency limit imposed by the coupling capacitors.

The bypass capacitors are needed because they bypass the emitters to ground. Without them, the voltage gain of each stage would be lost. These bypass capacitors also place a lower limit on the frequency response. In other words, as the frequency keeps decreasing, we eventually reach a point at which the bypass capacitors no longer look like ac shorts. At this frequency, the voltage gain starts to decrease because of the local feedback, and the overall gain of the amplifier drops significantly.

If you are interested in amplifying ac signals with frequencies greater than approximately 10 Hz, the *RC*-coupled amplifier is usually suitable. For discrete

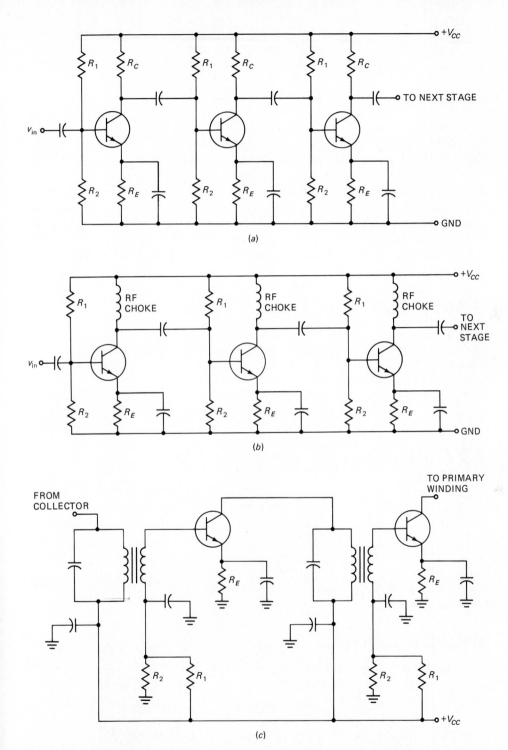

Fig. 8-11 *Amplifiers.* (a) *RC-coupled.* (b) *Impedance-coupled.* (c) *Transformer-coupled.*

circuits (those in which separate resistors, capacitors, and transistors are soldered together), it is the most convenient and least expensive way to build a multistage amplifier.

IMPEDANCE COUPLING

Sometimes you will see *impedance coupling* at higher frequencies. Here's the basic idea: Replace each R_C of Fig. 8-11a by an inductor, as shown in Fig. 8-11b. As the frequency increases, X_L approaches infinity and each inductor appears open. In other words, the inductors pass direct current but block alternating current. When used in this way, they are called *RF chokes.*

The advantage of impedance coupling is that no signal power is wasted in collector resistors; they have been replaced by inductors that have very small power dissipation. The disadvantage is that RF chokes are relatively expensive and their impedance drops off at lower frequencies. As a rule, impedance coupling is suitable only at *radio frequencies* (RF). These are frequencies above 20 kHz.

TRANSFORMER COUPLING

Figure 8-11c illustrates *transformer coupling.* R_1 and R_2 provide voltage-divider bias. A bypass capacitor is used on the bottom of each primary winding to produce an ac ground; this avoids the lead inductance of the connecting line that returns to the dc supply point. Similarly, a bypass capacitor is used on the bottom of each secondary winding to get an ac ground; this prevents signal power loss in the biasing resistors. The signal is transformer-coupled from one stage to the next.

At one time, transformers were widely used in *audio amplifiers* (20 Hz to 20 kHz), despite their high cost and bulkiness. Since the invention of the transistor, the emitter follower and Darlington amplifier have replaced transformers in most audio applications.

Transformers are still used in *RF amplifiers.* In AM radio receivers, for instance, RF signals have frequencies from 535 to 1605 kHz. In TV receivers, the frequencies are from 54 to 216 MHz (channels 2 through 13). At these higher frequencies, transformers are smaller and less expensive. Usually, a capacitor is shunted across one or both windings to get frequency resonance. For instance, each primary winding of Fig. 8-11c has a tuning capacitor across it. This allows us to filter out all frequencies except the resonant frequency and those near it. (This is the principle behind tuning in a radio station or TV channel.)

DIRECT COUPLING

Below 10 Hz (approximately), coupling and bypass capacitors become unreasonably large, electrically and physically. As an example, to bypass an emitter resistor of 100 Ω at 10 Hz, we need approximately 1590 μF. The lower the resistance or frequency, the worse the problem gets.

To break the low-frequency barrier, we can fall back on *direct coupling.* This means designing the stages without coupling and bypass capacitors, so that the direct current is coupled as well as the alternating current. As a result, there is no lower frequency limit; the amplifier enlarges signals no matter how low their frequency, including direct current or zero frequency. The next section tells you more about direct coupling.

8-5 DIRECT COUPLING

There are many possible direct-coupled designs. In this section, we look at a few elementary circuits to get the idea.

ONE-SUPPLY CIRCUIT

Figure 8-12 is a two-stage direct-coupled amplifier; no coupling or bypass capacitors are used. With a quiescent input voltage of +1.4 V, about 0.7 V is dropped across the first emitter diode, leaving +0.7 V across the 680-Ω resistor. This sets up approximately 1 mA of collector current. This 1 mA then produces a drop of 27 V across the collector resistor. Therefore, the first collector runs at about +3 V with respect to ground.

Allowing 0.7 V for the second emitter diode, we get 2.3 V across the 2.4-kΩ resistor. This results in approximately 1 mA of collector current, about a 24-V drop across the collector resistor, and around +6 V from the second collector to ground. Therefore, a quiescent input voltage of +1.4 V gives a quiescent output voltage of +6 V.

Because of the high β_{dc}, we can ignore the loading effect of the second base upon the first collector. Since the first stage is heavily swamped, we can ignore r'_e and calculate a voltage gain of

$$A_1 = -\frac{27,000}{680} \cong -40$$

The second stage has a voltage gain of

$$A_2 = -\frac{24,000}{2400} = -10$$

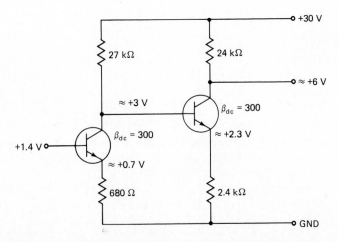

Fig. 8-12
Direct-coupled amplifier.

And the overall gain is

$$A = A_1 A_2 = (-40)(-10) = 400$$

The two-stage circuit will amplify an ac input voltage (any change in the input voltage) by a factor of 400. For instance, if the input voltage changes by $+5$ mV, the final output voltage changes by

$$v_{out} = Av_{in} = 400(5 \text{ mV}) = 2 \text{ V}$$

Because two inverting stages are involved, the final output goes from $+6$ to $+8$ V.

Here is the main disadvantage of direct coupling. Transistor characteristics like V_{BE} vary with temperature. This causes the collector currents and voltages to change. Because of the direct coupling, the voltage changes are coupled from one stage to the next, appearing at the final output as an amplified voltage change. This unwanted change is called *drift*. The trouble with drift is that you can't distinguish it from a genuine change produced by the input signal.

GROUND-REFERENCED INPUT

For the two-stage amplifier of Fig. 8-12 to work properly, we need a quiescent input voltage of $+1.4$ V. In most applications, it's necessary to have *ground-referenced* input, one where the quiescent input voltage is 0 V.

Figure 8-13 shows a ground-referenced input stage. This stage is a *pnp* Darlington connection with the input base returned to ground through the input signal source. Because of this, the first emitter is approximately $+0.7$ V above ground, and the second emitter is about $+1.4$ V above ground. The $+1.4$ V biases the second stage, which operates as previously described.

The quiescent V_{CE} of the first transistor is only 0.7 V, and the quiescent V_{CE} of the second transistor is only 1.4 V. Nevertheless, both transistors are operating in the active region because the $V_{CE(sat)}$ of low-power transistors is only about 0.1 V.

Fig. 8-13
Ground-referenced input.

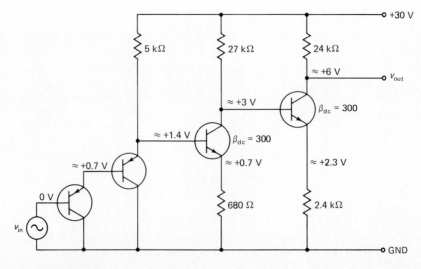

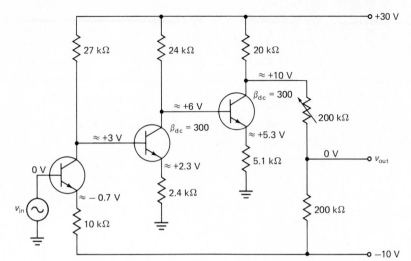

Fig. 8-14
*Split-supply
direct-coupled
amplifier.*

Furthermore, the input signal is typically in millivolts, which means that these transistors continue to operate in the active region when a small signal is present.

Remember the *pnp* ground-referenced input. It's used a lot in audio integrated circuits.

TWO-SUPPLY CIRCUIT

When a *split supply* is available (positive and negative voltages), we can reference both the input and the output to ground. Figure 8-14 is an example. The first stage is emitter-biased with an I_E around 1 mA. This produces about +3 V at the first collector. Subtracting the V_{BE} drop of the second emitter diode leaves +2.3 V at the second emitter.

The emitter current in the second stage is around 1 mA. This flows through the collector resistor, producing about +6 V from collector to ground. The final stage has +5.3 V across the emitter resistor, which gives about 1 mA of current. Therefore, the last collector has approximately +10 V to ground.

The output voltage divider references the output to ground. When the upper resistor is adjusted to 200 kΩ, the final output voltage is approximately 0 V. The adjustment allows us to eliminate errors caused by such factors as resistor tolerances or V_{BE} differences from one transistor to the next.

What is the overall voltage gain? The first stage has a gain of approximately 3, the second stage about 10, the third stage around 4, and the voltage divider about 0.5. Therefore,

$$A = 3(10)(4)(0.5) = 60$$

FINAL POINT

This gives you the basic idea behind direct coupling. We leave out all coupling and bypass capacitors. This allows us to couple both direct current and alternating current from one stage to the next. Because of this, the amplifier has no lower

Fig. 8-15
CE stage direct-coupled to CC stage.

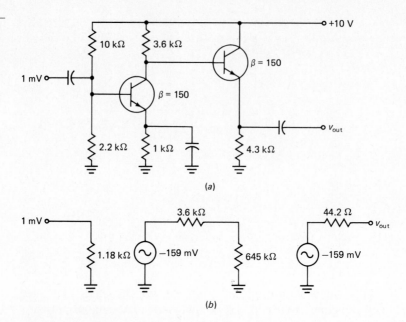

(a)

(b)

frequency limit; it amplifies all frequencies down to zero. Herein lies the strength and weakness of direct coupling: It's good to be able to amplify very low input frequencies, including dc; it's bad, however, to amplify unwanted inputs, such as very slow changes in supply voltage or transistor variations.

Later chapters introduce the *differential amplifier*, a two-transistor direct-coupled circuit that has become the backbone of linear integrated circuits. One reason the differential amplifier is so popular is that drift cancels out, at least partially. More is said about this later.

EXAMPLE 8-5

Figure 8-15a shows a CE amplifier coupled directly into an emitter follower. What is the output voltage?

SOLUTION

The first stage has been analyzed previously. It has the following properties:

$$A = -159$$
$$z_{in} = 1.18 \text{ k}\Omega$$
$$z_{out} = 3.6 \text{ k}\Omega$$
$$V_C = 6.04 \text{ V}$$

Notice that the second stage gets its dc base voltage from the collector of the first stage. Therefore, the dc emitter current in the second stage is

$$I_E = \frac{6.04 \text{ V} - 0.7 \text{ V}}{4.3 \text{ k}\Omega} = 1.24 \text{ mA}$$

and the ac resistance of the emitter diode is

$$r'_e = \frac{25 \text{ mV}}{1.24 \text{ mA}} = 20.2 \text{ }\Omega$$

Since there are no biasing resistors in the second stage, its input impedance is

$$z_{in} = \beta R_E = 150(4.3 \text{ k}\Omega) = 645 \text{ k}\Omega$$

The output impedance of the second stage is

$$z_{out} = 20.2 \text{ }\Omega + \frac{3.6 \text{ k}\Omega}{150} = 44.2 \text{ }\Omega$$

Figure 8-15*b* summarizes the calculations. The first stage looks like a stiff ac source to the second stage. This means that approximately all of the 159 mV at the collector of the first transistor reaches the base of the second transistor. For this reason, the emitter follower acts like a source of 159 mV with an output impedance of only 44.2 Ω.

8-6 IMPROVED VOLTAGE REGULATION

An emitter follower can improve the performance of a zener regulator. Figure 8-16*a* shows a *zener follower,* a circuit that combines a zener regulator and an emitter follower. Here is how it works. The zener voltage is the input to the base. Therefore, the dc output voltage is

$$V_{out} = V_Z - V_{BE} \qquad (8\text{-}10)$$

This output voltage is fixed, equal to the zener voltage less the V_{BE} drop of the transistor. If the supply voltage changes, the zener voltage remains approximately constant, and so does the output voltage. In other words, the circuit acts like a voltage regulator.

TWO ADVANTAGES

The zener follower has two advantages over an ordinary zener regulator. First, the direct current through R_S is the sum of the zener current and the base current, which equals

$$I_B = \frac{I_L}{\beta_{dc}} \qquad (8\text{-}11)$$

Since this base current is much smaller than the load current, a smaller zener diode can be used. For instance, if you are trying to supply amperes to a load resistor, an ordinary zener regulator requires a zener diode capable of handling amperes. On the other hand, with the improved regulator of Fig. 8-16*a*, the zener diode needs to handle only tens of milliamperes because of the reduction by β_{dc}.

In an ordinary zener regulator, the load resistor sees an output impedance of approximately R_Z, the zener impedance. But in the zener follower, the output

Fig. 8-16
Zener follower.
(a) *Circuit.*
(b) *Equivalent.*

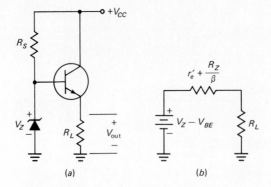

impedance is

$$z_{out} \cong r'_e + \frac{R_Z}{\beta} \tag{8-12}$$

This means that the equivalent output circuit looks like Fig. 8-16*b*. A circuit like this can hold the load voltage almost constant because the source looks stiff.

The two advantages of a zener follower, less load on the zener diode and lower output impedance, allow us to design stiff voltage regulators. The key idea is that the emitter follower increases the current-handling capability of a zener regulator. The zener follower increases the load current by a factor of β_{dc}.

SERIES REGULATORS

Figure 8-17 shows the zener follower as it is usually drawn. V_{in} is the unregulated input voltage. The zener voltage drives the base of an emitter follower; therefore, the dc output voltage is bootstrapped to within one V_{BE} drop of the zener voltage.

If you design a circuit like this, you've got to take the power dissipation of the transistor into account. It equals

$$P_D = V_{CE}I_C \tag{8-13}$$

In Fig. 8-17, the collector-emitter voltage is the difference between the input and output voltage:

$$V_{CE} = V_{in} - V_{out}$$

The collector current is approximately equal to the emitter current:

$$I_C \cong I_E$$

The zener follower is an example of a series voltage regulator. Since the collector-emitter terminals are in series with the load, the load current must pass through the transistor, and this is why it is called a *pass transistor*. Because of their simplicity, series regulators are widely used.

The main disadvantage of a series regulator is the power dissipation of the pass transistor. As long as the load current is not too large, the pass transistor does not get too hot. But when the load current is heavy, the pass transistor has to dissipate a lot of power, and this raises the internal temperature of the equipment. In some cases, a fan

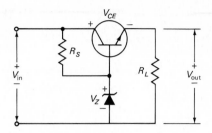

Fig. 8-17
Zener follower drawn to emphasize pass transistor.

may be needed to remove the heat. To avoid this, a designer may prefer a switching regulator (discussed in a later chapter).

TEMPERATURE EFFECTS

Now is a good time to mention the effect that temperature has on V_{BE}. When the emitter temperature increases, V_{BE} decreases. Data sheets usually give information on how much V_{BE} changes with temperature. The change in V_{BE} depends on the collector current, the particular transistor, and other factors. A useful approximation for the change is this: V_{BE} decreases 2 mV for each degree Celsius rise. For example, suppose that V_{BE} is 0.7 V for an emitter temperature of 25°C. If the emitter temperature rises to 75°C, V_{BE} decreases 100 mV to a new value of 0.6 V.

Why is this important? The output voltage of a zener regulator is

$$V_{out} = V_Z - V_{BE}$$

When the temperature rises from 25° to 75°C, V_{BE} decreases approximately 100 mV. If the zener diode has a zero temperature coefficient, the regulated output voltage decreases approximately 0.1 V. This is a relatively small change, but you must be aware of these effects if you are designing circuits. For more information on the temperature dependence of V_{BE}, consult data sheets for the particular transistor you are using.

EXAMPLE 8-6

In Fig. 8-18, the pass transistor has β_{dc} of 80. Calculate the current through the zener diode.

SOLUTION

The current through the series-limiting resistor is

$$I_S = \frac{20 \text{ V} - 10 \text{ V}}{680 \text{ }\Omega} = 14.7 \text{ mA}$$

Next, find the base current. The load voltage is

$$V_{out} = V_Z - V_{BE} = 10 \text{ V} - 0.7 \text{ V} = 9.3 \text{ V}$$

Fig. 8-18

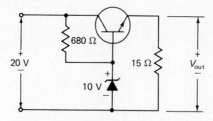

The load current is

$$I_E = \frac{V_{\text{out}}}{R_L} = \frac{9.3 \text{ V}}{15 \text{ }\Omega} = 0.62 \text{ A}$$

The base current is

$$I_B = \frac{I_E}{\beta_{\text{dc}}} = \frac{0.62 \text{ A}}{80} = 7.75 \text{ mA}$$

The zener current is

$$I_Z = 14.7 \text{ mA} - 7.75 \text{ mA} = 6.95 \text{ mA}$$

Notice how much smaller this zener current is than the load current. This is the whole point of the circuit. A simple zener regulator capable of handling milliamperes is cascaded to an emitter follower to get a load current in hundreds of milliamperes.

Incidentally, we use β_{dc} in this calculation because we are after the dc or total current through the zener diode.

EXAMPLE 8-7

What is the power dissipation of the transistor in Fig. 8-18? If $R_Z = 7 \text{ }\Omega$ and $\beta = 100$, what is the output impedance seen by the load resistor?

SOLUTION

The power dissipation of the pass transistor is

$$P_D = (20 \text{ V} - 9.3 \text{ V})(0.62 \text{ A}) = 6.63 \text{ W}$$

This is not too much power. In some applications where the load current is considerably larger, the power dissipation of the pass transistor can become very high.

In the preceding example, we calculated an emitter current of 0.62 A. So, the ac resistance of the emitter diode is

$$r'_e = \frac{25 \text{ mV}}{0.62 \text{ A}} = 0.04 \text{ }\Omega$$

Using Eq. (8-12), the output impedance is approximately

$$z_{out} \cong 0.04 \ \Omega + \frac{7}{100} = 0.11 \ \Omega$$

This implies a stiff source for all load resistances greater than 11 Ω.

We use β to calculate z_{out} because output impedance is an ac quantity that tells us what happens when currents and voltages change. For instance, if the load resistance changes from 15 to 14 Ω, the load current changes from 0.62 to 0.664 A, an increase of 0.044 A. In this case, the load voltage decreases by

$$\Delta V_L = (0.044 \ \text{A})(0.11 \ \Omega) = 4.84 \ \text{mV}$$

8-7 THE COMMON-BASE AMPLIFIER

Figure 8-19*a* shows a *common-base* (CB) amplifier. Since the base is grounded, this circuit is also called a *grounded-base* amplifier. The *Q* point is set by emitter bias, which we immediately recognize when the dc equivalent circuit is drawn as shown in Fig. 8-19*b*. Therefore, the dc emitter current is given by

$$I_E = \frac{V_{EE} - V_{BE}}{R_E} \tag{8-14}$$

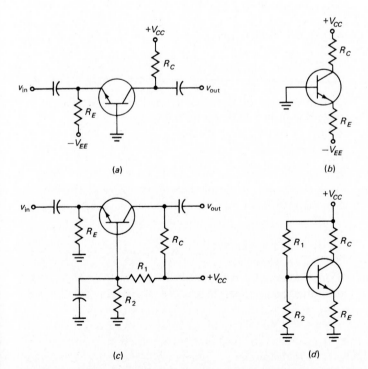

(a)

(b)

(c)

(d)

Fig. 8-19 *CB amplifier.* (a) *Split supply.* (b) *Emitter-biased dc equivalent circuit.* (c) *Single supply.* (d) *Voltage-divider-biased dc equivalent circuit.*

Figure 8-19*c* shows a voltage-divider-biased CB amplifier. You can recognize the voltage-divider bias by drawing the dc equivalent circuit, as shown in Fig. 8-19*d*.

In either amplifier, the base is ac ground. The input signal drives the emitter, and the output signal is taken from the collector. Figure 8-20*a* shows the ac equivalent circuit of a CB amplifier during the positive half cycle of input voltage. The biasing resistors are omitted because they have a negligible effect on input impedance. The input impedance of a CB amplifier is closely equal to

$$z_{\text{in}} \cong r'_e \qquad (8\text{-}15)$$

The output voltage is given by

$$v_{\text{out}} = i_c R_C$$

This is in phase with the input voltage. Since the input voltage equals

$$v_{\text{in}} = i_e r'_e$$

the voltage gain is

$$\frac{v_{\text{out}}}{v_{\text{in}}} = \frac{i_c R_C}{i_e r'_e}$$

Because $i_c \cong i_e$, we can rewrite the equation as

$$A \cong \frac{R_C}{r'_e} \qquad (8\text{-}16)$$

This says that the voltage gain has the same magnitude as it would in an unswamped CE amplifier; only the phase is different. For instance, if $R_C = 2500\ \Omega$ and $r'_e = 25\ \Omega$, then a CB amplifier has a voltage gain of 100, whereas a CE amplifier has a voltage gain of -100.

Ideally, the collector current source of Fig. 8-20*a* has an infinite internal impedance. Therefore, the output impedance of a CB amplifier is

$$z_{\text{out}} \cong R_C \qquad (8\text{-}17)$$

Figure 8-20*b* shows the ac model of a CB amplifier. The big difference between this and the CE amplifier is the extremely low input impedance. One reason why the CB amplifier is not used as much is its low input impedance. The ac source driving the CB amplifier sees

$$z_{\text{in}} \cong r'_e$$

which can be quite low. For instance, if $I_E = 1$ mA, the input impedance of a CB amplifier is only 25 Ω. Unless the ac source is stiff, most of the signal is lost across the source resistance.

The input impedance of a CB amplifier is so low that it overloads most signal sources. Because of this, a discrete CB amplifier is not used too often at low frequencies; it is mainly used in high-frequency applications (above 10 MHz) where low source impedances are common. In integrated circuits (discussed in a later chapter), the CB amplifier is widely used as part of a differential amplifier.

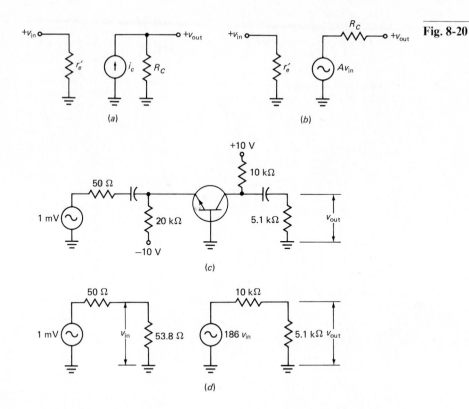

Fig. 8-20

EXAMPLE 8-8

What is the ac output voltage in Fig. 8-20c?

SOLUTION

The dc emitter current is

$$I_E = \frac{10 \text{ V} - 0.7 \text{ V}}{20 \text{ k}\Omega} = 0.465 \text{ mA}$$

and the ac emitter resistance is

$$r'_e = \frac{25 \text{ mV}}{0.465 \text{ mA}} = 53.8 \text{ }\Omega$$

Therefore,

$$z_{in} = 53.8 \text{ }\Omega$$

The unloaded voltage gain is

$$A = \frac{10{,}000}{53.8} = 186$$

and the output impedance is

$$z_{out} = 10 \text{ k}\Omega$$

Figure 8-20d shows the ac equivalent circuit. The input voltage is

$$v_{in} = \frac{53.8 \ \Omega}{50 \ \Omega + 53.8 \ \Omega}(1 \text{ mV}) = 0.518 \text{ mV}$$

The unloaded output voltage is

$$Av_{in} = 186(0.518 \text{ mV}) = 96.3 \text{ mV}$$

The actual output voltage is

$$v_{out} = \frac{5.1 \text{ k}\Omega}{10 \text{ k}\Omega + 5.1 \text{ k}\Omega}(96.3 \text{ mV}) = 32.5 \text{ mV}$$

PROBLEMS

Straightforward

8-1. In Fig. 8-21, $h_{FE} = 160$. Calculate the following dc quantities: V_B, V_E, V_C, I_E, I_C, and I_B.

Fig. 8-21

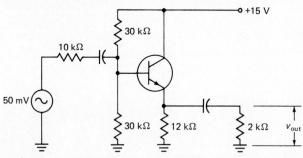

8-2. Draw the dc load line for Fig. 8-21.

8-3. If $\beta = 125$ in Fig. 8-21, what is the output voltage?

8-4. In Fig. 8-21, $h_{fe} = 150$. If you look at the emitter with a dc-coupled oscilloscope, what do you see? The answer should have a dc and an ac voltage.

8-5. If $\beta_{dc} = 80$ for both transistors, work out the following dc quantities for each stage of Fig. 8-22: V_B, V_E, V_C, I_E, I_C, and I_B.

Fig. 8-22

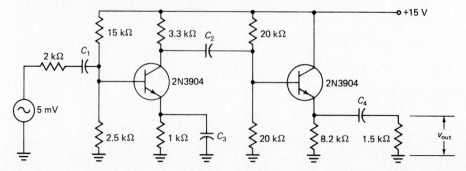

8-6. Draw the dc load line for each stage in Fig. 8-22.

8-7. In Fig. 8-22, if each transistor has a $\beta = 175$, what is the output voltage?

8-8. If $h_{fe} = 90$ for both transistors in Fig. 8-22, what does the output voltage equal?

8-9. In Fig. 8-23, each transistor has $\beta_{dc} = 120$. What is the dc collector current in the second transistor? In the first transistor? The base current of the first transistor?

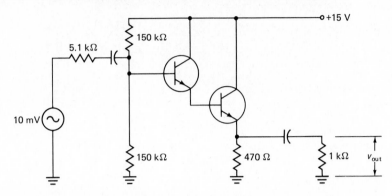

Fig. 8-23

8-10. If both transistors of Fig. 8-23 have $\beta = 100$, what does the output voltage equal?

8-11. In Fig. 8-23, the first transistor has an h_{fe} of 150 and the second transistor has an h_{fe} of 90. What is the ac output voltage?

8-12. In Fig. 8-24, each transistor has a $\beta_{dc} = 115$. Work out the following dc quantities for each stage: V_B, V_E, V_C, I_E, I_C, and I_B.

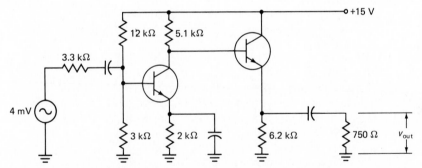

Fig. 8-24

8-13. If $\beta = 135$ for each transistor of Fig. 8-24, what does the output voltage equal?

8-14. If the first transistor of Fig. 8-24 has $\beta = 250$ and the second transistor has $\beta = 200$, what does v_{out} equal?

8-15. What is the output voltage in Fig. 8-25? If $\beta_{dc} = 120$, what does the zener current equal? What is the output impedance of the zener follower if $R_Z = 6\ \Omega$?

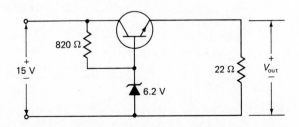

Fig. 8-25

8-16. A Darlington transistor is substituted for the transistor of Fig. 8-25. What is the output voltage?

8-17. In Fig. 8-26, $h_{FE} = 75$. Calculate V_B, V_E, V_C, I_E, I_C, and I_B.

Fig. 8-26

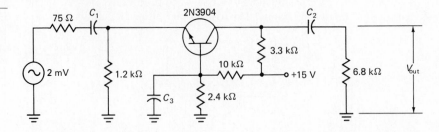

8-18. What does v_{out} equal in Fig. 8-26?

Troubleshooting

8-19. In Fig. 8-21, you decide to look at v_{be} with an oscilloscope. After touching the signal lead of the probe to the base and the common lead of the probe to the emitter, you discover that the circuit is not working. Why?

8-20. Does the ac output voltage of Fig. 8-22 increase, decrease, or remain the same for each of these troubles:
  a. C_3 opens
  b. Collector-emitter short in second transistor
  c. 3.9-kΩ resistor mistakenly used instead of 3.3-kΩ resistor
  d. C_1 shorts

8-21. Capacitor C_3 opens in Fig. 8-22. What does the ac output voltage equal if each transistor has a β of 100?

8-22. The output voltage of Fig. 8-24 is only 100 mV. If you measure an ac voltage of approximately 0.9 V at the first collector, is the trouble in the first or second stage?

8-23. The output voltage is very low in Fig. 8-25. Which of the following is a possible cause:
  a. Collector-emitter terminals shorted
  b. Zener diode open
  c. Collector-emitter terminals open
  d. 820-Ω resistor open

Design

8-24. Design an emitter follower to meet these specifications: $V_{CC} = 15$ V, $I_C = 2$ mA, $β = 100$, $z_{in} \geq 10$ kΩ, and $z_{out} \leq 112$ Ω.

8-25. Design a Darlington amplifier using these data: $V_{CC} = 20$ V, $z_{in} \geq 100$ kΩ, and $z_{out} \leq 10$ Ω.

8-26. Design a zener follower to meet these specifications: $V_{CC} = 15$ V, $V_{out} = 5.5$ V, and $I_L = 0.5$ A.

Challenging

8-27. Figure 8-27a shows a circuit called a *phase splitter.* It has two outputs. Draw the dc load line. Calculate the following dc quantities: V_B, V_E, V_C, I_E, I_C, and I_B.

8-28. If an input signal with a peak-to-peak value of 5 mV drives the circuit of Fig. 8-27a, what are the two ac output voltages? What do you think this circuit is supposed to do?

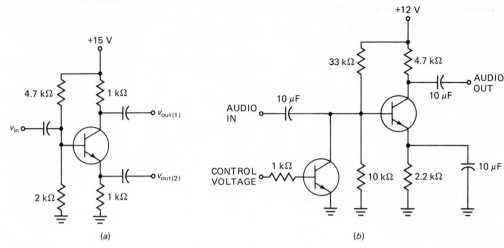

Fig. 8-27

8-29. Figure 8-27*b* shows a *squelch* circuit. The control voltage can be low or high, 0 V or +5 V. If the audio input voltage is 10 mV peak-to-peak, what is the audio output voltage when the control voltage is low? When the control voltage is high? What do you think the purpose of the circuit is?

Computer

8-30. The STOP and CONT commands allow you to set breakpoints in a program and then to resume operation. For instance, here is a program with a breakpoint in line 80:

```
10 PRINT "ENTER RS": INPUT RS
20 PRINT "ENTER R1": INPUT R1
30 PRINT "ENTER R2": INPUT R2
40 PRINT "ENTER AC BETA": INPUT BETA
50 PRINT "ENTER R PRIME E": INPUT RPE
60 Y = RS + R1 + R2
70 Z = 1/Y
80 STOP
90 ZOUT = RPE + Z/BETA
100 PRINT "ZOUT =": PRINT ZOUT
```

After you input the requested data, the computer stops at line 80 and displays

BREAK AT 80

It then waits until you enter CONT, which stands for continue. Then the program continues at line 90. Breakpoints are mainly used in *debugging* programs, which means locating errors.

The program has a bug (trouble) in it. To debug it, you have to correct the error. What is wrong with the program, and how can you fix it?

8-31. The inputs to an emitter-follower program are R_S, R_1, R_2, R_E, V_{CC}, and β. Write a program that calculates the input and output impedance of an emitter follower.

8-32. Write a program that calculates the input and output impedance of a Darlington amplifier.

H Parameters

The hybrid (*h*) parameters are an advanced mathematical approach to linear transistor circuit analysis. They represent the ultimate tool for finding the exact voltage gain, input impedance, and output impedance of a transistor amplifier. Because using the *h* parameters requires a lot of time-consuming calculations, this approach is sensible only if two conditions are satisfied: First, you are doing design work that requires the most accurate answers you can get, and second, you have access to a computer. The computer is not absolutely necessary, but it certainly eliminates the boredom and human error that accompany *h*-parameter analysis and design.

9-1 FOUR-PARAMETER SYSTEMS

Thevenin's and Norton's theorems can be applied to two-port networks like Fig. 9-1. A voltage v_1 is across the input port. This sets up a conventional current i_1 into the port. Conventional current is used throughout this chapter because the *h*-parameter method is used primarily by scientists and engineers. An ac output voltage v_2 appears across the output port, along with an ac output current i_2. By definition, the positive direction of conventional current is into a port. If a conventional current comes out of a port, the current is negative.

Z PARAMETERS

If a circuit contains only linear elements, then Thevenin's theorem can be applied to both ports to get the ac model shown in Fig. 9-1*b*. Looking into either port, we see an impedance in series with a voltage source. The Kirchhoff equations for this ac model are

$$v_1 = z_{11}i_1 + z_{12}i_2$$
$$v_2 = z_{21}i_1 + z_{22}i_2$$

In this system of analysis, the coefficients are called *z parameters*.

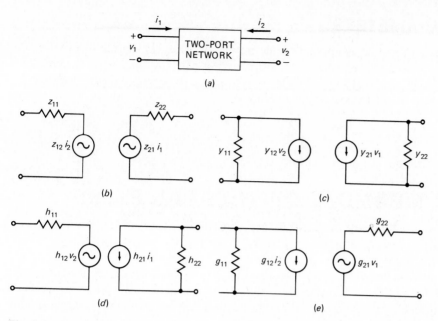

Fig. 9-1 (a) *Two-port network.* (b) *z parameters.* (c) *y parameters.* (d) *h parameters.* (e) *g parameters.*

Y PARAMETERS

If we apply Norton's theorem to both ports of a linear network, we get the equivalent circuit shown in Fig. 9-1c. This time, each port consists of an admittance shunted by a current source. The Kirchhoff equations for this ac model are

$$i_1 = y_{11}v_1 + y_{12}v_2$$
$$i_2 = y_{21}v_1 + y_{22}v_2$$

The coefficients are called *y parameters.* This system of analysis has been applied to transistors operating at high frequencies.

H PARAMETERS

If we apply Thevenin's theorem to the input port and Norton's theorem to the output port, we get the hybrid model shown in Fig. 9-1d. The input side contains an impedance h_{11} in series with a voltage source $h_{12}v_2$. The output side contains a current source $h_{21}i_1$ shunted by an admittance h_{22}. The Kirchhoff equations for the hybrid model are

$$v_1 = h_{11}i_1 + h_{12}v_2$$
$$i_2 = h_{21}i_1 + h_{22}v_2$$

The coefficients in these equations are called hybrid parameters, or simply *h parameters.* This system of analysis has been used to analyze CE, CC, and CB amplifiers operating at low frequencies.

G PARAMETERS

If Norton's theorem is applied to the input port and Thevenin's theorem to the output port, we get the ac model shown in Fig. 9-1e. On the input side, an admittance is shunted by a current source. On the output side, a voltage source is in series with an impedance. The Kirchhoff equations are

$$i_1 = g_{11}v_1 + g_{12}i_2$$
$$v_2 = g_{21}v_1 + g_{22}i_2$$

The coefficients in these equations are called *g parameters.*

9-2 MEANING OF *H* PARAMETERS

Of the four systems of analysis just described, the *h* parameters are best suited for analyzing transistor amplifiers operating at lower frequencies. For the rest of this chapter, therefore, we will concentrate on the *h* parameters.

Figure 9-2a shows the *hybrid model* with port voltages and currents. Voltages are considered positive when they have the plus-minus polarity shown. Also, currents are considered positive when they enter the ports as shown. As indicated earlier, the Kirchhoff equations for the hybrid model are

$$v_1 = h_{11}i_1 + h_{12}v_2 \tag{9-1}$$
$$i_2 = h_{21}i_1 + h_{22}v_2 \tag{9-2}$$

INPUT IMPEDANCE h_{11}

To discover the meaning of h_{11} and h_{21}, we proceed as follows. Suppose there is an ac short across the output terminals. Then $v_2 = 0$, and the hybrid equations reduce to

$$v_1 = h_{11}i_1 \tag{9-3}$$
$$i_2 = h_{21}i_1 \tag{9-4}$$

Solving the first equation, we get

$$h_{11} = \frac{v_1}{i_1} \qquad \text{(output shorted)} \tag{9-5}$$

Since it has the dimensions of volts divided by amperes, h_{11} is an impedance. For instance, if $v_1 = 35$ mV and $i_1 = 0.01$ mA,

$$h_{11} = \frac{35 \text{ mV}}{0.01 \text{ mA}} = 3.5 \text{ k}\Omega$$

So, h_{11} is the *input impedance* of a network when the output is shorted, as shown in Fig. 9-2b.

CURRENT GAIN h_{21}

Next, we can solve Eq. (9-4) to get

$$h_{21} = \frac{i_2}{i_1} \qquad \text{(output shorted)} \tag{9-6}$$

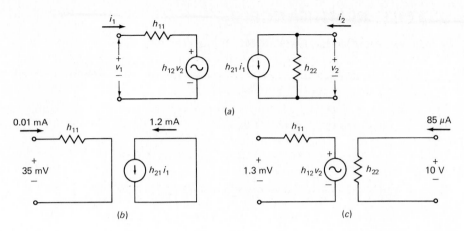

Fig. 9-2
(a) *Hybrid model.*
(b) *Shorted output.* (c) *Open input.*

Since it is the ratio of output current to input current, h_{21} is called the *current gain* with the output shorted. If $i_2 = 1.2$ mA and $i_1 = 0.01$ mA, then

$$h_{21} = \frac{1.2 \text{ mA}}{0.01 \text{ mA}} = 120$$

Therefore, h_{21} is the current gain of a network when the output is shorted (see Fig. 9-2*b*).

REVERSE VOLTAGE GAIN h_{12}

What is the meaning of h_{12} and h_{22}? If the input terminals are open, then $i_1 = 0$ and Eqs. (9-1) and (9-2) simplify to

$$v_1 = h_{12}v_2 \tag{9-7}$$
$$i_2 = h_{22}v_2 \tag{9-8}$$

Solving the first equation gives

$$h_{12} = \frac{v_1}{v_2} \qquad \text{(input open)} \tag{9-9}$$

Because it is the ratio of input voltage to output voltage, h_{12} is called the *reverse voltage gain* with the input open. In other words, if we deliberately drive the output port with a signal generator and measure the signal feeding back to the input side, we can calculate the reverse voltage gain with the input open.

For instance, suppose that the voltage driving the output port is 10 V and the voltage appearing at the input port is only 1.3 mV. Then,

$$h_{12} = \frac{1.3 \text{ mV}}{10 \text{ V}} = 1.3(10^{-4})$$

As you can see, the reverse voltage gain is very small. This means that the circuit does not work too well in the reverse direction. Remember, h_{12} is the reverse voltage gain with the input open, as shown in Fig. 9-2*c*.

OUTPUT ADMITTANCE h_{22}

Finally, we can solve Eq. (9-8) to get

$$h_{22} = \frac{i_2}{v_2} \quad \text{(input open)} \tag{9-10}$$

This is the ratio of output current to output voltage; therefore, h_{22} is the *output admittance* with the input open. As an example, if $i_2 = 85\ \mu\text{A}$ and $v_2 = 10\ \text{V}$, then

$$h_{22} = \frac{85\ \mu\text{A}}{10\ \text{V}} = 8.5\ \mu\text{S}$$

(Note: S stands for siemens or mhos, the reciprocal of ohms.) In Fig. 9-2c, h_{22} is an admittance.

MEASURING *H* PARAMETERS

Table 9-1 summarizes the meaning of each *h* parameter and the required condition. As you can see, h_{11} is the input impedance with the output shorted, h_{21} is the current gain with the output shorted, h_{12} is the reverse voltage gain with the input open, and h_{22} is the output admittance with the input open.

Because it is experimentally easy to put an ac short across the output of a transistor amplifier or to put an ac open at the input, manufacturers usually measure and specify the small-signal characteristics of a transistor with *h* parameters. For instance, the *h* parameters of a 2N3904 in a CE connection with 1 mA of collector current are as follows:

$$h_{11} = 3.5\ \text{k}\Omega$$
$$h_{12} = 1.3(10^{-4})$$
$$h_{21} = 120$$
$$h_{22} = 8.5\ \mu\text{S}$$

Incidentally, h_{22} equals the slope of the collector curves seen on a curve tracer. Pick any two points on the almost horizontal part of a collector curve; the ratio of the change in current to the change in voltage equals h_{22}. The more horizontal the collector curves are, the smaller the value of h_{22}, which is equivalent to saying that the collector current source has a higher impedance.

9-3 ANALYSIS FORMULAS

Figure 9-3a shows a voltage source v_S with an impedance r_S driving a two-port network whose output is connected to load resistance r_L. Impedance r_S is the

Table 9-1. *H* **Parameters**

Parameter	Meaning	Equation	Condition
h_{11}	Input impedance	v_1/i_1	Output shorted
h_{12}	Reverse voltage gain	v_1/v_2	Input open
h_{21}	Current gain	i_2/i_1	Output shorted
h_{22}	Output admittance	i_2/v_2	Input open

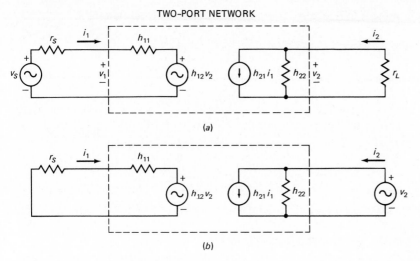

TWO-PORT NETWORK

(a)

(b)

Fig. 9-3 (a) *Two-port network with source and load resistances.* (b) *Driving the output side to find output impedance.*

Thevenin ac resistance driving the input terminals; load resistance r_L is the equivalent ac load resistance connected to the output terminals. If we know the h parameters of a transistor, we can calculate the exact current gain, voltage gain, input impedance, and output impedance. In the example of this section, we will use the h parameters of a 2N3904 in a CE connection with a quiescent current of 1 mA:

$$h_{11} = 3.5 \text{ k}\Omega$$
$$h_{12} = 1.3(10^{-4})$$
$$h_{21} = 120$$
$$h_{22} = 8.5 \ \mu\text{S}$$

CURRENT GAIN

In Fig. 9-3a,

$$A_i = \frac{i_2}{i_1} \qquad (9\text{-}11)$$

where A_i = current gain
$\quad i_2$ = ac output current
$\quad i_1$ = ac input current

Notice that i_1 and i_2 are the currents under load conditions; these current values are different from the values we got earlier with the output shorted. With Eq. (9-2), we can rewrite Eq. (9-11) as

$$A_i = \frac{h_{21}i_1 + h_{22}v_2}{i_1} = h_{21} + h_{22}\frac{v_2}{i_1}$$

Look at Fig. 9-3a and realize that

$$v_2 = -i_2 r_L$$

When this is substituted,

$$A_i = h_{21} - h_{22}\frac{i_2 r_L}{i_1} = h_{21} - A_i h_{22} r_L$$

Solving for A_i, we get

$$A_i = \frac{h_{21}}{1 + h_{22} r_L} \qquad (9\text{-}12)$$

For instance, if $h_{21} = 120$, $h_{22} = 8.5\ \mu S$, and $r_L = 3.6\ k\Omega$, then

$$A_i = \frac{120}{1 + 8.5(10^{-6})(3.6)(10^3)} = 116$$

The h values used in this example are the h parameters of a 2N3904 with 1 mA of dc collector current. As you can see, the current gain of a 2N3904 under load is slightly less than the current gain with a shorted output (116 versus 120).

What we are doing now is taking the output impedance of the current source into account. In previous chapters, we treated the collector current source as ideal. Now we are including the source impedance by using the output admittance h_{22}. The reciprocal of h_{22} is

$$\frac{1}{h_{22}} = \frac{1}{8.5\ \mu S} = 118\ k\Omega$$

This means that an output admittance of 8.5 μS is the same as an output impedance of 118 kΩ. This impedance shunts some of the current out of the current source. This is why the current gain is slightly less.

VOLTAGE GAIN

In Fig. 9-3a,

$$A_v = \frac{v_2}{v_1}$$

where A_v = voltage gain
   v_2 = ac output voltage
   v_1 = ac input voltage

With Eq. (9-1),

$$A_v = \frac{v_2}{h_{11}i_1 + h_{12}v_2} = \frac{-i_2 r_L}{h_{11}i_1 - h_{12}i_2 r_L}$$

Dividing the numerator and denominator by i_2 gives

$$A_v = \frac{-r_L}{h_{11}/A_i - h_{12}r_L}$$

Using Eq. (9-12), we can rearrange the foregoing to get

$$A_v = \frac{-h_{21}r_L}{h_{11} + (h_{11}h_{22} - h_{12}h_{21})r_L} \qquad (9\text{-}13)$$

For instance, with the *h* parameters of a 2N3904, given earlier, and a load resistance of 3.6 kΩ,

$$h_{11}h_{22} - h_{12}h_{21} = 3.5(10^3)8.5(10^{-6}) - 1.3(10^{-4})120$$
$$= 0.0142$$

and

$$A_v = \frac{-120(3600)}{3500 + (0.0142)(3600)} = -121.7$$

The minus sign indicates phase inversion.

INPUT IMPEDANCE

The input impedance of a loaded two-port network is

$$z_{in} = \frac{v_1}{i_1} = \frac{h_{11}i_1 + h_{12}v_2}{i_1} = h_{11} + \frac{h_{12}v_2}{i_1}$$

Since $v_2 = -i_2 r_L$,

$$z_{in} = h_{11} - \frac{h_{12}i_2 r_L}{i_1} = h_{11} - A_i h_{12} r_L$$

Using Eq. (9-12), we can rearrange the foregoing to get

$$z_{in} = h_{11} - \frac{h_{12}h_{21} r_L}{1 + h_{22} r_L} \tag{9-14}$$

For example, with the *h* parameters of the 2N3904, given earlier, and a load resistance of 3.6 kΩ,

$$z_{in} = 3.5 \text{ k}\Omega - \frac{1.3(10^{-4})(120)(3.6 \text{ k}\Omega)}{1 + 8.5(10^{-6})(3600)} = 3.45 \text{ k}\Omega$$

OUTPUT IMPEDANCE

To get the output impedance, reduce the source voltage to zero as shown in Fig. 9-3b. Then drive the output terminals with a signal of v_2. The ratio of v_2 to i_2 is the output impedance of the two-port network. The output impedance is

$$z_{out} = \frac{v_2}{i_2} = \frac{v_2}{h_{21}i_1 + h_{22}v_2} \tag{9-15}$$

On the input side, Ohm's law gives

$$i_1 = \frac{-h_{12}v_2}{r_S + h_{11}}$$

When this is substituted in Eq. (9-15) and the equation is rearranged, we get

$$z_{out} = \frac{r_S + h_{11}}{(r_S + h_{11})h_{22} - h_{12}h_{21}} \tag{9-16}$$

With the *h* parameters given earlier and a source resistance of 1 kΩ,

$$z_{out} = \frac{1 \text{ k}\Omega + 3.5 \text{ k}\Omega}{(4.5 \text{ k}\Omega)(8.5)(10^{-6}) - 1.3(10^{-4})(120)} = 199 \text{ k}\Omega$$

9-4 CE ANALYSIS

The *h* parameters of transistors can be listed as

$$h_i = h_{11}$$
$$h_r = h_{12}$$
$$h_f = h_{21}$$
$$h_o = h_{22}$$

where h_i = input impedance with output shorted
  h_r = reverse voltage gain with input open
  h_f = forward current gain with output shorted
  h_o = output admittance with input open

To remember this, notice that the subscript is the first letter of the description:

$$i = \text{input}$$
$$r = \text{reverse}$$
$$f = \text{forward}$$
$$o = \text{output}$$

The *h* parameters of a transistor depend on which connection is used: CE, CC, or CB. Because of this, the letter *e* is included for CE connection, *c* for CC connection, and *b* for CB connection. Table 9-2 summarizes the notation commonly used for transistor *h* parameters. As you can see, CE parameters are h_{ie}, h_{re}, h_{fe}, and h_{oe}. For instance, the data sheet of a 2N3904 lists the following *h* parameters for a quiescent collector current of 1 mA:

$$h_{ie} = 3.5 \text{ k}\Omega$$
$$h_{re} = 1.3(10^{-4})$$
$$h_{fe} = 120$$
$$h_{oe} = 8.5 \text{ }\mu\text{S}$$

Table 9-2. Relations

General	CE	CC	CB
h_{11}	h_{ie}	h_{ic}	h_{ib}
h_{12}	h_{re}	h_{rc}	h_{rb}
h_{21}	h_{fe}	h_{fc}	h_{fb}
h_{22}	h_{oe}	h_{oc}	h_{ob}

FORMULAS

For a CE amplifier, the *h* formulas derived earlier are usually written as follows:

$$A_i = \frac{h_{fe}}{1 + h_{oe}r_L} \tag{9-17}$$

$$A_v = \frac{-h_{fe}r_L}{h_{ie} + (h_{ie}h_{oe} - h_{re}h_{fe})r_L} \tag{9-18}$$

$$z_{in} = h_{ie} - \frac{h_{re}h_{fe}r_L}{1 + h_{oe}r_L} \tag{9-19}$$

$$z_{out} = \frac{r_S + h_{ie}}{(r_S + h_{ie})h_{oe} - h_{re}h_{fe}} \tag{9-20}$$

VARIATION OF *H* PARAMETERS

The *h* parameters vary with the *Q* point. For instance, Fig. 9-4*a* shows the typical value of h_{ie} for a 2N3904. As indicated, h_{ie} decreases as the quiescent collector current increases. At 1 mA, h_{ie} is approximately 3.5 kΩ.

In Sec. 5-4, we discussed the Early effect. Recall the basic idea. As the collector voltage increases, the base current decreases. This means that the output circuit has an effect on the input circuit. The reverse voltage gain h_{re} is a measure of the Early effect. Figure 9-4*b* shows the variation in h_{re} for a 2N3904. Notice how the reverse voltage gain reaches a minimum at approximately 2 mA. Above and below this value of collector current, h_{re} increases, indicating more Early effect.

As mentioned in Chap. 7, h_{fe} is the same as *β*. Figure 9-4*c* shows the variation in h_{fe}. Notice how it increases when I_C varies from 0.1 to 10 mA. Although this is not shown, h_{fe} reaches a maximum at some higher current and then decreases.

Finally, Fig. 9-4*d* shows the variation of h_{oe} with quiescent collector current. As you can see, the output admittance h_{oe} increases at higher levels of current. This is equivalent to saying that the output impedance of the current source decreases when the current increases.

AC RESISTANCE OF THE EMITTER DIODE

We already know that $β = h_{fe}$. Furthermore, the parameter h_{ie} is the input impedance of a CE transistor with the output ac shorted. In our earlier analysis of a CE amplifier, we found that $z_{in(base)}$ was equal to $βr'_e$. Therefore,

$$h_{ie} \cong βr'_e = h_{fe}r'_e$$

or

$$r'_e \cong \frac{h_{ie}}{h_{fe}} \tag{9-21}$$

This equation is useful because it relates r'_e to the *h* parameters, quantities found on the data sheet. As an example, a 2N3904 has $h_{ie} = 3.5$ kΩ and $h_{fe} = 120$ at $I_C =$

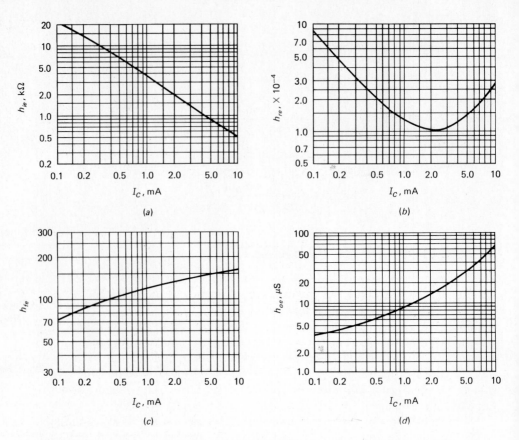

Fig. 9-4 *Parameter graphs.* (a) *Input impedance.* (b) *Reverse voltage gain.* (c) *Forward current gain.* (d) *Output admittance.*

1 mA. Therefore,

$$r'_e = \frac{3.5 \text{ k}\Omega}{120} = 29.2 \ \Omega$$

This is slightly larger than the ideal value of 25 Ω, the difference being due to the nonrectangular junction of a 2N3904.

Up to now, we have estimated r'_e using 25 mV/I_E. Equation (9-21) gives us a more accurate way to calculate r'_e using the h parameters listed on the data sheet of a small-signal transistor. This allows us to get more accurate answers using the simplified methods of earlier chapters. For instance, if $I_C = 1$ mA and $R_C = 5$ kΩ, the ideal voltage gain is

$$A_v = -\frac{5000 \ \Omega}{25 \ \Omega} = -200$$

We can improve the answer by using Eq. (9-21) to get an r'_e of 29.2 Ω. Now, the more

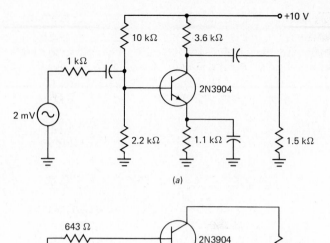

Fig. 9-5
(a) *CE amplifier.*
(b) *Ac equivalent circuit.*

accurate voltage gain is

$$A_v = -\frac{5000\ \Omega}{29.2\ \Omega} = -171$$

Using Eq. (9-18), we can calculate an exact voltage gain of

$$A_v = \frac{-120(5000)}{3500 + (0.0142)(5000)} = -168$$

EXAMPLE 9-1

Calculate the voltage gain, input impedance, and output impedance in Fig. 9-5a using h parameters.

SOLUTION

The dc emitter current is approximately 1 mA; therefore, the typical h parameters of a 2N3904 are

$$h_{ie} = 3.5\ \text{k}\Omega$$
$$h_{fe} = 120$$
$$h_{re} = 1.3(10^{-4})$$
$$h_{oe} = 8.5\ \mu\text{S}$$

These apply only to the transistor itself. For this reason, the standard approach in using the h-parameter method is to thevenize the circuit driving the base, as shown in Fig. 9-5b; notice that r_S equals 643 Ω. Likewise, the ac load resistance r_L is the parallel combination of 3.6 kΩ and 1.5 kΩ, which equals 1.06 kΩ.

Now, the transistor of Fig. 9-5b acts like a two-port network with the foregoing h parameters. Therefore, we can calculate the voltage gain using Eq. (9-18):

$$A_v = \frac{-120(1060)}{3500 + (0.0142)(1060)} = -36.2$$

From Eq. (9-19),

$$z_{in} = 3.5 \text{ k}\Omega - \frac{1.3(10^{-4})(120)(1.06 \text{ k}\Omega)}{1 + 8.5(10^{-6})(1060)} = 3.48 \text{ k}\Omega$$

From Eq. (9-20),

$$z_{out} = \frac{643 \ \Omega + 3500 \ \Omega}{(4143)(8.5)(10^{-6}) - 1.3(10^{-4})(120)} = 211 \text{ k}\Omega$$

9-5 CC ANALYSIS

To analyze an emitter follower, we need to use the h parameters of the CC connection: h_{ic}, h_{rc}, and h_{oc}. When these are substituted into the formulas derived earlier, we get the following:

$$A_i = \frac{h_{fc}}{1 + h_{oc}r_L} \tag{9-22}$$

$$A_v = \frac{-h_{fc}r_L}{h_{ic} + (h_{ic}h_{oc} - h_{rc}h_{fc})r_L} \tag{9-23}$$

$$z_{in} = h_{ic} - \frac{h_{rc}h_{fc}r_L}{1 + h_{oc}r_L} \tag{9-24}$$

$$z_{out} = \frac{r_S + h_{ic}}{(r_S + h_{ic})h_{oc} - h_{rc}h_{fc}} \tag{9-25}$$

With these formulas, we can calculate the exact voltage gain, input impedance, and output impedance of an emitter follower.

There is one slight problem. Data sheets usually list only the h parameters for the CE connection, because this is the most common configuration. Because of this, we have to use the following conversion formulas, which are derived in advanced literature:

$$h_{ic} = h_{ie} \tag{9-26}$$

$$h_{rc} = 1 - h_{re} \tag{9-27}$$

$$h_{fc} = -(1 + h_{fe}) \tag{9-28}$$

$$h_{oc} = h_{oe} \tag{9-29}$$

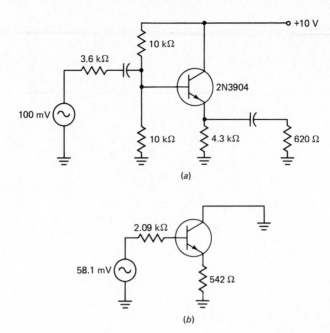

Fig. 9-6
(a) *CC amplifier.*
(b) *Ac equivalent
circuit.*

With these formulas we can work out the CC parameters of a transistor, given the CE parameters.

EXAMPLE 9-2

Calculate the exact current gain, voltage gain, input impedance, and output impedance for the emitter follower of Fig. 9-6a.

SOLUTION

The 2N3904 has approximately 1 mA of quiescent collector current; therefore, it has these typical CE parameters:

$$h_{ie} = 3.5 \text{ k}\Omega$$
$$h_{re} = 1.3(10^{-4})$$
$$h_{fe} = 120$$
$$h_{oe} = 8.5 \ \mu\text{S}$$

With Eqs. (9-26) through (9-29), we can convert to the CC parameters:

$$h_{ic} = 3.5 \text{ k}\Omega$$
$$h_{rc} = 1 - 1.3(10^{-4}) \cong 1$$
$$h_{fc} = -(1 + 120) = -121$$
$$h_{oc} = 8.5 \ \mu\text{S}$$

After thevenizing the base circuit and combining the output resistors, we can draw the ac equivalent circuit shown in Fig. 9-6b. Using Eqs. (9-22) through (9-25), we can calculate the following:

$$A_i = \frac{-121}{1 + 8.5(10^{-6})(542)} = -120$$

$$h_{ic}h_{oc} - h_{rc}h_{fc} = 3.5(10^3)(8.5)(10^{-6}) - (-121) = 121$$

$$A_v = \frac{(121)(542)}{3500 + (121)(542)} = 0.949$$

$$z_{in} = 3.5 \text{ k}\Omega - \frac{(-121)(542)}{1 + 8.5(10^{-6})(542)} = 68.8 \text{ k}\Omega$$

$$z_{out} = \frac{2.09 \text{ k}\Omega + 3.5 \text{ k}\Omega}{(5590)(8.5)(10^{-6}) + 121} = 46.2 \text{ }\Omega$$

9-6 CB ANALYSIS

When analyzing CB amplifiers, we need to use the CB parameters: h_{ib}, h_{rb}, h_{fb}, and h_{ob}. The basic formulas become

$$A_i = \frac{h_{fb}}{1 + h_{ob}r_L} \tag{9-30}$$

$$A_v = \frac{-h_{fb}r_L}{h_{ib} + (h_{ib}h_{ob} - h_{rb}h_{fb})r_L} \tag{9-31}$$

$$z_{in} = h_{ib} - \frac{h_{rb}h_{fb}r_L}{1 + h_{ob}r_L} \tag{9-32}$$

$$z_{out} = \frac{r_S + h_{ib}}{(r_S + h_{ib})h_{ob} - h_{rb}h_{fb}} \tag{9-33}$$

To convert from CE parameters to CB parameters, we need to use the following conversion formulas, derived in advanced literature:

$$D = (1 + h_{fe})(1 - h_{re}) + h_{ie}h_{oe} \tag{9-34}$$

$$h_{ib} = \frac{h_{ie}}{D} \tag{9-35}$$

$$h_{rb} = \frac{h_{ie}h_{oe} - h_{re}(1 + h_{fe})}{D} \tag{9-36}$$

$$h_{fb} = \frac{-h_{fe}(1 - h_{re}) - h_{ie}h_{oe}}{D} \tag{9-37}$$

$$h_{ob} = \frac{h_{oe}}{D} \tag{9-38}$$

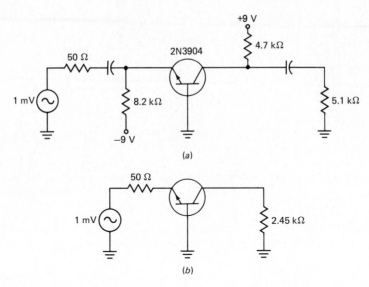

Fig. 9-7
(a) *CB amplifier.*
(b) *Ac equivalent circuit.*

EXAMPLE 9-3

Calculate A_i, A_v, z_{in} and z_{out} for the CB amplifier of Fig. 9-7a.

SOLUTION

The quiescent collector current is approximately 1 mA. Therefore, we can use the CE parameters given earlier. From Eq. (9-34),

$$D = (1 + 120)(1 - 0.00013) + 3500(8.5)10^{-6} = 121$$

From Eqs. (9-35) through (9-38),

$$h_{ib} = \frac{3500}{121} = 28.9 \ \Omega$$

$$h_{rb} = \frac{3500(8.5)(10^{-6}) - 1.3(10^{-4})(1 + 120)}{121} = 1.16(10^{-4})$$

$$h_{fb} = \frac{-120(1 - 0.00013) - 3500(8.5)(10^{-6})}{121} = -0.992$$

$$h_{ob} = \frac{8.5 \ \mu S}{121} = 0.0702 \ \mu S$$

By thevenizing the input side and combining the load resistors, we get the ac equivalent circuit of Fig. 9-7b. Now, using Eqs. (9-30) through (9-33),

$$A_i = \frac{-0.992}{1 + 7.02(10^{-8})(2450)} = -0.992$$

$$h_{ib}h_{ob} - h_{rb}h_{fb} = 28.9(7.02)(10^{-8}) - 1.16(10^{-4})(-0.992)$$
$$= 1.17(10^{-4})$$

$$A_v = \frac{(0.992)(2450)}{28.9 + (1.17)(10^{-4})(2450)} = 83.3$$

$$z_{in} = 28.9 \ \Omega - \frac{1.16(10^{-4})(-0.992)(2450)}{1 + 7.02(10^{-8})(2450)} = 29.2 \ \Omega$$

$$z_{out} = \frac{50 \ \Omega + 28.9 \ \Omega}{(78.9)(7.02)(10^{-8}) - 1.16(10^{-4})(-0.992)} = 654 \ k\Omega$$

9-7 PRACTICAL OBSERVATIONS

Look at how much mathematics is involved when you use the h parameters. The approach is time-consuming and prone to calculation error. For this reason, most people do not use h parameters for everyday troubleshooting and design. Unless you have a computer, the h parameters take too long to work with.

If you have a computer at your disposal, there is another problem. Data sheets do not always supply the h parameters of a transistor. If they supply parameter graphs like Fig. 9-4a through d, you must remember that these are only typical h parameters, values that are somewhere between the minimum and maximum parameters of the transistor type. For instance, besides the typical values given earlier, a 2N3904 has these minimum and maximum values listed on its data sheet for a quiescent collector current of 1 mA:

Parameter	Min	Max
h_{ie}	1 kΩ	10 kΩ
h_{re}	0.5(10^{-4})	8(10^{-4})
h_{fe}	100	400
h_{oe}	1 μS	40 μS

Look at how large these parameter spreads are. The data sheet does not guarantee that all parameters reach minimum or maximum at the same time. For this reason, it's impossible to know the exact h parameters of a transistor used in mass production, or even the worst-case parameters. In other words, we have exact formulas, but they are useless without an exact set of h parameters. For mass production, many designers use the typical h parameters listed on the data sheet and include some form of negative feedback to stabilize the transistor amplifier. (A later chapter discusses negative feedback.)

PROBLEMS

Straightforward

9-1. A 2N4401 has these h parameters for a quiescent collector current of 2 mA:

$$h_{ie} = 3 \ k\Omega$$
$$h_{re} = 0.45(10^{-4})$$
$$h_{fe} = 225$$
$$h_{oe} = 23 \ \mu S$$

If $r_S = 600 \ \Omega$ and $r_L = 2 \ k\Omega$, calculate A_i, A_v, z_{in}, and z_{out} for a CE stage.

9-2. A 2N3904 has a quiescent collector current of 5 mA. If it is used in a CE amplifier with $r_S = 430\ \Omega$ and $r_L = 1\ k\Omega$, what are the values of A_i, A_v, z_{in}, and z_{out}?

9-3. Use the parameters graph of Fig. 9-4 to work out the value of r'_e for the following collector currents: 0.1 mA, 0.2 mA, 0.5 mA, 1 mA, 2 mA, 5 mA, and 10 mA.

9-4. What are the A_i and A_v for the second stage of Fig. 9-8?

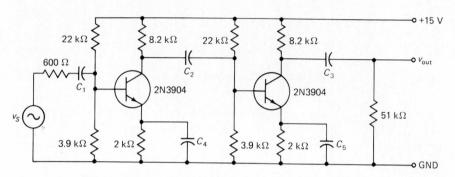

Fig. 9-8

9-5. In Fig. 9-8, what are the A_i and A_v of the first stage?

9-6. What is the z_{in} of the second transistor in Fig. 9-8? The z_{out} of the transistor?

9-7. In Fig. 9-9, calculate the A_i and A_v for the second stage.

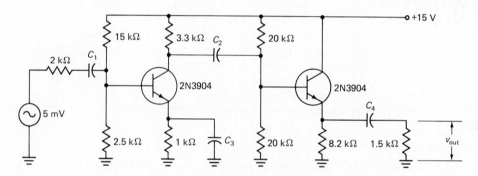

Fig. 9-9

9-8. What are the z_{in} and z_{out} of the second transistor in Fig. 9-9?

9-9. What are the CC parameters of a 2N3904 when the quiescent collector current is 2 mA?

9-10. In Fig. 9-10, what are the A_i and A_v of the CB amplifier?

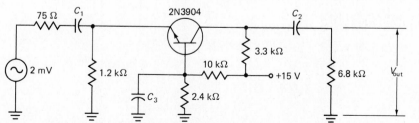

Fig. 9-10

9-11. What are the CB parameters of a 2N3904 when the quiescent collector current is 5 mA?

Troubleshooting

9-12. All three terminals of the second transistor are shorted in Fig. 9-8. What is the dc collector voltage to ground? The dc base voltage? Is there any ac signal at the final load?

9-13. There is no ac signal at all across the final load resistor of Fig. 9-8. Which of the following is a possible cause:

    a. C_1 shorted

    b. C_4 open

    c. C_2 open

    d. C_5 open

    e. C_3 shorted

9-14. In Fig. 9-10, the ac output voltage is zero. Which of the following is a possible trouble:

    a. C_1 open

    b. 1.2 kΩ shorted

    c. C_3 shorted

    d. 12 kΩ shorted

Design

9-15. Design a CE amplifier using the h parameters of a 2N3904 to meet these specifications: $V_{CC} = 15$ V, $I_C = 2$ mA, and $A_v = 100$.

9-16. Redesign the circuit of Fig. 9-5a to get an A_v of 50.

Challenging

9-17. Use h parameters to calculate the overall voltage gain of Fig. 9-8.

9-18. In Fig. 9-8, calculate the overall current gain of the two-stage amplifier.

9-19. Work out the CC parameters of a 2N3904 for each of these collector currents: 0.1 mA, 0.2 mA, 0.5 mA, 1 mA, 2 mA, 5 mA, and 10 mA.

9-20. A Darlington amplifier uses a pair of 2N3904s. If the first 2N3904 has an I_C of 0.1 mA and the second has an I_C of 10 mA, what is the input impedance looking into the first base if the load resistance seen by the second emitter is 100 Ω?

Computer

9-21. The IF . . . THEN statement allows you to make conditional jumps. For instance, here is part of a program:

```
10 PRINT "ENTER HIE": INPUT HIE
20 PRINT "ENTER HRE": INPUT HRE
30 PRINT "ENTER HFE": INPUT HFE
40 PRINT "ENTER HOE": INPUT HOE
50 PRINT "1. CE ANALYSIS"
60 PRINT "2. CC ANALYSIS"
70 PRINT "3. CB ANALYSIS"
80 PRINT "ENTER CHOICE"
90 INPUT C
100 IF C = 2 THEN GOTO 1000
110 IF C = 3 THEN GOTO 2000
120 STOP
```

After you key in the CE parameters, the computer prints a menu on the screen that allows you to choose the type of analysis. When the computer reaches line 100, it will branch to line 1000 if and only if C equals 2. If C does not equal 2, the program falls through to line 110. This time, the computer will branch to line 2000 if $C = 3$; otherwise, it falls through to line 120.

Do you see the point? The IF . . . THEN statement sets up a conditional jump. If certain conditions are satisfied, the jump takes place; otherwise, the program falls through to the next line.

Answer the following questions about the foregoing program:
a. If you choose CE analysis, does a jump occur?
b. If you choose CC analysis, does the computer jump to line 1000?
c. If you choose CB analysis, does the computer jump to line 1000?

9-22. Write a program similar to the foregoing program that does the following: Starting at line 1000 is a program that converts CE parameters to CC parameters. This program ends with a GOTO 110 statement. Also, starting at line 2000 is another program that converts CE parameters to CB parameters; this program ends with a GOTO 120 statement.

9-23. Write a program that calculates the A_i, A_v, z_{in}, and z_{out} of a CE amplifier.

10

Class A and B Power Amplifiers

After several stages of voltage gain, the signal swing uses up the entire load line. Any further gain has to be current gain. Stated another way, in the later stages of a multistage amplifier, the emphasis changes from voltage gain to power gain. In these later stages, the collector currents are much larger because the load resistances are smaller. In a typical AM radio, for instance, the final load resistance is 3.2 Ω, the impedance of the loudspeaker. Therefore, the final stage has to produce enough current to drive this low impedance.

As mentioned in Chap. 5, small-signal transistors have a power rating of less than half a watt and power transistors have a power rating of more than half a watt. Small-signal transistors are typically used near the front end of systems because the signal power is low, and power transistors are used near the end of systems because the signal power is high.

This chapter discusses ac load lines, ac output compliance, classes of operation, and other topics related to power amplifiers.

10-1 THE AC LOAD LINE OF A CE AMPLIFIER

Every amplifier sees two loads: a dc load and an ac load. Because of this, every amplifier has two load lines: a dc load line and an ac load line. In earlier chapters, we used the dc load line to analyze biasing circuits. In this chapter, we will use the ac load line to analyze large-signal operation.

DC AND AC LOAD LINES

The CE amplifier of Fig. 10-1a has the dc equivalent circuit shown in Fig. 10-1b. With this dc equivalent circuit we derived the dc load line of Fig. 10-1c. Recall that the dc saturation current is $V_{CC}/(R_C + R_E)$ and the dc cutoff voltage is V_{CC}.

When a signal drives the transistor of Fig. 10-1a, the capacitors appear like ac shorts. This is why the source and load resistances seen by the transistor are different.

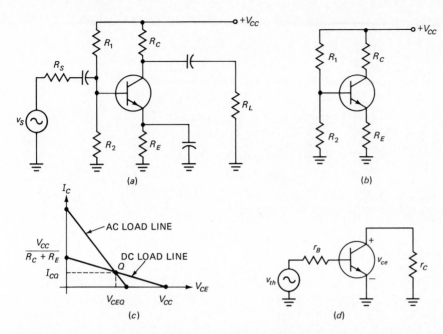

Fig. 10-1 (a) *CE amplifier.* (b) *Dc equivalent circuit.* (c) *Dc and ac load lines.* (d) *Ac equivalent circuit.*

In other words, the ac Thevenin resistance driving the base is

$$r_B = R_S \parallel R_1 \parallel R_2$$

and the ac load resistance seen by the collector is

$$r_C = R_C \parallel R_L$$

Figure 10-1*d* shows the ac equivalent circuit. This circuit produces the ac load line of Fig. 10-1*c*. When no signal is present, the transistor operates at the *Q* point shown in Fig. 10-1*c*. When a signal is present, the operating point swings along the ac load line rather than the dc load line because the ac load resistance is different from the dc load resistance.

Incidentally, to keep the *Q* point distinct in the discussion that follows, we will label the quiescent collector current as I_{CQ} and the quiescent collector-emitter voltage as V_{CEQ} (see Fig. 10-1*c*).

AC SATURATION AND CUTOFF

As shown in Fig. 10-1*c*, the saturation and cutoff points on the ac load line are different from those on the dc load line. Here is how to get the intercepts of the ac load line. In Fig. 10-1*d*, we can sum ac voltages around the collector loop to get

$$v_{ce} + i_c r_C = 0$$

or

$$i_c = -\frac{v_{ce}}{r_C} \tag{10-1}$$

Fig. 10-2
*Ac load line for a
CE amplifier.*

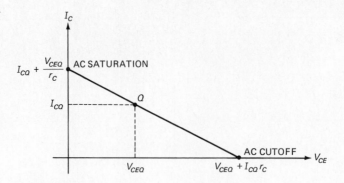

The ac collector current is given by

$$i_c = \Delta I_C = I_C - I_{CQ}$$

and the ac collector voltage is

$$v_{ce} = \Delta V_{CE} = V_{CE} - V_{CEQ}$$

Substituting these expressions into Eq. (10-1) and rearranging gives

$$I_C = I_{CQ} + \frac{V_{CEQ}}{r_C} - \frac{V_{CE}}{r_C} \qquad (10\text{-}2)$$

This is the equation of the ac load line. We can find the intercepts in the usual way. When the transistor goes into saturation, V_{CE} is zero, and Eq. (10-2) gives

$$I_{C(\text{sat})} = I_{CQ} + \frac{V_{CEQ}}{r_C} \qquad \text{(upper end)} \qquad (10\text{-}3)$$

where $I_{C(\text{sat})}$ = ac saturation current
$\quad I_{CQ}$ = dc collector current
$\quad V_{CEQ}$ = dc collector-emitter voltage
$\quad r_C$ = ac resistance seen by collector

When the transistor goes into cutoff, I_C equals zero, and we get an ac cutoff voltage of

$$V_{CE(\text{cut})} = V_{CEQ} + I_{CQ}r_C \qquad \text{(lower end)} \qquad (10\text{-}4)$$

Figure 10-2 shows the ac load line with its saturation current and cutoff voltage. This is called the *ac load line* because it represents all possible ac operating points. At any instant during the ac cycle, the operating point of the transistor is somewhere along the ac load line; the exact point is determined by the amount of change from the Q point.

AC OUTPUT COMPLIANCE

The ac load line is a visual aid for understanding large-signal operation. During the positive half cycle of ac source voltage, the collector voltage swings from the Q point toward saturation. On the negative half cycle, the collector voltage swings from the Q point toward cutoff. For a large enough ac signal, clipping can occur on either or both signal peaks.

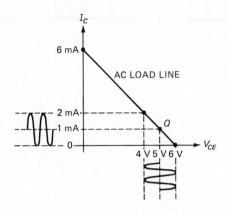

Fig. 10-3
Ac output compliance.

   The *ac output compliance* is the maximum unclipped peak-to-peak ac voltage that an amplifier can produce. For instance, in Fig. 10-3 the ac output compliance is 2 V. If we try to get more than 2 V peak to peak, the output signal will be clipped.

   Once we know the ac output compliance of an amplifier, we know its large-signal limit. From now on, we will symbolize the ac output compliance of an amplifier as *PP*, a reminder that it is the maximum unclipped peak-to-peak voltage that an amplifier can produce. In Fig. 10-3, the amplifier has a PP of 2 V.

   Since the ac cutoff voltage is $V_{CEQ} + I_{CQ}r_c$, the maximum positive swing from the Q point is

$$V_{CEQ} + I_{CQ}r_c - V_{CEQ} = I_{CQ}r_c$$

Since the ac saturation voltage is ideally zero, the maximum negative swing from the Q point is

$$0 - V_{CEQ} = -V_{CEQ}$$

The ac output compliance of a CE amplifier, therefore, is given by the smaller of these two approximate values:

$$PP \cong 2I_{CQ}r_c \qquad (10\text{-}5)$$

or

$$PP \cong 2V_{CEQ} \qquad (10\text{-}6)$$

EXAMPLE 10-1

The 2N3904 of Fig. 10-4*a* has the following maximum ratings: $I_C = 200$ mA and $V_{CEO} = 40$ V. Calculate the ac output compliance. Also, show that neither of the transistor ratings is exceeded during the ac cycle.

SOLUTION

The voltage divider sets up a dc base voltage of

$$V_B = 1.8 \text{ V}$$

Fig. 10-4

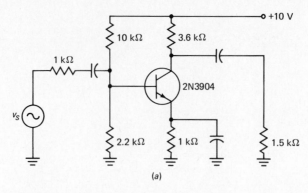

(a)

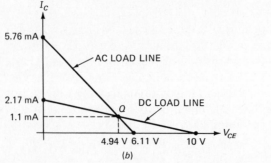

(b)

Therefore, the dc emitter current is

$$I_E = \frac{1.1 \text{ V}}{1 \text{ k}\Omega} = 1.1 \text{ mA}$$

Since collector current approximately equals emitter current, the quiescent collector current is

$$I_{CQ} \cong 1.1 \text{ mA}$$

This current produces a dc collector-emitter voltage of

$$V_{CE} = 10 \text{ V} - (1.1 \text{ mA})(4600 \text{ }\Omega) = 4.94 \text{ V}$$

or

$$V_{CEQ} = 4.94 \text{ V}$$

The dc saturation current is

$$\frac{V_{CC}}{R_C + R_E} = \frac{10 \text{ V}}{4.6 \text{ k}\Omega} = 2.17 \text{ mA}$$

and the dc cutoff voltage is 10 V. Figure 10-4*b* shows the dc load line.

Next, visualize the ac equivalent circuit. You can see that the ac load resistance is

$$r_C = 3.6 \text{ k}\Omega \parallel 1.5 \text{ k}\Omega = 1.06 \text{ k}\Omega$$

With Eq. (10-3), calculate the ac saturation current:

$$I_{C(\text{sat})} = 1.1 \text{ mA} + \frac{4.94 \text{ V}}{1.06 \text{ k}\Omega} = 5.76 \text{ mA}$$

With Eq. (10-4), the ac cutoff voltage is

$$V_{CE(\text{cut})} = 4.94 \text{ V} + (1.1 \text{ mA})(1.06 \text{ k}\Omega) = 6.11 \text{ V}$$

These intercepts are shown on the ac load line of Fig. 10-4b.

Now we have the whole story. A glance at Fig. 10-4b tells what the Q point is, the dc saturation and cutoff values, the ac saturation and cutoff values, and the ac output compliance. The maximum positive voltage swing is

$$I_{CQ}r_C = (1.1 \text{ mA})(1.06 \text{ k}\Omega) = 1.17 \text{ V}$$

This is smaller than the maximum negative swing of

$$-V_{CEQ} = -4.94 \text{ V}$$

Therefore, the ac output compliance is

$$PP = 2(1.17 \text{ V}) = 2.34 \text{ V}$$

This is the maximum unclipped peak-to-peak voltage that the CE amplifier of Fig. 10-4a can produce.

Now for the transistor maximum ratings. The largest collector current at any point in the ac cycle is less than 5.76 mA, the ac saturation current in Fig. 10-4b. Therefore, we aren't even close to the 2N3904's maximum current rating of 200 mA. Also, the largest collector-emitter voltage at any time during the ac cycle is less than 6.11 V, the ac cutoff voltage (see Fig. 10-4b); this is well below the breakdown voltage of 40 V.

10-2 AC LOAD LINES OF OTHER AMPLIFIERS

The emitter follower, CB amplifier, and swamped amplifier have their own ac load lines. This section examines each because the ac load line is the key to calculating the ac output compliance.

EMITTER FOLLOWER

Figure 10-5a shows an emitter follower. Since the ac output signal comes out of the emitter, the effective ac load resistance is

$$r_E = R_E \parallel R_L$$

This is the ac resistance that actually loads the emitter follower under signal conditions.

Fig. 10-5
(a) *Emitter
follower.* (b) *Ac
load line.*

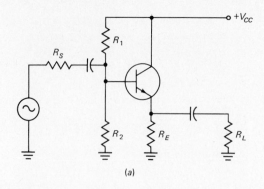

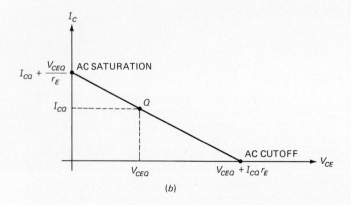

By a derivation almost identical to that given for the CE amplifier, we can prove that the ac saturation current is

$$I_{C(\text{sat})} = I_{CQ} + \frac{V_{CEQ}}{r_E} \tag{10-7}$$

and the ac cutoff voltage is

$$V_{CE(\text{cut})} = V_{CEQ} + I_{CQ} r_E \tag{10-8}$$

Figure 10-5*b* shows the ac load line of an emitter follower. Notice that the formulas for ac saturation current and cutoff voltage are identical to those given previously, except that r_E is used in the place of r_C. This makes sense because the ac load resistance now is r_E instead of r_C. Equations (10-7) and (10-8) let you know if you are exceeding the maximum current rating or breakdown voltage of the transistor.

The ac output compliance of an emitter follower is the smaller of

$$PP \cong 2I_{CQ} r_E \tag{10-9}$$

or

$$PP \cong 2V_{CEQ} \tag{10-10}$$

You will find these formulas useful when you need to know the maximum unclipped peak-to-peak output signal that an emitter follower can produce.

CB AMPLIFIER

The CB amplifier has an ac load resistance of

$$r_C = R_C \parallel R_L$$

The ac load line of a CB amplifier is approximately the same as that for a CE amplifier. For this reason, we can use Fig. 10-2 when analyzing the large-signal operation of a CB amplifier. Also, the ac output compliance is approximately the same as that of a CE amplifier.

SWAMPED AMPLIFIER

In the swamped amplifier of Fig. 10-6a, the ac load resistance seen by the transistor is $r_C + r_E$. The derivation for the ac load line is almost identical to that given earlier. This is why the ac saturation current is

$$I_{C(\text{sat})} = I_{CQ} + \frac{V_{CEQ}}{r_C + r_E} \qquad (10\text{-}11)$$

and the ac cutoff voltage is

$$V_{CE(\text{cut})} = V_{CEQ} + I_{CQ}(r_C + r_E) \qquad (10\text{-}12)$$

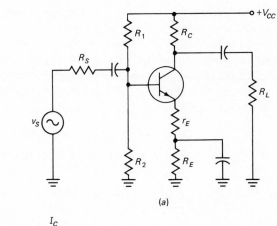

Fig. 10-6
(a) *Swamped amplifier.* (b) *Ac load line.*

(a)

(b)

Figure 10-6*b* shows the ac load line. Instead of using only r_C, we now use $r_C + r_E$ because this is the effective ac load resistance seen by the transistor. With Eqs. (10-11) and (10-12), you check to see if the transistor ratings are exceeded during the ac cycle.

The ac output voltage appears across r_C; the ac feedback voltage across r_E is used only for swamping. Because the total ac voltage appears across $r_C + r_E$, the ac output voltage equals $r_C/(r_C + r_E)$ times the total ac voltage. Therefore, the ac output compliance of a swamped amplifier is the smaller of

$$PP \cong 2I_{CQ}r_C \tag{10-13}$$

or

$$PP \cong 2V_{CEQ}\frac{r_C}{r_C + r_E} \tag{10-14}$$

MAXIMUM AC OUTPUT COMPLIANCE

In earlier chapters, we set the Q point near the middle of the dc load line. This was done to keep things simple. We can increase the ac output compliance if we set the Q higher than the center of the dc load line. Figure 10-7 illustrates why this is so. Q_1 is the Q point at the center of the dc load line.

Q_2 is a Q point further up the dc load line. As you can see, the higher Q point results in a larger unclipped ac output voltage. Therefore, if you are designing a large-signal amplifier and want to get maximum ac output compliance, locate the Q point above the center of the dc load line.

What you are trying to do in your design is to get equal voltage swing in both directions, as shown in Fig. 10-8. This allows maximum swing along the ac load line for each half cycle and produces maximum ac output compliance. To get equal voltage swing in either direction, you must satisfy the following relations for each of the basic amplifiers:

$$I_{CQ}r_C = V_{CEQ} \quad \text{(CE stage)} \tag{10-15}$$

$$I_{CQ}r_E = V_{CEQ} \quad \text{(CC stage)} \tag{10-16}$$

Fig. 10-7
Increasing the ac output compliance.

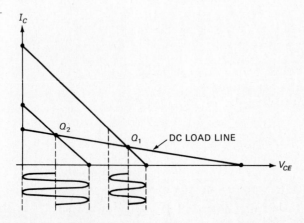

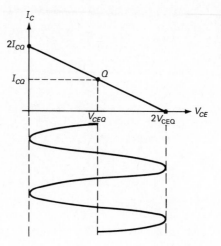

Fig. 10-8
Optimum Q point for maximum output.

$$I_{CQ}r_C = V_{CEQ}\frac{r_C}{r_C + r_E} \qquad \text{(swamped)} \qquad (10\text{-}17)$$

Most designers use a *trial-and-error* approach here. Try a collector current, see if the equation is approximately satisfied, then try again until the answer is close enough. With trial and error (also known as successive approximations), you can sneak up on the optimum Q point. (Other design alternatives include a graphical solution and a computer solution.)

EXAMPLE 10-2

A swamped amplifier has $I_{CQ} = 5$ mA, $V_{CEQ} = 10$ V, $R_C = 1$ kΩ, $R_L = 3$ kΩ, and $r_E = 120$ Ω. Calculate the ac saturation current, the ac cutoff voltage, and the ac output compliance.

SOLUTION

The ac resistance seen by the collector is

$$r_C = 1 \text{ k}\Omega \parallel 3 \text{ k}\Omega = 750 \text{ }\Omega$$

The total ac resistance seen by the transistor is

$$r_C + r_E = 750 \text{ }\Omega + 120 \text{ }\Omega = 870 \text{ }\Omega$$

Since $I_{CQ} = 5$ mA and $V_{CEQ} = 10$ V, the ac saturation current is

$$I_{C(\text{sat})} = 5 \text{ mA} + \frac{10 \text{ V}}{870 \text{ }\Omega} = 16.5 \text{ mA}$$

and the ac cutoff voltage is

$$V_{CE(\text{cut})} = 10 \text{ V} + (5 \text{ mA})(870 \text{ }\Omega) = 14.4 \text{ V}$$

The ac output compliance is the smaller of

$$PP = 2(5 \text{ mA})(750 \text{ }\Omega) = 7.5 \text{ V}$$

or

$$PP = 2(10 \text{ V}) \frac{750}{870} = 17.2 \text{ V}$$

Therefore, the ac output compliance is the smaller of the two, 7.5 V peak to peak.

EXAMPLE 10-3

Locate the optimum Q point in Fig. 10-4a to get maximum ac output compliance.

SOLUTION

In Fig. 10-4b, we can see that the optimum Q point lies somewhere between 1.1 mA and 2.17 mA. Pick a current halfway between, approximately

$$I_{CQ} = 1.64 \text{ mA}$$

The maximum positive swing is

$$I_{CQ}r_C = (1.64 \text{ mA})(1.06 \text{ k}\Omega) = 1.74 \text{ V}$$

The maximum negative swing is

$$V_{CEQ} = 10 \text{ V} - (1.64 \text{ mA})(4.6 \text{ k}\Omega) = 2.46 \text{ V}$$

The positive swing is still smaller than the negative swing; therefore, try a higher Q point.

For the second trial, choose I_{CQ} halfway between 1.64 mA and 2.17 mA, or

$$I_{CQ} = 1.91 \text{ mA}$$

The maximum positive swing is

$$I_{CQ}r_C = (1.91 \text{ mA})(1.06 \text{ k}\Omega) = 2.02 \text{ V}$$

The maximum negative swing is

$$V_{CEQ} = 10 \text{ V} - (1.91 \text{ mA})(4.6 \text{ k}\Omega) = 1.21 \text{ V}$$

The positive swing is now larger than the negative swing, so the Q point must be lower down the dc load line.

For the third trial, choose an I_{CQ} halfway between 1.91 mA and 1.64 mA, or

$$I_{CQ} = 1.78 \text{ mA}$$

which gives a maximum positive swing of

$$I_{CQ}r_C = (1.78 \text{ mA})(1.06 \text{ k}\Omega) = 1.89 \text{ V}$$

and a maximum negative swing of

$$V_{CEQ} = 10 \text{ V} - (1.78 \text{ mA})(4.6 \text{ k}\Omega) = 1.81 \text{ V}$$

Now the positive and negative swings are almost equal. This means that the optimum Q point has a dc collector current of approximately 1.78 mA and a collector-emitter voltage of 1.81 V. With this Q point, the amplifier has a maximum ac output compliance of approximately

$$PP = 2(1.81 \text{ V}) = 3.62 \text{ V}$$

Originally, the amplifier had a PP of 2.34 V, so there is a distinct improvement.

Finding the optimum Q point is not always necessary. In small-signal amplifiers, for instance, the center of the dc load line is just fine because only a small part of the ac load line is used. For convenience, we will continue to set the Q point at the center of the dc load line. Then we can check the ac output compliance to see if the peak-to-peak swing is large enough for our application. If not, then we can redesign the amplifier by using a smaller R_E to get maximum ac output compliance.

10-3 CLASS A OPERATION

Class A operation means that the transistor operates in the active region at all times. This implies that collector current flows for 360° of the ac cycle. In this section, we will discuss some properties of a class A amplifier needed for troubleshooting and design.

LOADED VOLTAGE GAIN

In the CE amplifier of Fig. 10-9*a*, an ac voltage v_{in} drives the base, producing an ac output voltage v_{out}. The unloaded voltage gain is

$$A = -\frac{R_C}{r'_e}$$

In earlier chapters, we thevenized the output circuit to find the ac voltage across R_L.
 But there is another approach we can use here. Since the ac resistance seen by the collector is

$$r_C = R_C \parallel R_L$$

we can calculate the *loaded voltage gain* directly by using

$$A_v = -\frac{r_C}{r'_e} \tag{10-18}$$

This alternative formula for voltage gain allows us to calculate the effects of R_L without thevenizing the output circuit. For instance, if $R_C = 10 \text{ k}\Omega$, $R_L = 30 \text{ k}\Omega$,

Fig. 10-9
(a) *CE amplifier.*
(b) *Load power.*
(c) *Transistor power dissipation.*

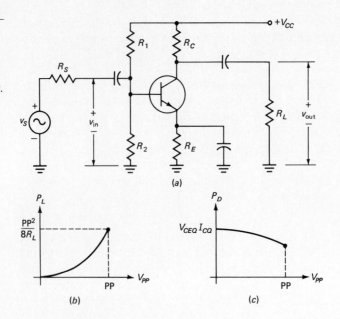

(a)

(b)

(c)

and $r'_e = 50\ \Omega$, then

$$A_v = -\frac{10\ \text{k}\Omega\ \|\ 30\ \text{k}\Omega}{50\ \Omega} = -150$$

CURRENT GAIN

In Fig. 10-9a, the *current gain* of the transistor is the ratio of ac collector current to ac base current. In symbols,

$$A_i = \frac{i_c}{i_b} \tag{10-19}$$

where A_i = current gain
i_c = ac collector current
i_b = ac base current

As discussed in Chap. 9, A_i depends on the output impedance of the collector current source and the load resistance; however, in most circuits, you can use the following approximation with negligible error:

$$A_i \cong \beta \tag{10-20}$$

POWER GAIN

In Fig. 10-9a, the ac input power to the base is

$$p_{\text{in}} = v_{\text{in}} i_b$$

The ac output power from the collector is

$$p_{\text{out}} = -v_{\text{out}} i_c$$

where the minus sign is needed because of the phase inversion. The ratio $p_{\text{out}}/p_{\text{in}}$ is called the *power gain* and is symbolized by A_p. Taking the ratio of p_{out} to p_{in}, we get

$$A_p = \frac{p_{\text{out}}}{p_{\text{in}}} = -\frac{v_{\text{out}} i_c}{v_{\text{in}} i_b}$$

Since $A_v = v_{\text{out}}/v_{\text{in}}$ and $A_i = i_c/i_b$,

$$A_p = -A_v A_i \qquad\qquad (10\text{-}21)$$

where A_p = power gain
A_v = voltage gain
A_i = current gain

The equation makes sense. It says that power gain equals the negative product of voltage gain and current gain.

For instance, if a CE amplifier has $r_C = 7500\ \Omega$, $r'_e = 50\ \Omega$, and $\beta = 125$, then the voltage gain is

$$A_v = -\frac{7500\ \Omega}{50\ \Omega} = -150$$

The current gain is

$$A_i = 125$$

The power gain is

$$A_p = -(-150)(125) = 18{,}750$$

This means that an ac input power of $1\ \mu\text{W}$ results in an ac output power of $18{,}750\ \mu\text{W}$ or $18.75\ \text{mW}$.

LOAD POWER

The load on an amplifier may be a loudspeaker, a motor, or some other device. It is important to know how much ac power reaches the load resistor. In Fig. 10-9a, the ac power into the load resistor R_L is

$$P_L = \frac{V_L^2}{R_L} \qquad\qquad (10\text{-}22)$$

where P_L = ac load power
V_L = rms load voltage
R_L = load resistance

This is a convenient equation to use when you measure the ac load voltage with a voltmeter because the typical voltmeter is calibrated in rms values.

Often you look at the ac output voltage with an oscilloscope. In this case, it is convenient to have a formula that uses peak-to-peak voltage instead of rms voltage. Since

$$V_L = 0.707 V_P$$

and

$$V_P = \frac{V_{PP}}{2}$$

we can write

$$V_L = 0.707 V_P = \frac{0.707 V_{PP}}{2}$$

Substitute this into Eq. (10-22) and you get

$$P_L = \frac{V_{PP}^2}{8R_L} \tag{10-23}$$

You will find this useful when you measure the peak-to-peak voltage with an oscilloscope.

MAXIMUM AC LOAD POWER

What is the maximum ac load power you can get from a CE amplifier operated class A? The ac output compliance PP equals the maximum unclipped output voltage. Therefore, we can rewrite Eq. (10-23) as

$$P_{L(\text{max})} = \frac{\text{PP}^2}{8R_L} \tag{10-24}$$

This is the maximum ac load power that a class A amplifier can produce without clipping.

Figure 10-9*b* shows how load power varies with the peak-to-peak load voltage. This is a parabolic curve because power is directly proportional to the square of voltage. As you can see, the maximum load power occurs when the peak-to-peak load voltage equals the ac output compliance.

TRANSISTOR POWER DISSIPATION

When no signal drives an amplifier, the *power dissipation* of the transistor equals the product of dc voltage and current:

$$P_{DQ} = V_{CEQ} I_{CQ} \tag{10-25}$$

where P_{DQ} = quiescent power dissipation
V_{CEQ} = quiescent collector-emitter voltage
I_{CQ} = quiescent collector current

This power dissipation must not exceed the power rating of the transistor. If it does, you run the risk of damaging the transistor. For instance, if $V_{CEQ} = 10$ V and $I_{CQ} = 5$ mA, then

$$P_{DQ} = (10 \text{ V})(5 \text{ mA}) = 50 \text{ mW}$$

A 2N3904 has a power rating of 310 mW for an ambient temperature of 25°C.

Therefore, a 2N3904 would have no problem dissipating 50 mW of quiescent power when the ambient temperature is 25°C.

Figure 10-9c shows how transistor power dissipation varies with the peak-to-peak load voltage. P_D is maximum when there is no input signal. It decreases when the peak-to-peak load voltage increases. In the worst case, the transistor must have a power rating that is greater than P_{DQ}, the quiescent dissipation. In symbols,

$$P_{D(max)} = P_{DQ} \qquad (10\text{-}26)$$

Therefore, a designer must make sure that P_{DQ} is less than the power rating of the transistor being used, because P_{DQ} represents the worst case.

Equation (10-26) is true only for class A operation. That is, it is only in class A operation that the worst-case dissipation of the transistor occurs under no-signal conditions. For the other classes of operation studied later, more transistor power dissipation occurs when the signal is present.

CURRENT DRAIN

Whoever designs the power supply has to know how much current is required by different stages. In an amplifier like Fig. 10-9a, the dc voltage source V_{CC} must supply direct current to the voltage divider and the collector circuit. Assuming a stiff voltage divider, the voltage divider produces a *dc current drain* of

$$I_1 = \frac{V_{CC}}{R_1 + R_2} \qquad (10\text{-}27)$$

In the collector circuit, the dc current drain is

$$I_2 = I_{CQ} \qquad (10\text{-}28)$$

In a class A amplifier, the sinusoidal variations in collector current average to zero. Therefore, whether the ac signal is present or not, the dc source must supply an average current of

$$I_S = I_1 + I_2 \qquad (10\text{-}29)$$

This is the total dc current drain. The dc source voltage multiplied by the dc current drain gives the total dc power supplied to an amplifier:

$$P_S = V_{CC}I_S \qquad (10\text{-}30)$$

STAGE EFFICIENCY

Sometimes we want to compare the efficiency of one design with that of another. In this case, it helps to talk about the *stage efficiency,* given by

$$\eta = \frac{P_{L(max)}}{P_S} \times 100\% \qquad (10\text{-}31)$$

where η = stage efficiency
$P_{L(max)}$ = maximum ac load power
P_S = dc input power

Table 10-1. Class A Formulas

Quantity	Formula	Comment
$I_{C(\text{sat})}$	$I_{CQ} + V_{CEQ}/(r_C + r_E)$	Applies to all stages
$V_{CE(\text{cut})}$	$V_{CEQ} + I_{CQ}(r_C + r_E)$	Applies to all stages
PP	$2I_{CQ}r_C$ or $2V_{CEQ}$	Use the smaller. Applies to CE and CB stages
PP	$2I_{CQ}r_E$ or $2V_{CEQ}$	Use the smaller. Applies to CC stages
PP	Eqs. (10–13) and (10–14)	Swamped amplifier
P_L	$V_L{}^2/R_L$	Voltage in rms volts
P_L	$V_{PP}{}^2/8R_L$	Voltage in peak-to-peak volts
$P_{L(\text{max})}$	$PP^2/8R_L$	Maximum undistorted output power
P_{DQ}	$V_{CEQ}I_{CQ}$	Maximum transistor power dissipation
P_S	$V_{CC}I_S$	Supply power
η	$P_{L(\text{max})}/P_S$	Stage efficiency; multiply by 100%

As an example, if $P_{L(\text{max})} = 50$ mW and $P_S = 400$ mW, the efficiency of the stage is

$$\eta = \frac{50 \text{ mW}}{400 \text{ mW}} \times 100\% = 12.5\%$$

This implies that 12.5 percent of the dc input power reaches the output in the form of ac load power.

CONCLUSION

Table 10-1 summarizes the more important formulas for class A operation. This will be helpful when you are troubleshooting and designing class A amplifiers. The first entry is the collector saturation current. Notice that this can be applied to all stages: CE, CC, CB, and swamped. For instance, in a CE stage, r_E is zero and the formula reduces to $I_{CQ} + V_{CEQ}/r_C$. Similarly, in a CC stage, r_C is zero and we get $I_{CQ} + V_{CEQ}/r_E$.

The ac output compliance is listed for CE, CC, and CB stages. The formula for the swamped amplifier is too complicated, and so you will have to refer to Eqs. (10-13) and (10-14). You should be able to understand the remaining entries. If you have any difficulty, refer to the original derivation and discussion.

EXAMPLE 10-4

Figure 10-10a is the CE amplifier analyzed earlier; Fig. 10-10b shows the dc and ac load lines. If $\beta = 150$, calculate A_v, A_i, A_p, $P_{L(\text{max})}$, I_S, P_S, and β.

SOLUTION

Earlier we calculated an r_C of 1.06 kΩ and an r'_e of 22.7 Ω. So, the loaded voltage gain is

$$A_v = -\frac{1060 \ \Omega}{22.7 \ \Omega} = -46.7$$

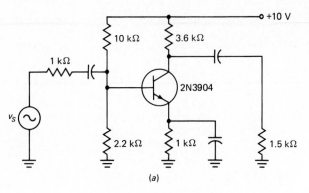

Fig. 10-10

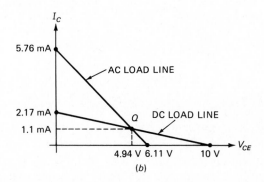

The current gain equals

$$A_i = 150$$

The power gain is

$$A_p = (46.7)(150) = 7005$$

In Fig. 10-10b, the ac output compliance is

$$PP = 2(6.11 \text{ V} - 4.94 \text{ V}) = 2.34 \text{ V}$$

This means that the largest unclipped peak-to-peak load voltage is 2.34 V. Therefore, the maximum load power is

$$P_{L(\text{max})} = \frac{(2.34 \text{ V})^2}{8(1500 \ \Omega)} = 456 \ \mu\text{W}$$

At the Q point of Fig. 10-10b, $V_{CEQ} = 4.94$ V and $I_{CQ} = 1.1$ mA, and so

$$P_{D(\text{max})} = (4.94 \text{ V})(1.1 \text{ mA}) = 5.43 \text{ mW}$$

The dc current drain through the biasing resistors of Fig. 10-10a is

$$I_1 = \frac{10 \text{ V}}{10 \text{ k}\Omega + 2.2 \text{ k}\Omega} = 0.82 \text{ mA}$$

and the dc current drain of the collector is

$$I_2 = 1.1 \text{ mA}$$

So, the total dc current drain is

$$I_S = 0.82 \text{ mA} + 1.1 \text{ mA} = 1.92 \text{ mA}$$

The dc input power is

$$P_S = (10 \text{ V})(1.92 \text{ mA}) = 19.2 \text{ mW}$$

Finally, the stage efficiency is

$$\eta = \frac{456 \ \mu\text{W}}{19.2 \ \text{mW}} \times 100\% = 2.38\%$$

10-4 CLASS B OPERATION

Class A is the common way to run a transistor in linear circuits because it leads to the simplest and most stable biasing circuits. But class A is not the most efficient way to operate a transistor. In some applications, like battery-powered systems, current drain and stage efficiency become important considerations in the design. For this reason, a number of other classes of operation have evolved.

Class B operation of a transistor means that collector current flows for only 180° of the ac cycle. This implies that the *Q* point is located approximately at cutoff on both the dc and ac load lines. The advantage of class B operation is lower transistor power dissipation and reduced current drain.

PUSH-PULL CIRCUIT

When a transistor operates class B, it clips off half a cycle. To avoid the resulting distortion, therefore, you have to use two transistors in a *push-pull* arrangement. This means that one transistor conducts during one half cycle, and the other transistor conducts during the other half cycle. With push-pull circuits, it is possible to build class B amplifiers that have low distortion, large load power, and high efficiency.

Figure 10-11*a* shows one way to connect a class B push-pull emitter follower. All we have done here is to connect an *npn* emitter follower and a *pnp* emitter follower in a complementary or push-pull arrangement. To understand what is going on, let's begin the analysis with the dc equivalent circuit of Fig. 10-11*b*. The designer selects biasing resistors to set the *Q* point at cutoff. This biases the emitter diode of each transistor between 0.6 and 0.7 V, whatever voltage is needed to just barely turn off the emitter diode. Ideally,

$$I_{CQ} = 0$$

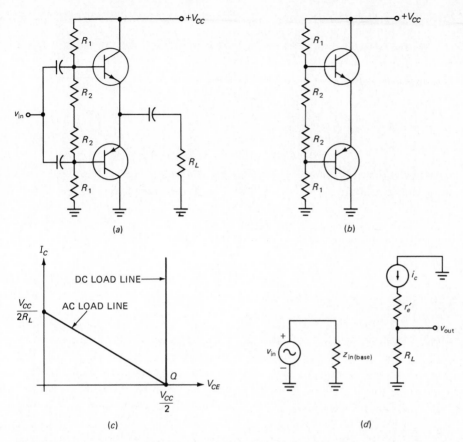

Fig. 10-11   (a) *Class B push-pull emitter follower.* (b) *Dc equivalent circuit.* (c) *Load lines.* (d) *Ac equivalent circuit.*

Notice the symmetry of the circuit. Because the biasing resistors are equal, each emitter diode is biased with the same voltage. As a result, half the supply voltage is dropped across each transistor. That is,

$$V_{CEQ} = \frac{V_{CC}}{2}$$

DC LOAD LINE

Since there is no dc resistance in the collector or emitter circuits of Fig. 10-11*b*, the dc saturation current is infinite. This means that the dc load line is vertical, as shown in Fig. 10-11*c*. If this seems to you like a dangerous situation, you are quite right. The most difficult thing about designing a class B amplifier is setting up a stable Q point at cutoff. Any significant decrease in V_{BE} with temperature can move the Q point up the dc load line to dangerously high currents. For the moment, however, assume that the Q point is rock solid at cutoff, as shown in Fig. 10-11*c*.

AC LOAD LINE

The ac load line derived earlier still applies. For an emitter follower, the ac saturation current is

$$I_{C(sat)} = I_{CQ} + \frac{V_{CEQ}}{r_E}$$

and the ac cutoff voltage is

$$V_{CE(cut)} = V_{CEQ} + I_{CQ}r_E$$

In the class B emitter follower of Fig. 10-11a, $I_{CQ} = 0$, $V_{CEQ} = V_{CC}/2$, and $r_E = R_L$. Therefore, the ac saturation current and cutoff voltage reduce to

$$I_{C(sat)} = \frac{V_{CC}}{2R_L} \qquad (10\text{-}32)$$

and

$$V_{CE(cut)} = \frac{V_{CC}}{2} \qquad (10\text{-}33)$$

Figure 10-11c shows the ac load line. When either transistor is conducting, that transistor's operating point swings upward along the ac load line; the operating point of the other transistor remains at cutoff. The voltage swing of the conducting transistor can go all the way from cutoff to saturation. On the alternate half cycle, the other transistor does the same thing. This means that the ac output compliance of a class B push-pull amplifier is higher than class A because it now equals

$$PP \cong V_{CC} \qquad (10\text{-}34)$$

Given a 10-V supply, we can build a class B push-pull emitter follower with an ac output compliance of 10 V.

AC ANALYSIS

Figure 10-11d shows the ac equivalent of the conducting transistor. This is almost identical to the class A emitter follower. The loaded voltage gain is

$$A_v = \frac{R_L}{R_L + r_e'} \qquad (10\text{-}35)$$

The loaded input impedance of the base is

$$z_{in(base)} \cong \beta(R_L + r_e') \qquad (10\text{-}36)$$

and the output impedance is

$$z_{out} = r_e' + \frac{r_B}{\beta} \qquad (10\text{-}37)$$

The current gain A_i is still approximately equal to β, and the power gain is

$$A_p = A_v A_i \qquad (10\text{-}38)$$

OVERALL ACTION

Now we have a good idea of what the circuit of Fig. 10-11*a* does. On the positive half cycle of input voltage, the upper transistor conducts and the lower one cuts off. The upper transistor acts like an ordinary emitter follower, so that the output voltage approximately equals the input voltage. As usual, the output impedance is very low because of the emitter follower.

On the negative half cycle of input voltage, the upper transistor cuts off and the lower transistor conducts. The lower transistor acts like an ordinary emitter follower and produces a load voltage approximately equal to the input voltage.

The overall action is now clear. The upper transistor handles the positive half cycle of input voltage, and the lower transistor takes care of the negative half cycle. During either half cycle, the source sees a high input impedance looking into either base, and the load sees a low output impedance.

CROSSOVER DISTORTION

Figure 10-12*a* shows the ac equivalent circuit of a class B push-pull emitter follower. Suppose that no bias at all is applied to the emitter diodes. Then the incoming ac voltage has to rise to about 0.7 V to overcome the barrier potential. Because of this, no current flows through Q_1 when the signal is less than 0.7 V. The action is similar on the other half cycle; no current flows in Q_2 until the ac input voltage is more negative than -0.7 V. For this reason, if no bias is applied to the emitter diodes, the output of a class B push-pull emitter follower looks like Fig. 10-12*b*.

The signal is distorted. Because of clipping action between half cycles, it no longer

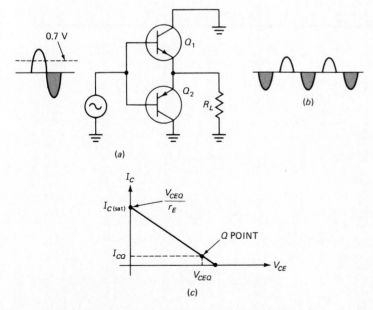

Fig. 10-12 (a) *Ac equivalent circuit of class B amplifier.* (b) *Crossover distortion.* (c) *Ac load line with trickle bias.*

is a sine wave. Since the clipping occurs between the time one transistor cuts off and the time the other one comes on, we call it *crossover distortion*. To eliminate crossover distortion, we need to apply a slight forward bias to each emitter diode. This means locating the Q point slightly above cutoff, as shown in Fig. 10-12c. As a guide, an I_{CQ} from 1 to 5 percent of $I_{C(sat)}$ is enough to eliminate crossover distortion.

Strictly speaking, we have class AB operation. This means that collector current flows in each transistor for more than 180° but less than 360°. Since the operation is closer to class B than to class A, however, most people refer to the circuit as a class B amplifier.

NONLINEAR DISTORTION

As discussed earlier, a large-signal class A amplifier elongates one half cycle and squashes the other. One cure for this is swamping, which reduces the *nonlinear distortion* to acceptable levels. The class B push-pull emitter follower reduces the distortion even further because both half cycles are identical in shape. Although some nonlinear distortion still remains, it is much less than with class A.

The reason for the lower distortion is that all even harmonics cancel out. Harmonics are multiples of the input frequency. For instance, if $f_{in} = 1$ kHz, the second harmonic is 2 kHz, the third harmonic is 3 kHz, and so on. A large-signal class A amplifier produces all harmonics: f_{in}, $2f_{in}$, $3f_{in}$, $4f_{in}$, $5f_{in}$, and so forth. A class B push-pull amplifier produces only the odd harmonics: f_{in}, $3f_{in}$, $5f_{in}$, etc. Because of this, the distortion is less with class B push-pull amplifiers. (Chapter 22 gives a complete discussion of harmonics and explains why even harmonics cancel out with push-pull operation.)

10-5 POWER FORMULAS FOR CLASS B

The load power, transistor dissipation, current drain, and stage efficiency of a class B push-pull emitter follower are quite different from those of a class A amplifier. Whether you are troubleshooting or designing class B amplifiers, it will help to know the following power formulas.

LOAD POWER

The ac load power of a class B push-pull amplifier is given by

$$P_L = \frac{V_{PP}^2}{8R_L} \tag{10-39}$$

where P_L = ac load power
V_{PP} = peak-to-peak load voltage
R_L = load resistance

You can use this equation when you measure the peak-to-peak load voltage with an oscilloscope.

Next, let us find the maximum load power. Figure 10-13a shows the ideal ac load line of a class B push-pull emitter follower. It is ideal because it ignores $V_{CE(sat)}$ and

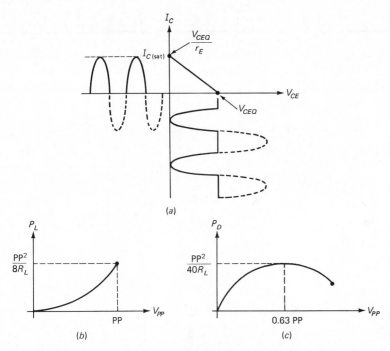

Fig. 10-13 (a) *Class B current and voltage.* (b) *Load power.* (c) *Transistor power dissipation.*

I_{CQ}. In a real amplifier, the ac saturation point does not quite touch the vertical axis, and the Q point is slightly above the cutoff. Figure 10-13a illustrates the maximum unclipped current and voltage waveforms we can get with one transistor of a class B push-pull emitter follower; the other transistor produces the dotted half cycle. Since the ac output compliance equals the peak-to-peak voltage, the maximum load power is

$$P_{L(\text{max})} = \frac{\text{PP}^2}{8R_L} \tag{10-40}$$

In Fig. 10-13a, PP equals $2V_{CEQ}$. Therefore, an alternative formula is

$$P_{L(\text{max})} = \frac{V_{CEQ}^2}{2R_L} \tag{10-41}$$

Figure 10-13b shows how load power varies with peak-to-peak load voltage. There are no surprises here. Load power increases to a maximum when the peak-to-peak load voltage equals the ac output compliance.

TRANSISTOR POWER DISSIPATION

Under no-signal conditions, the transistors of a class B push-pull amplifier are idling because only a small trickle of current passes through them. For this reason, the power dissipation in each transistor is very small. When a signal is present, however, the transistors have large swings in current, causing much more power dissipation.

The transistor power dissipation depends on how much of the ac load line is used. In the worst case, the dissipation reaches a maximum when 63 percent of the ac load line is used. Appendix 1 proves that the maximum transistor power dissipation is

$$P_{D(max)} = \frac{PP^2}{40R_L} \tag{10-42}$$

Figure 10-13*c* shows how the transistor power dissipation varies with peak-to-peak load voltage. As you can see, P_D reaches a maximum when the peak-to-peak load voltage is 0.63PP. Further increases in signal level cause the transistor dissipation to decrease. Since the worst-case power dissipation is $PP^2/40R_L$, each transistor in the class B amplifier must have a power rating greater than $PP^2/40R_L$.

CURRENT DRAIN

The dc current drain of a class B push-pull amplifier like Fig. 10-11a is

$$I_S = I_1 + I_2 \tag{10-43}$$

where I_1 = dc current through biasing resistors
I_2 = dc current through the upper collector

When no signal is present, $I_2 = I_{CQ}$, and the current drain is small. But when a signal is present, the current drain increases because the upper collector current becomes large.

If the entire ac load line is used, then the upper transistor has a half sine wave of current through it with a peak value of

$$I_{C(sat)} = \frac{V_{CEQ}}{R_L}$$

As discussed in Chap. 3, the average or dc value of a half-wave signal is

$$I_2 = 0.318 I_{C(sat)}$$

or

$$I_2 = \frac{0.318 V_{CEQ}}{R_L} \tag{10-44}$$

This allows you to calculate the maximum collector current drain.

The dc power supplied to the circuit is

$$P_S = V_{CC} I_S \tag{10-45}$$

This applies to any class B push-pull amplifier with a single power supply V_{CC}. Under no-signal conditions, the dc power is small because the current drain is minimum. But when a signal uses the entire ac load line, the dc power supplied to the circuit reaches a maximum.

Table 10-2. Class B Formulas

Quantity	Formula	Comment
$I_{C(sat)}$	$I_{CQ} + V_{CC}/2R_L$	Emitter follower, single supply
$V_{CE(cut)}$	$V_{CC}/2$	Emitter follower, single supply
PP	V_{CC}	Emitter follower, single supply
P_L	V_L^2/R_L	Voltage in rms volts
P_L	$V_{PP}^2/8R_L$	Voltage in peak-to-peak volts
$P_{L(max)}$	$PP^2/8R_L$	Maximum undistorted output power
$P_{D(max)}$	$PP^2/40R_L$	Maximum transistor power dissipation
P_S	$V_{CC}I_S$	Supply power
η	$P_{L(max)}/P_{S(max)}$	Stage efficiency; multiply by 100%

STAGE EFFICIENCY

The stage efficiency is

$$\eta = \frac{P_{L(max)}}{P_{S(max)}} \times 100\% \qquad (10\text{-}46)$$

As will be shown in a later example, class B has higher stage efficiency than class A because it produces much more output power with less dc power from the supply. In fact, it can be shown that a class B push-pull stage has a maximum efficiency of 78.5 percent. A class A stage has a maximum efficiency of either 25 percent (*RC* coupled) or 50 percent (transformer coupled). In either case, class B is more efficient.

CONCLUSION

Table 10-2 summarizes the more important formulas for class B operation. The entries are self-explanatory. If you have any difficulty, refer to the original derivation and discussion for the entry.

EXAMPLE 10-5

Figure 10-14 shows the ac load line of a class B push-pull emitter follower. If $R_L = 100\ \Omega$, calculate the ac output compliance and the maximum load power.

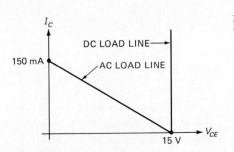

Fig. 10-14

SOLUTION

In Fig. 10-14, the maximum voltage swing during each half cycle is 15 V. Therefore, the ac output compliance is

$$PP = 2(15 \text{ V}) = 30 \text{ V}$$

This is the maximum unclipped peak-to-peak voltage the amplifier can produce. The maximum load power is

$$P_{L(max)} = \frac{(30 \text{ V})^2}{8(100 \text{ }\Omega)} = 1.13 \text{ W}$$

EXAMPLE 10-6

A class B push-pull amplifier has a supply voltage of 30 V, a biasing current of 1 mA, and a quiescent collector current of 1 mA. If the amplifier has the ac load line of Fig. 10-14, what are the no-signal current drain, the full-signal current drain, and the stage efficiency?

SOLUTION

The no-signal current drain is

$$I_S = 1 \text{ mA} + 1 \text{ mA} = 2 \text{ mA}$$

Under full-signal conditions, the entire ac load line is used, and the average collector current in the upper transistor increases to

$$I_2 = 0.318(150 \text{ mA}) = 47.7 \text{ mA}$$

Therefore, the full-signal current drain is

$$I_S = 1 \text{ mA} + 47.7 \text{ mA} = 48.7 \text{ mA}$$

Finally, let us calculate the stage efficiency. The maximum ac load power was calculated in Example 10-5:

$$P_L = 1.13 \text{ W}$$

The maximum dc power supplied to the stage is

$$P_S = (30 \text{ V})(48.7 \text{ mA}) = 1.46 \text{ W}$$

So, the stage efficiency is

$$\eta = \frac{1.13 \text{ W}}{1.46 \text{ W}} = 77.4\%$$

Notice how high this efficiency is compared with class A. This is one of the reasons the class B push-pull circuit is popular near the end of a system. The higher efficiency implies more load power than class A.

10-6 BIASING A CLASS B AMPLIFIER

As mentioned earlier, the hardest thing about designing a class B amplifier is setting up a stable Q point near cutoff. This section discusses the problem and its solution.

VOLTAGE-DIVIDER BIAS

Figure 10-15a shows voltage-divider bias for a class B push-pull circuit. The two transistors have to be complementary, meaning that they have similar V_{BE} curves, maximum ratings, etc. For instance, the 2N3904 and 2N3906 are complementary, the first being an *npn* transistor and the second a *pnp*; they have similar V_{BE} curves, maximum ratings, and so on. Complementary pairs like these are available for almost any class B push-pull design.

In Fig. 10-15a, the collector and emitter currents are approximately equal. Because of the series connection of complementary transistors, each transistor drops half the supply voltage. To avoid crossover distortion, we set the Q point slightly above cutoff, with the correct V_{BE} somewhere between 0.6 and 0.7 V, depending on the transistor type, the temperature, and other factors. Data sheets indicate that an increase of 60 mV in V_{BE} produces 10 times as much emitter current. Because of this, it is extremely difficult to find standard resistors that can produce the correct V_{BE}. Almost always, an adjustable resistor is needed to set the correct Q point.

But an adjustable resistor does not solve the temperature problem. As discussed in Chap. 8, for a given collector current, V_{BE} decreases approximately 2 mV per degree rise. In other words, the V_{BE} required to set up a particular collector current decreases as the temperature increases. In Fig. 10-15a, the voltage dividers produce a stiff drive for each emitter diode. Therefore, as the temperature increases, the fixed voltage on each emitter diode forces the collector current to increase. For instance, if the

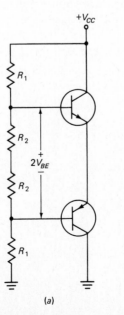
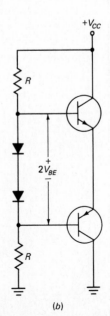

(a) (b)

Fig. 10-15
(a) *Voltage-divider bias for class B.*
(b) *Diode bias.*

required V_{BE} decreases 60 mV, the collector current increases by a factor of 10 because the fixed bias is 60 mV too high.

The ultimate danger is *thermal runaway.* When the temperature increases, collector current increases, and this is equivalent to the Q point moving up along the vertical dc load line. As the Q point moves toward higher collector currents, the temperature of the transistor increases, further reducing the correct V_{BE}. This escalating situation means that the Q point may "run away" by rising along the dc load line until excessive power destroys the transistor. Whether or not thermal runaway takes place depends on the thermal properties of the transistor, how it is cooled, and the type of heat sink used (discussed later).

DIODE BIAS

One way to avoid thermal runaway is with *diode bias,* shown in Fig. 10-15*b.* The idea is to use compensating diodes to provide the biasing voltage to the emitter diodes. For this to work, the diode curves must match the V_{BE} curves of the transistors. Then, any increase in temperature reduces the biasing voltage developed by the compensating diodes. For instance, assume that a biasing voltage of 0.65 V sets up 2 mA of quiescent collector current. If the temperature rises 30°C, the voltage across each compensating diode drops about 60 mV. Since the required V_{BE} also decreases by approximately 60 mV, the quiescent collector current remains at approximately 2 mA.

CURRENT MIRROR

Diode bias is based on the concept of the *current mirror,* a circuit technique widely used in linear integrated circuits. In Fig. 10-16*a,* the base current is much smaller than the current through the resistor and the diode. For this reason, the resistor current and the diode current are approximately equal. If the diode curve is identical to the V_{BE} curve of the transistor, the diode current equals the emitter current. Since collector current almost equals emitter current, we arrive at the following conclusion: The collector current approximately equals the current through the biasing resistor. In symbols,

$$I_C \cong I_R \tag{10-47}$$

Fig. 10-16
(a) *npn current mirror.* (b) *pnp current mirror.*

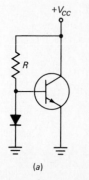

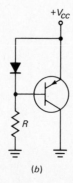

(a)   (b)

This is an important result. It means that we can set the collector current by controlling the resistor current. Think of the circuit as a mirror; the current through the resistor is reflected into the collector circuit. This is why the circuit of Fig. 10-16*a* is called a *current mirror.*

Figure 10-16*b* shows a *pnp* current mirror. By a similar line of reasoning, the current through the collector approximately equals the current through the biasing resistor. If the V_{BE} curve of the transistor matches the diode curve, the collector current approximately equals the resistor current.

Diode bias of a class B push-pull emitter follower (Fig. 10-15*b*) relies on two current mirrors. The upper half is an *npn* current mirror, and the lower half is a *pnp* current mirror. For diode bias to be immune to changes in temperature, the diode curves must match the V_{BE} curves of the transistors over a wide temperature range. This is not easily done with discrete circuits because of tolerances. But diode bias is easy to implement with integrated circuits because the diode and transistors are on the same chip, which means that they have almost identical characteristics.

EXAMPLE 10-7

What is the quiescent collector current in Fig. 10-17*a*?

SOLUTION

We have an *npn* mirror in series with a *pnp* mirror. The current through the biasing resistors is

$$I_R = \frac{30 \text{ V} - 1.4 \text{ V}}{2(4.7 \text{ k}\Omega)} = 3.04 \text{ mA}$$

Assuming that the diode curves match the V_{BE} curves, the collector current in each transistor is approximately 3.04 mA.

EXAMPLE 10-8

What is the collector current in Fig. 10-17*b*?

SOLUTION

Recall collector-feedback bias. When the base resistor decreases to zero, a collector-feedback-biased transistor acts like a diode. This is what we have in the circuit of Fig. 10-17*b*. Instead of using ordinary diodes, we use transistors connected as diodes. The current through the resistors is

$$I_R = \frac{30 \text{ V} - 1.4 \text{ V}}{2 \text{ k}\Omega} = 14.3 \text{ mA}$$

Therefore, the collector current in each transistor is approximately 14.3 mA.

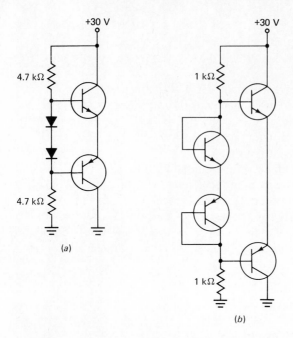

Fig. 10-17
(a) *Diode bias.*
(b) *Transistors
connected as
diodes.*

The reason for using transistors connected as diodes is because it is easier to match the diode curves with the V_{BE} curves when the same transistor type is used for the diode and the transistor. This is always done with integrated circuits.

10-7 CLASS B DRIVER

In the initial discussion of the class B push-pull emitter follower, capacitors were used to couple the ac signal into the amplifier. This is not the best way to drive a class B amplifier. It is easier to use a direct-coupled CE *driver,* as shown in Fig. 10-18*a*. Transistor Q_2 is a current source that sets up the dc biasing current through the diodes. By adjusting R_2, we can control the dc emitter current through R_4; this means that Q_2 sources direct current through the compensating diodes. Because of the current mirrors, the same value of quiescent current exists in the collectors of Q_3 and Q_4.

When an ac signal drives the input, Q_2 acts like a swamped amplifier. The amplified and inverted ac signal at the Q_2 collector drives the bases of Q_3 and Q_4. On the positive half cycle, Q_3 conducts and Q_4 cuts off. On the negative half cycle, Q_3 cuts off and Q_4 conducts. Because the output coupling capacitor is an ac short, the ac signal is coupled into the load resistance.

Figure 10-18*b* shows the ac equivalent circuit of the CE driver. The diodes (transistors connected as diodes) are replaced by their ac emitter resistances. In any practical circuit, r_e' is at least 100 times smaller then R_3; therefore, the ac equivalent circuit simplifies to Fig. 10-18*c*. Now, we can see that the driver stage is a swamped amplifier with an unloaded voltage gain of

$$A = -\frac{R_3}{R_4}$$

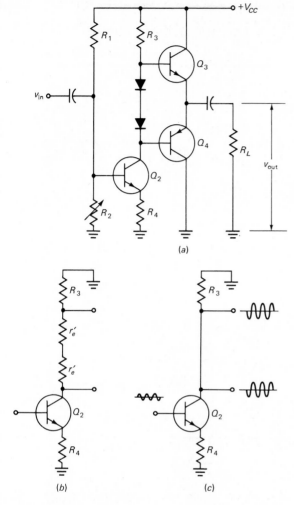

Fig. 10-18　*Driver for class B push-pull emitter follower.* (a) *Circuit.* (b) *Ac equivalent circuit.* (c) *Simplified equivalent circuit.*

Usually, the $z_{in(base)}$ of the class B transistors is very high, so that the loaded voltage gain of the driver stage is almost equal to its unloaded voltage gain.

EXAMPLE 10-9

Figure 10-19 is an example of a complete amplifier. It has three stages: a small-signal amplifier (Q_1), a large-signal class A amplifier (Q_2), and a class B push-pull emitter follower (Q_3 and Q_4). The approximate dc voltages of all nodes are listed. Calculate the quiescent current through Q_3 and Q_4. If $\beta = 120$, what is the loaded voltage gain of the driver stage?

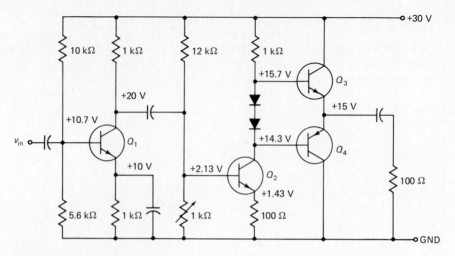

Fig. 10-19 *Complete amplifier including small-signal CE stage, class B driver, and push-pull output stage.*

SOLUTION

The emitter resistor of the driver stage has $+1.43$ V across it. Therefore, the dc emitter current in Q_2 is

$$I_E = \frac{1.43 \text{ V}}{100 \ \Omega} = 14.3 \text{ mA}$$

The Q_2 collector current is approximately 14.3 mA. Since this is the biasing current for the current mirrors, the quiescent collector currents of Q_3 and Q_4 approximately equal 14.3 mA:

$$I_{CQ} = 14.3 \text{ mA}$$

The input impedance looking into the base of the conducting transistor is

$$z_{\text{in(base)}} \cong 120(100 \ \Omega) = 12 \text{ k}\Omega$$

In the ac equivalent circuit, this input impedance is in parallel with the collector resistor of Q_2. Because of the swamping resistor, the loaded voltage gain of the driver stage is approximately $-r_C/r_E$:

$$A_v \cong -\frac{1 \text{ k}\Omega \parallel 12 \text{ k}\Omega}{100 \ \Omega} = -9.23$$

This is almost equal to the unloaded voltage gain:

$$A \cong -\frac{1 \text{ k}\Omega}{100 \ \Omega} = -10$$

10-8 OTHER CLASS B AMPLIFIERS

The class B push-pull emitter follower is the most commonly used class B circuit. It has the advantages of low distortion, large ac output compliance, and high stage efficiency. But there are other class B amplifiers that are worth knowing about.

SPLIT SUPPLY

When a *split supply* (equal positive and negative voltages) is available, both the input and the output can be referenced to ground, as shown in Fig. 10-20. Because the supplies are equal and opposite, each transistor has a V_{CEQ} of V_{CC}. This implies that the quiescent output voltage is zero. This is why the signal can be direct-coupled to the load resistor. Also, the quiescent voltage between the compensating diodes is zero, which makes this a convenient input point for applications that need a ground-referenced input.

As far as the ac signal is concerned, the diodes act like small resistances of r'_e. Since the $z_{in(base)}$ of each transistor is very high, almost all the ac input signal passes through the diodes to the bases of the class B transistors.

Another advantage of a split-supply circuit is its large ac output compliance. Since each transistor has a V_{CEQ} of V_{CC}, the ac output compliance is

$$PP = 2V_{CC}$$

This high ac output compliance means that the circuit can produce more undistorted load power.

THERMISTOR COMPENSATION

Instead of using current mirrors for compensating a class B push-pull emitter follower, we can use *thermistors* (resistors that decrease in value when temperature increases). Here is the idea behind thermistor compensation. In Fig. 10-21, the circled resistors are thermistors. We can select a room-temperature value R_2 to set the Q point slightly above cutoff. Then as the temperature increases, the required V_{BE}

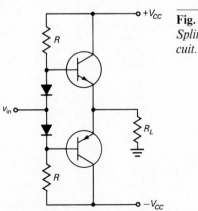

Fig. 10-20
Split-supply circuit.

Fig. 10-21
*Thermistors
compensate for
temperature
changes.*

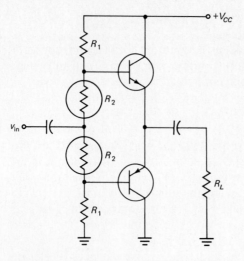

decreases approximately 2 mV per degree. Since the resistance of the thermistors also decreases, less voltage is applied to the emitter diodes. If the correct thermistor is used, it can approximately compensate for temperature increases.

DARLINGTON AND SZIKLAI

If the class B push-pull emitter follower is not stiff enough for the load resistance, we can use Darlington pairs, as shown in Fig. 10-22a. As previously discussed, each

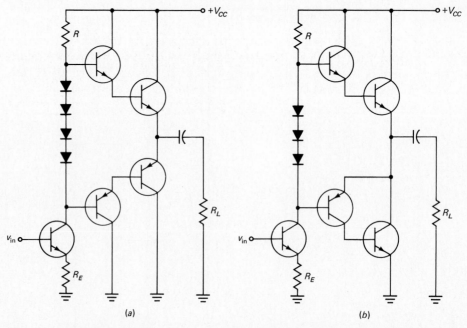

Fig. 10-22 (a) *Darlington pairs increase load power.* (b) *Darlington and Sziklai output stage.*

Darlington pair acts like a single transistor with a very high current gain. On account of this, the input impedance of the base increases and the output impedance of the emitter decreases. Since each Darlington pair has two V_{BE} drops, we need to use four compensating diodes, as shown. A circuit like this can produce a large ac load power.

Sometimes it is easier to design a class B push-pull amplifier with the same type of output transistor, either *npn* or *pnp*. Figure 10-22b shows a class B push-pull circuit with a Darlington pair on top and a *Sziklai pair* on the bottom. The Sziklai pair, sometimes called a complementary Darlington, acts like a single *pnp* transistor with a very high current gain. Notice that only three compensating diodes are needed. But this is not the main advantage. The best thing about this circuit is that both output transistors are *npn*. This is convenient from the design standpoint because it is easier to match power transistors if they are of the same type.

TRANSFORMER-COUPLED CE AMPLIFIER

Complementary transistors have made transformers obsolete in most audio applications. Nevertheless, you still may encounter the class B push-pull CE amplifier shown in Fig. 10-23. Notice that both transistors are *npn*. A single diode is used to bias these transistors slightly above cutoff. If the diode curve approximately matches the V_{BE} curves of the transistors, the quiescent collector current will not change too much with temperature.

The ac input signal is transformer-coupled into the bases. Because of the way a transformer works, the signals driving the bases are equal in amplitude but opposite in phase. As a result, the positive half cycle turns on the upper transistor and cuts off

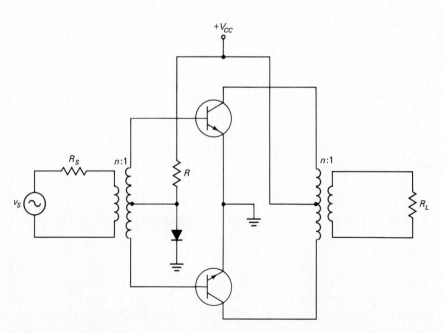

Fig. 10-23
Transformer-coupled push-pull amplifier.

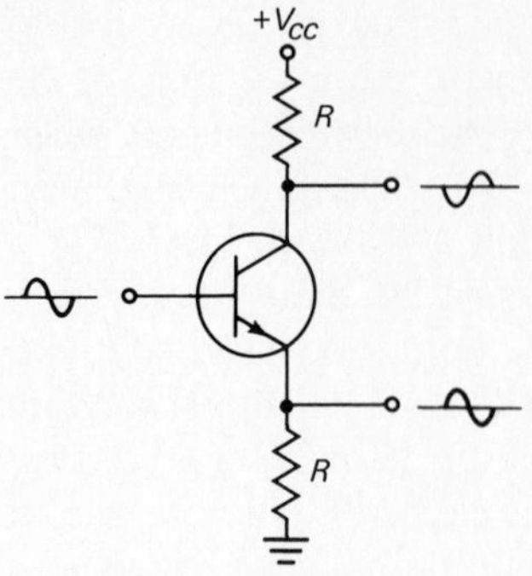

Fig. 10-24
Phase splitter.

the lower transistor. Conversely, the negative half cycle cuts off the upper transistor and turns on the lower transistor.

During the positive half cycle, the upper transistor conducts through the upper half winding of the output transformer. During the negative half cycle, the lower transistor conducts through the lower half winding. In either case, the ac signal is transformer-coupled to the load resistor.

PHASE SPLITTER

The two bases of Fig. 10-23 receive ac signals that are 180° out of phase with each other. This is necessary because both transistors are of the same type (*npn*). A transformer is an expensive and bulky way to produce two out-of-phase signals. It is more practical to use a *phase splitter* like Fig. 10-24 for the input driver.

Notice that the phase splitter is a heavily swamped amplifier. Because the emitter and collector resistors are equal in value, the phase splitter has an unloaded voltage gain of 1. Furthermore, the bootstrapped emitter produces an in-phase signal, while the collector produces an out-of-phase signal. Therefore, the output signals are equal in amplitude and opposite in phase, precisely the kind of drive needed for the circuit of Fig. 10-23. In other words, we can replace the input transformer by a phase splitter.

Remember the phase splitter. It is very useful whenever you need to drive a circuit that requires two equal and out-of-phase input signals.

10-9 TRANSISTOR POWER RATING

The temperature at the collector junction places a limit on the allowable power dissipation P_D. Depending on the transistor type, a *junction* temperature in the range of 150°C to 200°C will destroy the transistor. Data sheets specify this maximum junction temperature as $T_{J(max)}$. For instance, the data sheet of a 2N3904 gives a $T_{J(max)}$ of 150°C; the data sheet of a 2N3719 specifies a $T_{J(max)}$ of 200°C.

AMBIENT TEMPERATURE

The heat produced at the junction passes through the transistor case (metal or plastic housing) and radiates to the surrounding air. The temperature of this air, known as the *ambient* temperature, is normally around 25°C, but it can be much higher on hot days. Also, the ambient temperature may be much higher inside a piece of electronic equipment.

DERATING FACTOR

Data sheets often specify the $P_{D(max)}$ of a transistor at an ambient temperature of 25°C. For instance, the 2N1936 has a $P_{D(max)}$ of 4 W for a T_A of 25°C. This means that a 2N1936 used in a class A amplifier can have a quiescent power dissipation as high as 4 W. As long as the ambient temperature is 25°C or less, the transistor is within its specified power rating.

What do you do if the ambient temperature is greater than 25°C? You have to derate (reduce) the power rating. Data sheets sometimes include a *derating curve* like Fig. 10-25. As you can see, the power rating decreases when the ambient temperature increases. For instance, at an ambient temperature of 100°C, the power rating is 2 W. Notice that the power rating decreases linearly with temperature.

Some data sheets do not give a derating curve like Fig. 10-25. Instead, they list a *derating factor D*. For instance, the derating factor of a 2N1936 is 26.7 mW/°C. This means that you have to subtract 26.7 mW for each degree the ambient temperature is above 25°C. In symbols,

$$\Delta P = D(T_A - 25°C) \tag{10-48}$$

where ΔP = decrease in power rating
D = derating factor
T_A = ambient temperature

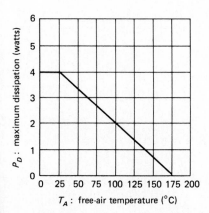

Fig. 10-25 *Power derating curve for ambient temperature.*

For example, if the ambient temperature rises to 75°C, you have to reduce the power rating by

$$\Delta P = 26.7 \text{ mW} \times (75 - 25) = 1.34 \text{ W}$$

Since the power rating is 4 W at 25°C, the new power rating is

$$P_{D(\text{max})} = 4 \text{ W} - 1.34 \text{ W} = 2.66 \text{ W}$$

This agrees with the derating curve of Fig. 10-25.

Whether you get the reduced power rating from a derating curve like Fig. 10-25 or from a formula like Eq. (10-48), the important thing to be aware of is the reduction in power rating as the ambient temperature increases. Just because a circuit works well at 25°C, doesn't mean it will perform well over a large temperature range. When you design circuits, therefore, you must take the operating temperature range into account by derating all transistors for the highest expected ambient temperature.

HEAT SINKS

One way to increase the power rating of a transistor is to get rid of the heat faster. This is the purpose of a *heat sink* (a mass of metal). If we increase the surface area of the transistor case, we allow the heat to escape more easily into the surrounding air. For instance, Fig. 10-26a shows one type of heat sink. When this is pushed on to the transistor case, heat radiates more quickly because of the increased surface area of the fins.

Figure 10-26b shows another approach. This is the outline of a *power-tab* transistor. A metal tab provides a path out of the transistor for heat. This metal tab can be fastened to the chassis of electronics equipment. Because the chassis is a massive heat sink, heat can easily escape from the transistor to the chassis.

Large power transistors like Fig. 10-26c have the *collector connected directly to the case* to let heat escape as easily as possible. The transistor case is then fastened to the chassis. To prevent the collector from shorting to chassis ground, a thin mica washer is used between the transistor case and the chassis. The important idea here is that

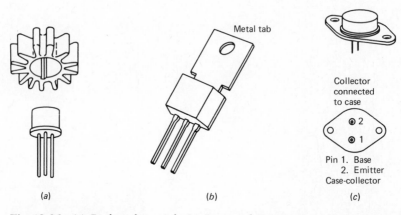

Fig. 10-26   (a) *Push-on heat sink.* (b) *Power-tab transistor.* (c) *Power transistor with collector connected to case.*

heat can leave the transistor more rapidly, which means that the transistor has a higher power rating at the same ambient temperature. Sometimes the transistor is fastened to a large heat sink with fins; this is even more efficient in removing heat from the transistor.

CASE TEMPERATURE

When heat flows out of a transistor, it passes through the case of the transistor and into the heat sink, which then radiates the heat into the surrounding air. The temperature of the transistor case T_C will be slightly higher than the temperature of the heat sink T_S, which in turn is slightly higher than the ambient temperature T_A.

The data sheets of large power transistors give derating curves for the case temperature rather than the ambient temperature. For instance, Fig. 10-27 shows the derating curve of a 2N5877. The power rating is 150 W at a case temperature of 25°C; then it decreases linearly with temperature until it reaches zero for a case temperature of 200°C.

Sometimes you get a derating factor instead of a derating curve. In this case, you can use the following equation to calculate the reduction in power rating:

$$\Delta P = D(T_C - 25°C) \tag{10-49}$$

where ΔP = decrease in power rating
$\quad\quad D$ = derating factor
$\quad\quad T_C$ = case temperature

THERMAL ANALYSIS

To use the derating curve of a large power transistor, you need to know what the case temperature will be in the worst case. Then you can derate the transistor to arrive at its maximum power rating. To calculate the case temperature, you need to know something about thermodynamics, the study of heat flow.

Thermal resistance θ is the resistance to heat flow between two temperature points. For instance, Fig. 10-28a shows the case temperature, the sink temperature, and the ambient temperature. Heat flows from the case of the transistor to the heat

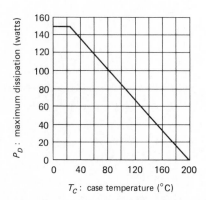

Fig. 10-27 *Power derating curve for case temperature.*

Fig. 10-28
*Thermal resist-
ances.*

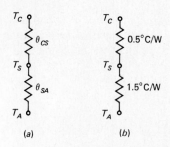

(a)   (b)

sink and on to the surrounding air. As this heat flows from the case to the heat sink, it encounters the thermal resistance θ_{CS}. When the heat passes from the sink to the surrounding air, it passes through a thermal resistance θ_{SA}. As a guide, θ_{CS} is from 0.2 to 1°C/W, and θ_{SA} is from 1 to 100°C/W, depending on the size of the heat sink, number of fins, finish, and other such factors. For instance, if the data sheet of a heat sink lists $\theta_{CS} = 0.5°C/W$ and $\theta_{SA} = 1.5°C/W$, then the thermal resistances are as shown in Fig. 10-28b.

The transistor power dissipation P_D is the same as the rate at which heat flows out of the transistor. In thermodynamics, the rate of heat flow is analogous to current, thermal resistance to resistance, and temperature difference to voltage:

$$P_D \rightarrow \text{current}$$
$$\theta \rightarrow \text{resistance}$$
$$T_1 - T_2 \rightarrow \text{voltage}$$

where T_1 and T_2 are the temperatures of any two points. Using this analogy, Ohm's law for thermodynamics can be written as

$$P_D = \frac{T_1 - T_2}{\theta} \tag{10-50}$$

The thermal resistances of Fig. 10-28a are in series and can be added to get the total thermal resistance between the case and the surrounding air:

$$\theta_{CA} = \theta_{CS} + \theta_{SA}$$

So, we can rewrite Eq. (10-50) as

$$P_D = \frac{T_C - T_A}{\theta_{CS} + \theta_{SA}}$$

Solving for case temperature gives

$$T_C = T_A + P_D(\theta_{CS} + \theta_{SA}) \tag{10-51}$$

where T_C = case temperature
T_A = ambient temperature
P_D = transistor power dissipation
θ_{CS} = thermal resistance between case and sink
θ_{SA} = thermal resistance between sink and surrounding air

This is the key formula needed to calculate the case temperature of a power transistor.

EXAMPLE 10-10

A circuit must operate over an ambient temperature range of 0° to 70°C. A 2N5877 and a heat sink have the following thermal resistances: $\theta_{CS} = 0.5°C/W$ and $\theta_{SA} = 1.5°C/W$. If the transistor has a maximum power dissipation of 30 W, what is the maximum case temperature of the transistor? Using the derating curve of Fig. 10-27, determine what the power rating of the 2N5877 is at this maximum case temperature.

SOLUTION

The highest case temperature occurs when the ambient temperature is 70°C. With Eq. (10-51),

$$T_C = 70°C + (30 \text{ W})(0.5°C/W + 1.5°C/W) = 130°C$$

This tells us that the case temperature is 130°C when the transistor is dissipating 30 W. From the derating curve of Fig. 10-27, the power rating of the 2N5877 is

$$P_{D(\text{max})} = 60 \text{ W}$$

In summary, the highest ambient temperature is 70°C and the highest case temperature is 130°C. The transistor power dissipation of 30 W is still well within the maximum power rating of 60 W at the highest temperature.

PROBLEMS

Straightforward

10-1. If $\beta_{dc} = 100$ in Fig. 10-29a, draw the ac load line and calculate the ac output compliance.

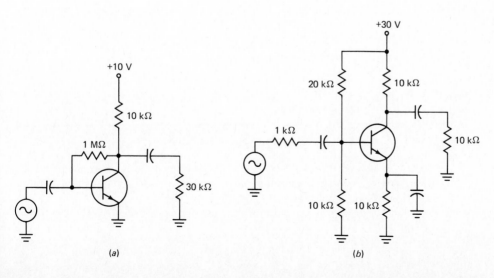

Fig. 10-29

(a)

(b)

Fig. 10-29
(Continued)

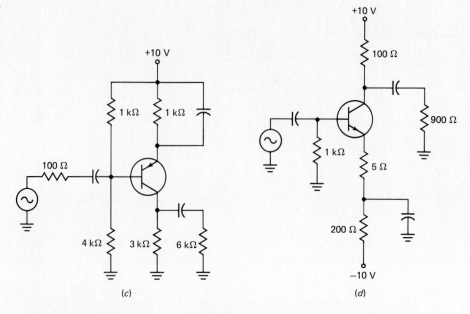

(c)

(d)

10-2. In Fig. 10-29b, draw the ac load line and calculate the ac output compliance.

10-3. What is the ac output compliance for Fig. 10-29c?

10-4. Draw the ac load line for Fig. 10-29d. Calculate the ac output compliance.

10-5. What is the ac output compliance for the first stage of Fig. 10-30? Draw the ac load line for the second stage.

Fig. 10-30

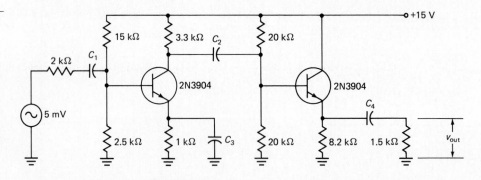

10-6. If $\beta = 125$ in Fig. 10-29b, calculate A_v, A_i, A_p, $P_{L(max)}$, P_{DQ}, I_S, P_S, and η.

10-7. Repeat Prob. 10-6 for Fig. 10-29d.

10-8. In Fig. 10-30, calculate the total dc current drain.

10-9. What is the efficiency of the second stage in Fig. 10-30?

10-10. The ac load line of a class B push-pull emitter follower has an ac saturation current of 250 mA and an ac cutoff voltage of 10 V. What is the ac output compliance? If the load resistance is 50 Ω, what is the maximum load power? What is the maximum transistor power dissipation?

10-11. Prove that the maximum transistor power dissipation of a class B push-pull amplifier is given by

$$P_{D(max)} = 0.2 P_{L(max)}$$

10-12. Draw the ac load line for Fig. 10-31*a*. What is the ac output compliance? The maximum load power? In the worst case, what is the maximum transistor power dissipation?

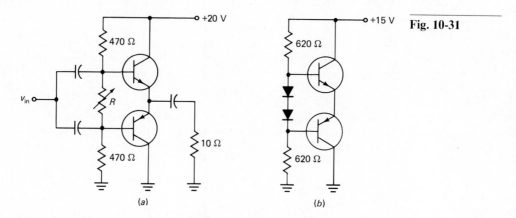

Fig. 10-31

(a)   (b)

10-13. In Fig. 10-31*a*, *R* is adjusted to get a V_{BE} of 0.68 V and an I_{CQ} of 20 mA. What is the no-signal current drain of the stage? The full-signal current drain? The stage efficiency?

10-14. After adjusting *R* in Fig. 10-31*a*, $V_{BE} = 0.66$ V and $I_{CQ} = 5$ mA. If the temperature of the transistor rises from 25°C to 55°C, what is the new value of I_{CQ}?

10-15. In Fig. 10-31*b*, what is the current through the biasing resistors? If the diode curves match the V_{BE} curves, what does I_{CQ} equal?

10-16. The supply voltage of Fig. 10-31*b* changes from 15 V to 25 V. What does I_{CQ} equal?

10-17. In Fig. 10-32, what is the value of *R* that produces $V_{CEQ} = 10$ V for each output transistor? (Use 0.7 V for the drop of the compensating diodes.)

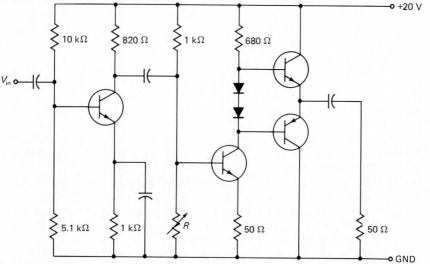

Fig. 10-32

10-18. The amplifier of Fig. 10-32 has equal drops across the output transistors. Work out the voltage for every node in the amplifier.

10-19. Calculate the approximate $P_{L(max)}$ and $P_{D(max)}$ for the output stage of Fig. 10-32.

10-20. What does I_{CQ} equal in the output stage of Fig. 10-32?

10-21. What is the maximum dc current drain of Fig. 10-32?

10-22. A 2N3904 has a power rating of 310 mW for an ambient temperature of 25°C. If the derating factor is 2.81 mW/°C, what is the power rating when the ambient temperature is 70°C?

10-23. A transistor has the derating curve of Fig. 10-25. If the ambient temperature range is from 0 to 70°C, what is the power rating in the worst case?

10-24. The data sheet of a 2N3055 lists a power rating of 115 W for a case temperature of 25°C. If the derating factor is 0.657 W/°C, what is $P_{D(max)}$ when the case temperature is 90°C?

10-25. A circuit operates over an ambient temperature range of 0 to 80°C. A transistor and heat sink have these thermal resistances: $\theta_{CS} = 0.3$°C/W and $\theta_{SA} = 2.3$°C/W. If the transistor power dissipation is 40 W, what is the maximum case temperature?

Troubleshooting

10-26. You have just built a circuit like Fig. 10-31*a*. You adjust R to get an I_{CQ} of 20 mA. Five minutes later, you recheck the circuit and discover that the upper transistor has been destroyed. Explain what has happened and how you can fix the problem.

10-27. The amplifier of Fig. 10-30 is not working. Furthermore, the current drain is higher than it should be, because an ammeter in series with the 15-V supply reads approximately 5.6 mA. Which of these can cause the trouble:

   a. C_1 shorted

   b. C_3 shorted

   c. Collector-emitter terminals shorted in first stage

   d. C_4 open

10-28. Someone is trying to get diode bias with the circuit of Fig. 10-31*b* using two 1N914s, a 2N3904, and a 2N3906. If I_{CQ} is 25 mA, what is wrong?

10-29. Under full-signal conditions, the peak-to-peak voltage across the 50-Ω load of Fig. 10-32 is zero. Which of the following can be the cause:

   a. Input coupling capacitor is shorted

   b. Supply voltage is only 15 V

   c. Load resistor is open

   d. Load resistor is shorted

Design

10-30. Locate the optimum Q point for Fig. 10-29*b* to get maximum ac output compliance.

10-31. Redesign the voltage divider of the second stage in Fig. 10-30 to get maximum ac output compliance.

10-32. In Fig. 10-31*b*, the diode curves match the V_{BE} curves. Select resistances to set up an I_{CQ} of 5 mA.

10-33. Design an amplifier like Fig. 10-32 to meet these specifications: $V_{CC} = 9$ V and $R_L = 3.2$ Ω.

Challenging

10-34. Prove that the maximum stage efficiency for a class B push-pull emitter follower is 78.6 percent.

10-35. The class B push-pull emitter follower of Fig. 10-33*a* does not use diode bias or any other kind of bias. Explain why there is no crossover distortion.

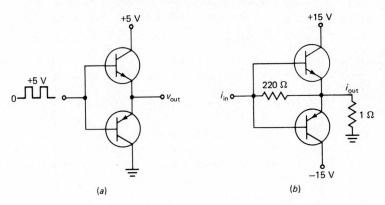

Fig. 10-33

(a)

(b)

10-36. The circuit shown in Fig. 10-33*b* is called a current booster. If the transistors turn on when $V_{BE} = 0.7$, what is the source current at this point? If $\beta_{dc} = 80$, what is the load current when the source current is 5 mA? Can you figure out what the circuit is used for?

Computer

10-37. The variables in typical microcomputers must start with a letter, A through Z. The second character may be a letter or a number 1 through 9. Examples of valid variables are P, D3, R4, and PS. It's all right to use more than two characters, but the computer recognizes only the first two characters. For this reason, the computer cannot tell the difference between ICQ and ICSAT. Which of the following are valid variables?
   a. PS
   b. 5WQ
   c. R
   d. ICQ
   e. VCEQ

10-38. A subroutine is a smaller program tucked away inside a larger program. Subroutines can be used over and over again during the execution of the larger program. The GOSUB and RETURN statements allow you to use a subroutine at any time you want. For instance, here is a program that works out the ac saturation current and cutoff voltage of a CE amplifier:

```
10 PRINT "ENTER ICQ": INPUT ICQ
20 PRINT "ENTER VCEQ": INPUT VCEQ
30 PRINT "ENTER RC": INPUT RC
40 PRINT "ENTER RL": INPUT RL
50 GOSUB 1000
60 PRINT "THE SATURATION CURRENT IS": PRINT ISAT
70 PRINT "THE CUTOFF VOLTAGE IS": PRINT VOFF
80 STOP
     .
     .
     .
1000 Y = 1/RC + 1/RL
1010 R = 1/Y
```

1020 ISAT = ICQ + VCEQ/R
1030 VOFF = VCEQ + ICQ*R
1040 RETURN

After you input I_{CQ}, V_{CEQ}, R_C, and R_L, the program encounters the GOSUB 1000, which tells the computer to go to the subroutine located at line 1000. Lines 1000 through 1040 are the subroutine. A subroutine always ends with a RETURN statement; this tells the computer to return to the main program. When the computer returns, it goes to line 60.

What does the computer calculate in line 1020? Line 1030? What does the computer print on the screen?

10-39. Here is a subroutine:

2000 PS = ICQ*R
2010 NS = VCEQ
2020 IF PS < NS THEN GOTO 2050
2030 PP = 2*NS
2040 GOTO 2060
2050 PP = 2*PS
2060 RETURN

What does line 2000 calculate? Line 2030? If PS is smaller than NS, what is the value of PP when the subroutine ends?

10-40. Write a subroutine similar to the one in Prob. 10-38 that calculates the ac output compliance of a class A emitter follower. Start the subroutine at line 3000.

10-41. Write a program that inputs the power rating at 25°C, the derating factor, and the ambient temperature. The program should end by printing out the new power rating.

Class C and Other Amplifiers

A class C amplifier can produce more load power than a class B amplifier. To amplify a sine wave, however, it has to be tuned to the sinusoidal frequency. Because of this, the tuned class C amplifier is a narrowband circuit; it can amplify only the resonant frequency and those frequencies near it. To avoid the need for large inductors and capacitors in the resonant circuit, class C amplifiers always operate at radio frequencies (RF). These are frequencies above 20 kHz. Therefore, even though it is the most efficient of all classes, class C is useful only for narrowband RF applications.

In addition to class C, this chapter also discusses other classes of operation, such as classes D, E, F, and S. Class D is important in transmitters; they send out RF signals that are picked up by receivers. Class S is important in switching regulators (discussed in Chap. 19).

11-1 CLASS C OPERATION

Class C operation means that the collector current flows for less than 180° of the ac cycle. This implies that the collector current of a class C amplifier is highly nonsinusoidal because current flows in pulses. To avoid the distortion that would occur with a purely resistive load, a class C amplifier always drives a resonant tank circuit. This results in a sinusoidal output voltage.

TUNED AMPLIFIER

Figure 11-1a shows one way to build a class C amplifier. The resonant tank circuit is tuned to the frequency of the input signal. When the circuit has a high quality factor (Q), *parallel resonance* occurs at approximately

$$f_r \cong \frac{1}{2\pi\sqrt{LC}} \qquad (11\text{-}1)$$

where f_r = resonant frequency
L = inductance
C = capacitance

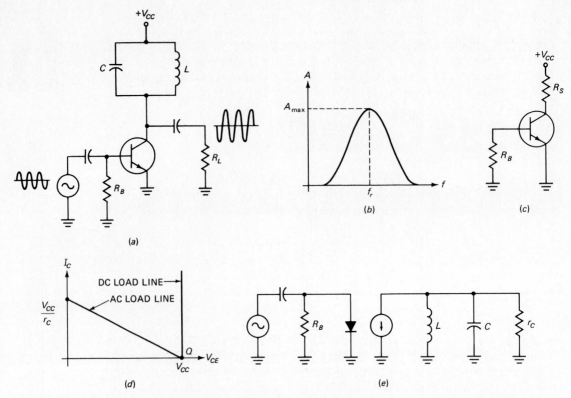

Fig. 11-1 (a) *Tuned class C amplifier.* (b) *Frequency response.* (c) *Dc equivalent circuit.* (d) *Load lines.* (e) *Ac equivalent circuit for Q greater than 10.*

At the resonant frequency, the impedance of the parallel resonant circuit is very high and is purely resistive. (This approximation assumes that the Q of the circuit is greater than 10, a condition usually satisfied in tuned RF circuits.) When the circuit is tuned to the resonant frequency, the voltage across R_L is maximum and sinusoidal.

Figure 11-1b shows the variation of voltage gain with frequency. As you can see, the voltage gain reaches a maximum value A_{max} when the frequency is f_r. Above and below this resonant frequency, the voltage gain decreases. The higher the Q of the circuit, the faster the gain drops off on either side of resonance.

NO BIAS

Figure 11-1c is the dc equivalent circuit. Notice that there is no bias applied to the transistor. Therefore, its Q point is at cutoff on the dc load line. Since there is no dc bias, V_{BE} is zero. This means that no collector current can flow until the input signal is greater than approximately 0.7 V. Also, notice that the dc collector resistance is R_S. This is the dc resistance of the RF inductor, typically a few ohms.

LOAD LINES

Since R_S is so small, the dc load line appears to be almost vertical, as shown in Fig. 11-1d. There is no danger of thermal runaway because the transistor has no current aside from leakage. The Q point is located at cutoff with no risk of thermal runaway.

The ac load line derived in Chap. 10 still applies. For a CE amplifier,

$$I_{C(\text{sat})} = I_{CQ} + \frac{V_{CEQ}}{r_C}$$

and

$$V_{CE(\text{cut})} = V_{CEQ} + I_{CQ}r_C$$

In the class C amplifier of Fig. 11-1a, $I_{CQ} = 0$ and $V_{CEQ} = V_{CC}$. Therefore, the foregoing equations reduce to

$$I_{C(\text{sat})} = \frac{V_{CC}}{r_C} \tag{11-2}$$

and

$$V_{CE(\text{cut})} = V_{CC} \tag{11-3}$$

Figure 11-1d shows the ac load line. When the transistor is conducting, its operating point swings upward along the ac load line. As before, r_C is the ac resistance seen by the collector. Therefore, the ac saturation current in a class C amplifier is V_{CC}/r_C, and the maximum voltage swing is V_{CC}.

AC EQUIVALENT CIRCUIT

When the Q of a resonant circuit is greater than 10, we can use the approximate ac equivalent circuit shown in Fig. 11-1e. In this equivalent circuit, the series resistance of the inductor is lumped into the collector resistance. In a class C amplifier, the input capacitor is part of a negative dc clamper. On the input side of a class C amplifier, therefore, the signal is negatively clamped. On the output side, the collector current source drives a parallel resonant tank circuit. At resonance, the peak-to-peak load voltage reaches a maximum.

As discussed in basic courses, the *bandwidth* of a resonant circuit is

$$B = f_2 - f_1 \tag{11-4}$$

where f_1 = lower half-power frequency
f_2 = upper half-power frequency

The bandwidth is related to the resonant frequency and the circuit Q as follows:

$$B = \frac{f_r}{Q} \tag{11-5}$$

where B = bandwidth
f_r = resonant frequency
Q = quality factor of the overall circuit

This means that a large Q produces a small bandwidth, equivalent to sharp tuning. Class C amplifiers almost always have a circuit Q that is greater than 10. This means that the bandwidth is less than 10 percent of the resonant frequency. For this reason, class C amplifiers are called *narrowband amplifiers*. The output of a narrowband amplifier is a large sinusoidal voltage at resonance with a rapid dropoff above and below resonance.

CURRENT DIP AT RESONANCE

Usually, the Q of a tuned circuit is greater than 10, which allows us to use the approximate ac equivalent circuit of Fig. 11-1e. In this equivalent circuit, the series resistance of the inductor is lumped into the collector resistance r_C. This means that an ideal inductor is in parallel with an ideal capacitor. When this circuit is resonant, the ac load impedance seen by the collector current source is purely resistive, and the collector current is minimum. Above and below resonance, the ac load impedance decreases and the collector current increases.

For example, suppose the resonant frequency is 5 MHz. When the input frequency is 5 MHz, the circuit is resonant and the collector current is minimum. If the input frequency is less than 5 MHz, the tank appears inductive and the collector current increases. Similarly, if the input frequency is greater than 5 MHz, the tank appears capacitive and the collector current increases.

One way to tune a resonant tank to the input frequency is to look for a *dip in the direct current* supplied to the circuit. In Fig. 11-1a, we can connect a dc ammeter in series with the V_{CC} supply. When the tank is adjusted to resonance, the ammeter reading will dip to a minimum value. This indicates that the circuit is resonant at the input frequency.

AC COLLECTOR RESISTANCE

Any coil or inductor has some series resistance R_S. Although schematic diagrams never show this series resistance as a separate component, it is important to remember that it does exist, as indicated in Fig. 11-2a. The Q of the inductor is given by

$$Q_L = \frac{X_L}{R_S} \tag{11-6}$$

where Q_L = quality factor of coil
X_L = inductive reactance
R_S = coil resistance

Remember that this is the Q of the coil only. The overall circuit has a lower Q because it includes the effect of load resistance as well as coil resistance.

As discussed in basic ac courses, the series resistance of the inductor can be replaced by a *parallel resistance* R_P, as shown in Fig. 11-2b. This equivalent resistance is given by

$$R_P = Q_L X_L \tag{11-7}$$

When Q_L is greater than 10, this formula has an error of less than 1 percent.

In Fig. 11-2b, it is important to realize that all the losses in the coil are now being

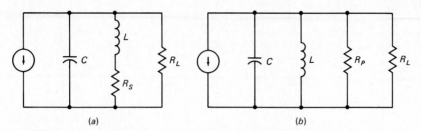

represented by the parallel resistance R_P; series resistance R_S no longer exists in the equivalent circuit. Because of this, X_L cancels X_C at resonance, leaving only R_P in parallel with R_L. This means that the ac resistance seen by the collector at resonance is

$$r_C = R_P \parallel R_L \tag{11-8}$$

The *Q of the overall circuit* is given by

$$Q = \frac{r_C}{X_L} \tag{11-9}$$

This circuit Q is lower than Q_L, the coil Q.

In practical class C amplifiers, the Q of the coil is typically 50 or more, and the Q of the circuit is 10 or more. Since the overall Q is 10 or more, the operation is narrowband. Also, since the Q of the coil is 50 or more, most of the ac load power is delivered to the load resistor, with only a small amount of power wasted in the coil resistance.

DC CLAMPING

Let's take a closer look at the dc clamping on the input side. In Fig. 11-3a, the input signal charges the coupling capacitor to approximately V_P with the polarity shown. On the positive half cycles, the emitter diode conducts briefly at the peaks; this replaces any capacitor charge lost during the cycle. On the negative half cycles, the only discharge path is through R_B. As long as the period T of the input signal is much smaller than the time constant $R_B C$, the capacitor loses only a small amount of its charge.

To replace the lost capacitor charge, the base voltage must swing slightly above 0.7 V to turn on the emitter diode briefly at each positive peak (see Fig. 11-3a). The conduction angle of base and collector current, therefore, is much less than 180°. This is why the collector current is a train of narrow pulses like Fig. 11-3b.

DUTY CYCLE

The brief turnon of the emitter diode at each positive peak produces narrow pulses of collector current. With pulses like these, it is convenient to use the *duty cycle*, defined as

$$D = \frac{W}{T} \tag{11-10}$$

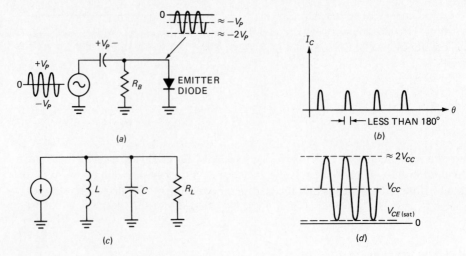

Fig. 11-3 (a) *Negative clamping at the base.* (b) *Narrow pulses of collector current.* (c) *Ac equivalent circuit.* (d) *Collector voltage.*

where D = duty cycle
$\quad W$ = width of pulse
$\quad T$ = period of pulses

For instance, if an oscilloscope shows a pulse width of 0.2 μs and a period of 1.6 μs, the duty cycle is

$$D = \frac{0.2\ \mu s}{1.6\ \mu s} = 0.125$$

This is equivalent to 12.5%.

FILTERING THE HARMONICS

As discussed earlier, any nonsinusoidal waveform is equivalent to a fundamental frequency f, a second harmonic $2f$, a third harmonic $3f$, and so on. In Fig. 11-3c, the collector current source drives the tank circuit with the nonsinusoidal current of Fig. 11-3b. If the tank is resonant at the fundamental frequency f, then all the harmonics are filtered out, and the load voltage is a sine wave at the fundamental frequency f, as shown in Fig. 11-3d. As discussed earlier, the maximum voltage swing along the ac load line is approximately V_{CC}. Therefore, under full-signal conditions, the load voltage swings from approximately $V_{CE(sat)}$ to $2V_{CC}$. Since $V_{CE(sat)}$ is close to zero, the ac output compliance of a class C amplifier is

$$PP \cong 2V_{CC} \tag{11-11}$$

The class C amplifier is rather unusual. First it clamps the input signal negatively to get highly distorted current pulses. Then it uses a high-Q resonant circuit to restore the fundamental frequency. Why this approach? Mainly to improve stage efficiency. The absence of biasing resistors means that there is less current drain. Furthermore, because of the narrow current pulses, the transistor power dissipation is lower than with class A or class B. The overall effect is less current drain, which means a higher stage efficiency. As will be discussed later, class C efficiency can approach 100 percent.

TROUBLESHOOTING

There is an excellent troubleshooting test that you can use on a class C amplifier. Since the amplifier has a negative clamped signal on the input side, you can use a dc voltmeter to measure the average voltage across the emitter diode. If the circuit is working correctly, you should read a negative voltage approximately equal to the peak of the input signal. (You can see that this is true by looking at Fig. 11-3a; the dc voltage is approximately $-V_p$.) If you try this test on a class C amplifier, be sure to use a high-impedance voltmeter to avoid loading the circuit and changing the time constant.

The voltmeter test just described is useful when an oscilloscope is not handy. If you have an oscilloscope, however, an even better test is to look across the emitter diode. You should see a negatively clamped waveform when the circuit is working properly.

EXAMPLE 11-1

Explain the waveforms shown in Fig. 11-4.

SOLUTION

The ac source signal has a peak-to-peak value of 10 V. Since it is referenced to ground, it has an average value of 0 V. The signal is negatively clamped at the base of the transistor. The dc base voltage is -4.3 V because the base voltage has to swing to approximately $+0.7$ V to turn on the emitter diode at each positive peak. Notice that the peak-to-peak value of the clamped signal is 10 V, the same as the source signal.

At the collector the signal is inverted because of the CE connection. The dc or average voltage of the collector waveform is $+15$ V, the supply voltage. (We know from earlier discussions that the Q point is located at cutoff, which implies a quiescent voltage of $+V_{CC}$.)

Across the load resistor we get the same inverted signal, except that it is referenced to ground. Why? Because a capacitor passes ac but blocks dc. For this reason, the ac voltage but not the dc gets through to the load resistor.

EXAMPLE 11-2

In Fig. 11-4, the Q of the coil is 50. Calculate the resonant frequency, ac saturation current, ac cutoff voltage, bandwidth, and ac output compliance.

Fig. 11-4
*Tuned class C
amplifier.*

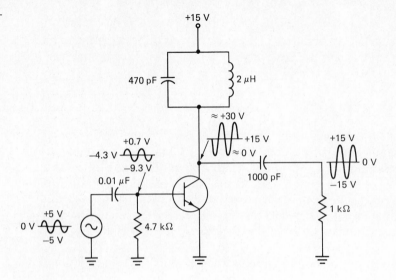

SOLUTION

The resonant frequency is

$$f_r = \frac{1}{2\pi\sqrt{(2\ \mu H)(470\ pF)}} = 5.19\ \text{MHz}$$

Therefore, the inductive reactance is

$$X_L = 2\pi(5.19\ \text{MHz})(2\ \mu H) = 65.2\ \Omega$$

From Eq. (11-7),

$$R_P = 50(65.2\ \Omega) = 3.26\ \text{k}\Omega$$

The ac load resistance is the parallel equivalent of R_P and R_L:

$$r_C = 3.26\ \text{k}\Omega \parallel 1\ \text{k}\Omega = 765\ \Omega$$

The ac saturation current is

$$I_{C(\text{sat})} = \frac{15\ \text{V}}{765\ \Omega} = 19.6\ \text{mA}$$

and the ac cutoff voltage is

$$V_{CE(\text{cut})} = 15\ \text{V}$$

The Q of the overall circuit is

$$Q = \frac{765\ \Omega}{65.2} = 11.7$$

and the bandwidth is

$$B = \frac{5.19 \text{ kHz}}{11.7} = 0.444 \text{ kHz}$$

With Eq. (11-11), the ac output compliance is

$$PP \cong 2(15 \text{ V}) = 30 \text{ V}$$

11-2 POWER RELATIONS FOR CLASS C

The load power, transistor dissipation, current drain, and stage efficiency of a class C amplifier are different from those of class A and class B amplifiers. Because the conduction angle is less than 180°, the mathematical analysis of power relations in class C amplifiers becomes very complicated, beyond the level of this book. This section will briefly describe the power relations in a class C amplifier without derivation.[1]

LOAD POWER

The *ac load power* of a class C amplifier is given by

$$P_L = \frac{V_{PP}^2}{8R_L} \tag{11-12}$$

where P_L = ac load power
V_{PP} = peak-to-peak load voltage
R_L = load resistance

This is useful when you measure the load voltage with an oscilloscope.

Maximum load power occurs when the entire ac load line is used. Since PP is the maximum unclipped value of V_{PP}, we can write the maximum load power in terms of ac output compliance:

$$P_{L(\text{max})} = \frac{PP^2}{8R_L} \tag{11-13}$$

Class C amplifiers are almost always driven hard enough to use the entire ac load line. This results in maximum load power and maximum stage efficiency.

TRANSISTOR POWER DISSIPATION

Figure 11-5a shows the ideal collector-emitter voltage in a class C transistor amplifier. Because of the resonant tank circuit, all harmonics are filtered to get a sinusoidal voltage with a fundamental frequency f_r. Since the maximum voltage is approximately $2V_{CC}$, the transistor must have a V_{CEO} rating greater than $2V_{CC}$.

---

[1]For a mathematical analysis of class C amplifiers, see H. L. Krauss, C. W. Bostian, and F. H. Raab, *Solid State Radio Engineering*, John Wiley & Sons, New York, pp. 394–428.

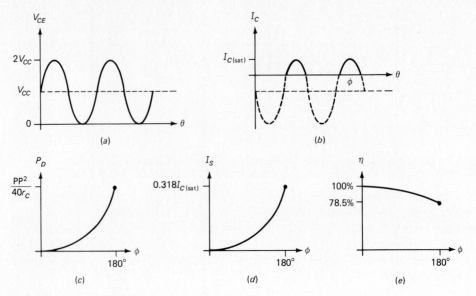

Fig. 11-5 (a) *Ideal collector voltage.* (b) *Collector current.* (c) *Transistor power dissipation.* (d) *Direct current drain.* (e) *Stage efficiency.*

Figure 11-5*b* shows the collector current for a class C amplifier. The conduction angle ϕ is less than 180°. Notice that the collector current reaches a maximum value of $I_{C(\text{sat})}$. The transistor must have a peak current rating greater than this. The dotted parts of the cycle represent the off time of the transistor.

Using calculus, it is possible to derive the *power dissipation* of the transistor. This power dissipation varies with the conduction angle, as shown in Fig. 11-5*c*. Notice that the power dissipation increases with conduction angle up to 180°, the class B case. At this point, the worst-case power dissipation of the transistor is $PP^2/40r_C$. A safe design guideline is to use transistors with a power rating greater than $PP^2/40r_C$. Under normal drive conditions, the conduction angle will be less than 180° and the transistor will be well within its power rating.

CURRENT DRAIN

In Fig. 11-5*b*, the dc or average value of the collector current depends on the conduction angle. For a conduction angle of 180°, the average current is $0.318I_{C(\text{sat})}$. For smaller conduction angles, the average current is less than this, as shown in Fig. 11-5*d*. This dc or average current is the only *current drain* in a class C amplifier.

The dc power supplied to the circuit is

$$P_S = V_{CC}I_S \qquad (11\text{-}14)$$

where P_S = dc power from supply
$\quad V_{CC}$ = supply voltage
$\quad I_S$ = dc current drain

The power dissipations are in the load, transistor, and coil. Ignoring the small ac signal power into the amplifier,

$$P_S = P_L + P_D + P_{(coil)} \qquad (11\text{-}15)$$

where P_S = dc power from supply
  P_L = ac load power
  P_D = power dissipation in transistor
  $P_{(coil)}$ = power loss in coil

Equation (11-15) says that whatever dc power goes into the circuit must come out in the form of load power or of power wasted in the transistor and coil.

STAGE EFFICIENCY

The *stage efficiency* of a class C amplifier is given by

$$\eta = \frac{P_{L(max)}}{P_S} \times 100\% \qquad (11\text{-}16)$$

In a class C amplifier, most of the dc power from the supply is converted into ac load power; the transistor and coil losses are small enough to ignore. For this reason, a class C amplifier has high stage efficiency.

Figure 11-5*e* shows how the optimum stage efficiency varies with conduction angle. When the angle is 180°, the stage efficiency is 78.5 percent, the theoretical maximum for a class B amplifier. When the conduction angle decreases, the stage efficiency increases. As indicated, class C has a maximum efficiency of 100 percent, approached at very small conduction angles.

FULL-SIGNAL DRIVE

To attain high efficiency, the ac input signal is made large enough to drive the class C amplifier over the entire ac load line. Then the output voltage has a peak-to-peak swing of approximately $2V_{CC}$. In this case, the class C amplifier is the most efficient of all classes because it delivers more load power for a given supply power than any other class. But remember, you get this efficiency only within a narrow bandwidth.

CONCLUSION

Table 11-1 summarizes the more important formulas for a class C tuned RF amplifier. It is especially important to note that the Q in these formulas includes the coil and load. This circuit Q is lower than the coil Q. In most tuned RF amplifiers, the circuit Q is greater than 10 to ensure a narrowband amplifier. Also note the maximum transistor power dissipation has a safety factor.

EXAMPLE 11-3

Reanalyze Fig. 11-4 to determine the maximum load power and the coil losses. If the transistor power dissipation is 7.5 mW, what are the current drain and stage efficiency?

Table 11-1. Class C Formulas

Quantity	Formula	Comment
f_r	$1/2\pi\sqrt{LC}$	Resonant frequency of tuned RF stage
B	f_r/Q	Use circuit Q; includes load and coil
Q	r_C/X_L	Use equivalent resistance of load and coil
$I_{C(sat)}$	V_{CC}/r_C	Use parallel resistance of load and coil
$V_{CE(cut)}$	V_{CC}	Cutoff voltage
PP	$2V_{CC}$	Ac output compliance
P_L	V_L^2/R_L	Voltage in rms volts
P_L	$V_{PP}^2/8R_L$	Voltage in peak-to-peak volts
$P_{L(max)}$	$PP^2/8R_L$	Maximum undistorted output power
$P_{D(max)}$	$PP^2/40r_C$	Includes safety factor; see text
P_S	$V_{CC}I_S$	Supply power
η	$P_{L(max)}/P_S$	Stage efficiency; multiply by 100%

SOLUTION

Since V_{CC} is 15 V,

$$PP = 2(15 \text{ V}) = 30 \text{ V}$$

and

$$P_{L(max)} = \frac{(30 \text{ V})^2}{8(1000 \text{ }\Omega)} = 113 \text{ mW}$$

The parallel resistance of the coil is 3.26 kΩ, found earlier. Therefore, the power lost to the coil is

$$P_{(coil)} = \frac{(30 \text{ V})^2}{8(3260 \text{ }\Omega)} = 34.5 \text{ mW}$$

With Eq. (11-15),

$$P_S = 113 \text{ mW} + 7.5 \text{ mW} + 34.5 \text{ mW} = 155 \text{ mW}$$

The current drain is

$$I_S = \frac{155 \text{ mW}}{15 \text{ V}} = 10.3 \text{ mA}$$

and the stage efficiency is

$$\eta = \frac{113 \text{ mW}}{155 \text{ mW}} = 72.9\%$$

11-3 FREQUENCY MULTIPLIERS

In addition to its use for efficient power amplification, a tuned class C circuit can be used as a *frequency multiplier*. The idea is to tune the resonant tank to a harmonic or multiple of the input frequency. Figure 11-6a shows narrow current pulses driving a

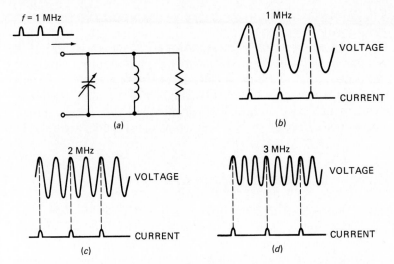

Fig. 11-6 *Frequency multipliers.* (a) *Current pulses drive resonant tank.* (b) *Tuned to fundamental.* (c) *Tuned to second harmonic.* (d) *Tuned to third harmonic.*

tuned circuit. These pulses have a fundamental frequency of 1 MHz. In an ordinary tuned class C amplifier, we tune the resonant circuit to the fundamental frequency. Then each current pulse recharges the capacitor once per output cycle (Fig. 11-6b).

Suppose we change the resonant frequency of the tank to 2 MHz. This is the second multiple or second harmonic of 1 MHz. The capacitor and inductor will now exchange energy at a 2-MHz rate and produce a load voltage of 2 MHz, as shown in Fig. 11-6c. In this case, the 1-MHz current pulses recharge the capacitor on every other output cycle. This offsets the coil and load power losses.

If we tune the tank to 3 MHz, the third harmonic of 1 MHz, we get a 3-MHz output signal (Fig. 11-6d). Now the 1-MHz current pulses recharge the capacitor on every third output cycle. As long as the Q of the resonant circuit is high, the load voltage looks almost like a perfect sine wave.

Figure 11-7 shows one way to build a frequency multiplier. The input signal has a frequency of f. This signal is negatively clamped at the base. The resulting collector current is a train of narrow current pulses with a fundamental frequency f. By tuning

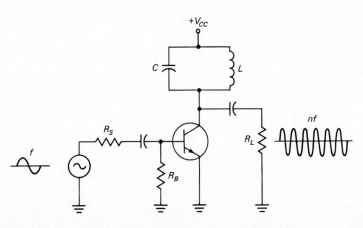

Fig. 11-7
Frequency multiplier.

the resonant tank to the *n*th harmonic, we get a load voltage whose frequency is *nf*. (Note: *n* is an integer or whole number.)

In Fig. 11-7, the current pulses recharge the capacitor every *n*th output cycle. For this reason, the load power decreases when we tune to higher harmonics. The higher *n* is, the lower the load power. Because of the decreasing efficiency at higher harmonics, a tuned class C frequency multiplier like Fig. 11-7 is normally used only for lower harmonics, like the second or third. For *n* greater than 3, we usually turn to more efficient semiconductor devices, such as the step-recovery diode (discussed in Chap. 4).

11-4 CLASS D OPERATION

The *class D amplifier* is widely used in transmitters because of its high efficiency. The basic idea behind a class D amplifier is to use transistors as switches instead of current sources. Since the power dissipation of a switch is ideally zero, the stage efficiency of a class D amplifier approaches 100 percent. In a class D amplifier, two push-pull transistor switches are used to produce a square wave, which is then filtered to recover the fundamental frequency.

CIRCUIT

Figure 11-8*a* shows one way to build a class D amplifier. This is a push-pull CE connection of two complementary transistors. An RF transformer couples the input signal to each base. On the positive half cycle of input voltage, the upper transistor cuts off and the lower one saturates. On the negative half cycle, the upper transistor saturates and the lower one cuts off. Because of this, the voltage driving the tuned circuit alternates between 0 and $+V_{CC}$.

DOT CONVENTION

In case you haven't seen dots used with transformers before, here is their meaning. Each dotted end of a transformer winding has the same phase as the other dotted ends. In Fig. 11-8*a*, on the positive half cycle of input voltage, the dotted end of the primary is positive. At the same time, the dotted end of the upper secondary winding is positive, and the dotted end of the lower secondary winding is positive. This is why the upper transistor cuts off and the lower one saturates.

TRANSISTORS ACT AS SWITCHES

The reason a class D amplifier is very efficient is because each transistor is saturated for almost 180° of the cycle. This means that each transistor acts like a switch rather than like a current source. When the transistor is saturated, the power dissipation is equal to

$$P_D = V_{CE(\text{sat})} I_{C(\text{sat})} \tag{11-17}$$

This is low because $V_{CE(\text{sat})}$ is near zero. Similarly, when the transistor is cut off, P_D is ideally zero. Therefore, the average transistor power dissipation over the cycle is very small and the stage efficiency approaches 100 percent.

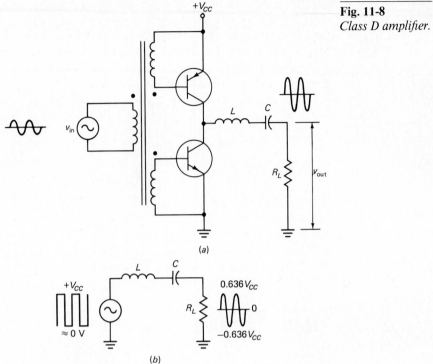

Fig. 11-8
Class D amplifier.

AC EQUIVALENT CIRCUIT

Figure 11-8*b* shows the ac equivalent circuit for the output side. The drive to the series resonant circuit is a square wave with a peak-to-peak swing of V_{CC}. This nonsinusoidal drive contains a fundamental frequency f and harmonics. As shown in Chap. 22, this square-wave collector voltage can be expressed as

$$v_c = 0.636V_{CC}\left(\sin \theta + \frac{\sin 3\theta}{3} + \frac{\sin 5\theta}{5} + \cdots\right)$$

Provided that the Q is high, the series resonant circuit transmits the fundamental frequency while blocking the harmonics. The result is a load voltage that is almost a perfect sine wave, given by

$$v_{out} = 0.636V_{CC} \sin \theta$$

The peak value of the sine wave is ideally $0.636V_{CC}$.

11-5 CLASS S OPERATION

As mentioned in Chap. 8, a series voltage regulator is popular because of its simplicity. Its main disadvantage is the power dissipation in the pass transistor. When the load current is heavy, most designers prefer to use a switching regulator, even though it is more complicated. Switching regulators are based on *class S operation*, described in this section.

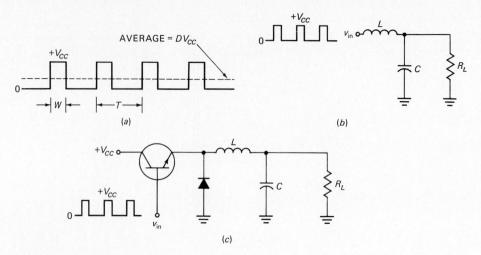

Fig. 11-9 (a) *Average value of rectangular pulses.* (b) *Rectangular pulses drive LC filter to produce a dc output voltage.* (c) *Class S amplifier.*

INPUT SIGNAL

In class S operation, a string of pulses like Fig. 11-9a is used as the input signal. The pulses have a width W and a period T. Therefore, the duty cycle D equals W/T. From basic mathematics, the average value of a rectangular waveform equals the area under a pulse divided by the period:

$$\text{Average} = \frac{\text{area}}{\text{period}}$$

In Fig. 11-9a, the average value is

$$\text{Average} = \frac{WV_{CC}}{T}$$

Since $D = W/T$, we can rewrite the equation as

$$V_{dc} = DV_{CC} \qquad (11\text{-}18)$$

This says that the average value of a string of rectangular pulses equals the duty cycle times the supply voltage. For instance, if $D = 0.1$ and $V_{CC} = 20$ V, then $V_{dc} = 2$ V.

LC FILTER

Although it is obsolete in most power supplies with a ripple frequency of 120 Hz, the *LC filter* is widely used in class S amplifiers because they operate at much higher frequencies, like 20 kHz. At these higher frequencies, an *LC* filter is an efficient way to pass the dc component and block the ac component.

Figure 11-9b shows a rectangular waveform driving an *LC* filter. If X_L is much greater than X_C at the fundamental frequency, the ac variations in the signal will be dropped across the inductor. Therefore, the output across the load resistor is a dc voltage with a small ripple. The dc voltage equals DV_{CC} because this is the average value of the input signal.

As an example, suppose that $W = 10 \ \mu s$, $T = 50 \ \mu s$, $V_{CC} = 15$ V, $L = 10$ mH, and $C = 1.5 \ \mu F$. Then the duty cycle is

$$D = \frac{10 \ \mu s}{50 \ \mu s} = 0.2$$

and the fundamental frequency is

$$f = \frac{1}{50 \ \mu s} = 20 \text{ kHz}$$

This fundamental frequency is often called the *switching frequency*. The inductive reactance is

$$X_L = 2 \ \pi(20 \text{ kHz})(10 \text{ mH}) = 1.26 \text{ k}\Omega$$

and the capacitive reactance is

$$X_C = \frac{1}{2 \ \pi(20 \text{ kHz})(1.5 \ \mu F)} = 5.31 \ \Omega$$

Since X_L is much greater than X_C, most of the ac variations are dropped across the inductor. The final output voltage is therefore a dc voltage of

$$V_{dc} = (0.2)(15 \text{ V}) = 3 \text{ V}$$

with a small ripple.

CIRCUIT

Figure 11-9c shows one way to build a class S amplifier. The transistor is an emitter follower driven by a train of pulses. Because of the V_{BE} drop, the voltage driving the LC filter is a train of pulses with an amplitude of $V_{CC} - V_{BE}$. If X_L is much greater than X_C at the switching frequency, the output is a dc voltage of

$$V_{dc} = D(V_{CC} - V_{BE}) \tag{11-19}$$

The higher the duty cycle, the larger the dc output.

A switching regulator uses a class S amplifier. By varying the duty cycle, we can regulate the dc output. Furthermore, since the transistor is either cut off or saturated, its power dissipation is much lower than that in a series regulator. When saturated, the transistor dissipates

$$P_D = V_{CE(sat)}I_{dc} \tag{11-20}$$

Since $V_{CE(sat)}$ is near zero, the power dissipation is very low.

This means that the switching regulator has two advantages over a series regulator. First, the transistor does not have to be as large, which simplifies the heat sinking. Second, the power supply runs cooler because less heat is dissipated.

INDUCTIVE KICKBACK

Notice the diode in Fig. 11-9c, included because of *inductive kickback.* As you know, the current through an inductor cannot change instantaneously. When the transistor suddenly cuts off, the extra diode continues to provide a path for the current through the inductor. Without the diode, the self-induced voltage across the inductor (inductive kickback) would produce a reverse voltage large enough to break down the emitter diode.

Remember this idea. Whenever you use a transistor as a switch that drives an inductor, you must add a diode to provide a path for inductor current when the transistor is cut off. Otherwise, the inductive kickback will probably destroy the transistor.

11-6 OTHER CLASSES OF OPERATION

Classes A, B, C, D, and S are the most important classes of operation. Classes A and B are useful in untuned or wideband circuits. Classes C and D are found in tuned or narrowband applications. Class S is mainly used in switching regulators. Besides these classes of operation, there are two others that will be mentioned briefly in this section.

CLASS E

Figure 11-10a shows a *class E amplifier.* The RF choke is an inductor whose reactance is so large that it can be approximated as an ac open. The word "choke" reminds us that it stops all alternating current through it. A sinusoidal signal is transformer-coupled into the base of the transistor. This signal is large enough to produce saturation for approximately 180° of the ac cycle. When the transistor goes into cutoff, capacitor C_1 charges. Inductance L and capacitance C_2 are tuned to produce resonance with C_1. This results in a final output voltage that is sinusoidal. With a class E amplifier, it is possible to optimize values and arrive at a stage efficiency near 100 percent.

CLASS F

Figure 11-10b is a *class F amplifier.* Capacitor C_2 is a coupling capacitor. Tank L_3 and C_3 is tuned to the third harmonic, while tank L_1 and C_1 is tuned to the fundamental frequency. For instance, if $f = 5$ MHz, then the third-harmonic tank is tuned to 15 MHz and the first-harmonic tank is tuned to 5 MHz.

Since the output tank is resonant at the fundamental frequency, the voltage across the load resistor is approximately a sine wave at the fundamental frequency. Furthermore, because the third-harmonic tank is in series with the load, the signal across the transistor is the sum of the first and third harmonics, as shown in Fig. 11-10c. As indicated, the sum of the harmonics produces a collector-emitter voltage that is more like a square wave then like a sine wave. For this reason, the transistor tends to act like a switch, which implies lower power dissipation and higher stage efficiency.

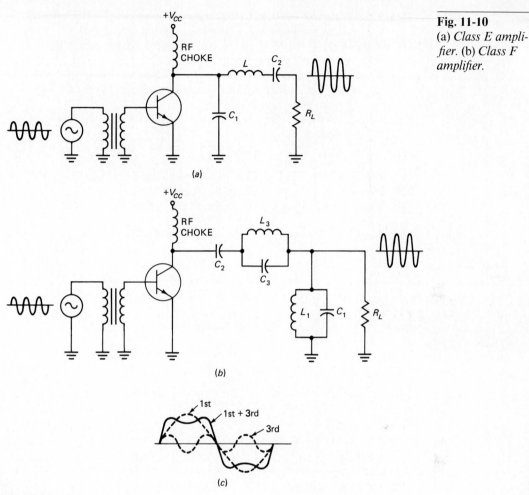

Fig. 11-10
(a) *Class E ampli-
fier.* (b) *Class F
amplifier.*

(a)

(b)

(c)

PROBLEMS

Straightforward

11-1. The RF choke of Fig. 11-11 acts like an ac open at the resonant frequency of the tank. If the Q of the coil is 60, calculate the resonant frequency, ac saturation current, ac cutoff voltage, bandwidth, and ac output compliance.

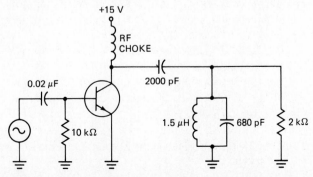

Fig. 11-11

11-2. What is the time constant of the negative clamper in Fig. 11-11? What is the period of the input signal if the tank is tuned to the input frequency?

11-3. If the input signal of Fig. 11-11 has a peak-to-peak value of 10 V, what is the dc voltage across the emitter diode?

11-4. If the Q of the coil in Fig. 11-11 is 75, what are the maximum load power and the maximum coil power? If the transistor power dissipation is 5 mW, what are the current drain and the stage efficiency?

11-5. If the tank of Fig. 11-11 is tuned to the third harmonic of the input signal, what is the input frequency?

11-6. In Fig. 11-11, what is the capacitive reactance of the 2000-pF capacitor at the resonant frequency of the tank?

11-7. If the RF choke in Fig. 11-11 has an inductance of 2 mH, what does its inductive reactance equal at the resonant frequency?

11-8. The transistor of Fig. 11-11 conducts for 30° of the ac cycle. What does the duty cycle equal?

11-9. Figure 11-12 shows a design variation of the class D amplifier; it uses two *npn* transistors instead of a complementary pair. What is the resonant frequency?

Fig. 11-12

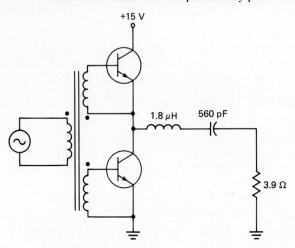

11-10. The transistors of a class D amplifier have a $V_{CE(sat)}$ of 1 V. If the $I_{C(sat)}$ is 1.92 A, what is the power dissipation of a transistor when it is conducting? If the leakage current is zero, what is the average transistor power dissipation over the ac cycle?

11-11. In Fig. 11-12, what is the peak-to-peak ac load voltage? The load power?

11-12. A rectangular waveform like Fig. 11-9a has $W = 15\ \mu s$, $T = 56\ \mu s$, and $V_{CC} = 15$ V. What is the dc value?

11-13. In the class S amplifier of Fig. 11-13, the input has a pulse width of 18 μs. If the switching frequency is 20 kHz, what is the dc load voltage?

Fig. 11-13

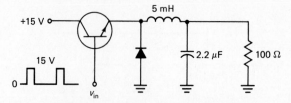

11-14. In Fig. 11-13, calculate the X_L and X_C of the LC filter for a switching frequency of 20 kHz.

11-15. If the switching frequency of Fig. 11-13 is 20 kHz and the pulse width is 19 μs, what is the current through the diode when the transistor is cut off?

Troubleshooting

11-16. In Fig. 11-11, the dc voltage between the base and ground is zero. Give some of the possible causes for this.

11-17. The load voltage of Fig. 11-12 is zero. An oscilloscope shows a peak-to-peak voltage of 15 V driving the left end of the inductor. Which of the following is a possible trouble:
  a. Inductor shorted
  b. Inductor open
  c. Capacitor shorted
  d. Capacitor open
  e. Load resistor shorted
  f. Load resistor open

11-18. You read no dc voltage across the load resistor of Fig. 11-13. Which of the following is a possible cause:
  a. Diode open
  b. Inductor shorted
  c. Inductor open
  d. Capacitor shorted
  e. Capacitor open

Design

11-19. Select a base resistor in Fig. 11-11 that will produce a discharging time constant of 100 μs.

11-20. Redesign the class C amplifier of Fig. 11-11 to have a resonant frequency of 3.5 MHz and a bandwidth less than 350 kHz. Assume a Q_{coil} of 50.

11-21. Design a class C amplifier similar to Fig. 11-11 to meet these specifications: PP = 20 V, f_r = 1 MHz, Q_{coil} = 60, R_L = 1.8 kΩ, and B = 100 kHz.

11-22. Select values of L and C that will give a resonant frequency of 1.5 MHz in Fig. 11-12. The bandwidth should be less than 150 kHz. Assume a Q_{coil} of 55.

Challenging

11-23. Figure 11-14 shows transformer coupling to the final load resistor. What is the reflected load resistance seen by the collector? If the coil has an X_L of 100 Ω and a Q_{coil} of 50, what is the Q of the circuit?

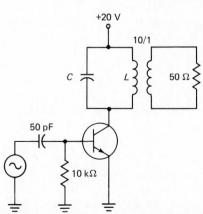

Fig. 11-14

11-24. What is the maximum load power in Fig. 11-14?

Computer

11-25. The ON . . . GOSUB statement lets you branch to a subroutine. For instance, here is a program:

10 PRINT "1. CLASS C"
20 PRINT "2. CLASS D"
30 PRINT "3. CLASS S"
40 PRINT: PRINT "ENTER CHOICE"
50 INPUT X
60 ON X GOSUB 1000, 2000, 3000
70 STOP

Variable X can be 1, 2, or 3. When line 60 is executed, the program goes to the first, second, or third subroutine. If X = 2, what line does the computer branch to? If you select class S, what line does the computer branch to?

11-26. The SQR function allows you to take the square root of a quantity. Here is a program using the SQR function:

10 PRINT "ENTER L": INPUT L
20 PRINT "ENTER C": INPUT C
30 X = SQR(L*C)
40 DEN = 2*3.1416*X
50 FR = 1/DEN
60 PRINT "THE RESONANT FREQUENCY IS ": PRINT FR

What happens in line 30? What does this program print out?

11-27. Given inputs of L, C, Q_{coil}, and R_L, write a program that prints out the resonant frequency and bandwidth of a parallel tank circuit.

JFETs

The *bipolar* transistor is the backbone of linear electronics. Its operation relies on two types of charge, holes and electrons. This is why it is called bipolar. For many linear applications, the bipolar transistor is the best choice. But there are some applications to which a *unipolar* transistor is better suited. The operation of a unipolar transistor depends on only one type of charge, either holes or electrons.

The *junction field-effect transistor,* abbreviated JFET, is an example of a unipolar transistor. This chapter discusses JFET fundamentals, biasing, amplifiers, and analog switches. JFETs are voltage-controlled devices, in contrast to bipolar transistors, which are current-controlled.

12-1 BASIC IDEAS

Figure 12-1a shows an *n*-channel JFET, a silicon bar of *n*-type semiconductor material with two islands of *p*-type material embedded in the sides. The lower end of the device is called the *source* because free electrons enter the device at this point. The upper end is known as the *drain* because free electrons leave from here. The two *p* regions are internally connected and are called the *gate*. Notice the small space between the *p* regions. It is through this narrow channel that the free electrons must pass as they move from the source to the drain. The width of the channel is important because it determines how much current there is through the JFET.

SCHEMATIC SYMBOL

Figure 12-1b shows the schematic symbol for an *n-channel* JFET. As a memory aid, visualize the thin vertical line as the *n* channel; the source and drain connect to this line. Since the gate and the channel form a *pn* diode, the gate arrow points toward the channel.

There is also a *p-channel* JFET. It consists of a silicon bar of *p* material with islands of *n*-type material embedded in the sides. The schematic symbol for a *p*-channel JFET is similar to that for an *n*-channel JFET, except that the arrow points outward. In the remainder of this chapter, we will emphasize the *n*-channel JFET. The action

Fig. 12-1
(a) *n-channel
JFET.* (b) *Schematic symbol.*

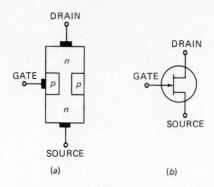

(a)   (b)

of a *p*-channel JFET is complementary, which means that all voltages and currents are reversed.

BIASING A JFET

Figure 12-2*a* shows the normal polarities for biasing an *n*-channel JFET. A positive V_{DD} supply is connected between the drain and the source. This sets up a flow of free electrons from the source to the drain. Since the electrons must pass through the channel, the drain current depends on the channel width.

Notice that a negative V_{GG} supply is connected between the gate and the source. This is standard for all JFET applications. The gate of a JFET is always reverse-biased to prevent gate current. This reverse bias sets up depletion layers around the *p* regions, as shown in Fig. 12-2*b*, and this makes the conducting channel narrower. The more negative the gate voltage is, the narrower the channel becomes because the depletion layers get closer together.

GATE-SOURCE CUTOFF VOLTAGE

When the gate voltage is negative enough, the depletion layers touch and the conducting channel pinches off (disappears). In this case, the drain current is cut off. The gate voltage that produces cutoff is symbolized $V_{GS(off)}$. For instance, a 2N5951 has a typical $V_{GS(off)}$ of -3.5 V. Data sheets for JFETs sometimes call $V_{GS(off)}$ the *pinchoff voltage.*

Fig. 12-2
(a) *Biasing a
JFET.* (b) *Depletion layers control
width of channel.*

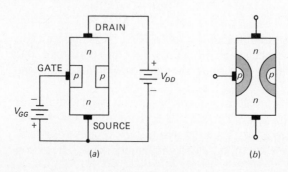

(a)   (b)

GATE LEAKAGE CURRENT

Since the gate-source junction is a reverse-biased silicon diode, only a very small reverse current flows through it. Ideally, gate current is zero. As a result, all the free electrons from the source go to the drain. This is equivalent to saying that drain current equals source current. For this reason, the only significant current in a JFET is the source-to-drain current, designated I_D; usually this is just called the drain current.

HIGH INPUT RESISTANCE

One big difference between a JFET and a bipolar transistor is the input impedance at low frequencies. Because the gate draws almost no reverse current, the input resistance of a JFET is well into the tens or hundreds of megohms. Therefore, in applications in which a very high input resistance is needed, a JFET is preferred to a bipolar transistor. The price paid for high input resistance is less control over output current. In other words, with a JFET it takes larger changes in input voltage to produce changes in output current. For this reason, a JFET amplifier has much less voltage gain than a bipolar amplifier.

DRAIN CURVES

Figure 12-3 shows a set of JFET drain curves. Notice the similarity to bipolar curves. There is a saturation region, an active region, a breakdown region, and a cutoff region. Let us start with $V_{GS} = 0$, known as the shorted-gate condition because it is equivalent to shorting the gate to the source. When $V_{GS} = 0$ the drain current rises rapidly until V_{DS} reaches 4 V. Beyond this value of V_{DS}, the drain current levels off and becomes almost horizontal. Between 4 and 30 V, the drain current is almost constant and the JFET acts like a current source of approximately 10 mA. When V_{DS} exceeds 30 V, the JFET breaks down. So, the active region of the JFET is between 4 and 30 V.

The subscripts of I_{DSS} stand for drain to source with shorted gate. I_{DSS} is the maximum drain current a JFET can produce. In Fig. 12-3, $I_{DSS} = 10$ mA at

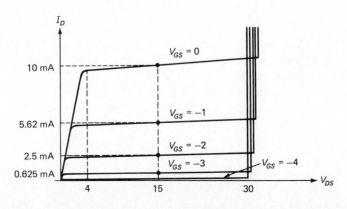

Fig. 12-3
Drain curves.

$V_{DS} = 15$ V. Because the JFET drain curves are almost horizontal, I_{DSS} is approximately 10 mA in the active region.

By making the gate voltage more negative, we can reduce the drain current. When $V_{GS} = -1$ V in Fig. 12-3, the drain current decreases to 5.62 mA. When $V_{GS} = -2$ V, the drain current equals 2.5 mA. When $V_{GS} = -3$ V, the drain current decreases to 0.625 mA. And when $V_{GS} = -4$ V, the drain current is approximately zero. The bottom curve represents the cutoff region, and so $V_{GS(off)}$ is -4 V.

Notice the saturation region of Fig. 12-3. When the JFET is saturated, V_{DS} is between 0 and 4 V, depending on the load line. Notice that the highest saturation voltage (4 V) is equal to the magnitude of gate-source cutoff voltage (-4 V). This is a property of all JFETs; it allows us to use $V_{GS(off)}$ as an estimate of the maximum saturation voltage. For instance, a 2N5457 has a $V_{GS(off)}$ of -2 V. Therefore, the maximum V_{DS} in the saturation region is approximately 2 V.

TRANSCONDUCTANCE CURVE

The *transconductance curve* of a JFET is a graph of output current versus input voltage, I_D as a function of V_{GS}. By reading the values of I_D and V_{GS} in Fig. 12-3, we can plot the transconductance curve shown in Fig. 12-4a. In general, the transconductance curve of any JFET will look like Fig. 12-4b.

The transconductance curve in Fig. 12-4b is part of a parabola. It has an equation of

$$I_D = I_{DSS}\left[1 - \frac{V_{GS}}{V_{GS(off)}}\right]^2 \tag{12-1}$$

This equation applies to any JFET because it is derived from the action of the depletion layers.[1]

Many data sheets do not show drain curves or transconductance curves. Instead, they list the values of I_{DSS} and $V_{GS(off)}$. By substituting these values into Eq. (12-1), you can calculate the drain current for any gate voltage. For instance, if a JFET has an I_{DSS} of 4 mA and a $V_{GS(off)}$ of -2 V, then substitute into Eq. (12-1) to get

$$I_D = 0.004\left(1 - \frac{V_{GS}}{-2}\right)^2$$

or

$$I_D = 0.004\left(1 + \frac{V_{GS}}{2}\right)^2$$

With this you can calculate the drain current for any gate-source voltage.

Figure 12-4c shows a normalized transconductance curve for any JFET. You can plot your own transconductance curve using the three inner points shown in Fig. 12-4c. For instance, suppose that $I_{DSS} = 12$ mA and $V_{GS(off)} = -4$ V. When $V_{GS} = -1$ V,

$$I_D = \frac{9}{16}\,(12\text{ mA}) = 6.75\text{ mA}$$

[1] If you are interested in the mathematical derivation of this equation, see L. J. Sevin, *Field-effect Transistors*, McGraw-Hill Book Company, New York, 1965, pp. 1–23.

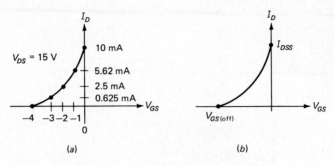

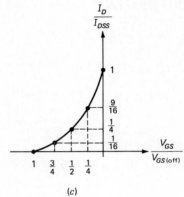

Fig. 12-4
Transconductance curves.

When $V_{GS} = -2$ V,

$$I_D = \frac{1}{4}\ (12\ \text{mA}) = 3\ \text{mA}$$

When $V_{GS} = -3$ V,

$$I_D = \frac{1}{16}\ (12\ \text{mA}) = 0.75\ \text{mA}$$

Now you can plot five known points and connect a smooth curve through them. (The data sheet gives two points, saturation and cutoff, and you have calculated three more.) This is how you get the transconductance curve if the data sheet of a JFET does not show it.

One final point: Square-law is another name for parabolic. This is why JFETs are often called *square-law* devices. For reasons that will be given later, the square-law property gives the JFET another advantage over the bipolar transistor.

EXAMPLE 12-1

Gate leakage current is symbolized I_{GSS}. If the data sheet of a 2N5951 lists a typical I_{GSS} of 5 pA at 20 V, what is the dc input resistance?

SOLUTION

The input resistance is

$$R_{GS} = \frac{20 \text{ V}}{5 \text{ pA}} = 4(10^{12}) \text{ }\Omega$$

Here you see the major advantage of a JFET over a bipolar transistor. The dc input resistance is enormously high. Because of this, the JFET is useful at the front end of electronic voltmeters, oscilloscopes, and other instruments that need a high input resistance to avoid loading the circuit being tested.

EXAMPLE 12-2

The data sheet of a 2N5951 lists these typical values: $I_{DSS} = 10$ mA and $V_{GS(off)} = -3.5$ V. Calculate the drain current for $V_{GS} = -1$ V, -2 V, and -3 V.

SOLUTION

When $V_{GS} = -1$ V, Eq. (12-1) gives

$$I_D = 0.01 \left(1 - \frac{-1}{-3.5}\right)^2 = 5.1 \text{ mA}$$

When $V_{GS} = -2$ V,

$$I_D = 0.01 \left(1 - \frac{-2}{-3.5}\right)^2 = 1.84 \text{ mA}$$

When $V_{GS} = -3$ V,

$$I_D = 0.01 \left(1 - \frac{-3}{-3.5}\right)^2 = 0.204 \text{ mA}$$

12-2 GATE BIAS

Figure 12-5*a* is an example of *gate bias* (similar to base bias of a bipolar transistor). Figure 12-5*b* shows a simplified drawing. This is the worst possible way to set up the *Q* point of a linear JFET amplifier. Here is the reason. There is considerable variation between the minimum and maximum values of JFET parameters. For example, here are the parameters for a 2N5459:

	I_{DSS}	$V_{GS(off)}$
Minimum	4 mA	−2 V
Maximum	16 mA	−8 V

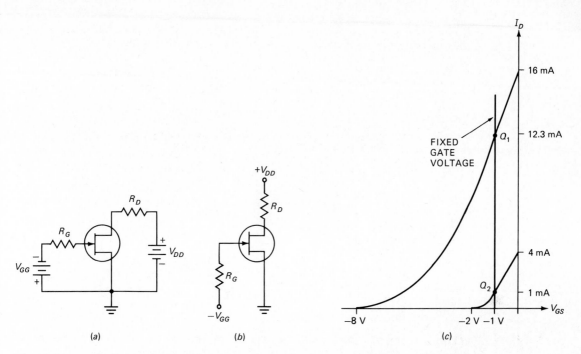

Fig. 12-5 (a) *Gate bias.* (b) *Simplified drawing.* (c) *Q point varies excessively with gate bias.*

This implies that the minimum and maximum transconductance curves are displaced as shown in Fig. 12-5c. Gate bias applies a fixed voltage to the gate. This fixed gate voltage results in a Q point that is highly sensitive to the particular JFET used.

For instance, suppose that $V_{GS} = -1$ V, and we draw a vertical line at $V_{GS} = -1$ V, as shown in Fig. 12-5c. In mass production, a gate-biased circuit using the 2N5459 can have a Q point located anywhere between Q_1 and Q_2. The drain current for Q_1 is

$$I_D = 0.016 \left(1 - \frac{1}{8}\right)^2 = 12.3 \text{ mA}$$

whereas the drain current for Q_2 is only

$$I_D = 0.004 \left(1 - \frac{1}{2}\right)^2 = 1 \text{ mA}$$

The variation in drain current is so large that gate bias is out of the question for setting up a solid Q point.

12-3 SELF-BIAS

Figure 12-6a shows *self-bias,* another way to bias a JFET. Notice that only a drain supply is used; there is no gate supply. The idea is to use the voltage across the source resistor R_S to produce the gate-source reverse voltage. This is a form of local feedback

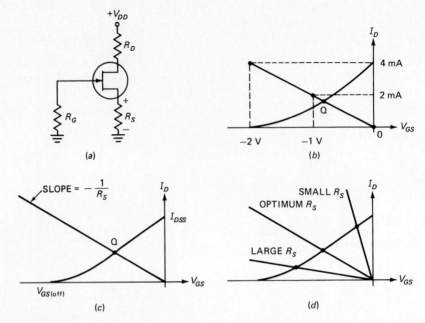

Fig. 12-6 (a) *Self-bias.* (b) *Drawing the self-bias line.* (c) *Self-bias line intersects transconductance curve at the Q point.* (d) *Effect of different biasing resistors.*

similar to that used with bipolar transistors. Recall how this feedback works. If the drain current increases, the voltage drop across R_S increases because the $I_D R_S$ product increases. This increases the gate-source reverse voltage, which makes the channel narrower and reduces the drain current. The overall effect is to partially offset the original increase in drain current.

Similarly, if drain current decreases, the gate-source reverse voltage decreases and the channel gets wider. This allows more free electrons through, and the drain current increases. This partially offsets the original decrease in drain current.

GATE-SOURCE VOLTAGE

Since the gate is reverse-biased in Fig. 12-6a, negligible gate current flows through R_G, and so the gate voltage with respect to ground is zero:

$$V_G = 0$$

The source voltage to ground equals the product of the drain current and the source resistance:

$$V_S = I_D R_S$$

The gate-source voltage is the difference between the gate voltage and the source voltage:

$$V_{GS} = V_G - V_S = 0 - I_D R_S$$

or

$$V_{GS} = -I_D R_S \qquad (12\text{-}2)$$

This means that the gate-source voltage equals the negative of the voltage across the source resistor. So, the greater the drain current, the more negative the gate-source voltage becomes.

SELF-BIAS LINE

Rearranging Eq. (12-2), we get

$$I_D = \frac{-V_{GS}}{R_S} \qquad (12\text{-}3)$$

The graph of this equation is called the *self-bias line.* For instance, if $R_S = 500\ \Omega$, then the equation of the self-bias line is

$$I_D = \frac{-V_{GS}}{500\ \Omega}$$

When $V_{GS} = 0$,

$$I_D = \frac{0}{500\ \Omega} = 0$$

When $V_{GS} = -1$ V, the drain current is

$$I_D = \frac{1\ \text{V}}{500\ \Omega} = 2\ \text{mA}$$

When $V_{GS} = -2$ V,

$$I_D = \frac{2\ \text{V}}{500\ \Omega} = 4\ \text{mA}$$

As you can see, drain current increases linearly with changes in V_{GS}.

Figure 12-6b shows a transconductance curve with an I_{DSS} of 4 mA and a $V_{GS(\text{off})}$ of -2 V. If we plot I_D and V_{GS} for a source resistance of 500 Ω, we get the self-bias line shown in Fig. 12-6b. The Q point is the point of intersection between the transconductance curve and the self-bias line. Notice that the drain current at the Q point is slightly less than 2 mA.

Any self-biased circuit has a transconductance curve and a self-bias line, as shown in Fig. 12-6c. The slope of the self-bias line is $-1/R_S$ because this line is the graph of Ohm's law for resistance R_S. The only way to satisfy both Ohm's law and the transconductance curve is with the Q point, located at the point of intersection.

EFFECT OF SOURCE RESISTANCE

Figure 12-6d shows how the Q point changes when the source resistance changes. When R_S is large, the Q point is far down the transconductance curve and the drain

Fig. 12-7
Transconductance curve.

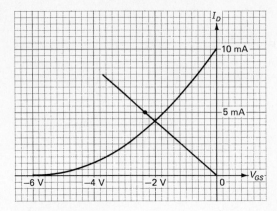

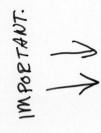

current is small. When R_S is small, the Q point is far up the transconductance curve and the drain current is large. In between, there is an optimum value of R_S that sets up a Q point near the middle of the transconductance curve.

GRAPHICAL ANALYSIS

When the data sheet of a JFET includes a transconductance curve (sometimes called the transfer characteristic), you can find the Q point of a self-biased JFET as follows:

1. Select any convenient value of drain current.
2. Multiply the assumed drain current by R_S.
3. Plot the assumed current and corresponding V_{GS}.
4. Draw a line through the plotted point and the origin.
5. Read the coordinates of the Q point.

For instance, suppose R_S is 470 Ω and the data sheet of a JFET gives the transconductance curve of Fig. 12-7. The first step is to assume a convenient value of drain current. Let us assume half of I_{DSS}, or 5 mA. Then the corresponding voltage drop across R_S is

$$V_S = (5 \text{ mA})(470 \text{ } \Omega) = 2.35 \text{ V}$$

which means that $V_{GS} = -2.35$ V. After plotting the point and drawing the self-bias line, we can read the following approximate values at the Q point:

$$I_D = 4.3 \text{ mA}$$
$$V_{GS} = -2 \text{ V}$$

GRAPHICAL DESIGN

If you are designing a self-bias circuit, draw a self-bias line so that it intersects the transconductance curve somewhere near the middle of the transconductance curve. Then read the coordinates of the Q point. The ratio of voltage to current gives you the design value of R_S:

$$R_S = \frac{-V_{GS}}{I_D} \tag{12-4}$$

CONVENIENT DESIGN

Here is another way to select the value of R_S for a self-bias circuit. In Fig. 12-8a, a self-bias line is drawn through the point with coordinates I_{DSS} and $V_{GS(off)}$. Every point on the self-bias line must satisfy Ohm's law. Therefore, using the coordinates of the upper point, we can calculate a source resistance of

$$R_S = \frac{-V_{GS(off)}}{I_{DSS}} \tag{12-5}$$

If you satisfy this design equation, you have a self-biased JFET with a Q point somewhere near the middle of the transconductance curve.

For example, if $V_{GS(off)} = -6$ V and $I_{DSS} = 10$ mA,

$$R_S = \frac{6\text{ V}}{10\text{ mA}} = 600\text{ }\Omega$$

The nearest standard value is 620 Ω. Figure 12-8b shows the transconductance curve and the self-bias line for a source resistance of 600 Ω. The Q point is near the middle of the transconductance curve.

TOLERANCES

Because of the wide variations in I_{DSS} and $V_{GS(off)}$, you have to strike a compromise for the value of source resistance to be used in mass production. If the data sheet of a JFET includes both minimum and maximum transconductance curves, you can draw a self-bias line through both curves to get a Q point that is near the middle of either transconductance curve.

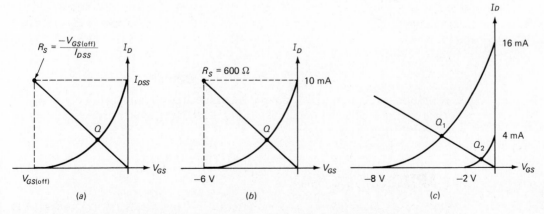

Fig. 12-8   (a) *One way to calculate biasing resistor.* (b) *An example.* (c) *Minimum and maximum transconductance curves.*

For example, suppose a data sheet shows the minimum and maximum transconductance curves of Fig. 12-8c. Draw a self-bias line that approximately goes through the center of each transconductance curve. Next, read the voltage and current for either Q point. Finally, calculate the optimum R_S with $-V_{GS}/I_D$.

If the data sheet does not show the minimum and maximum transconductance curves, use Eq. (12-5) to calculate the minimum R_S and the maximum R_S. For instance, the data sheet of a 2N5486 lists these parameters:

	I_{DSS}	$V_{GS(off)}$
Minimum	8 mA	-2 V
Maximum	20 mA	-6 V

Using Eq. (12-5),

$$R_S = \frac{2 \text{ V}}{8 \text{ mA}} = 250 \text{ }\Omega$$

and

$$R_S = \frac{6 \text{ V}}{20 \text{ mA}} = 300 \text{ }\Omega$$

The average of the two is 275 Ω. The nearest standard value is 270 Ω, which would produce a Q point near the middle of either transconductance curve.

EXAMPLE 12-3

Figure 12-9 shows the transconductance curve of a 2N5457. If this JFET is used in a self-biased circuit, what are the quiescent current and voltage for an R_S of 100 Ω? For an R_S of 1 kΩ? What source resistance do you get with Eq. (12-5)?

SOLUTION

Assume a drain current of 5 mA. Using Eq. (12-2),

$$V_{GS} = (-5 \text{ mA})(100 \text{ }\Omega) = -0.5 \text{ V}$$

Plot this point and draw a line between it and the origin. If you do this correctly, you get the upper self-bias line shown in Fig. 12-9. Next, read the coordinates of Q_1 to get the quiescent values:

$$I_D \cong 3.33 \text{ mA}$$
$$V_{GS} \cong -0.32 \text{ V}$$

For an R_S of 1 kΩ, assume a drain current of 2.5 mA. Then

$$V_{GS} = (-2.5 \text{ mA})(1 \text{ k}\Omega) = -2.5 \text{ V}$$

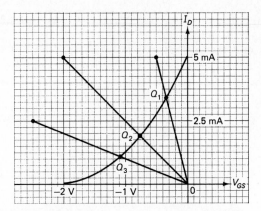

Fig. 12-9
Transconductance curve with three self-bias lines.

Plotting the point and drawing the self-bias line gives us the lower line in Fig. 12-9. The coordinates of Q_3 are

$$I_D \cong 1.2 \text{ mA}$$
$$V_{GS} \cong -1.1 \text{ V}$$

The middle self-bias line is what you get with Eq. (12-5), the formula for quick design. At Q_2, the coordinates are

$$I_D \cong 1.9 \text{ mA}$$
$$V_{GS} \cong -0.75 \text{ V}$$

which implies a source resistance of

$$R_S = \frac{0.75 \text{ V}}{1.9 \text{ mA}} = 395 \ \Omega$$

12-4 VOLTAGE-DIVIDER AND SOURCE BIAS

Self-bias is one way to stabilize the Q point. In this section, we discuss two more biasing methods, both of which are similar to methods used with bipolar transistors.

VOLTAGE-DIVIDER BIAS

Figure 12-10a shows one of the better ways to bias a JFET. The idea is similar to the *voltage-divider bias* used with a bipolar transistor. The Thevenin voltage V_{TH} applied to the gate is

$$V_{TH} = \frac{R_2}{R_1 + R_2} V_{DD} \qquad (12\text{-}6)$$

This is the dc voltage from the gate to ground. Because of V_{GS}, the voltage from the source to ground is

$$V_S = V_{TH} - V_{GS} \qquad (12\text{-}7)$$

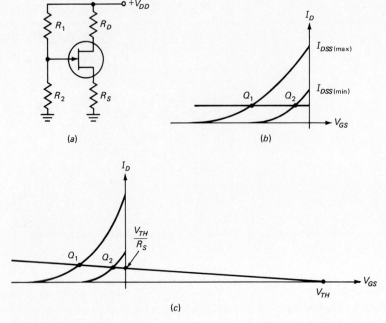

Fig. 12-10 (a) *Voltage-divider bias.* (b) *Drain current is ideally constant.* (c) *Drain current increases slightly.*

Therefore, the drain current equals

$$I_D = \frac{V_{TH} - V_{GS}}{R_S} \qquad (12\text{-}8)$$

and the dc voltage from the drain to ground is

$$V_D = V_{DD} - I_D R_D \qquad (12\text{-}9)$$

If V_{TH} is large enough to swamp out V_{GS} in Eq. (12-8), the drain current is approximately constant for any JFET, as shown in Fig. 12-10*b*.

But there is a problem. In a bipolar transistor, V_{BE} is approximately 0.7 V, with only minor variations from one transistor to the next. In a JFET, however, V_{GS} can vary several volts from one JFET to another. With typical supply voltages, it is difficult to make V_{TH} large enough to swamp out V_{GS}. For this reason, voltage-divider bias is less effective with JFETs than with bipolars.

If we plot Eq. (12-8), we get the bias line shown in Fig. 12-10*c*. Notice how the drain current increases slightly going from Q_2 to Q_1. The larger V_{TH} is, the more horizontal the bias line becomes. But there is a limit to how big we can make V_{TH}. Therefore, although it is a big improvement, voltage-divider bias still falls short of providing the solid kind of Q point we are looking for.

SOURCE BIAS

Figure 12-11*a* shows *source bias* (similar to emitter bias). The idea is to swamp out the variations in V_{GS}. Since most of V_{SS} appears across R_S, the drain current is

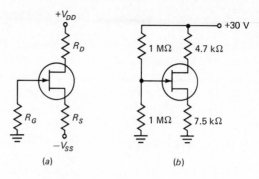

Fig. 12-11
(a) *Source bias.*
(b) *Example of voltage-divider bias.*

(a)

(b)

roughly equal to V_{SS}/R_S. The exact value is given by

$$I_D = \frac{V_{SS} - V_{GS}}{R_S} \qquad (12\text{-}10)$$

For source bias to work well, V_{SS} must be much greater than V_{GS}. However, a typical range for V_{GS} is from -1 to -5 V, so you can see that perfect swamping is not possible with typical supply voltages.

EXAMPLE 12-4

In Fig. 12-11b, if the minimum V_{GS} is -1 V, what is the drain current? If the maximum V_{GS} is -5 V, what is the drain current? Calculate the drain voltage to ground.

SOLUTION

The voltage divider produces a Thevenin voltage of 15 V. The minimum drain current is

$$I_D = \frac{15 \text{ V} - (-1 \text{ V})}{7.5 \text{ k}\Omega} = 2.13 \text{ mA}$$

The maximum drain current is

$$I_D = \frac{15 \text{ V} - (-5 \text{ V})}{7.5 \text{ k}\Omega} = 2.67 \text{ mA}$$

When $I_D = 2.13$ mA, the drain voltage is

$$V_D = 30 \text{ V} - (2.13 \text{ mA})(4.7 \text{ k}\Omega) = 20 \text{ V}$$

When $I_D = 2.67$ mA, the drain voltage is

$$V_D = 30 \text{ V} - (2.67 \text{ mA}) (4.7 \text{ k}\Omega) = 17.5 \text{ V}$$

Between Q_2 and Q_1, there is a slight increase in drain current and a slight decrease in drain voltage. Our conclusion is that voltage-divider bias works fairly well in maintaining a fixed Q point.

12-5 CURRENT-SOURCE BIAS

There is a way to get a solid Q point with JFETs. We need to produce a drain current that is independent of V_{GS}. Voltage-divider bias and source bias attempt to do this by swamping out the variations in V_{GS}. But, as already mentioned, it is difficult to completely swamp out V_{GS} with typical supply voltages because V_{GS} may be several volts. This section discusses two circuits that truly swamp out V_{GS}.

TWO SUPPLIES

When positive and negative supplies are available, you can use *current-source bias,* shown in Fig. 12-12a. Since the bipolar transistor is emitter-biased, its collector current is given by

$$I_C = \frac{V_{EE} - V_{BE}}{R_E} \tag{12-11}$$

Because the bipolar transistor acts like a dc current source, it forces the JFET drain current to equal the bipolar collector current:

$$I_D = I_C$$

Figure 12-12b illustrates how effective current-source bias is. Since I_C is constant, both Q points have the same value of drain current. The current source effectively wipes out the influence of V_{GS}. Although V_{GS} is different for each Q point, it no longer influences the value of drain current.

ONE SUPPLY

When only a positive supply is available, you can use a circuit like Fig. 12-13a to set up a constant drain current. In this case, the bipolar transistor is voltage-divider-biased. Assuming a stiff voltage divider, the emitter and collector currents are constant for all bipolar transistors. This forces the JFET drain current to equal the bipolar collector current.

Fig. 12-12
(a) *Current-source bias.* (b) *Drain current is constant.*

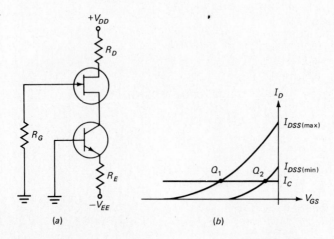

(a)   (b)

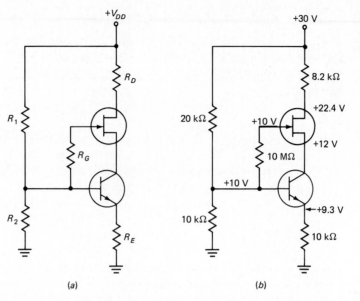

Fig. 12-13
*Single-supply
current-source
bias.*

(a)　　　　　　(b)

EXAMPLE 12-5

If $V_{GS} = -2$ V, calculate all currents and voltages in Fig. 12-13b.

SOLUTION

The voltage divider produces a Thevenin voltage of 10 V. Therefore, the emitter current is

$$I_E = \frac{10 \text{ V} - 0.7 \text{ V}}{10 \text{ k}\Omega} = 0.93 \text{ mA}$$

So the drain current is approximately 0.93 mA. The dc voltage from the drain to ground is

$$V_D = 30 \text{ V} - (0.93 \text{ mA})(8.2 \text{ k}\Omega) = 22.4 \text{ V}$$

Since the gate current can be ignored, the dc voltage from the gate to ground equals the Thevenin voltage out of the divider:

$$V_G = 10 \text{ V}$$

Because $V_{GS} = -2$ V, the dc voltage from the source to ground is

$$V_S = 10 \text{ V} - (-2 \text{ V}) = 12 \text{ V}$$

12-6 TRANSCONDUCTANCE

Before analyzing JFET amplifiers, we need to discuss an ac quantity called the *transconductance*, designated g_m. In symbols, transconductance is given by

$$g_m = \frac{\Delta I_D}{\Delta V_{GS}} \qquad (12\text{-}12)$$

Because the changes in I_D and V_{GS} are equivalent to ac current and voltage, Eq. (12-12) can be written as

$$g_m = \frac{i_d}{v_{gs}} \qquad (12\text{-}13)$$

This is a small-signal formula, exact for infinitesimal changes. We will use Eq. (12-13) as an approximation for g_m whenever the peak-to-peak value of i_d is less than 10 percent of the drain current at the Q point.

If the peak-to-peak values are $i_d = 0.2$ mA and $v_{gs} = 0.1$ V, then

$$g_m = \frac{0.2 \text{ mA}}{0.1 \text{ V}} = 2(10^{-3}) \text{ S} = 2000 \text{ } \mu\text{S}$$

Note: S is the symbol for the unit *siemens,* formerly referred to as the *mho*. The unit represents conductance, the ratio of current to voltage.

Most data sheets continue to use mho instead of siemens. They also use the symbol g_{fs} for g_m. As an example, the data sheet of a 2N5451 lists a typical g_{fs} of 2000 μmhos for a drain current of 1 mA. This is identical to saying that the 2N5451 has a typical g_m of 2000 μS at 1 mA.

GRAPHICAL MEANING

Figure 12-14*a* brings out the meaning of g_m in terms of the transconductance curve. Between points A and B, a change in V_{GS} produces a change in I_D. The ratio of the change in I_D to the change in V_{GS} equals the value of g_m between A and B. If we select another pair of points further up the curve at C and D, we get more of a change in I_D for a given change in V_{GS}. Therefore, g_m has a larger value further up the curve. In a nutshell, g_m tells us how much control gate voltage has over drain current. The higher g_m is, the more effective gate voltage is in controlling drain current.

Data sheets for JFETs usually include a graph that shows how g_m varies with quiescent drain current. Therefore, once we figure out how much dc drain current a JFET amplifier has, we can look up the value of g_m for this value of quiescent drain current.

SIMPLE JFET MODEL

Figure 12-14*b* shows a simple ac equivalent circuit for a JFET. A very high resistance R_{GS} is between the gate and the source. This is well into the tens or hundreds of megohms. The drain of a JFET acts like a current source with a value of $g_m v_{gs}$. If we know the g_m and the v_{gs}, we can calculate the ac drain current.

This model is a first approximation because it does not include the internal resistance r_{ds} of the current source, capacitances inside the JFET, and so on. At low frequencies, we can use this simple ac model for troubleshooting and preliminary design.

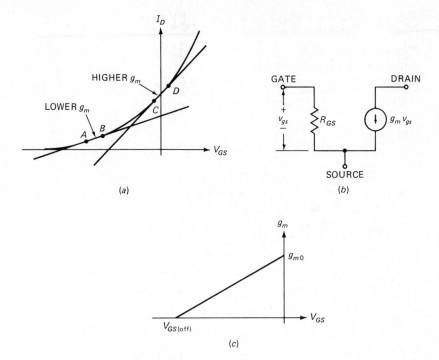

Fig. 12-14 (a) *Transconductance increases at higher drain current.* (b) *Ac equivalent circuit of JFET.* (c) *Variation of transconductance with gate-source voltage.*

VALUE OF TRANSCONDUCTANCE

When $V_{GS} = 0$, g_m has its maximum value. This maximum value is designated g_{m0}, or g_{fs0} on data sheets. In Appendix 1, we derive this useful relation between the g_m at any Q point and the maximum g_m:

$$g_m = g_{m0}\left[1 - \frac{V_{GS}}{V_{GS(\text{off})}}\right] \tag{12-14}$$

Notice that g_m decreases linearly when V_{GS} becomes more negative, as shown in Fig. 12-14c. This property is useful in automatic gain control, which will be discussed later.

AN ACCURATE VALUE OF $V_{GS(\text{off})}$

Appendix 1 derives this formula:

$$V_{GS(\text{off})} = \frac{-2I_{DSS}}{g_{m0}} \tag{12-15}$$

The quantity $V_{GS(\text{off})}$ is very difficult to measure accurately. On the other hand, I_{DSS} and g_{m0} are easy to measure with high accuracy. Therefore, the standard approach is to measure I_{DSS} and g_{m0}, then calculate $V_{GS(\text{off})}$.

12-7 THE CS AMPLIFIER

Figure 12-15*a* shows a common-source (CS) amplifier. When a small ac signal is coupled into the gate, it produces variations in gate-source voltage. This produces a sinusoidal drain current. Since an alternating current flows through the drain resistor, we get an amplified ac voltage at the output.

PHASE INVERSION

An increase in gate-source voltage produces more drain current, which means that the drain voltage is decreasing. Since the positive half cycle of input voltage produces the negative half cycle of output voltage, we get phase inversion in a CS amplifier.

VOLTAGE GAIN

Figure 12-15*b* shows the ac equivalent of a CS amplifier. On the input side, R_1 is in parallel with R_2; the internal R_{GS} is high enough to ignore. The ac output voltage is

$$v_{\text{out}} = -g_m v_{gs} R_D$$

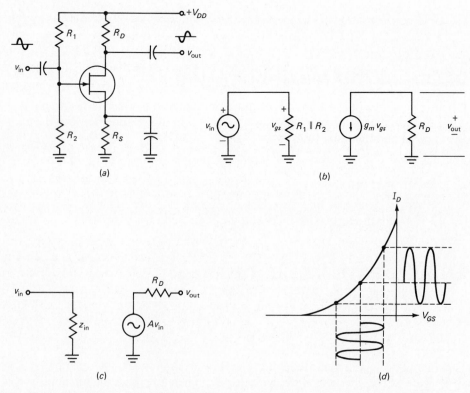

(a)

(b)

(c)

(d)

Fig. 12-15 (a) *A common-source amplifier.* (b) *Ac equivalent circuit.* (c) *Another equivalent circuit.* (d) *Distortion.*

where the minus sign indicates phase inversion. The ac input voltage equals

$$v_{in} = v_{gs}$$

because the ac source is connected directly between the gate-source terminals. Taking the ratio of the output to the input voltage gives

$$\frac{v_{out}}{v_{in}} = -g_m R_D$$

which can be written as

$$A = -g_m R_D \tag{12-16}$$

where A = unloaded voltage gain
g_m = transconductance
R_D = drain resistance

This is unloaded voltage gain because the output coupling capacitor is not connected to a load resistor.

AC MODEL OF CS STAGE

Figure 12-15c shows the ac model of a CS amplifier. Notice how similar it is to the ac model of a CE amplifier. To begin with, there is an input impedance z_{in}. At low frequencies, this is the equivalent resistance of the biasing resistors in the gate circuit. For instance, if voltage-divider bias is used, $z_{in} = R_1 \| R_2$. If self bias is used, $z_{in} = R_G$, and so on for the other forms of bias. As before, we can thevenize the output circuit to get an ac voltage source and an output impedance. For a CS amplifier this means an ac voltage source of Av_{in} and an output impedance of R_D.

DISTORTION

Because of the nonlinear transconductance curve, a JFET distorts large signals, as shown in Fig. 12-15d. Given a sinusoidal input voltage, we get a nonsinusoidal output current in which the positive half cycle is elongated and the negative half cycle is compressed. This type of distortion is called *square-law distortion* because the transconductance curve is parabolic. There is an important application of this type of distortion in frequency mixers (Chap. 23).

As far as linear amplifiers are concerned, distortion is undesirable. One way to minimize square-law distortion in JFET amplifiers is to keep the signal small. In that case only a small part of the transconductance curve is used, which means that the operation is approximately linear. As mentioned earlier, we consider the signal small when the peak-to-peak drain current is less than 10 percent of the quiescent drain current. This 10:1 rule keeps the distortion down to low enough levels for most applications. (Chapter 22 discusses distortion in more detail.)

SWAMPING RESISTOR

Sometimes a *swamping resistor* is added to the source resistance, as shown in Fig. 12-16a. Now the source is no longer at ac ground. The drain current through r_S

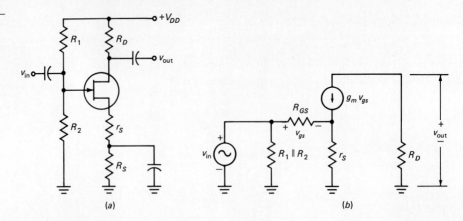

(a)   (b)

produces an ac voltage between the source and ground. If r_S is large enough, the local feedback can swamp out the nonlinearity of the transconductance curve. Then the voltage gain approaches an ideal value of $-R_D/r_S$, which is the ratio of drain resistance to unbypassed source resistance. (Notice the similarity to $-R_C/r_E$, the gain of a swamped bipolar amplifier.)

Here is the mathematical analysis. Figure 12-16b is the ac equivalent circuit. Since R_{GS} approaches infinity, all the ac drain current flows through r_S, producing a voltage drop of $g_m v_{gs} r_S$. Summing voltages around the input loop gives

$$v_{gs} + g_m v_{gs} r_S - v_{in} = 0$$

or

$$v_{in} = (1 + g_m r_S) v_{gs}$$

The ac output voltage is

$$v_{out} = -g_m v_{gs} R_D$$

Taking the ratio of output to input gives

$$\frac{v_{out}}{v_{in}} = \frac{-g_m R_D}{1 + g_m r_S}$$

which can be rewritten as

$$A = -\frac{R_D}{r_S + 1/g_m} \tag{12-17}$$

where A = unloaded voltage gain
        R_D = drain resistance
        r_S = swamping resistance
        g_m = transconductance

This tells us two things. First, the swamping resistor reduces the voltage gain. Second, changes in g_m from one JFET to the next have less effect on voltage gain. To swamp out the changes in g_m, r_S must be much greater than $1/g_m$. In this case, A approaches a value of $-R_D/r_S$.

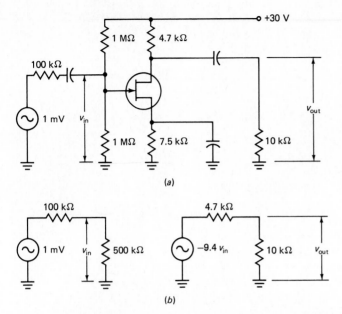

Fig. 12-17 (a) *CS amplifier with generator and load resistances.* (b) *Ac equivalent circuit.*

EXAMPLE 12-6

If $g_m = 2000\ \mu S$ for the JFET of Fig. 12-17a, what is the ac output voltage?

SOLUTION

The unloaded voltage gain is

$$A = (-2000\ \mu S)(4.7\ k\Omega) = -9.4$$

The input impedance of the amplifier is

$$z_{in} = 1\ M\Omega\ \|\ 1\ M\Omega = 500\ k\Omega$$

Therefore, we can visualize the ac circuit as shown in Fig. 12-17b. The input voltage divider reduces the signal at the gate to

$$v_{in} = \frac{500\ k\Omega}{600\ k\Omega} 1\ mV = 0.833\ mV$$

The Thevenin output voltage is

$$Av_{in} = -9.4(0.833\ mV) = -7.83\ mV$$

This is the unloaded output. The actual output is what appears across the 10-kΩ resistor:

$$v_{out} = \frac{10\ k\Omega}{14.7\ k\Omega} (-7.83\ mV) = -5.33\ mV$$

Notice how small the overall voltage gain is with a JFET amplifier. This is typical; JFETs produce much less voltage gain than bipolars.

EXAMPLE 12-7

If a swamping resistance of 1 kΩ is used in the preceding example, what is the output voltage?

SOLUTION

The reciprocal of g_m is

$$\frac{1}{g_m} = \frac{1}{2000\ \mu S} = 500\ \Omega$$

From Eq. (12-17), the unloaded voltage gain is

$$A = -\frac{4700}{1000 + 500} = -3.13$$

The ac input voltage to the gate is still 0.833 mV, but the Thevenin output voltage drops to

$$Av_{in} = (-3.13)(0.833\ \text{mV}) = -2.61\ \text{mV}$$

So, the final output voltage is

$$v_{out} = \frac{10\ \text{k}\Omega}{14.7\ \text{k}\Omega}(-2.61) = -1.78\ \text{mV}$$

12-8 THE CD AMPLIFIER

Figure 12-18*a* is a *common-drain* (CD) amplifier. It is similar to an emitter follower. An ac signal drives the gate, producing an ac drain current. This flows through the unbypassed source resistor and produces an ac output voltage that is approximately equal to and in phase with the input voltage. For this reason, the circuit is called a *source follower*. Because of its high input impedance, a source follower is often used at the front end of measuring instruments like electronic voltmeters and oscilloscopes.

VOLTAGE GAIN

Figure 12-18*b* shows the ac equivalent circuit of a source follower. To derive a formula for voltage gain, start by summing voltages around the input loop:

$$v_{gs} + g_m v_{gs} R_S - v_{in} = 0$$

or

$$v_{in} = (1 + g_m R_S) v_{gs}$$

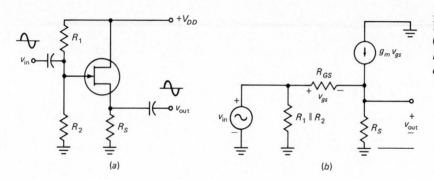

Fig. 12-18
(a) *Source fol-
lower.* (b) *Ac
equivalent circuit.*

The output voltage is

$$v_{\text{out}} = g_m v_{gs} R_S$$

Take the ratio of output to input to get

$$\frac{v_{\text{out}}}{v_{\text{in}}} = \frac{g_m R_S}{1 + g_m R_S}$$

which can be written as

$$A = \frac{R_S}{R_S + 1/g_m} \qquad (12\text{-}18)$$

When R_S is much greater than $1/g_m$, the unloaded voltage gain approaches 1.

LESS DISTORTION

The source follower has less distortion than a CS amplifier because of the unbypassed source resistor. When R_S is 10 times greater than $1/g_m$, the distortion is reduced by approximately a factor of 10. The source follower is inherently a low-distortion amplifier because the voltage gain approaches unity. If the voltage gain were exactly 1, there would be no distortion because the output would be a duplicate of the input.

OUTPUT IMPEDANCE

The unloaded voltage gain of a source follower is

$$\frac{v_{\text{out}}}{v_{\text{in}}} = \frac{R_S}{R_S + 1/g_m}$$

Therefore, the output voltage is

$$v_{\text{out}} = \frac{R_S}{R_S + 1/g_m} v_{\text{in}}$$

What does this remind you of? This is a voltage-divider equation. A voltage v_{in} drives two resistances, R_S and $1/g_m$, with output voltage taken across R_S. This means that the output side of a source follower appears as shown in Fig. 12-19a. The source resistor is driven by an ac source with an output impedance of

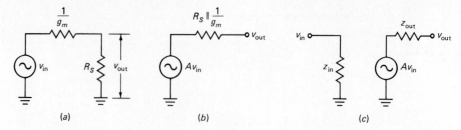

Fig. 12-19 (a) *Ac equivalent circuit for output of source follower.* (b) *Thevenin equivalent of source follower.* (c) *Another equivalent circuit.*

$$z_{out(source)} = \frac{1}{g_m} \qquad (12\text{-}19)$$

For instance, if g_m is 2000 μS, then $z_{out(source)}$ is 500 Ω. This is the output impedance looking back into the source terminal of the JFET.

If we thevenize the output circuit, then R_S is in parallel with $1/g_m$, and the output impedance of the stage is

$$z_{out} = R_S \parallel \frac{1}{g_m} \qquad (12\text{-}20)$$

Figure 12-19b shows the Thevenin output circuit. The Thevenin voltage is Av_{in}, and the output impedance is the equivalent parallel resistance of R_S and $1/g_m$. When R_S is much greater than $1/g_m$, the output impedance of the source follower approximately equals $1/g_m$.

AC MODEL

Figure 12-19c summarizes the source follower as far as the ac signal is concerned. Here you see an ac input voltage driving an input impedance. In a voltage-divider-biased stage, z_{in} is the parallel of R_1 and R_2. In a self-biased stage, z_{in} equals R_G, and so on for the other forms of bias. On the output side, there is an ac source Av_{in} and an output impedance z_{out}, which equals the R_S in parallel with $1/g_m$.

EXAMPLE 12-8

The JFET of Fig. 12-20a has a g_m of 2500 μS. If v_{in} is 5 mV, what does v_{out} equal?

SOLUTION

The reciprocal of transconductance is

$$\frac{1}{g_m} = \frac{1}{2500 \ \mu\text{S}} = 400 \ \Omega$$

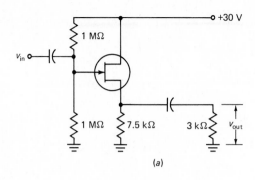

(a)

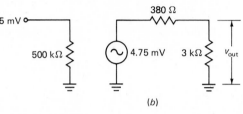

(b)

Fig. 12-20 (a) *Source follower drives load resistance.* (b) *Ac equivalent circuit.*

From Eq. (12-18), the unloaded voltage gain of the source follower is

$$A = \frac{7500}{7500 + 400} = 0.949$$

and the unloaded output voltage is

$$Av_{in} = (0.949)(5 \text{ mV}) = 4.75 \text{ mV}$$

The output impedance is

$$z_{out} = 7500 \text{ }\Omega \text{ || } 400 \text{ }\Omega = 380 \text{ }\Omega$$

Figure 12-20b shows the ac equivalent circuit of the source follower. An ac source of 4.75 mV is in series with an output impedance of 380 Ω. The ac voltage across the load resistor is

$$v_{out} = \frac{3000}{3380} \text{ } 4.75 \text{ mV} = 4.22 \text{ mV}$$

12-9 THE CG AMPLIFIER

Figure 12-21a shows a *common-gate* (CG) amplifier, and Fig. 12-21b is the ac equivalent circuit. The ac output voltage is

$$v_{out} = g_m v_{gs} R_D \tag{12-21}$$

Fig. 12-21
(a) *Common-gate
amplifier.* (b) *Ac
equivalent circuit.*

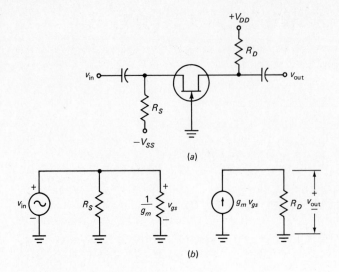

(a)

(b)

and the ac input voltage is

$$v_{in} = v_{gs}$$

The ratio of output to input is

$$\frac{v_{out}}{v_{in}} = \frac{g_m v_{gs} R_D}{v_{gs}} = g_m R_D$$

or

$$A = g_m R_D \qquad (12\text{-}22)$$

The ac input current to the JFET is

$$i_{in} = i_d = g_m v_{gs}$$

which can be rearranged as

$$\frac{v_{gs}}{i_{in}} = \frac{1}{g_m}$$

Since $v_{gs} = v_{in}$,

$$z_{in} = \frac{1}{g_m} \qquad (12\text{-}23)$$

For instance, if $g_m = 2000 \ \mu S$,

$$z_{in} = \frac{1}{2000 \ \mu S} = 500 \ \Omega$$

Therefore, the input impedance of a CG amplifier is low. This is in marked contrast to the CS and CD amplifiers, where the input impedance approaches infinity at low frequencies. Because of its small input impedance, the CG amplifier has only a few applications (discussed in Chap. 14).

12-10 THE JFET ANALOG SWITCH

One of the main applications of a JFET is *switching*. The idea is to use only two points on the load line: cutoff and saturation. When the JFET is cut off, it's like an open switch. When it's saturated, it's like a closed switch.

LOAD LINE

Figure 12-22a shows a JFET with a grounded source. When V_{GS} is zero, the JFET is saturated and operates at the upper end of the load line, as shown in Fig. 12-22b. When V_{GS} is equal to or more negative than $V_{GS(off)}$, the JFET is cut off and operates at the lower end of the load line. Ideally, the JFET acts like a closed switch when saturated and like an open switch when cut off.

The region between saturation and cutoff is not used in switching operation. In other words, only two points on the load line are used: saturation and cutoff. We get this two-state operation by applying either a zero gate or a large negative gate voltage.

DC ON-STATE RESISTANCE

The switching action is not perfect because a JFET has a small resistance when it's saturated. The *static* or *dc on-state resistance* is defined as the ratio of total drain voltage to total drain current:

$$r_{DS(on)} = \frac{V_{DS}}{I_D} \qquad (12\text{-}24)$$

where $r_{DS(on)}$ = dc resistance in saturation region
V_{DS} = dc drain-source voltage
I_D = dc drain current

For instance, if the saturation point of Fig. 12-22b has $V_{DS} = 0.1$ V and $I_D = 0.8$ mA, then

$$r_{DS(on)} = \frac{0.1 \text{ V}}{0.8 \text{ mA}} = 125 \text{ }\Omega$$

This means that the JFET has a dc resistance of 125 Ω.

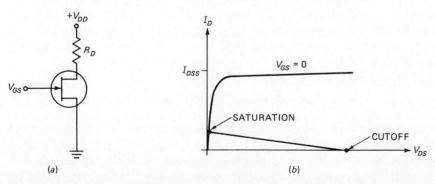

Fig. 12-22   (a) *JFET switching circuit.* (b) *Load line intersects well below knee of drain curve.*

Dc resistance $r_{DS(\text{on})}$ is useful when you are testing a JFET with an ohmmeter. On the lower ranges, where the internal voltage is 1.5 V, you can use an ohmmeter to measure the approximate value of $r_{DS(\text{on})}$. Remember to connect a jumper between the gate and the source to ensure that V_{GS} is zero. For instance, the data sheet of a 2N5951 lists a minimum $r_{DS(\text{on})}$ of 100 Ω and a maximum $r_{DS(\text{on})}$ of 500 Ω for $V_{GS} = 0$. An ohmmeter connected between the drain and source of a 2N5951 should read between 100 and 500 Ω when the gate is jumpered to the source.

AC ON-STATE RESISTANCE

When used as a switch, the JFET is normally saturated well below the knee of the drain curve, as shown in Fig. 12-22b. For this reason, the drain current is much smaller than I_{DSS}. If an ac signal is applied to a JFET when it is saturated, it appears as an ac resistance, given by

$$r_{ds(\text{on})} = \frac{\Delta V_{DS}}{\Delta I_D}$$

(12-25)

where $r_{ds(\text{on})}$ = ac resistance
ΔV_{DS} = change in drain-source voltage
ΔI_D = change in drain current

This resistance, known as the *small-signal on-state resistance,* equals the reciprocal slope of the drain curve. The steeper the saturation region, the smaller the ac resistance.

SHUNT SWITCH

Figure 12-23a is a *shunt switch.* The idea is to either transmit or block a small ac input signal. Voltage v_{in} is small, typically less than 100 mV. When the control voltage V_{con} is zero, the JFET is saturated. Since the JFET is equivalent to a closed switch, v_{out} is ideally zero. When V_{con} is more negative than $V_{GS(\text{off})}$, the JFET is like an open switch and v_{out} equals v_{in}. As long as the input voltage is small, the JFET remains cut off even though the input voltage goes negative during part of the cycle.

The switching is not perfect. Because of the ac resistance of a saturated JFET, the ac equivalent circuit appears as shown in Fig. 12-23b. When the switch is open, all the input signal reaches the output. But when the switch is closed, some of the input signal still reaches the output because $r_{ds(\text{on})}$ is not zero. For the switching to be effective, $r_{ds(\text{on})}$ must be much smaller than R_D. For instance, a 2N3970 has an $r_{ds(\text{on})}$ of 30 Ω. If R_D is 4.7 kΩ, then

$$v_{\text{out}} = \frac{30}{4730} v_{\text{in}} = 0.00634 v_{\text{in}}$$

This means that less than 1 percent of the input signal reaches the output when the JFET is saturated.

Incidentally, the shunt switch is sometimes called an *analog switch* because it is used to switch an ac signal on or off. The word "analog" means continuous or

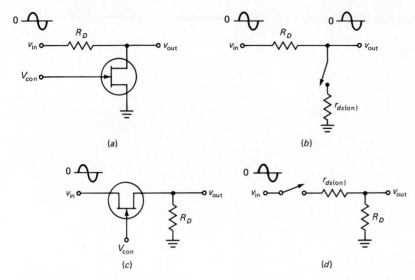

Fig. 12-23 (a) *Shunt switch.* (b) *Ac equivalent circuit.* (c) *Series switch.* (d) *Ac equivalent circuit.*

smooth. The input signal is called an analog signal because it varies continuously from its minimum to its maximum value.

SERIES SWITCH

Figure 12-23*c* shows a *series switch.* When V_{con} is zero, the JFET is equivalent to a closed switch. In this case, the output approximately equals the input. When V_{con} is equal to or more negative than $V_{GS(off)}$, the JFET is like an open switch and v_{out} is approximately zero.

Figure 12-23*d* is the ac equivalent circuit. When the switch is open, no signal reaches the output. When the switch is closed, most of the signal reaches the output, provided $r_{ds(on)}$ is much smaller than R_D. You will see both kinds of JFET switches used in practice. The series switch is used more often because it has a better on-off ratio. The series switch is another example of an analog switch, a device that either transmits or blocks an ac signal.

MULTIPLEXING

"Multiplex" means "many into one." Figure 12-24 shows an *analog multiplexer,* a circuit that steers one of the input signals to the output line. When the control signals (V_1, V_2, and V_3) are more negative than $V_{GS(off)}$, all input signals are blocked. By making any control voltage equal to zero, we can transmit one of the inputs to the output. For instance, if V_1 is zero, we get a sinusoidal output. If V_2 is zero, we get a triangular output. And if V_3 is zero, we get a square-wave output. Normally, only one of the control signals is zero.

Fig. 12-24
Analog multi-plexer.

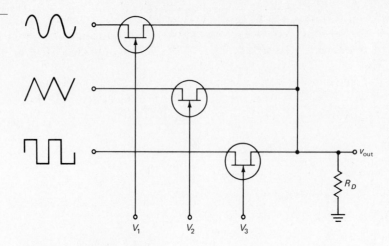

JFET CHOPPERS

As discussed earlier, we can build a direct-coupled amplifier by leaving out the coupling and bypass capacitors, and connecting the output of each stage directly to the input of the next stage. In this way, direct current is coupled, as well as alternating current. But the major disadvantage of direct coupling is *drift,* a slow shift in the final output voltage produced by supply and transistor variations.

Figure 12-25*a* shows the *chopper* method of building a dc amplifier. The input dc voltage is chopped by a switching circuit. This results in the square wave shown at the chopper output. The peak value of this square wave equals V_{DC}. Because the square wave is an ac signal, we can use a conventional ac amplifier, one with coupling capacitors between the stages. The amplified output can then be peak-detected to recover the dc signal.

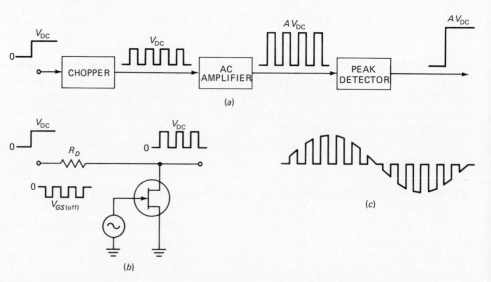

Fig. 12-25 (a) *Chopper-type amplifier.* (b) *FET chopper.* (c) *Output of FET chopper driven by low-frequency sinusoid.*

If we apply a square wave to the gate of a JFET analog switch, it becomes a chopper (see Fig. 12-25*b*). The gate square wave is negative-going, swinging from 0 V to at least $V_{GS(\text{off})}$. This alternately saturates and cuts off the JFET. Therefore, the output voltage is a square wave with a peak of V_{DC}.

If the input is a low-frequency ac signal, it gets chopped into the ac waveform of Fig. 12-25*c*. This chopped signal can now be amplified by an ac amplifier that is drift-free. Once the signal is large enough, it can be peak-detected to recover the original input signal.

12-11 OTHER JFET APPLICATIONS

In this section, we discuss some of the applications in which the JFET's properties give it a clear-cut advantage over the bipolar transistor.

BUFFER AMPLIFIER

Figure 12-26 shows a *buffer amplifier,* a stage that isolates the preceding stage from the following stage. Ideally, a buffer should have a high input impedance; if it does, almost all the Thevenin voltage from stage *A* appears at the buffer input. The buffer should also have a low output impedance. This ensures that all its output reaches the input of stage *B*.

The source follower is an excellent buffer amplifier because of its high input impedance (well into the megohms at low frequencies) and its low output impedance (typically a few hundred ohms). The high input impedance means light loading of the preceding stage. The low output impedance means that the buffer can drive heavy loads (small load resistances).

LOW-NOISE AMPLIFIER

Noise is any unwanted disturbance superimposed upon a useful signal. Noise interferes with the information contained in the signal; the greater the noise, the less the information. For instance, the noise in television receivers produces small white or black spots on the picture; severe noise can wipe out the picture. Similarly, the noise in radio receivers produces crackling and hissing, which sometimes completely masks voice or music. Noise is independent of the signal because it exists even when the signal is off.

Chapter 23 discusses the nature of noise, how it is produced, and how to reduce it. Any electronic device produces a certain amount of noise. The JFET is an outstanding low-noise device because it produces very little noise. This is especially important near the front end of receivers and other electronic equipment; subsequent stages

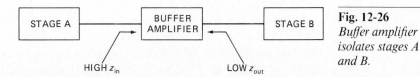

Fig. 12-26
Buffer amplifier isolates stages A and B.

Fig. 12-27
Drain curves are linear near origin.

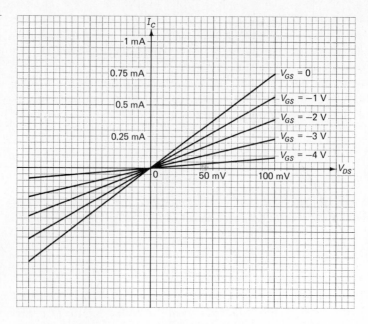

amplify front-end noise along with the signal. If we use a JFET amplifier at the front end, we get less amplified noise at the final output.

VOLTAGE-VARIABLE RESISTANCE

The saturation region of a JFET is also called the *ohmic region* because a saturated JFET acts like a resistance instead of a current source. In the ohmic region, $r_{ds(on)}$ can be controlled by V_{GS}. The more negative V_{GS} is, the larger $r_{ds(on)}$ becomes.

For instance, Fig. 12-27 shows the drain curves of a 2N5951 in the ohmic region. The small-signal resistance $r_{ds(on)}$ depends on the value of V_{GS}. You can calculate $r_{ds(on)}$ by taking the ratio of drain voltage to drain current. For instance, when $V_{GS} = 0$, $I_D = 0.75$ mA for $V_{DS} = 100$ mV. Therefore,

$$r_{ds(on)} = \frac{100 \text{ mV}}{0.75 \text{ mA}} = 133 \text{ } \Omega$$

For a more negative V_{GS}, $r_{ds(on)}$ becomes larger. When $V_{GS} = -2$ V, $I_D = 0.4$ mA and $V_{DS} = 100$ mV. So,

$$r_{ds(on)} = \frac{100 \text{ mV}}{0.4 \text{ mA}} = 250 \text{ } \Omega$$

Notice that the drain curves of Fig. 12-27 extend on both sides of the origin. This means that a JFET can be used as a *voltage-variable resistance* for small ac signals, typically those less than 100 mV. When used in this way, the JFET does not need a dc drain voltage from the supply. All that is required is an ac input signal. You will see many applications for a voltage-variable resistance in Chaps. 17 and 18.

AUTOMATIC GAIN CONTROL

When a receiver is tuned from a weak to a strong station, the loudspeaker will blare unless the volume is immediately decreased. Or the volume may change because of *fading,* a variation in signal strength caused by an electrical change in the path between the transmitting and receiving antennas. To counteract unwanted changes in volume, most receivers use *automatic gain control* (AGC).

This is where the JFET comes in. As shown earlier,

$$g_m = g_{m0}\left[1 - \frac{V_{GS}}{V_{GS(\text{off})}}\right]$$

This is a linear equation. When graphed, it results in Fig. 12-28*a*. For a JFET, g_m reaches a maximum value when $V_{GS} = 0$. As V_{GS} becomes more negative, the value of g_m decreases. Since a common-source amplifier has a voltage gain of

$$A = -g_m r_D$$

we can control the voltage gain by controlling the value of g_m.

Figure 12-28*b* shows how it's done. A JFET amplifier is near the front end of a receiver. It has a voltage gain of $-g_m r_D$. Subsequent stages amplify the JFET output. This amplified output goes into a negative peak detector that produces voltage V_{AGC}. This negative voltage returns to the JFET amplifier, where it is applied to the gate through a 10-kΩ resistor. When the receiver is tuned from a weak to a strong station, a larger signal is peak-detected and V_{AGC} is more negative; this reduces the gain of the JFET amplifier.

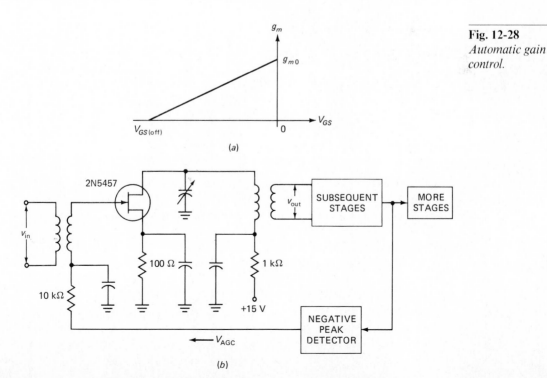

(a)

(b)

Fig. 12-28
Automatic gain control.

The overall effect of AGC is this: the final signal increases, but not nearly as much as it would without AGC. For instance, in some AGC systems an increase of 100 percent in the input signal results in an increase of less than 1 percent in the final output signal.

CASCODE AMPLIFIER

Figure 12-29a is an example of a *cascode amplifier,* a common-source amplifier driving a common-gate amplifier. Here's how it works. For simplicity, assume matched JFETs so that both have the same g_m. The CS amplifier has a gain of

$$A_1 = -g_m R_D$$

The input impedance of the CG amplifier is $1/g_m$; this is the drain resistance seen by the CS amplifier. Therefore,

$$A_1 = -g_m R_D = -g_m \frac{1}{g_m} = -1$$

The CG amplifier has a gain of

$$A_2 = g_m R_D$$

So, the overall gain of the two JFETs is

$$A = A_1 A_2 = -g_m R_D$$

This says that a cascode connection has the same voltage gain as a CS amplifier.

Fig. 12-29
(a) *Cascode amplifier.* (b) *FET current limiter.*

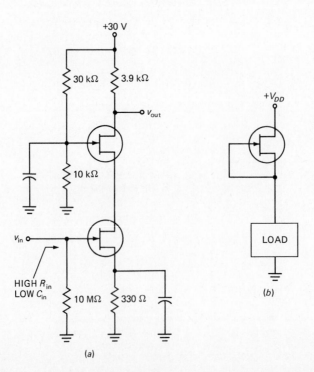

(a)

(b)

The main advantage of a cascode connection is its low input capacitance, which is considerably less than the input capacitance of a CS amplifier. Chapter 14 will explain why a cascode amplifier has a low input capacitance (it is related to the Miller effect). Until then, all you have to remember is that a cascode amplifier has the same voltage gain as a CS amplifier, but its input capacitance is very low.

CURRENT LIMITING

The JFET of Fig. 12-29*b* can protect a load against excessive current. For instance, suppose the normal load current is 1 mA. If $I_{DSS} = 10$ mA and $r_{DS(on)} = 200 \, \Omega$, then a normal load current of 1 mA means that the JFET is operating in the ohmic region with a voltage drop of only

$$V_{DS} = (1 \text{ mA})(200 \, \Omega) = 0.2 \text{ V}$$

Almost all the supply voltage therefore appears across the load. Now suppose the load shorts. Then load current tries to increase to an excessive level. The increased load current forces the JFET into the active region, where it limits the current to 10 mA. The JFET now acts like a current source and prevents excessive load current.

A manufacturer can tie the gate to the source and package the JFET as a two-terminal device. This is how constant-current diodes are made. Constant-current diodes, also called current-regulator diodes, were discussed in Chap. 4.

CONCLUSION

Look at Table 12-1. Some of the terms are new and will be discussed in later chapters. The JFET buffer has the advantage of high input impedance and low output impedance. This is why the JFET is the natural choice at the front end of voltmeters,

Table 12-1. FET Applications

Application	Main Advantage	Uses
Buffer	High $z_{in'}$, low z_{out}	General purpose, measuring equipment, receivers
RF amplifier	Low noise	FM tuners, communication equipment
Mixer	Low intermodulation distortion	FM and TV receivers, communication equipment
AGC amplifier	Ease of gain control	Receivers, signal generators
Cascode amplifier	Low input capacitance	Measuring instruments, test equipment
Chopper	No drift	Dc amplifiers, guidance control systems
Voltage-variable resistor	Voltage-controlled	Operational amplifiers, organ, tone controls
Low-frequency amplifier	Small coupling capacitor	Hearing aids, inductive transducers
Oscillator	Minimum frequency drift	Frequency standards, receivers
MOS digital circuit	Small size	Large-scale integration, computers, memories

oscilloscopes, and other such devices, where you need high input resistances (10 MΩ or more). As a guide, the input resistance at the gate of a JFET is from 100 to more than 10,000 MΩ.

When a JFET is used as a small-signal amplifier, the output voltage from it is linearly related to the input because only a small part of the transconductance curve is used. Near the front end of TV and radio receivers, the signals are small; therefore, JFETs are often used as RF amplifiers.

But with larger signals, more of the transconductance curve is used, resulting in square-law distortion. This nonlinear distortion is unwanted in an amplifier. But in a mixer (discussed in Chap. 23), square-law distortion has a tremendous advantage; this is why the JFET is preferred to the bipolar for FM and TV mixer applications.

As indicated in Table 12-1, JFETs are also useful in AGC amplifiers, cascode amplifiers, choppers, and voltage-variable resistors. The last two entries, oscillators and MOS digital circuits, are discussed in later chapters.

PROBLEMS

Straightforward

12-1. At room temperature the 2N4220 (an *n*-channel JFET) has a reverse gate current of 0.1 nA for a reverse gate voltage of 15 V. Calculate the resistance from gate to source.

12-2. If a JFET has the drain curves of Fig. 12-30*a*, what does I_{DSS} equal? What is the maximum V_{DS} in the saturation region? Over what voltage range of V_{DS} does the JFET act like a current source?

Fig. 12-30

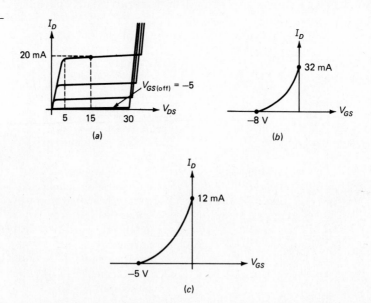

(a)

(b)

(c)

12-3. Write the transconductance equation for the JFET whose curve is shown in Fig. 12-30*b*. How much drain current is there when $V_{GS} = -4$ V? When $V_{GS} = -2$ V?

12-4. If a JFET has a square-law curve like Fig. 12-30*c*, how much drain current is there when $V_{GS} = -1$ V?

refer to R 330.

12-5. Figure 12-31*a* shows a self-biased circuit, and Fig. 12-31*b* is the transconductance curve of the JFET. What are the quiescent values of I_D and V_{GS}? Calculate the dc voltage between the drain and ground.

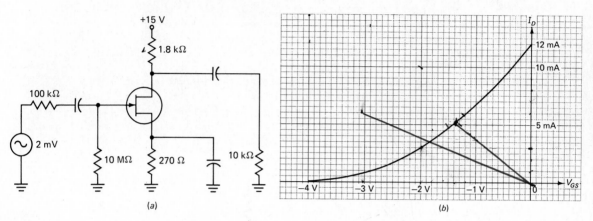

(a)   (b)

Fig. 12-31

12-6. If R_S is changed from 270 Ω to 510 Ω in Fig. 12-31*a*, what are the quiescent values of I_D and V_{GS}? (Use the transconductance curve of Fig. 12-31*b*.) What are the following dc voltages to ground: V_G, V_S, and V_D?

12-7. In Fig. 12-32*a*, V_{GS} is −1 V. What does the quiescent drain current equal? What do the following dc voltages equal: V_G, V_S, and V_D?

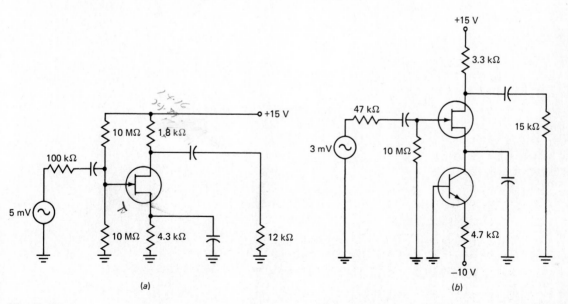

(a)   (b)

Fig. 12-32

12-8. V_{GS} can be as low as −0.5 V and as high as −2 V in Fig. 12-32*a*. Calculate the minimum and maximum values of dc drain current. What are the minimum and maximum values of V_D?

12-9. What is the quiescent drain current in Fig. 12-32*b*? The value of V_D?

12-10. When V_{GS} changes from -2.1 to -2 V, the drain current changes from 1 to 1.3 mA. What is the value of g_m in this region?

12-11. A JFET has the transconductance curve shown in Fig. 12-31*b*. Work out the approximate value of g_m when $V_{GS} = 0$. What is the value of g_m when $V_{GS} = -2$ V?

12-12. A JFET has a g_{m0} of 5000 μS and a $V_{GS(\text{off})}$ of -4 V. Calculate g_m at -1 V and at -3 V.

12-13. A JFET has an I_{DSS} of 10 mA and a g_{m0} of 10,000 μS. Calculate its $V_{GS(\text{off})}$.

12-14. The ac input voltage has a peak-to-peak value of 2 mV in Fig. 12-31*a*. If $g_m = 3850$ μS, what is the ac voltage across the load resistor?

12-15. If $g_m = 2100$ μS in Fig. 12-32*a*, what is the ac voltage across the load resistor?

12-16. The g_m equals 2500 μS in Fig. 12-32*b*. What is the ac load voltage?

12-17. In Fig. 12-33, the first JFET has a g_m of 2850 μS, and the second JFET has a g_m of 4275 μS. If v_{in} is 1 mV, what does v_{out} equal?

Fig. 12-33

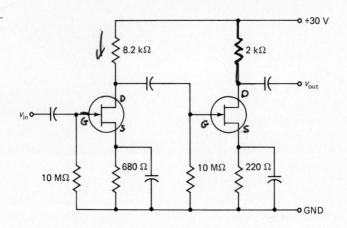

12-18. The JFET of Fig. 12-34 has a g_m of 2500 μS, and the bipolar transistor has a β of 150. If v_{in} equals 1 mV, answer the following:

a. What is the output impedance of the first stage?

b. What is the ac input voltage to the second stage?

c. What is the final output voltage?

Fig. 12-34

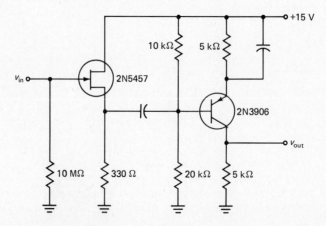

12-19. The JFET of Fig. 12-35a has an $r_{ds(on)}$ of 120 Ω. If v_{in} equals 20 mV, what does v_{out} equal when V_{con} is zero? When V_{con} is more negative than $V_{GS(off)}$?

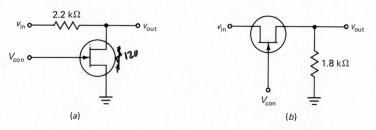

Fig. 12-35

(a) (b)

12-20. In Fig. 12-35b, $r_{ds(on)}$ is 200 Ω. If v_{in} is 50 mV, what does v_{out} equal when V_{con} is zero? When V_{con} is more negative than $V_{GS(off)}$?

Troubleshooting

12-21. You are testing a JFET whose data sheet lists an $r_{DS(on)}$ of 200 Ω for a V_{GS} of zero. You connect a jumper between the gate and the source. What should an ohmmeter read, approximately, when it is connected between the drain and the source? If a battery of 1.5 V is used to reverse-bias the gate, should the ohmmeter reading increase or decrease?

12-22. In Fig. 12-32a, you measure a dc drain-to-ground voltage of 2 V. Which of the following is the trouble:
   a. Source bypass capacitor open
   b. Source bypass capacitor shorted
   c. Input coupling capacitor open
   d. Input coupling capacitor shorted

12-23. All dc voltages are normal in Fig. 12-33. The ac output voltage is lower than it should be. Which of the following is the trouble:
   a. Input coupling capacitor open
   b. Source bypass capacitor open in first stage
   c. Gate resistor shorted in second stage
   d. Drain resistor shorted in second stage

12-24. In Fig. 12-34, the emitter bypass capacitor opens in the second stage. What happens to the dc voltage at the output? What happens to the ac voltage?

Design

12-25. A JFET has the transconductance curve shown in Fig. 12-31b. Select a source resistance that will satisfy Eq. (12-5).

12-26. Using the transconductance curve of Fig. 12-31b, redesign the circuit of Fig. 12-31a to meet these specifications: input impedance, 6.8 MΩ; load resistor, 8.2 kΩ; quiescent I_D, 3 mA; and quiescent V_D, 8 V.

12-27. Redesign the circuit of Fig. 12-32b to get a quiescent drain current of approximately 1.5 mA and a quiescent drain-to-ground voltage of 2.7 V.

12-28. Design a two-stage amplifier with a source-follower input and a bipolar second stage, similar to Fig. 12-34. Use the transconductance curve of Fig. 12-31b for the JFET and use a β of 120 for the 2N3906. The two-stage amplifier should meet these specifications: input impedance, 4.7 MΩ; output impedance, 1 kΩ; and unloaded voltage gain, 20.

Challenging

12-29. Instead of a graphical approach, it is possible to apply trial and error to Eq. (12-1) to converge on the quiescent I_D and V_{GS} for a self-bias circuit. If the JFET of Fig. 12-31a has an I_{DSS} of 8 mA and a $V_{GS(off)}$ of -3.5 V, work out the quiescent I_D and V_{GS} by trial and error using Eq. (12-1).

12-30. Figure 12-36 shows a simple FET dc voltmeter. The zero adjust is set as in any voltmeter; the calibrate adjust is set periodically to give full-scale deflection when $V_{in} = 2.5$ V. A calibrate adjustment like this takes care of variations from one FET to another and FET aging effects.

    a. The current through the 510-Ω resistor equals 4 mA. How much dc voltage is there from the source to ground?

    b. If no current flows through the ammeter, what voltage does the wiper tap off the zero adjust?

    c. If an input voltage of 2.5 V produces 1 mA deflection, how much deflection does 1.25 V produce (assume linearity)?

    d. The MPF102 has an I_{GSS} of 2 nA for a V_{GS} of 15 V. What is the input impedance of the voltmeter?

Fig. 12-36

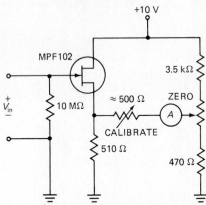

12-31. A JFET can be used as a current limiter to protect a load from excessive current. In Fig. 12-37a, the JFET has an I_{DSS} of 16 mA and an $r_{DS(on)}$ of 200 Ω. If the load accidentally shorts, what are the load current and the voltage across the JFET? If the load has a resistance of 10 kΩ, what are the load current and the voltage across the JFET?

Fig. 12-37

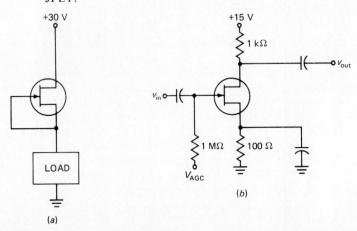

12-32. Figure 12-37*b* shows part of an AGC amplifier. Figure 12-31*b* is the transconductance curve. What is the unloaded voltage gain for each of these:
  a. $V_{AGC} = 0$
  b. $V_{AGC} = -1$ V
  c. $V_{AGC} = -2$ V
  d. $V_{AGC} = -3$ V
  e. $V_{AGC} = -3.5$ V

Computer

12-33. The FOR . . . NEXT statement is a two-line command that allows you to repeat part of a program as many times as desired. For instance, here is a program:

```
10 FOR X = 1 TO 7
20 PRINT X
30 NEXT X
40 STOP
```

When line 10 is executed the first time, X equals 1. Line 20 then prints a 1 on the screen. Line 30 forces the program to jump back to line 10 for the next X value, which is 2. The program keeps looping and produces the integer values of X: 1, 2, 3, . . . , 7. Each time line 20 is executed, the next higher value of X is printed on the screen. Finally, when X is 7, the computer prints a 7 on the screen. When line 30 is executed, X = 8 and the program falls through to line 40.

Write a program that prints all values of X between 1 and 1000.

12-34. Here is a program:

```
10 IDSS = 0.012
20 VGS(OFF) = -4
30 FOR X = 0 TO 4
40 Y = 1 + X/VGS(OFF)
50 Z = Y*Y
60 I = IDSS*Z
70 PRINT I
80 NEXT X
```

What values does the program print? (You may use Fig. 12-31.)

12-35. Write a program that inputs I_{DSS} and $V_{GS(off)}$, then prints out the source resistance given by Eq. (12-5).

12-36. Write a program for a source follower like Fig. 12-18*a* that prints out the unloaded voltage gain, input impedance of the stage, and output impedance of the stage. The inputs to the program are R_1, R_2, R_S, and g_m.

MOSFETs

The *metal-oxide semiconductor* FET, or MOSFET, has a source, gate, and drain. Unlike in a JFET, however, the gate is insulated from the channel. Because of this, gate current is extremely small whether the gate is positive or negative. The MOSFET is sometimes referred to as an IGFET, which stands for *insulated-gate* FET. This chapter discusses MOSFET fundamentals, biasing, amplifiers, and switching circuits.

13-1 THE DEPLETION-TYPE MOSFET

Figure 13-1 shows an *n*-channel MOSFET, a conducting bar of *n* material with a *p* region on the right and an insulated gate on the left. Free electrons can flow from the source to the drain through the *n* material. The *p* region is called the *substrate* (or body); it physically reduces the conducting path to a narrow channel. Electrons flowing from source to drain must pass through this narrow channel.

A thin layer of *silicon dioxide* (SiO_2) is deposited on the left side of the channel. Silicon dioxide is the same as glass, which is an insulator. In a MOSFET the gate is metallic. Because the gate is insulated from the channel, negligible gate current flows even when the gate voltage is positive. The *pn* diode that exists in a JFET has been eliminated in the MOSFET.

DEPLETION MODE

Figure 13-2*a* shows a MOSFET with a negative gate. The V_{DD} supply forces free electrons to flow from source to drain. These electrons flow through the narrow channel on the left of the *p* substrate.

As before, the gate voltage can control the width of the channel. The more negative the gate voltage, the smaller the drain current. When the gate voltage is negative enough, the drain current is cut off. Therefore, with negative gate voltage, the operation of a MOSFET is similar to that of a JFET. Because the action with a negative gate depends on depleting the channel of free electrons, we call negative-gate operation the *depletion mode*.

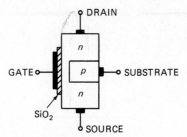

Fig. 13-1
Depletion-type MOSFET.

ENHANCEMENT MODE

Since the gate of a MOSFET is insulated from the channel, we can apply a positive voltage to the gate, as shown in Fig. 13-2*b*. The positive gate voltage increases the number of free electrons flowing through the channel. The more positive the gate voltage, the greater the conduction from source to drain. Operation of the MOSFET with a positive gate voltage depends on enhancing channel conductivity. For this reason, positive-gate operation (Fig. 13-2*b*) is called the *enhancement mode.*

Because of the insulating layer, negligible gate current flows in either mode of operation. The input resistance of a MOSFET is incredibly high, typically from 10,000 MΩ to over 10,000,000 MΩ.

DRAIN CURVES

Figure 13-3*a* shows typical drain curves for an *n*-channel MOSFET, along with a dc load line for a common-source circuit. Notice that the upper curves have a positive V_{GS} and the lower curves have a negative V_{GS}. The bottom drain curve is for $V_{GS} = V_{GS(off)}$. Along this curve, the drain current is approximately zero. When V_{GS} is between $V_{GS(off)}$ and zero, we get depletion-mode operation. On the other hand, V_{GS} greater than zero gives enhancement-mode operation.

TRANSCONDUCTANCE CURVE

Figure 13-3*b* is the transconductance curve of a MOSFET. I_{DSS} still represents the drain current with a shorted gate. But now the curve extends to the right of the origin,

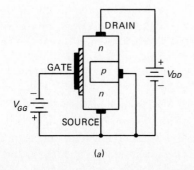

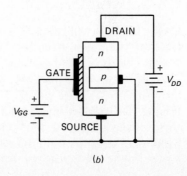

Fig. 13-2
(a) *Depletion mode.* (b) *Enhancement mode.*

(a)   (b)

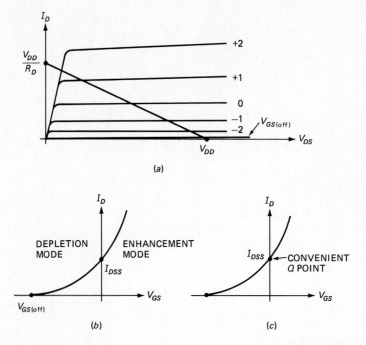

Fig. 13-3 (a) *Drain curves.* (b) *Transconductance curve.* (c) *Convenient Q point.*

as shown. The relation between drain current and gate-source voltage is still parabolic, and so we can use the square-law equation described in Chap. 12:

$$I_D = I_{DSS}\left[1 - \frac{V_{GS}}{V_{GS(\text{off})}}\right]^2 \qquad (13\text{-}1)$$

This is identical to the square-law equation of a JFET. The value of V_{GS}, however, can now be positive or negative.

MOSFETs with a transconductance curve like Fig. 13-3*b* are easier to bias than JFETs because we can use the Q point shown in Fig. 13-3*c*. With this Q point, $V_{GS} = 0$ and $I_D = I_{DSS}$. Setting up a V_{GS} of zero is easy; it requires no dc voltage on the gate.

NORMALLY ON

The MOSFET just described is called a *depletion-type* MOSFET; it can have drain current in either the depletion mode or the enhancement mode. Since this kind of MOSFET conducts when $V_{GS} = 0$, it also is known as a *normally on* MOSFET.

SCHEMATIC SYMBOL

Figure 13-4*a* shows the schematic symbol for a depletion-type MOSFET. Just to the right of the gate is the thin vertical line representing the channel. The drain lead comes out the top of the channel, and the source lead connects to the bottom. The arrow on the *p* substrate points to the *n* material. In some applications, a voltage can

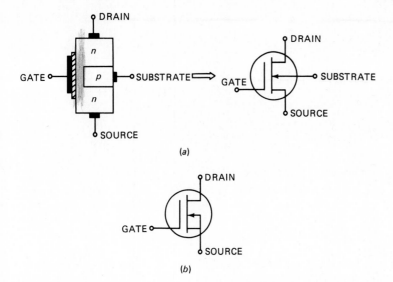

Fig. 13-4 *Schematic symbols for n-channel depletion-type MOSFET.*

be applied to the substrate for added control of drain current. For this reason, some MOSFETs have four external leads.

In most applications the substrate is connected to the source. Usually, the manufacturer internally connects the substrate to the source. This results in a three-terminal device whose schematic symbol is shown in Fig. 13-4*b*.

There is also a *p*-channel depletion-type MOSFET. It consists of a bar of *p* material with an *n* region on the right and an insulated gate on the left. The schematic symbol of a *p*-channel MOSFET is similar to that of an *n*-channel MOSFET, except that the arrow points outward. In the remainder of this chapter, we emphasize the *n*-channel MOSFET. The action of a *p*-channel MOSFET is complementary, meaning that all voltages and currents are reversed.

13-2 BIASING DEPLETION-TYPE MOSFETs

Because depletion-type MOSFETs can operate in the depletion mode, all the biasing methods discussed for JFETs can be used. These include gate bias, self-bias, voltage-divider bias, and current-source bias. In addition to these biasing methods, you have another option with depletion-type MOSFETs.

Since a depletion-type MOSFET can operate in either the depletion or the enhancement mode, we can set its Q point at $V_{GS} = 0$, as shown in Fig. 13-5*a*. Then, an ac input signal to the gate can produce variations above and below the Q point. Being able to use zero V_{GS} is an advantage when it comes to biasing. It permits the unique biasing circuit of Fig. 13-5*b*. This simple circuit has no applied gate or source voltage. Therefore, $V_{GS} = 0$ and $I_D = I_{DSS}$. The dc drain voltage is

$$V_{DS} = V_{DD} - I_{DSS}R_D \tag{13-2}$$

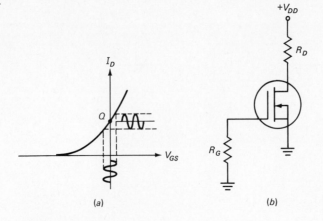

The *zero bias* of Fig. 13-5a is unique with depletion-type MOSFETs; it will not work with a bipolar transistor or a JFET.

13-3 DEPLETION-TYPE MOSFET APPLICATIONS

After the depletion-type MOSFET is biased to a Q point, it can amplify small signals just as a JFET can. MOSFET amplifiers are similar to JFET amplifiers, so that most of the ac analysis of the preceding chapter applies. For instance, a CS MOSFET amplifier has an unloaded voltage gain of $-g_m R_D$, a MOSFET source follower has an output impedance of $1/g_m$, and so on.

If the input impedance of a JFET is not high enough, you can use a MOSFET. It makes an almost ideal buffer amplifier because the insulated gate means that the input resistance approaches infinity. Furthermore, MOSFETs have excellent low-noise properties, a definite advantage for any stage near the front end of a system, where the signal is weak. As with a JFET, the g_m of a MOSFET can be controlled by changing the dc gate voltage. Because of this, MOSFETs can also serve as AGC amplifiers.

Some MOSFETs are *dual-gate* devices. This means that they have two separate gates, like the dual-gate depletion-type MOSFET shown in Fig. 13-6a. One use for a device like this is to build a cascode amplifier like Fig. 13-6b. For convenience, the circuit uses zero bias on each gate. The ac input signal drives the lower gate. The upper gate is grounded. Because of its internal structure, the dual-gate MOSFET is equivalent to one MOSFET driving another MOSFET, as shown in Fig. 13-6c. Here you can recognize the cascode connection; the lower half acts like a CS amplifier, and the upper half acts like a CG amplifier. As a result, the cascode amplifier has an unloaded voltage gain of $-g_m R_D$.

The important thing to remember is that the dual-gate MOSFET is a convenient way to build a cascode amplifier. This type of amplifier is useful at high frequencies because of its low input capacitance. The next chapter tells you why a cascode amplifier has much lower input capacitance than an ordinary CS amplifier.

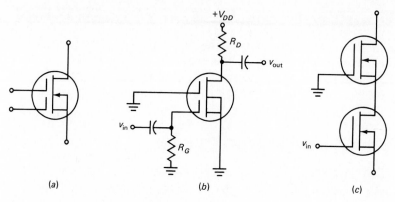

Fig. 13-6 (a) *Dual-gate MOSFET.* (b) *Cascode amplifier using dual-gate MOSFET.* (c) *Equivalent circuit.*

EXAMPLE 13-1

The 3N201 of Fig. 13-7 has a g_m of 10,000 μS. Calculate the unloaded voltage gain, input impedance, and output impedance at low frequencies.

SOLUTION

The unloaded voltage gain is

$$A = -10,000 \ \mu S \times 1.8 \ k\Omega = -18$$

At low frequencies the input impedance is 1 MΩ, the value of the gate return resistor. The output impedance approximately is 1.8 kΩ, the value of the drain resistor.

At higher frequencies, the input and output impedances decrease because of capacitive effects, to be discussed in the next chapter. The cascode amplifier is important because its capacitive effects are much smaller than those of other kinds of amplifiers. A common application for the cascode amplifier is in RF circuits.

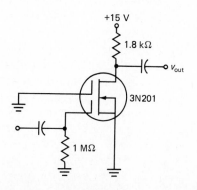

Fig. 13-7
Cascode amplifier.

13-4 THE ENHANCEMENT-TYPE MOSFET

By changing the internal structure of an *n*-channel MOSFET, we can produce a new kind of MOSFET that conducts only in the enhancement mode. This type of MOSFET is widely used in microprocessors and computer memories because it acts like a normally off switch. To get drain current, you have to apply a positive gate voltage.

CREATING THE INVERSION LAYER

Figure 13-8*a* shows an *n*-channel *enhancement-type* MOSFET. The substrate extends all the way to the silicon dioxide; physically, there no longer is an *n* channel between the source and the drain.

How does it work? Figure 13-8*b* shows normal biasing polarities. When $V_{GS} = 0$, the V_{DD} supply tries to force free electrons from source to drain, but the *p* substrate has only a few thermally produced conduction-band electrons. Aside from these minority carriers and some surface leakage, the current between source and drain is zero. For this reason, an enhancement-type MOSFET is also called a *normally off* MOSFET.

The gate and the *p* substrate are like two plates of a capacitor separated by a dielectric (SiO$_2$). When the gate is positive, it induces negative charges in the *p* substrate. In other words, the positive gate attracts free electrons from the source into the lower left corner of the *p* region. When the gate is positive enough, it can attract enough free electrons to form a thin layer of electrons between the source and the drain.

Put another way, a positive gate voltage draws free electrons into the *p* substrate. These free electrons recombine with some of the holes adjacent to the silicon dioxide. When the gate voltage is positive enough, all the holes touching the silicon dioxide are filled, and free electrons begin to flow from the source to the drain. The effect is equivalent to creating a thin layer of *n*-type material next to the silicon dioxide. This layer of free electrons is called an *n-type inversion layer.*

THRESHOLD VOLTAGE

The minimum V_{GS} that creates the *n*-type inversion layer is designated the *threshold voltage* $V_{GS(th)}$. When V_{GS} is less than $V_{GS(th)}$, zero current flows from source to drain.

Fig. 13-8
Enhancement-type MOSFET.

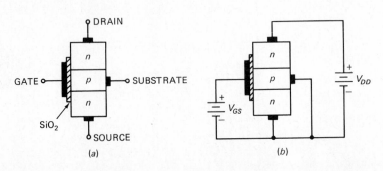

(a)   (b)

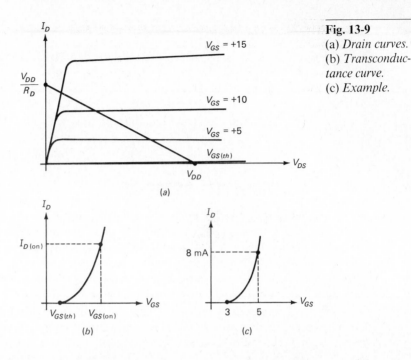

Fig. 13-9
(a) *Drain curves.*
(b) *Transconduc-
tance curve.*
(c) *Example.*

But when V_{GS} is greater than $V_{GS(th)}$, an *n*-type inversion layer connects the source to the drain, and we get current.

$V_{GS(th)}$ can vary from less than 1 V to more than 5 V, depending on the particular device being used. The 3N169 is an example of an enhancement-type MOSFET. It has a maximum threshold voltage of 1.5 V.

DRAIN CURVES

Figure 13-9*a* shows a set of drain curves for an *n*-channel enhancement-type MOSFET, along with a dc load line for a common-source circuit. The lowest curve is the $V_{GS(th)}$ curve. When V_{GS} is less than $V_{GS(th)}$, the drain current is extremely small. When V_{GS} is greater than $V_{GS(th)}$, significant drain current flows, with the amount depending on the value of V_{GS}.

TRANSCONDUCTANCE CURVE

Figure 13-9*b* is the transconductance curve. The curve is parabolic or square-law. The vertex of the parabola is at $V_{GS(th)}$. Because of this, the equation for the parabola is different from before. It now equals

$$I_D = K[V_{GS} - V_{GS(th)}]^2 \qquad (13\text{-}3)$$

where K is a constant that depends on the particular MOSFET.

Data sheets usually give the coordinates for one point on the transconductance curve, as shown in Fig. 13-9*b*. After you substitute $I_{D(on)}$, $V_{GS(on)}$, and $V_{GS(th)}$ into Eq. (13-3), you can solve for the value of K. For instance, if an enhancement-type MOSFET has $I_{D(on)} = 8$ mA, $V_{GS(on)} = 5$ V, and $V_{GS(th)} = 3$ V, its transconductance curve looks like Fig. 13-9c. When you substitute these values into Eq. (13-3),

$$0.008 = K(5 - 3)^2 = 4K$$

or

$$K = 0.002$$

Therefore, the transconductance equation of Fig. 13-9c is

$$I_D = 0.002(V_{GS} - 3)^2$$

SCHEMATIC SYMBOL

When $V_{GS} = 0$, the enhancement-type MOSFET is off because there is no conducting channel between source and drain. The schematic symbol of Fig. 13-10a has a broken channel line to indicate this normally off condition. As you know, a gate voltage greater than the threshold voltage creates an *n*-type inversion layer that connects the source to the drain. The arrow points to this inversion layer, which acts like an *n* channel when the device is conducting.

There is also a *p*-channel enhancement-type MOSFET. The schematic symbol is similar, except that the arrow points outward, as shown in Fig. 13-10b. With a *p*-channel enhancement-type MOSFET, all voltages and currents are complementary to those of the *n*-channel enhancement-type MOSFET.

MAXIMUM GATE-SOURCE VOLTAGE

Both depletion-type and enhancement-type MOSFETs have a thin layer of silicon dioxide, an insulator that prevents gate current for positive as well as negative gate voltages. This insulating layer is kept as thin as possible to give the gate more control over drain current. Because the insulating layer is so thin, it is easily destroyed by excessive gate-source voltage. For instance, a 2N3796 has a $V_{GS(max)}$ rating of ±30 V. If the gate-source voltage becomes more positive than $+30$ V or more negative than -30 V, you can throw away the MOSFET because the thin insulating layer has been destroyed.

Aside from directly applying an excessive V_{GS}, you can destroy the thin insulating layer in more subtle ways. If you remove or insert a MOSFET into a circuit while the power is on, transient voltages caused by inductive kickback and other effects may exceed the $V_{GS(max)}$ rating. This will wipe out the MOSFET. Even picking up a MOSFET may deposit enough static charge to exceed the $V_{GS(max)}$ rating. This is the

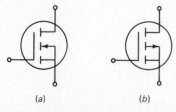

(a)   (b)

Fig. 13-10   *Symbols for enhancement-type MOSFETs.* (a) *n-channel.* (b) *p-channel.*

reason why MOSFETs are often shipped with a wire ring around the leads. You remove the ring after the MOSFET is connected in the circuit.

Some MOSFETs are protected by built-in zener diodes in parallel with the gate and the source. The zener voltage is less than the $V_{GS(max)}$ rating. Therefore, the zener diode breaks down before any damage to the thin insulating layer occurs. The disadvantage of these internal zener diodes is that they reduce the MOSFET's high input resistance. Nevertheless, the tradeoff is often worth it in many applications because expensive MOSFETs are easily destroyed without zener protection.

13-5 BIASING ENHANCEMENT-TYPE MOSFETs

With enhancement-type MOSFETs, V_{GS} has to be greater than $V_{GS(th)}$ to get current. This eliminates self-bias, current-source bias, and zero bias, because all these will have depletion-mode operation. That leaves gate bias and voltage-divider bias. Both of these will work with enhancement-type MOSFETs because both types of bias can produce the enhancement mode. In addition to gate bias and voltage-divider bias, there is one more method of biasing enhancement-type MOSFETs.

Figure 13-11*a* shows *drain-feedback bias,* a type of bias you can use only with enhancement-type MOSFETs. When the MOSFET is conducting, it has a drain current of $I_{D(on)}$ and a drain voltage of $V_{DS(on)}$. Since the gate current is approximately zero, no voltage appears across R_G. Therefore, $V_{GS} = V_{DS(on)}$. As with collector-feedback bias, the circuit of Fig. 13-11*a* tends to compensate for changes in FET characteristics. If $I_{D(on)}$ tries to increase for some reason, $V_{DS(on)}$ decreases. This reduces V_{GS}, which partially offsets the original increase in $I_{D(on)}$.

Figure 13-11*b* shows the Q point on the transconductance curve. It has coordinates of $I_{D(on)}$ and $V_{DS(on)}$. Data sheets for enhancement-type MOSFETs usually give a value of $I_{D(on)}$ for $V_{GS} = V_{DS(on)}$. This helps in setting up the Q point. In design, all you do is select a value of R_D that sets up the specified V_{DS}. As a formula,

$$R_D = \frac{V_{DD} - V_{DS(on)}}{I_{D(on)}} \qquad (13\text{-}4)$$

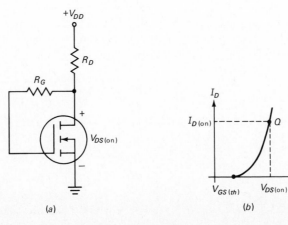

(a)   (b)

Fig. 13-11
Drain-feedback bias. (a) *Circuit.* (b) *Quiescent point.*

Fig. 13-12

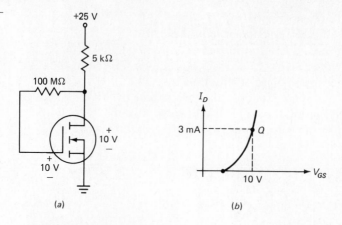

(a)   (b)

For instance, suppose the data sheet of an enhancement-type MOSFET specifies $I_{D(on)} = 3$ mA and $V_{DS(on)} = 10$ V. If $V_{DD} = 25$ V, you can select an R_D of 5 kΩ, as shown in Fig. 13-12a. When $I_{D(on)}$ is 3 mA, $V_{DS(on)} = 10$ V. Therefore, the enhancement-type MOSFET is operating at its specified Q point (Fig. 13-12b).

DC AMPLIFIER

A *dc amplifier* is one that can operate all the way down to zero frequency without a loss of gain. One way to build a dc amplifier, or dc amp, is to leave out all coupling and bypass capacitors.

Figure 13-13 shows a dc amp using MOSFETs. The input stage is a depletion-type MOSFET with zero bias. The second and third stages use enhancement-type MOSFETs; each gate gets its V_{GS} from the drain of the preceding stage. The design of Fig. 13-13 uses MOSFETs with drain currents of 3 mA. For this reason, each drain runs at +10 V with respect to ground. We tap the final output voltage between the 100-kΩ resistors. Since the lower resistor is returned to −10 V, the quiescent output

Fig. 13-13
Three-stage dc amplifier.

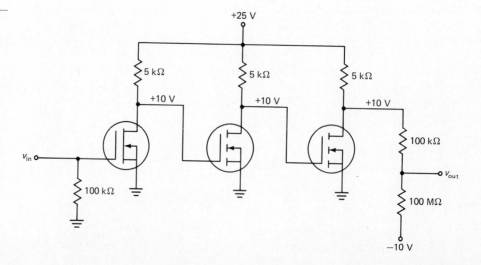

Table 13-1. FET Biasing Circuits

Method	JFET	D MOSFET	E MOSFET
Gate bias	Yes	Yes	Yes
Self-bias	Yes	Yes	No
Voltage-divider bias	Yes	Yes	Yes
Current-source bias	Yes	Yes	No
Zero bias	No	Yes	No
Drain-feedback bias	No	No	Yes

voltage is 0 V. When an ac signal drives the amplifier, regardless of how low its frequency, we get an amplified output voltage.

There are other ways of designing dc amplifiers. The beauty of Fig. 13-11 is its simplicity.

SUMMARY OF BIAS

By examining the types of bias used with JFETs, we can figure out which ones will work with MOSFETs. Table 13-1 summarizes all biasing circuits discussed so far. For convenience, abbreviations are used for depletion-type (D) MOSFETs and enhancement-type (E) MOSFETs. As you can see, self-bias and current-source bias will not work with an E MOSFET. Zero bias works only with depletion-type MOSFETs. Drain-feedback bias works only with enhancement-type MOSFETs.

13-6 ENHANCEMENT-TYPE MOSFET APPLICATIONS

Computers use integrated circuits with thousands of transistors. These ICs work remarkably well, despite the variations in temperature and in transistor parameters. How is this possible? The answer is *two-state* design, using only two points on the load line of each transistor. When used this way, the transistor acts like a switch rather than a current source. Circuits using transistor switches are called switching circuits, digital circuits, logic circuits, and so on. On the other hand, circuits using transistor current sources are called linear circuits, analog circuits, and so on.

The enhancement-type MOSFET has made its greatest impact in digital circuits. One reason is its low power consumption. Another is the small amount of space it takes on a chip. In other words, a manufacturer can put many more MOS transistors on a chip than bipolar transistors. This is the reason MOSFETs are used in large-scale integration for microprocessors, memories, and other devices requiring thousands of transistors on a chip.

This section discusses some of the applications for enhancement-type MOSFETs, especially in switching and digital circuits.

SAMPLE-AND-HOLD AMPLIFIER

Like the JFET, the MOSFET can act as a switch, either in shunt or in series with the load. The enhancement-type MOSFET is particularly convenient in switching

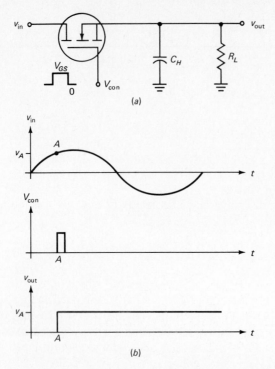

Fig. 13-14   (a) *Sample-and-hold amplifier.* (b) *Waveforms of input, control, and output.*

applications because it is normally off. Figure 13-14*a* shows a useful circuit called a *sample-and-hold* amplifier. When V_{con} is high, the MOSFET turns on, and the capacitor charges to the value of input voltage. The charging time constant is very short because $r_{ds(on)}$ is small. When V_{con} goes low, the MOSFET opens, and the capacitor begins to discharge through the load resistor. If the discharging time constant is very large, the capacitor can hold its charge for a long time.

In many applications we need a dc output equal to the input voltage at a particular instant. For example, suppose we want the value of input voltage at point A in Fig. 13-14*b*. If we apply a narrow V_{con} at point A, the v_{out} of the sample-and-hold amplifier can charge to approximately v_A as shown. When V_{con} returns to zero, the MOSFET opens, and the input voltage can no longer affect the value of output voltage. Given a long time constant, the output voltage holds at v_A for an indefinite period (see Fig. 13-14*b*).

When V_{con} is high in Fig. 13-14*a*, the circuit is *sampling* the input and the capacitor charges to the approximate value of input voltage. When V_{con} returns low, the circuit goes into a *hold* condition because the capacitor stores the sampled value of input voltage. Remember the basic idea of a sample-and-hold amplifier; you will see it used a lot with analog-to-digital converters (computer circuits).

ACTIVE LOAD

Figure 13-15*a* shows a MOSFET driver and a passive load (resistor R_D). In this switching circuit, v_{in} is either low or high, and the MOSFET acts like a switch that is

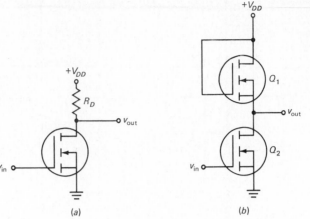

Fig. 13-15
(a) R_D is passive
load. (b) Q_2 is
active load.

either off or on. When v_{in} is low, the MOSFET is cut off and v_{out} equals the supply voltage. On the other hand, when v_{in} is high, the MOSFET conducts heavily and v_{out} drops to a low value.

Figure 13-15b shows a MOSFET driver (the lower MOSFET) and an *active load* (the upper MOSFET). Because of the drain-feedback bias, the upper MOSFET is always conducting. By deliberate design, this upper MOSFET has an $r_{DS(on)}$ at least 10 times greater than the $r_{DS(on)}$ of the lower MOSFET. For this reason, the upper MOSFET acts like a resistor and the lower MOSFET acts like a switch.

Using an MOS driver and an MOS load leads to much smaller integrated circuits because MOSFETs take up less room on a chip than resistors. This is why MOS technology dominates in computer applications; it allows you to get many more circuits on a chip.

The main thing to remember is the idea of active loading, using one active device as the load on another. Active loading is also possible with bipolar transistors; this is discussed further in Chap. 15.

CMOS INVERTER

We can build *complementary* MOS (CMOS) circuits with *p*-channel and *n*-channel MOSFETs. One of the most important of all is the CMOS inverter shown in Fig. 13-16a. Notice that Q_1 is a *p*-channel device and Q_2 is an *n*-channel device. This circuit is analogous to the class B push-pull bipolar amplifier of Fig. 13-16b. When one device is on, the other is off, and vice versa.

For instance, when v_{in} is low in Fig. 13-16a, the lower MOSFET is off but the upper one is on. Therefore, the output voltage is high. On the other hand, when v_{in} is high, the lower MOSFET is on and the upper one is off. In this case, the output voltage is low. Since the output voltage is always opposite in phase to the input, the circuit is called an *inverter*.

The CMOS inverter can be modified to build other complementary-type circuits. The key advantage in using CMOS design is its extremely *low power consumption*. Because both devices are in series, the current is determined by the leakage in the off device, which is typically in nanoamperes. This means that the total power dissipa-

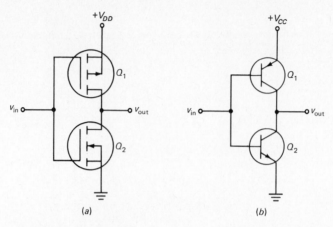

Fig. 13-16 *Complementary circuits.* (a) *CMOS inverter.* (b) *Bipolar inverter.*

tion of the circuit is in nanowatts. This low power consumption is the main reason CMOS circuits are popular in pocket calculators, digital wristwatches, and satellites.

13-7 VMOS

Figure 13-17*a* shows the structure of an enhancement-type MOSFET in an integrated circuit. The source is on the left, the gate in the middle, and the drain on the right. Free electrons flow horizontally from the source to the drain when V_{GS} is greater than the threshold voltage. This conventional structure limits the maximum current because the free electrons must flow along the narrow inversion layer, symbolized by the dashed line. Because the channel is so narrow, conventional MOS devices have small drain currents, which implies low power ratings (typically less than 1 W).

VERTICAL CHANNEL

Figure 13-17*b* shows the structure of *vertical* MOS (VMOS). Notice that it has two sources at the top. These are usually connected. Furthermore, the substrate now acts like the drain. When V_{GS} is greater than the threshold voltage, free electrons flow vertically downward from the two sources to the drain. Because the conducting channel is much wider along both sides of the V groove, the current can be much larger. The overall effect is an enhancement-type MOSFET that can handle much larger currents and voltages than a conventional MOSFET.

Prior to the invention of the VMOS transistor, MOSFETs could not compete with the power ratings of large bipolar transistors. But now, the VMOS offers a new type of MOSFET that is better than the bipolar transistor in many applications requiring high load power, including audio amplifiers, RF amplifiers, and so on.

LACK OF THERMAL RUNAWAY

One major advantage VMOS transistors have over bipolar transistors is the *lack of thermal runaway.* As you recall, an increase in device temperature lowers the V_{BE} of a bipolar transistor. This increases the collector current, which raises the temperature

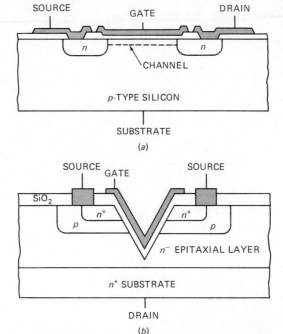

Fig. 13-17
(a) *Conventional MOSFET struc- ture.* (b) *VMOS structure.*

further. If the heat sinking is inadequate, the bipolar transistor can go into thermal runaway and be destroyed by the excessive power dissipation.

A VMOS transistor, on the other hand, has a negative thermal coefficient. As the device temperature increases, the drain current decreases, which reduces power dissipation. Because of this, the VMOS transistor cannot go into thermal runaway, and this is a big advantage in any power amplifier.

PARALLEL CONNECTION

Bipolar transistors cannot be connected in parallel to increase load power because their V_{BE} drops do not match closely enough. If you try to connect them in parallel, *current hogging* occurs (the one with the lower V_{BE} drop has more collector current).

Because of their negative temperature coefficients, two VMOS transistors can be connected in parallel to increase the load power. If one of the parallel VMOS transistors tries to hog the current, its negative temperature coefficient reduces the current through it, so that approximately equal currents flow through the parallel VMOS transistors.

FASTER SWITCHING SPEED

When a small-signal bipolar transistor is used as saturated switch, conservative design calls for a base current that is approximately one-tenth the saturated collector current. Since most transistors have a β_{dc} greater than 10, the excess base current guarantees saturation from one transistor to the next. But the excess base current also does something else that we have not mentioned until now. Extra carriers are stored in the base region of a saturated transistor. When the transistor tries to come out of

Fig. 13-18
Class C amplifier using VMOS transistor.

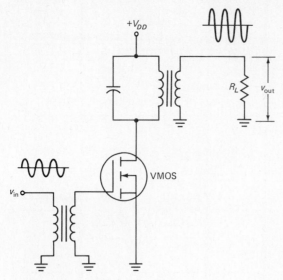

saturation, there is a small delay called the *saturation delay time* t_s (also called *storage time*). For instance, the storage time of a 2N3713 is 0.3 μs. This means that it takes approximately 0.3 μs for a 2N3713 to come out of saturation after the base drive is removed.

Another advantage the VMOS transistor has over the bipolar transistor is the *lack of storage time*. Because no extra charges are stored in VMOS when it is conducting, it can come out of saturation almost immediately. Typically, a VMOS transistor can shut off amperes of current in tens of nanoseconds. This is from 10 to 100 times faster than a comparable bipolar transistor. Therefore, the VMOS transistor finds numerous applications in high-speed switching circuits, switching regulators, and so on.

CLASS C AMPLIFIER

Figure 13-18 shows a VMOS class C amplifier. Because of the threshold voltage, conduction does not occur until the input signal swings above $V_{GS(th)}$. This results in class C operation, since the conduction angle is less than 180°. As before, the current pulses are filtered by the tank circuit to get an amplified sinusoidal output voltage. The output transformer couples the RF signal into the load.

INTERFACING

Digital ICs are low-power devices because they can supply only small load currents. *Interfacing* means using some kind of buffer between a low-power device (often a digital IC) and a high-power load (such as a relay, motor, or incandescent lamp). The VMOS transistor is an excellent device for interfacing digital ICs to high-power loads. As shown in Fig. 13-19, a digital IC drives the gate of a VMOS. When the digital output is low, the VMOS is off. When the digital output is high, the VMOS acts like a closed switch, and maximum current flows through the load. Interfacing digital ICs (such as CMOS, MOS, or TTL) to high-power loads is one of the most important applications for the VMOS transistor.

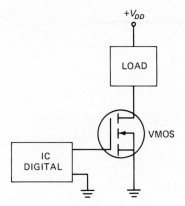

Fig. 13-19 *VMOS transistor interfaces a low-power digital IC with a high-power load.*

EXAMPLE 13-2

Figure 13-20 shows part of a robot. The VN0300M is a VMOS transistor with an $r_{DS(on)}$ of 1.2 Ω. It has a maximum current rating of 700 mA and a breakdown voltage of 30 V. Explain what the circuit does.

SOLUTION

The whole point of the circuit is to interface a CMOS inverter with a relay. Typically, a relay takes tens to hundreds of milliamperes, too heavy a load for a digital IC like a CMOS inverter. The VN0300M has a maximum current rating of 700 mA, more than enough to supply the relay current.

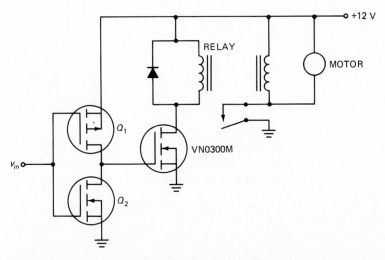

Fig. 13-20 *Part of a robot: CMOS inverter, VMOS interface, relay, and motor.*

When v_{in} is low, the CMOS inverter (Q_1 and Q_2) has a high output, which turns on the VMOS. Since the VMOS has an $r_{DS(on)}$ of only 1.2 Ω, it effectively shorts the lower end of the relay to ground. The relay points then close, and the motor starts turning. The motor continues to run as long as v_{in} is low. When v_{in} goes high, the CMOS inverter output goes low. The VMOS transistor then shuts off, the relay opens, and the motor stops running.

Robotics, the science of robots, is a field emerging in electronics. It heralds the second industrial revolution. Modern robots combine microprocessors, interfacing circuits, and mechanical devices. Because of this, we can now build machines that have fingers and toes as well as brains. VMOS transistors such as the VN0300M are ideal for interfacing digital devices and high-power loads in robotic systems.

PROBLEMS

Straightforward

13-1. An *n*-channel depletion-type MOSFET has an I_{DSS} of 8 mA and a $V_{GS(off)}$ of −4 V. How much drain current is there when $V_{GS} = -1$ V? When $V_{GS} = +1$ V?

13-2. A MOSFET has the transconductance curve shown in Fig. 13-21*a*. If it is used in the circuit of Fig. 13-21*b*, what is the dc voltage from the drain to ground?

Fig. 13-21

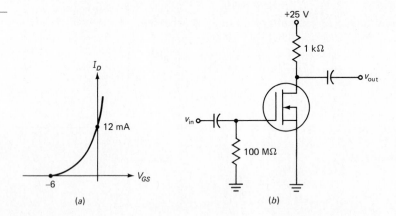

(a)   (b)

13-3. If the MOSFET of Fig. 13-21*b* has a g_m of 8000 μS, what is the input impedance of the amplifier? The unloaded voltage gain? The output impedance?

13-4. A dual-gate depletion-type MOSFET is used in a zero-biased cascode amplifier. If $R_G = 10$ MΩ, $R_D = 1.5$ kΩ, and $g_m = 7500$ μS, what is the input impedance? The unloaded voltage gain? The output impedance?

13-5. If an enhancement-type MOSFET has the transconductance curve shown in Fig. 13-22*a*, how much drain current is there when $V_{GS} = +5$ V?

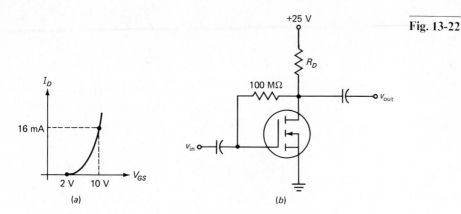

Fig. 13-22

(a)

(b)

13-6. The MOSFET of Fig. 13-22*b* has the transconductance curve of Fig. 13-22*a*. If $V_{DS} = +10$ V, what does V_{GS} equal? What is the value of R_D?

13-7. If g_m is 8500 μS in Fig. 13-22*b*, what is the unloaded voltage gain for an R_D of 910 Ω?

13-8. The upper MOSFET of Fig. 13-23*a* has an $r_{DS(on)}$ of 2 kΩ. When v_{in} is +3 V, the lower MOSFET has an $r_{DS(on)}$ of 150 Ω. What is the output voltage when v_{in} is zero? When v_{in} is +3 V?

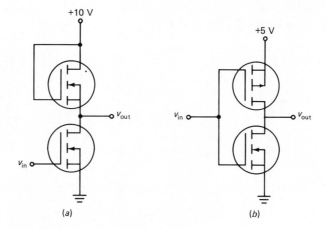

Fig. 13-23

(a)

(b)

13-9. In Fig. 13-23*b*, the threshold voltage is +2 V. What does v_{out} equal when v_{in} is zero? When v_{in} is +5 V?

13-10. In Fig. 13-20 the relay windings have a resistance of 150 Ω. If the VMOS has an $r_{DS(on)}$ of 1.2 Ω when conducting, how much current is there through the relay windings when v_{in} is low? When v_{in} is high?

Troubleshooting

13-11. The dc drain voltage is 0 V in Fig. 13-21*b*. Which of the following is a possible trouble:
   a. Drain resistor open
   b. Drain resistor shorted
   c. MOSFET open
   d. Gate resistor shorted

13-12. If the 100-MΩ resistor in Fig. 13-22*b* shorts, do the following increase, decrease, or stay the same?
   a. Dc drain voltage
   b. Dc gate current
   c. Ac output voltage
   d. Supply voltage

13-13. The motor of Fig. 13-20 runs continuously, no matter what the value of v_{in}. Which of the following is a possible trouble?
   a. VMOS transistor open
   b. Relay contacts fused (shorted)
   c. Diode shorted
   d. Q_1 open

Design

13-14. A load resistor of 4.7 kΩ is connected to the output coupling capacitor of Fig. 13-7. If $g_m = 10,000 \ \mu S$, select a new value of R_D to produce a loaded voltage gain of -10.

13-15. The enhancement-type MOSFETs of Fig. 13-13 are replaced by MOSFETs that have $I_{D(on)} = 2$ mA when $V_{GS} = 10$ V. Select new values of drain resistors so that all drain voltages are +10 V.

13-16. In Fig. 13-14*a*, the load resistance is 470 kΩ. Select a value of hold capacitance C_H to get a discharging time constant of approximately 0.15 s.

Challenging

13-17. All MOSFETs have a g_m of 2000 μS in Fig. 13-13. If v_{in} is 1 mV, what does v_{out} equal?

13-18. The MOSFET of Fig. 13-21*b* has the transconductance curve of Fig. 13-21*a*. What is the unloaded voltage gain?

13-19. The enhancement-type MOSFET of Fig. 13-14*a* has an $r_{ds(on)}$ of 50 Ω. The hold capacitance C_H equals 0.1 μF, and the load resistance is 1 MΩ. What pulse width should V_{CON} have for the capacitor to charge to 99 percent of the sampled voltage? For the discharging time constant to be at least 100 times longer than the period, what is the minimum input frequency?

Computer

13-20. Here is a program:

```
10 R = 1250
20 PRINT "THE RESISTANCE IS ";R
30 STOP
```

In line 20, the semicolon following the quotation marks tells the computer to print whatever follows the semicolon. In other words, after the program is run, the screen shows

THE RESISTANCE IS 1250

The semicolon can be used more than once. For instance, line 20 can be changed to

20 PRINT "THE RESISTANCE IS";R; " OHMS"

In this case, the execution of line 20 would print out

THE RESISTANCE IS 1250 OHMS

Look at this program:

```
10 PRINT "ENTER VDD": INPUT VDD
20 PRINT "ENTER IDSS": INPUT IDSS
30 PRINT "ENTER RD": INPUT RD
40 V = VDD − IDSS*RD
50 PRINT "THE DRAIN-SOURCE VOLTAGE IS";V; "VOLTS"
```

If $V_{DD} = 20$ V, $I_{DSS} = 5$ mA, and $R_D = 1.5$ kΩ, what does the computer screen display after the program is run?

13-21. The following program is for an enhancement-type MOSFET.

```
10 PRINT "ENTER ID": INPUT ID
20 PRINT "ENTER VGS": INPUT VGS
30 PRINT "ENTER TH": INPUT TH
40 A = (VGS − TH)*(VGS − TH)
50 K = ID/A
60 FOR VGS = TH TO TH + 5
70 ID = K*(VGS − TH)*(VGS − TH)
80 PRINT "ID =";ID; "WHEN VGS =";VGS
90 NEXT VGS
```

What does the program do in line 50? If $V_{GS(th)} = 2$ V, what is the largest V_{GS} used in line 70? How many print lines are there on the screen when the program is through running?

13-22. Write a program that prints out the input impedance, unloaded voltage gain, and output impedance of a dual-gate cascode amplifier like Fig. 13-6*b*. The inputs are R_G, g_m, and R_D. Use semicolons in your final PRINT statements.

13-23. Write a program that calculates the ac output voltage for the dc amplifier of Fig. 13-13. The inputs are g_m for each stage.

CHAPTER 14

Frequency Effects

The midband of an ac amplifier is the range of frequencies where capacitors have no effect, where only resistances appear in the ac equivalent circuit. In this chapter we will discuss the operation of amplifiers outside the midband. Below the midband, the voltage gain of an ac amplifier drops off because of the coupling and bypass capacitors. Above the midband, the voltage gain drops off because of internal transistor capacitance and stray wiring capacitance.

The chapter begins with a discussion of the lead network, the key to low-frequency effects. Next, we take up the lag network because it helps explain high-frequency effects. In the remainder of the chapter we discuss Miller's theorem, decibels, Bode plots, and other topics needed to understand frequency effects in amplifiers.

14-1 THE LEAD NETWORK

The *lead network* of Fig. 14-1a is the key to analyzing low-frequency effects in amplifiers. As you know, capacitive reactance is given by

$$X_C = \frac{1}{2\pi f C}$$

At very low frequencies, X_C approaches infinity. At very high frequencies, X_C approaches zero. A capacitor is equivalent to an open circuit at very low frequencies, and it is equivalent to a short circuit at very high frequencies.

Incidentally, the circuit is called a lead network because the output voltage leads the input voltage. We will discuss phase angle later in the chapter. For now, our main interest is the way in which the magnitude of the output voltage varies with frequency.

FREQUENCY RESPONSE

In Fig. 14-1a, voltages V_{in} and V_{out} indicate rms values. As we vary the frequency, the output voltage changes because of the reactance of the capacitor. This means that the voltage gain V_{out}/V_{in} is a function of frequency.

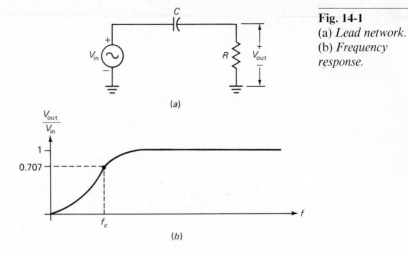

Fig. 14-1
(a) *Lead network.*
(b) *Frequency response.*

Figure 14-1*b* shows the *frequency response* (voltage gain versus frequency) of a lead network. At zero frequency, X_C is infinite. Therefore, the output voltage is zero and the voltage gain is zero. As the frequency increases, X_C decreases and the voltage gain increases. When the frequency is high enough, X_C is much smaller than R, and V_{out} approximately equals V_{in}. Therefore, the voltage gain of the lead network approaches 1 at higher frequencies, as shown in Fig. 14-1*b*.

CUTOFF FREQUENCY

The lead network of Fig. 14-1*a* is an ac voltage divider with an output voltage of

$$V_{out} = \frac{R}{\sqrt{R^2 + X_C^2}} V_{in}$$

which can be rewritten as

$$\frac{V_{out}}{V_{in}} = \frac{R}{\sqrt{R^2 + X_C^2}} \tag{14-1}$$

By graphing this equation versus frequency, we can get exact values for the frequency response of Fig. 14-1*b*.

The *cutoff frequency* (also called the critical frequency, break frequency, and corner frequency) is the frequency at which X_C equals R. In symbols,

$$X_C = R$$

or

$$\frac{1}{2\pi f C} = R$$

Solving for f gives

$$f = \frac{1}{2\pi RC}$$

To keep this frequency distinct, the subscript c (cutoff) is attached, and the equation is usually written as

$$f_c = \frac{1}{2\pi RC} \tag{14-2}$$

HALF-POWER POINT

At the cutoff frequency, $X_C = R$. If we substitute this into Eq. (14-1), we get

$$\frac{V_{out}}{V_{in}} = 0.707$$

This is the voltage gain at the cutoff frequency. In Fig. 14-1*b*, the cutoff point is sometimes called the *half-power point* because the load power is half of its maximum value. For instance, suppose V_{in} is 1 V and R is 1 Ω. Then the maximum load power is 1 W. At the cutoff frequency, the output voltage is 0.707 V, and the load power is

$$P = \frac{(0.707 \text{ V})^2}{1 \text{ }\Omega} = 0.5 \text{ W}$$

SOURCE RESISTANCE

Figure 14-2*a* shows a lead network with source resistance. The voltage gain is given by

$$\frac{V_{out}}{V_{in}} = \frac{R_L}{\sqrt{(R_S + R_L)^2 + X_C^2}} \tag{14-3}$$

In this case, the half-power point occurs when the capacitive reactance equals the total series resistance:

$$X_C = R_S + R_L \tag{14-4}$$

or

$$\frac{1}{2\pi fC} = R_S + R_L$$

Solving for the cutoff frequency gives

$$f_c = \frac{1}{2\pi(R_S + R_L)C} \tag{14-5}$$

Figure 14-2*b* shows the frequency response of a lead network with a source resistance. In the midband of the circuit, the capacitor acts like a short and the voltage gain is

$$\frac{V_{out}}{V_{in}} = \frac{R_L}{R_S + R_L}$$

which can be written as

$$A_{mid} = \frac{R_L}{R_S + R_L}$$

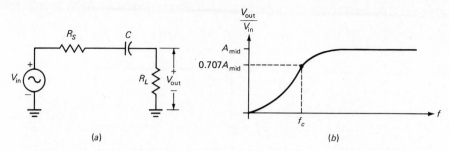

Fig. 14-2 (a) *Lead network with source and load resistances.* (b) *Frequency response.*

A_{mid} is the voltage gain in the midband, the range of frequencies at which the capacitor acts approximately like an ac short. Below the midband, the voltage gain drops off. Notice that it equals $0.707A_{mid}$ at the cutoff frequency.

STIFF COUPLING

In earlier chapters, we assumed *stiff coupling*. This means that all lead networks satisfied this rule:

$$X_C = 0.1(R_S + R_L) \tag{14-6}$$

at the lowest frequency to be coupled. Substituting this into Eq. (14-3) gives a voltage gain of

$$\frac{V_{out}}{V_{in}} = 0.995 \, A_{mid}$$

This shows you how effective stiff coupling is. At the lowest frequency to be coupled, the voltage gain is within half a percent of the midband gain.

If you compare Eq. (14-6) with Eq. (14-4), you can see that they differ by a factor of 10. This implies that the minimum frequency with stiff coupling is 10 times greater than the cutoff frequency. In symbols,

$$f_{min} = 10f_c \tag{14-7}$$

where f_{min} = lowest frequency with stiff coupling
f_c = cutoff frequency of lead network

This is useful because sometimes you know one frequency but not the other. For instance, if you measure a cutoff frequency of 200 Hz, then you can calculate a minimum frequency of 2000 Hz. Conversely, if you design an amplifier with a minimum frequency of 500 Hz, then you immediately know that it has a cutoff frequency of 50 Hz.

AMPLIFIER ANALYSIS

Figure 14-3a shows the CE amplifier analyzed in earlier chapters. It has an input coupling capacitor and an output coupling capacitor. What we want to do now is

Fig. 14-3
(a) *CE amplifier.*
(b) *Ac equivalent circuit.*

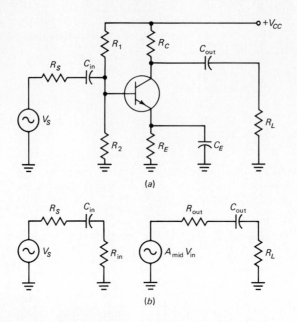

(a)

(b)

identify the input and output lead networks, so that we can easily calculate the cutoff frequencies.

The next section analyzes the effect of the emitter bypass capacitor. For now, let us eliminate it from consideration by assuming that it has infinite capacitance. This allows us to draw the ac equivalent circuit as shown in Fig. 14-3b. On the input side, R_{in} is the input impedance of the stage in the midband of the amplifier. It equals

$$R_{in} = R_1 \parallel R_2 \parallel \beta r'_e$$

On the output side, R_{out} is the output impedance of the stage in the midband of the amplifier. Ignoring the resistance of the collector current source,

$$R_{out} \cong R_C$$

Incidentally, R_{in} and R_{out} are identical to the z_{in} and z_{out} used in Chaps. 7 and 8. In these earlier chapters, we were analyzing amplifiers operating in the midband. In this frequency range, the input impedance z_{in} and the output impedance z_{out} are purely resistive. But outside the midband, z_{in} and z_{out} become complex variables because they include reactive effects. Now that we are discussing frequency effects, we will use R_{in} instead of z_{in} and R_{out} instead of z_{out}.

We have two lead networks, one on the input side and one on the output side. The input lead network has a cutoff frequency of

$$f_{in} = \frac{1}{2\pi(R_S + R_{in})(C_{in})} \qquad (14\text{-}8)$$

where f_{in} = cutoff frequency of input lead network
   R_S = source resistance
   R_{in} = input resistance of stage
   C_{in} = capacitance of input lead network

Similarly, the output lead network has a cutoff frequency of

$$f_{\text{out}} = \frac{1}{2\pi(R_{\text{out}} + R_L)(C_{\text{out}})} \tag{14-9}$$

where f_{out} = cutoff frequency of output lead network
R_{out} = output resistance of stage
R_L = load resistance
C_{out} = capacitance of output lead network

You can use Eqs. (14-8) and (14-9) to analyze any amplifier. These equations apply to emitter followers, JFET amplifiers, and other devices, provided that you can calculate the input and output resistances of the amplifier. The following examples give the idea of how this is done.

EXAMPLE 14-1

If $\beta = 150$, what are the cutoff frequencies of the input and output lead networks in Fig. 14-4a?

SOLUTION

We analyzed this CE amplifier in Example 7-3 and found that

$$R_{\text{in}} = 1.18 \text{ k}\Omega$$
$$R_{\text{out}} = 3.6 \text{ k}\Omega$$

With Eq. (14-8),

$$f_{\text{in}} = \frac{1}{2\pi(1 \text{ k}\Omega + 1.18 \text{ k}\Omega)(0.47 \text{ }\mu\text{F})} = 155 \text{ Hz}$$

With Eq. (14-9),

$$f_{\text{out}} = \frac{1}{2\pi(3.6 \text{ k}\Omega + 1.5 \text{ k}\Omega)(2.2 \text{ }\mu\text{F})} = 14.2 \text{ Hz}$$

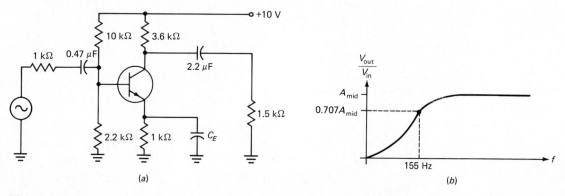

(a)

(b)

Fig. 14-4

Fig. 14-5

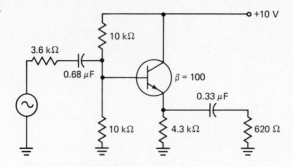

The input lead network has the higher cutoff frequency. Therefore, it causes the load power to drop to the half-power point when the frequency is 155 Hz, as shown in Fig. 14-4*b*. Whenever you have two or more lead networks, the one with the higher cutoff frequency is more important because it causes the *first break* in amplifier response. We call it the *dominant* lead cutoff frequency.

EXAMPLE 14-2

In Fig. 14-5, the emitter follower has these values: $R_{in} = 5 \text{ k}\Omega$ and $R_{out} = 45.9 \ \Omega$ (found in Example 8-3). Calculate the cutoff frequencies of the input and output lead networks.

SOLUTION

The cutoff frequency of the input lead network is

$$f_{in} = \frac{1}{2\pi(3.6 \text{ k}\Omega + 5 \text{ k}\Omega)(0.68 \ \mu\text{F})} = 27.2 \text{ Hz}$$

The cutoff frequency of the output lead network is

$$f_{out} = \frac{1}{2\pi(45.9 \ \Omega + 620 \ \Omega)(0.33 \ \mu\text{F})} = 724 \text{ Hz}$$

The output lead network has the higher cutoff frequency. This means that the load power drops to the half-power point when the frequency is 724 Hz. The frequency response of the amplifier looks like Fig. 14-4*b*, except that the cutoff frequency is at 724 Hz.

14-2 THE LAG NETWORK

The *lag network* of Fig. 14-6*a* is the key to analyzing high-frequency effects in amplifiers. At very low frequencies, X_C is large and the output voltage approximately equals the input voltage. At very high frequencies, X_C is small and the output voltage approaches zero. The circuit is called a lag network because the output voltage lags the input voltage.

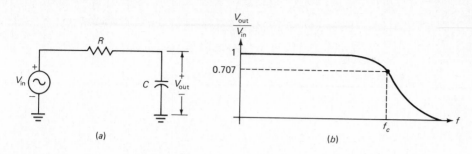

Fig. 14-6
(a) *Lag network.*
(b) *Frequency response.*

(a)

(b)

FREQUENCY RESPONSE

Figure 14-6b shows the frequency response of a lag network. The voltage gain is 1 at low frequencies. At the cutoff frequency, the voltage gain is 0.707. Beyond the cutoff frequency, the voltage gain continues to decrease; it approaches zero at infinite frequency.

CUTOFF FREQUENCY

The voltage gain of a lag network is

$$\frac{V_{out}}{V_{in}} = \frac{X_C}{\sqrt{R^2 + X_C^2}} \qquad (14\text{-}10)$$

By graphing this equation, you can get the exact values for the frequency response of Fig. 14-6b. The cutoff frequency is defined as the frequency at which

$$X_C = R$$

and is given by

$$f_c = \frac{1}{2\pi RC} \qquad (14\text{-}11)$$

where f_c = cutoff frequency of lag network
R = resistance of lag network
C = capacitance of lag network

LOAD RESISTANCE

Often, a capacitor is in parallel with the load resistor, as shown in Fig. 14-7a. At low frequencies, where the capacitor appears to be open, the circuit acts like a voltage divider with a midband gain of

$$A_{mid} = \frac{R_L}{R_S + R_L}$$

At higher frequencies, however, the capacitor begins to shunt alternating current away from the load. This causes the load voltage to drop off.

The simplest way to find the cutoff frequency is by thevenizing the circuit driving

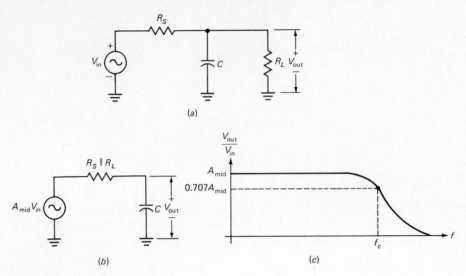

Fig. 14-7 (a) *Lag network with source and load resistances.* (b) *Equivalent circuit.* (c) *Frequency response.*

the capacitor. The Thevenin voltage is

$$V_{TH} = \frac{R_L}{R_S + R_L} V_{in}$$

or

$$V_{TH} = A_{mid} V_{in}$$

and the Thevenin resistance is

$$R_{TH} = R_S \parallel R_L$$

Figure 14-7*b* shows the Thevenin equivalent circuit. Notice that the equivalent circuit is a lag network. Therefore, it has a cutoff frequency of

$$f_c = \frac{1}{2\pi(R_S \parallel R_L)C} \tag{14-12}$$

As shown in Fig. 14-7*c*, the voltage gain is 0.707 A_{mid} at this cutoff frequency.

CURRENT SOURCE

Figure 14-8*a* shows a current source driving a parallel connection of R_C, R_L, and C. This is equivalent to a lag network. The easiest way to see this is to thevenize the circuit, as shown in Fig. 14-8*b*. Because the equivalent circuit is a lag network, the cutoff frequency is

$$f_c = \frac{1}{2\pi(R_C \parallel R_L)C} \tag{14-13}$$

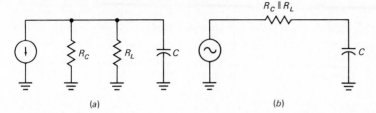

Fig. 14-8 (a) *Current source drives parallel components.* (b) *Equivalent circuit is lag network.*

This formula is important in high-frequency analysis of bipolar and FET amplifiers. More will be said about this later.

EMITTER BYPASS CAPACITOR

The emitter bypass capacitor causes the frequency response of an amplifier to break at a cutoff frequency, designated f_E. Therefore, an amplifier like Fig. 14-9a has three cutoff frequencies: f_{in}, f_{out}, and f_E. To isolate the effects of the emitter bypass capacitor, assume that the coupling capacitors have infinite capacitance. This means that the frequency response breaks at f_E, as shown in Fig. 14-9b, and that cutoff frequencies f_{in} and f_{out} are much lower than f_E.

In the midband of the amplifier, the emitter bypass capacitor appears like an ac short. This grounds the emitter and produces a loaded voltage gain of $-r_C/r'_e$, where

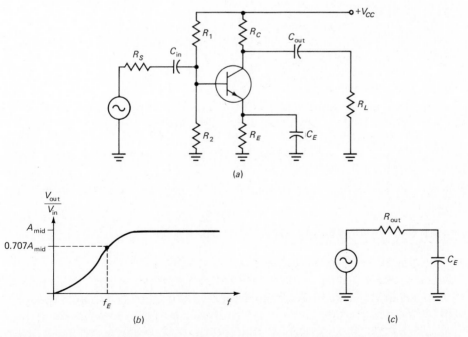

Fig. 14-9 (a) *CE amplifier.* (b) *Frequency response.* (c) *Thevenin circuit facing bypass capacitor.*

$r_C = R_C \parallel R_L$. Below the midband, however, the bypass capacitor no longer appears like a perfect ac short. Therefore, the voltage gain decreases, as shown in Fig. 14-9b.

The emitter circuit is equivalent to a lag network. You can see this by thevenizing the circuit driving C_E, as shown in Fig. 14-9c. In this equivalent circuit, R_{out} is the Thevenin resistance facing the capacitor. As derived in Chap. 8,

$$R_{out} \cong r'_e + \frac{R_S \parallel R_1 \parallel R_2}{\beta} \tag{14-14}$$

(*Note:* Because impedance becomes complex outside the midband, we use R_{out} instead of z_{out} to designate the Thevenin resistance facing the capacitor.) The cutoff frequency of the lag network is

$$f_E = \frac{1}{2\pi R_{out} C_E} \tag{14-15}$$

where f_E = cutoff frequency of emitter network
$\quad R_{out}$ = output resistance facing bypass capacitor
$\quad C_E$ = emitter bypass capacitance

EXAMPLE 14-3

If $\beta = 150$ in Fig. 14-10a, what is the cutoff frequency of the emitter bypass network?

SOLUTION

Earlier in Example 14-1, we calculated these cutoff frequencies for the lead networks:

$$f_{in} = 155 \text{ Hz}$$
$$f_{out} = 14.2 \text{ Hz}$$

To calculate the cutoff frequency of the emitter bypass network, we first need to get the output resistance facing the emitter bypass capacitor. If you analyze the voltage-divider bias, you can calculate $I_C = 1.1$ mA and $r'_e = 22.7 \ \Omega$. From Eq. (14-14),

$$R_{out} = 22.7 \ \Omega + \frac{1 \text{ k}\Omega \parallel 10 \text{ k}\Omega \parallel 2.2 \text{ k}\Omega}{150} = 27 \ \Omega$$

For a C_E of 10 μF, the emitter cutoff frequency is

$$f_E = \frac{1}{2\pi(27 \ \Omega)(10 \ \mu\text{F})} = 589 \text{ Hz}$$

In Example 7-3, we analyzed this amplifier and calculated an output voltage of 25 mV. This means that the amplifier has the frequency response shown in Fig. 14-10b. As you can see, the output voltage is 25 mV in the midband. The output voltage drops to 17.7 mV at 589 Hz, the most important of the cutoff

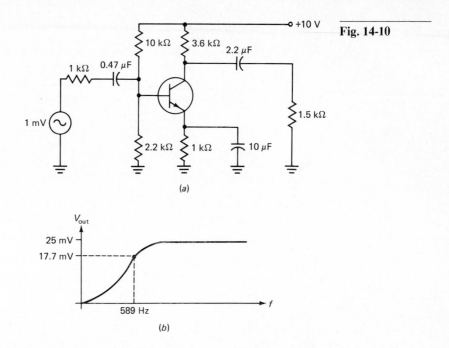

Fig. 14-10

(a)

(b)

frequencies. The cutoff frequencies of the lead networks are 14.2 Hz and 155 Hz, and so they have little effect until the frequency is much lower.

Given the three cutoff frequencies (f_{in}, f_{out}, and f_E), the highest one is called the *dominant* cutoff frequency because the amplifier response breaks first at this cutoff frequency. In this example, f_E is the dominant cutoff frequency.

EXAMPLE 14-4

Figure 14-11a shows a depletion-type MOSFET amplifier. If $g_m = 5000 \ \mu S$, what is the output voltage in the midband? If a stray capacitance of 20 pF is across the load, what is the cutoff frequency? Sketch the frequency response.

SOLUTION

The dashed lines used for the capacitor indicate that the capacitance is not a component. Instead, it represents internal or stray wiring capacitance. In the midband of the amplifier, this capacitance appears open, and the voltage gain is

$$A = -5000 \ \mu S \times 10 \ k\Omega = -50$$

Figure 14-11b shows the ac equivalent circuit for the collector. This is equivalent to a lag network with a cutoff frequency of

$$f_c = \frac{1}{2 \ \pi (10 \ k\Omega)(20 \ pF)} = 796 \ kHz$$

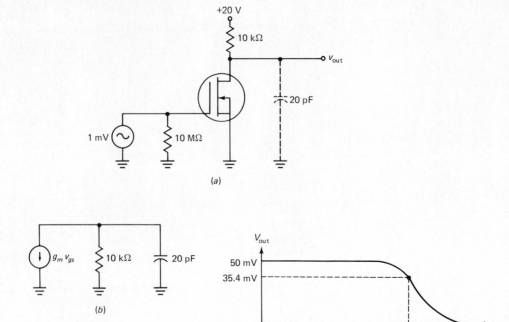

Fig. 14-11 (a) *MOSFET amplifier.* (b) *Ac equivalent circuit of output.* (c) *Frequency response.*

Figure 14-11c shows the frequency response. In the midband, the output voltage has an rms value of 50 mV. At the cutoff frequency, the output voltage drops to 0.707 of the midband value. Notice that the midband extends all the way to zero frequency because the amplifier is direct-coupled. In other words, there is no lower cutoff frequency.

14-3 MILLER'S THEOREM

Figure 14-12a shows an amplifier with a capacitor between the input and output terminals. The capacitor is sometimes called a *feedback capacitor* because the amplified output signal is fed back to the input. When *A* is large, this feedback significantly changes the input impedance of the amplifier.

MILLER EQUIVALENT CIRCUIT

A circuit like Fig. 14-12a is difficult to analyze because the feedback capacitor is part of the input and output circuits. Miller's theorem says that the original circuit can be replaced by the equivalent circuit of Fig. 14-12b. In the equivalent circuit, the input

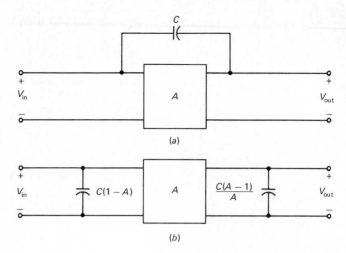

capacitance is

$$C_{\text{in(Miller)}} = C(1 - A) \qquad (14\text{-}16)$$

and the output capacitance is

$$C_{\text{out(Miller)}} = C\frac{A - 1}{A} \qquad (14\text{-}17)$$

The advantage of the Miller equivalent circuit is that it splits the feedback capacitor into two capacitors, one on the input side and the other on the output side. This simplifies the analysis because the input and output circuits are no longer coupled.

MATHEMATICAL PROOF

In Fig. 14-12a, the feedback capacitor has an alternating current given by

$$I_C = \frac{V_{\text{in}} - V_{\text{out}}}{-jX_C}$$

Since $V_{\text{out}} = AV_{\text{in}}$, we can rewrite the equation as

$$I_C = \frac{V_{\text{in}}(1 - A)}{-jX_C}$$

or

$$\frac{V_{\text{in}}}{I_C} = \frac{-jX_C}{1 - A} = \frac{-j1}{2\,\pi fC(1 - A)}$$

The ratio V_{in}/I_C is the impedance of the capacitor seen from the input side of the amplifier. Notice that the feedback capacitance is multiplied by $1 - A$. Because of this, the equivalent input capacitance is

$$C_{\text{in(Miller)}} = C(1 - A)$$

This input Miller capacitance appears in parallel with the input terminals of the amplifier (see Fig. 14-12b).

Similarly, the output capacitance can be derived as follows. The current through the capacitor is

$$I_C = \frac{V_{out} - V_{in}}{-jX_C} = \frac{(1 - 1/A)V_{out}}{-jX_C}$$

or

$$\frac{V_{out}}{I_C} = \frac{-jX_C}{(A - 1)/A} = \frac{-j1}{2\,\pi fC(A - 1)/A}$$

The ratio V_{out}/I_C is the impedance of the capacitor seen from the output terminals, and its effective capacitance is

$$C_{out(Miller)} = \frac{C(A - 1)}{A}$$

as shown in Fig. 14-12b. When A is large, this capacitance is approximately equal to C, the feedback capacitance.

INVERTING AMPLIFIER

The most important application of the Miller theorem is with an inverting amplifier. When an inverting amplifier is used, A is negative and the input Miller capacitance is larger than the feedback capacitance. The increase in the input capacitance is called the *Miller effect*.

For example, if $C = 5$ pF and $A = -120$, as shown in Fig. 14-13a, then,

$$C_{in(Miller)} = C(1 - A) = 5 \text{ pF}(121) = 605 \text{ pF}$$

This means that the equivalent input capacitance is 605 pF, shown in Fig. 14-13b. Because of the Miller effect, the input capacitance of the circuit is much larger than the feedback capacitance.

The output Miller capacitance is

$$A = \frac{5 \text{ pF}(-121)}{-120} \cong 5 \text{ pF}$$

As you see, a large voltage gain means the output Miller capacitance is approximately equal to the feedback capacitance as shown in Fig. 14-13b.

Fig. 14-13
Example of Miller theorem.

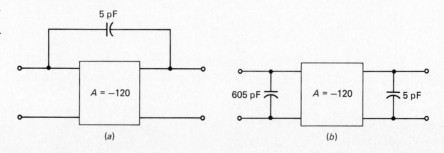

(a)   (b)

14-4 HIGH-FREQUENCY FET ANALYSIS

Figure 14-14*a* shows a signal generator V_G with an internal resistance R_G driving a FET amplifier with voltage-divider bias. At high frequencies, coupling and bypass capacitors act like ac shorts. For this reason, the ac equivalent circuit appears as shown in Fig. 14-14*b*. Resistance r_D is the ac resistance seen by the drain, the parallel combination of R_D and R_L:

$$r_D = R_D \parallel R_L$$

Resistance r_G is the ac Thevenin resistance facing the gate terminal of the FET. This resistance includes the biasing resistors in parallel with the generator resistance. For instance, with voltage-divider bias,

$$r_G = R_1 \parallel R_2 \parallel R_G$$

In the midband of the amplifier, the loaded voltage gain is

$$A = - g_m r_D \tag{14-18}$$

Above the midband, internal FET capacitances and stray wiring capacitances form lag networks that cause the voltage gain to decrease.

CAPACITANCES

The FET has *internal capacitances* between its three electrodes. C_{gs} is the internal capacitance between the gate and the source. C_{gd} is the capacitance between the gate

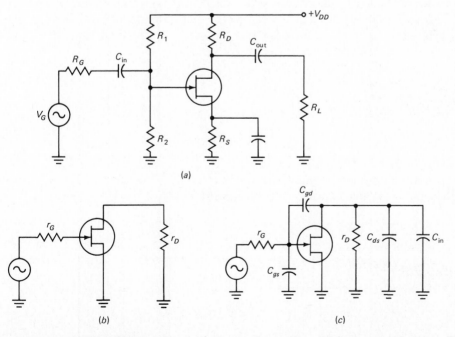

Fig. 14-14 (a) *FET amplifier.* (b) *Ac equivalent circuit in midband.* (c) *Ac equivalent circuit above midband.*

and the drain. C_{ds} is the capacitance between the drain and the source. Figure 14-14c shows these capacitances in the ac equivalent circuit.

When the output of a FET amplifier drives another stage, the input capacitance C_{in} of the next stage appears across the drain-ground terminals, as shown in Fig. 14-14c. This C_{in} includes the input capacitance of the next stage and the stray wiring capacitance, which is the capacitance between connecting wires and ground. As a guide, we will use 0.3 pF/in as a rough estimate for stray wiring capacitance. This means that each inch of connecting wire between the drain of the first stage and the gate of the second stage shunts 0.3 pF across the load in Fig. 14-14c. (This is why you should keep your leads as short as possible in high-frequency amplifiers.)

GATE LAG NETWORK

In Fig. 14-14c, C_{gd} is a feedback capacitor. With Miller's theorem,

$$C_{in(Miller)} = C_{gd}(1 - A)$$

or

$$C_{in(Miller)} = C_{gd}(1 + g_m r_D)$$

Figure 14-15 shows this input Miller capacitance.

In most cases, A is large enough to approximate the output Miller capacitance as equal to the feedback capacitance:

$$C_{out(Miller)} = C_{gd}$$

As shown in Fig. 14-15, the output Miller capacitance is in parallel with C_{ds} and C_{in}.

Capacitances in parallel add. Therefore, the FET amplifier of Fig. 14-15 has two lag networks, one on the gate side and the other on the drain side. The total capacitance in the gate circuit is

$$C_G = C_{gs} + C_{gd}(1 + g_m r_D) \tag{14-19}$$

and the cutoff frequency of the gate lag network is

$$f_G = \frac{1}{2\pi r_G C_G} \tag{14-20}$$

where f_G = gate cutoff frequency
r_G = ac resistance seen by gate
C_G = total capacitance in gate lag network

Fig. 14-15
Ac equivalent circuit of FET amplifier with Miller capacitances.

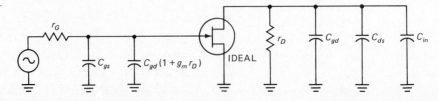

DRAIN LAG NETWORK

The drain acts like a current source driving ac resistance r_D in parallel with capacitances C_{gd}, C_{ds}, and C_{in}. The total capacitance in the drain circuit is

$$C_D = C_{gd} + C_{ds} + C_{in} \qquad (14\text{-}21)$$

and the cutoff frequency is

$$f_D = \frac{1}{2\pi r_D C_D} \qquad (14\text{-}22)$$

where f_D = drain cutoff frequency
r_D = ac resistance seen by drain
C_D = total capacitance in drain circuit

CAPACITANCES ON A DATA SHEET

Figure 14-16*a* shows the three FET capacitances: C_{gs}, C_{gd}, and C_{ds}. For convenience, the manufacturer measures FET capacitances under short-circuit conditions. For instance, C_{iss} is the input capacitance with an ac short across the output, as shown in Fig. 14-16*b*. Since C_{gd} is in parallel with C_{gs},

$$C_{iss} = C_{gs} + C_{gd} \qquad (14\text{-}23)$$

Data sheets also list C_{oss}, the capacitance looking back into the FET with an ac short across the input terminals (Fig. 14-16*c*). Because C_{ds} is in parallel with C_{gd},

$$C_{oss} = C_{ds} + C_{gd} \qquad (14\text{-}24)$$

One more capacitance shown on a data sheet is C_{rss}, the feedback capacitance:

$$C_{rss} = C_{gd} \qquad (14\text{-}25)$$

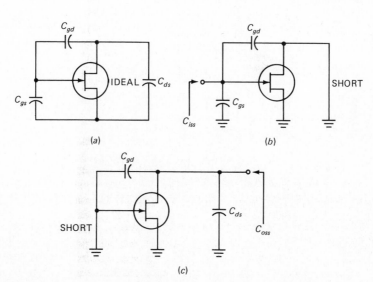

Fig. 14-16
Measuring FET capacitances.

Solving Eqs. (14-23) through (14-25), we get these useful formulas:

$$C_{gd} = C_{rss} \tag{14-26}$$
$$C_{gs} = C_{iss} - C_{rss} \tag{14-27}$$
$$C_{ds} = C_{oss} - C_{rss} \tag{14-28}$$

With these formulas we can calculate the capacitances needed to analyze the lag networks of a FET amplifier.

EXAMPLE 14-5

Figure 14-17a shows a FET amplifier. The MPF102 has these capacitances:

$$C_{iss} = 7 \text{ pF}$$
$$C_{oss} = 4 \text{ pF}$$
$$C_{rss} = 3 \text{ pF}$$

The stray wiring capacitance in the drain circuit is 4 pF. If $g_m = 4000 \ \mu S$, what are the cutoff frequencies of the gate and drain lag networks?

SOLUTION

From Eqs. (14-26) through (14-28),

$$C_{gd} = 3 \text{ pF}$$
$$C_{gs} = 7 \text{ pF} - 3 \text{ pF} = 4 \text{ pF}$$
$$C_{ds} = 4 \text{ pF} - 3 \text{ pF} = 1 \text{ pF}$$

Fig. 14-17

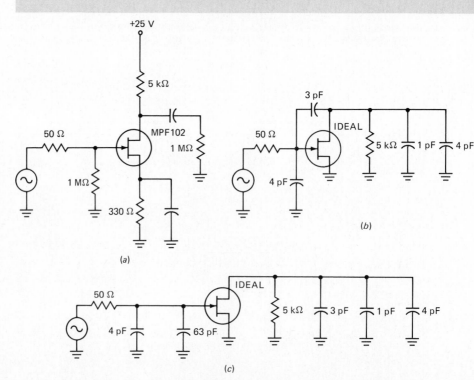

(a)

(b)

(c)

Figure 14-17*b* shows the ac equivalent circuit at higher frequencies, where the coupling and bypass capacitors act like ac shorts. In addition to the internal capacitances of the FET, there is a stray wiring capacitance of 4 pF in parallel with the drain resistance.

The midband voltage gain is

$$A = (-4000 \ \mu S)(5 \ k\Omega) = -20$$

The input Miller capacitance is

$$C_{in(Miller)} = (3 \ pF)(21) = 63 \ pF$$

and the output Miller capacitance is approximately

$$C_{out(Miller)} \cong 3 \ pF$$

Figure 14-17*c* shows the equivalent Miller circuit. Now the two lag networks are evident.

The gate lag network has these values:

$$r_G = 50 \ \Omega$$
$$C_G = 4 \ pF + 63 \ pF = 67 \ pF$$

Therefore, it has a cutoff frequency of

$$f_G = \frac{1}{2 \ \pi(50 \ \Omega)(67 \ pF)} = 47.5 \ MHz$$

The drain lag network has

$$r_D = 5 \ k\Omega$$
$$C_D = 3 \ pF + 1 \ pF + 4 \ pF = 8 \ pF$$

and the cutoff frequency is

$$f_D = \frac{1}{2 \ \pi(5 \ k\Omega)(8 \ pF)} = 3.98 \ MHz$$

The drain cutoff frequency is lower than the gate cutoff frequency. Therefore, the first break in high-frequency response is at 3.98 MHz. In other words, the drain lag network is dominant because it has a lower cutoff frequency than the gate lag network.

EXAMPLE 14-6

Figure 14-18 shows a cascode amplifier. As you know, this two-stage amplifier has a voltage gain of $-g_m R_D$. What is the input capacitance if the first stage has $C_{gs} = 4 \ pF$ and $C_{gd} = 3 \ pF$?

SOLUTION

The first stage (common-source) drives the second stage (common-gate). Since the input impedance of a common-gate amplifier is approximately $1/g_m$, the

Fig. 14-18
Cascode amplifier.

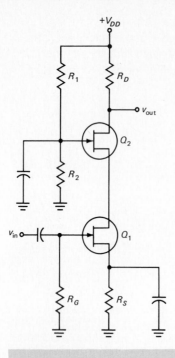

voltage gain of the first stage is

$$A_1 = -g_m r_D \cong (-g_m)\frac{1}{g_m} = -1$$

The first stage has a feedback capacitance of 3 pF and a voltage gain of -1; therefore, the input Miller capacitance is

$$C_{in(Miller)} = 3\ pF \times 2 = 6\ pF$$

Notice how low the Miller effect is. This is because the first stage has a voltage gain of only -1. The total input capacitance of the first stage is the sum of C_{gs} and the input Miller capacitance:

$$C_{in} = 4\ pF + 6\ pF = 10\ pF$$

The advantage of a cascode amplifier is its low input capacitance because the Miller effect is small. In general, the input capacitance of any cascode FET amplifier is

$$C_{in} = C_{gs} + 2C_{gd} \tag{14-29}$$

14-5 HIGH-FREQUENCY BIPOLAR ANALYSIS

Figure 14-19a shows a signal generator V_G with an internal resistance R_G driving a CE amplifier. Figure 14-19b shows the ac equivalent circuit in the midband of the

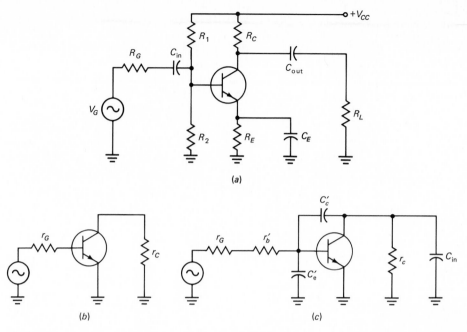

Fig. 14-19   (a) *Bipolar amplifier.* (b) *Midband equivalent circuit.* (c) *Above the midband.*

amplifier. Resistance r_G is the ac Thevenin resistance facing the base terminal of the transistor:

$$r_G = R_1 \parallel R_2 \parallel R_G$$

Resistance r_C is the ac resistance seen by the collector:

$$r_C = R_C \parallel R_L$$

ABOVE THE MIDBAND

C'_e is the capacitance of the emitter diode. C'_c is the capacitance of the collector diode. Figure 14-19c shows the ac equivalent circuit above the midband of the amplifier. Notice that C'_c is a feedback capacitor between the base and the collector. Also notice r'_b, the base-spreading resistance. This is included in high-frequency analysis because it is part of the base lag network.

BASE LAG NETWORK

To find the cutoff frequencies for a bipolar amplifier, we have to identify the lag networks in the base and the collector. The first step is to get the input Miller capacitance. The midband voltage gain from the base to the collector is

$$A = -\frac{r_C}{r'_e}$$

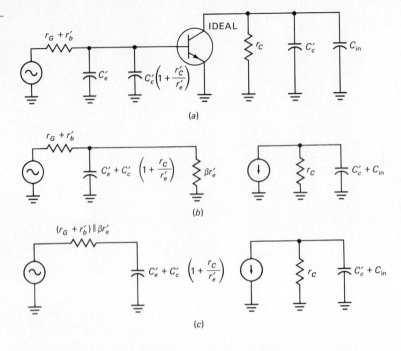

Fig. 14-20
*Ac equivalent
circuits of bipolar
amplifier.*

The input Miller capacitance therefore equals

$$C_{\text{in(Miller)}} = C'_c \left(1 + \frac{r_C}{r'_e}\right)$$

Figure 14-20a shows this input Miller capacitance.

The output Miller capacitance is approximately C'_c because the voltage gain A is ordinarily high in a CE amplifier. Figure 14-20a shows the output Miller capacitance in parallel with C_{in}, the input capacitance of the next stage.

The capacitances in Fig. 14-20a are in parallel. The total capacitance in the base circuit is

$$C_B = C'_e + C'_c \left(1 + \frac{r_C}{r'_e}\right) \tag{14-30}$$

and the total capacitance in the collector circuit is

$$C_C = C'_c + C_{\text{in}} \tag{14-31}$$

as shown in Fig. 14-20b. To get the base circuit into the form of a lag network, we have to thevenize the circuit driving the base capacitance to get Fig. 14-20c. Notice that the Thevenin resistance facing the base capacitance is

$$r_B = (r_G + r'_b) \parallel \beta r'_e \tag{14-32}$$

Look at Fig. 14-20c. Aside from the complicated collection of symbols, the circuit is simple; it contains two lag networks. The base lag network has a cutoff frequency of

$$f_B = \frac{1}{2 \pi r_B C_B} \tag{14-33}$$

where f_B = base cutoff frequency
r_B = Thevenin resistance facing base capacitance
C_B = total capacitance of base lag network

COLLECTOR LAG NETWORK

The collector circuit is another lag network. It has a cutoff frequency of

$$f_C = \frac{1}{2\,\pi r_C C_C} \qquad (14\text{-}34)$$

where f_C = collector cutoff frequency
r_C = ac resistance seen by collector
C_C = total capacitance in collector circuit

CAPACITANCES ON A DATA SHEET

There is no standard designation for C'_c. Data sheets may list it by any of the following equivalent symbols: C_c, C_{cb}, C_{ob}, and C_{obo}. For instance, the data sheet of a 2N2330 gives a C_{ob} of 10 pF. This is the value of C'_c to use in high-frequency analysis.

Capacitance C'_e is not usually listed on a data sheet because it is too difficult to measure directly. Instead, the manufacturer gives a value called the *current-gain-bandwidth product,* designated f_T. This is the frequency at which the current gain of a transistor drops to unity. You can calculate C'_e by using

$$C'_e = \frac{1}{2\,\pi f_T r'_e} \qquad (14\text{-}35)$$

(Appendix 1 derives this formula.)

EXAMPLE 14-7

The data sheet of a 2N3904 gives an f_T of 300 MHz at I_E = 10 mA. Work out the value of C'_e.

SOLUTION

Since I_E is 10 mA,

$$r'_e = \frac{25\text{ mV}}{10\text{ mA}} = 2.5\ \Omega$$

From Eq. (14-35), C'_e equals

$$C'_e = \frac{1}{2\,\pi(300\text{ MHz})(2.5\ \Omega)} = 212\text{ pF}$$

EXAMPLE 14-8

Suppose we use the 2N3904 of the preceding example in a CE amplifier with the following values: r_G = 1 kΩ, r'_b = 100 Ω, $\beta r'_e$ = 250 Ω, r_C = 1 kΩ,

$r'_e = 2.5 \, \Omega$, $C'_e = 212 \, \text{pF}$, $C'_c = 4 \, \text{pF}$, and $C_{in} = 5 \, \text{pF}$. Calculate the cutoff frequencies of a CE amplifier.

SOLUTION

In Fig. 14-20c, the Thevenin resistance facing the base is

$$r_B = (1000 \, \Omega + 100 \, \Omega) \parallel 250 \, \Omega = 204 \, \Omega$$

The voltage gain is

$$A = -\frac{1000}{2.5} = -400$$

and the input Miller capacitance is

$$C_{in(Miller)} = 4 \, \text{pF} \times 401 = 1604 \, \text{pF}$$

The total base capacitance is

$$C_B = 212 \, \text{pF} + 1604 \, \text{pF} = 1816 \, \text{pF}$$

So the critical frequency of the base lag network is

$$f_B = \frac{1}{2 \, \pi (204 \, \Omega)(1816 \, \text{pF})} = 430 \, \text{kHz}$$

In the collector circuit of Fig. 14-20c, the total capacitance is

$$C_C = 4 \, \text{pF} + 5 \, \text{pF} = 9 \, \text{pF}$$

Therefore, the cutoff frequency of the collector lag network is

$$f_C = \frac{1}{2 \, \pi (1 \, \text{k}\Omega)(9 \, \text{pF})} = 17.7 \, \text{MHz}$$

There is no question which lag network is dominant. The base lag network has a much lower cutoff frequency. Therefore, it causes the frequency response to break at 430 kHz. If you were trying to improve the high-frequency response of this amplifier, you would start with the base circuit because it has the lower cutoff frequency.

14-6 DECIBEL POWER GAIN

The power gain G of an amplifier is the ratio of output power to input power:

$$G = \frac{P_2}{P_1}$$

If the output power is 15 W and the input power is 0.5 W,

$$G = \frac{15 \, \text{W}}{0.5 \, \text{W}} = 30$$

This says that the output power is 30 times greater than the input power.

DECIBELS

The *decibel power gain* is defined as

$$G' = 10 \log G \qquad\qquad (14\text{-}36)$$

where G' = power gain in decibels
$\quad\log$ = logarithm to the base 10
$\quad\ G$ = power gain

If a circuit has a power gain of 100, its decibel power gain is

$$G' = 10 \log 100 = 20$$

G' is dimensionless, but to make sure it is not confused with ordinary power gain G, we will attach *decibel* (abbreviated dB) to all answers for G'. The preceding answer therefore is written as

$$G' = 20 \text{ dB}$$

When an answer is in decibels, we automatically know that it represents the decibel power gain and not the ordinary power gain.

3 dB FOR EACH FACTOR OF 2

Suppose the power gain is 2. Then the decibel power gain is

$$G' = 10 \log 2 = 3.01 \text{ dB}$$

If $G = 4$, then

$$G' = 10 \log 4 = 6.02 \text{ dB}$$

If $G = 8$, then

$$G' = 10 \log 8 = 9.03 \text{ dB}$$

Usually, 3.01 dB is rounded off to 3 dB, 6.02 dB to 6 dB, and 9.03 dB to 9 dB. Can you see the pattern? Every time the ordinary power gain doubles, the decibel power gain increases by approximately 3 dB, as shown in Table 14-1.

NEGATIVE DECIBELS

If the power gain is less than 1, there is a power loss (attenuation), and the decibel power gain is negative. For instance, if the output power is 1.5 W when the input

Table 14-1

G	G'
1	0 dB
2	3 dB
4	6 dB
8	9 dB
16	12 dB

Table 14-2

G	G'
1	0 dB
0.5	−3 dB
0.25	−6 dB
0.125	−9 dB
0.0625	−12 dB

power is 3 W, then

$$G = \frac{1.5 \text{ W}}{3 \text{ W}} = 0.5$$

and the decibel power gain is

$$G' = 10 \log 0.5 = -3.01 \text{ dB}$$

When the power gain is 0.25,

$$G' = 10 \log 0.25 = -6.02 \text{ dB}$$

If the power gain is 0.125, then

$$G' = 10 \log 0.125 = -9.03 \text{ dB}$$

Again, these are usually rounded off to −3 dB, −6 dB, and −9 dB. Each time the power gain decreases by a factor of 2, the decibel power gain decreases by approximately 3 dB, as indicated in Table 14-2.

10 dB FOR EACH FACTOR OF 10

Suppose the power gain is 10. Then the decibel power gain is

$$G' = 10 \log 10 = 10 \text{ dB}$$

If the power gain is 100, then

$$G' = 10 \log 100 = 20 \text{ dB}$$

If the power gain is 1000, then

$$G' = 10 \log 1000 = 30 \text{ dB}$$

The pattern emerging here is that each time the power gain increases by a factor of 10, the decibel power gain increases by 10 dB, as shown in Table 14-3.

Table 14-3

G	G'
1	0 dB
10	10 dB
100	20 dB
1,000	30 dB
10,000	40 dB

A similar result applies to power gain of less than 1. When the power gain is 0.1, the decibel power gain is −10 dB. When the power gain is 0.01, the decibel power gain is −20 dB. When the power gain is 0.001, the decibel power gain is −30 dB; and so on.

ORDINARY GAINS MULTIPLY

Figure 14-21*a* shows two stages in an amplifier. The first stage has an input power of P_1, an output power of P_2, and a power gain of

$$G_1 = \frac{P_2}{P_1}$$

The second stage has an input power of P_2, an output power of P_3, and a power gain of

$$G_2 = \frac{P_3}{P_2}$$

The total power gain of the two stages is

$$G = \frac{P_3}{P_1}$$

which is equal to

$$G = \frac{P_2}{P_1}\frac{P_3}{P_2}$$

or

$$G = G_1 G_2 \qquad (14\text{-}37)$$

This proves that the total power gain of cascaded stages equals the product of the stage gains. No matter how many stages there are, we can find the total power gain by

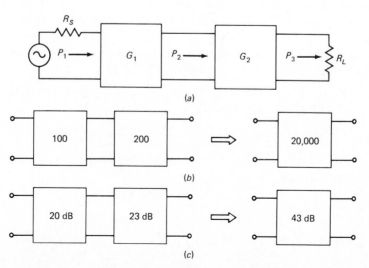

Fig. 14-21
Cascaded stages.

multiplying the individual stage gains. As an example, Fig. 14-21*b* shows a first-stage power gain of 100 and a second-stage power gain of 200. The total power gain is

$$G = 100 \times 200 = 20,000$$

DECIBEL GAINS ADD

Since the total power gain of two cascaded stages is

$$G = G_1 G_2$$

we can take the logarithm of both sides to get

$$\log G = \log G_1 G_2 = \log G_1 + \log G_2$$

Multiplying both sides by 10 gives

$$10 \log G = 10 \log G_1 + 10 \log G_2$$

which can be written as

$$G' = G_1' + G_2' \tag{14-38}$$

where G' = total decibel power gain
G_1' = decibel power gain of first stage
G_2' = decibel power gain of second stage

This equation says that the total power gain of two cascaded stages equals the sum of the decibel power gains for each stage. The same idea applies to any number of stages: Add the decibel power gains of the stages to find the total decibel power gain.

As an example, Fig. 14-21*c* shows the same two-stage amplifier as Fig. 14-21*b*, except that the gains are expressed in decibels. The total decibel power gain is

$$G' = 20 \text{ dB} + 23 \text{ dB} = 43 \text{ dB}$$

We can leave the answer like this, or convert it back to ordinary power gain as follows:

$$G = \text{antilog} \frac{G'}{10} = \text{antilog } 4.3 \cong 20,000$$

An answer in decibels has the advantage of being compact and easy to write. In this example it is easier to write 43 dB than 20,000.

1-mW REFERENCE

Although they are ordinarily used with the power gain, decibels are sometimes used to indicate the power level with respect to 1 mW. In this case, the label dBm is used instead of dB; the *m* at the end of dBm reminds you of the *m*illiwatt reference. The dBm formula is

$$P' = 10 \log \frac{P}{1 \text{ mW}} \tag{14-39}$$

where P' = power in dBm
P = power in watts

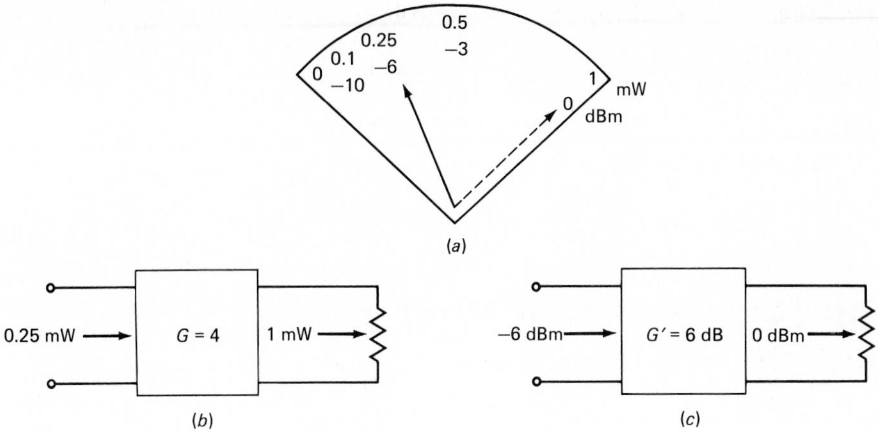

Fig. 14-22
Meaning of dBm.

(a)

(b)

(c)

For instance, if the power is 2 W, then

$$P' = 10 \log \frac{2 \text{ W}}{1 \text{ mW}} = 10 \log 2000 = 33 \text{ dBm}$$

SIMPLIFIED MEASUREMENTS

The advantage of using dBm is simpler power measurements. For example, some instruments have two scales like Fig. 14-22*a* to indicate power level. The upper scale is marked in milliwatts. Suppose we measure the input power and output power for the stage of Fig. 14-22*b*. Then the upper scale will indicate 0.25 mW (solid line) for the input power and 1 mW (dashed line) for the output power.

The lower scale of Fig. 14-22*a* is the dBm scale. As shown, 0 dBm is equivalent to 1 mW, −3 dBm to 0.5 mW, −6 dBm to 0.25 mW, and so on. If we use the dBm scale to measure power in Fig. 14-22*b*, we will read −6 dBm for the input power and 0 dBm for the output power of Fig. 14-22*c*. Since the needle moves from −6 dBm to 0 dBm, the amplifier has a decibel power gain of 6 dB.

14-7 DECIBEL VOLTAGE GAIN

Voltage measurements are more common than power measurements. Therefore, it is not surprising that decibels are also used to specify voltage gain. *Decibel voltage gain* is defined as

$$A' = 20 \log A \tag{14-40}$$

where A' = voltage gain in decibels
A = voltage gain

For example, if A is 40, then

$$A' = 20 \log 40 = 32 \text{ dB}$$

Table 14-4	
A	A'
1	0 dB
2	6 dB
4	12 dB
8	18 dB

Table 14-5	
A	A'
1	0 dB
0.5	−6 dB
0.25	−12 dB
0.125	−18 dB

FACTORS OF 2 AND 10

When $A = 2$,

$$A' = 20 \log 2 \cong 6 \text{ dB}$$

When $A = 4$,

$$A' = 20 \log 4 \cong 12 \text{ dB}$$

When $A = 8$,

$$A' = 20 \log 8 \cong 18 \text{ dB}$$

As you can see, when the voltage gain doubles, the decibel voltage gain increases by 6 dB (Table 14-4).

If the voltage gain decreases by a factor of 2, then the decibel voltage gain decreases by 6 dB, as shown in Table 14-5. Also, when the voltage gain increases by a factor of 10, the decibel voltage gain increases by 20 dB (Table 14-6). Notice that the changes in decibel voltage gain are exactly twice the changes in decibel power gain.

RELATION OF POWER AND VOLTAGE GAINS

In Fig. 14-23a the rms voltage across the amplifier input terminals equals V_1 and the rms voltage across the amplifier output terminals equals V_2. Therefore, the input power to the amplifier is

$$P_1 = \frac{V_1^2}{R_1}$$

and the output power is

$$P_2 = \frac{V_2^2}{R_2}$$

Table 14-6	
A	A'
1	0 dB
10	20 dB
100	40 dB
1000	60 dB

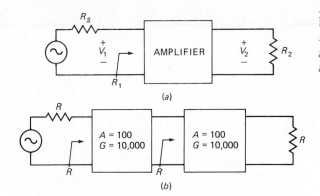

Fig. 14-23
*Relation of power
gain and voltage
gain.*

The power gain from input to output equals

$$G = \frac{P_2}{P_1} = \frac{V_2^2/R_2}{V_1^2/R_1} = \left(\frac{V_2}{V_1}\right)^2 \frac{R_1}{R_2}$$

Since the ratio V_2/V_1 is the voltage gain A, we can rewrite the power gain as

$$G = A^2 \frac{R_1}{R_2} \tag{14-41}$$

IMPEDANCE-MATCHED CASE

In many systems (for example, microwave and telephone) the input and load impedances are *matched,* meaning that $R_1 = R_2$. For this condition, Eq. (14-41) simplifies to

$$G = A^2 \tag{14-42}$$

This says that the power gain equals the square of the voltage gain. For instance, if $A = 100$, then $G = 10,000$. Figure 14-23b shows a system with these gains. Each block has an input impedance of R and a load impedance of R.

Taking the logarithm of both sides of Eq. (14-42) gives

$$\log G = \log A^2 = 2 \log A$$

Multiplying both sides by 10,

$$10 \log G = 10(2 \log A) = 20 \log A$$

which can be written as

$$G' = A' \tag{14-43}$$

This says that the decibel power gain equals the decibel voltage gain, a relation that is true for all impedance-matched systems.

NOT MATCHED

In most amplifiers the impedances are not matched because the load impedance does not equal the input impedance. In this case, G' does not equal A', and we must calculate each separately. In other words, we have to use

$$G' = 10 \log G$$

for the decibel power gain and

$$A' = 20 \log A$$

for the decibel voltage gain.

Both types of decibel gain are widely used. Decibel power gain predominates in communications, microwaves, and other systems in which power is important. Decibel voltage gain is preferred in areas of electronics in which measuring voltage is more convenient.

CASCADED STAGES

Since decibel voltage gain is based on logarithms, the additive property of cascaded stages holds. That is, the total decibel voltage gain of cascaded stages equals the sum of the individual decibel voltage gains. In symbols,

$$A' = A'_1 + A'_2 + A'_3 + \cdots \tag{14-44}$$

Also, many voltmeters have a decibel scale. When you measure the input and output voltages of an amplifier, the difference in the dB readings is the decibel voltage gain of the amplifier.

As an example, the two-stage amplifier of Fig. 14-24a has a first-stage voltage gain of 100 (40 dB) and a second-stage voltage gain of 200 (46 dB). The total voltage gain is

$$A = 100 \times 200 = 20,000$$

This is equivalent to a total decibel voltage gain of

$$A' = 40 \text{ dB} + 46 \text{ dB} = 86 \text{ dB}$$

If we use the decibel scale of a voltmeter when measuring input and output voltages, the needle will show an increase of 40 dB for the first stage, 46 dB for the second stage, and 86 dB for the entire amplifier.

Fig. 14-24
(a) *Ordinary gains multiply.*
(b) *Decibel gains add.*

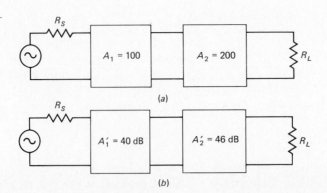

14-8 BODE PLOTS

In complex numbers, the voltage gain of a lag network is

$$\frac{V_{out}}{V_{in}} = \frac{-jX_C}{R - jX_C}$$

This can be converted to a magnitude of

$$\frac{V_{out}}{V_{in}} = \frac{X_C}{\sqrt{R^2 + X_C^2}} \tag{14-45}$$

and a phase angle of

$$\phi = -\arctan \frac{R}{X_C} \tag{14-46}$$

DECIBEL VOLTAGE GAIN

Equation (14-45) can be rewritten as

$$A = \frac{V_{out}}{V_{in}} = \frac{X_C}{\sqrt{R^2 + X_C^2}} = \frac{1}{\sqrt{1 + (R/X_C)^2}}$$

Since

$$\frac{R}{X_C} = 2\pi f RC = \frac{f}{f_c}$$

the voltage gain can be written as

$$A = \frac{1}{\sqrt{1 + (f/f_c)^2}}$$

The decibel voltage gain is

$$A' = 20 \log \frac{1}{\sqrt{1 + (f/f_c)^2}} \tag{14-47}$$

where A' = decibel voltage gain of lag network
    $\log$ = logarithm to base 10
     f = input frequency
    f_c = cutoff frequency of lag network

Using Eq. (14-47), we can calculate the decibel voltage gain of a lag network. For instance, when $f/f_c = 0.1$, Eq. (14-47) gives

$$A' = 20 \log \frac{1}{\sqrt{1 + (0.1)^2}} = -0.0432 \text{ dB} \cong 0 \text{ dB}$$

When $f/f_c = 1$, the decibel voltage gain is

$$A' = 20 \log \frac{1}{\sqrt{1 + (1)^2}} = -3.01 \text{ dB} \cong -3 \text{ dB}$$

When $f/f_c = 10$,

$$A' = 20 \log \frac{1}{\sqrt{1 + (10)^2}} \cong -20 \text{ dB}$$

When $f/f_c = 100$,

$$A' = 20 \log \frac{1}{\sqrt{1 + (100)^2}} = -40 \text{ dB}$$

When $f/f_c = 1000$,

$$A' = 20 \log \frac{1}{\sqrt{1 + (1000)^2}} = -60 \text{ dB}$$

The decibel voltage gains we have just calculated tell a story. Here is what we have found:

1. When the input frequency to a lag network is a *decade* (a factor of 10) below the cutoff frequency, the decibel voltage gain is approximately zero.
2. When the input frequency equals the cutoff frequency, the decibel voltage gain equals -3 dB (the half-power point).
3. When the input frequency is a decade above the cutoff frequency, the decibel voltage gain is -20 dB.
4. Beyond this point, each time the frequency increases by a decade, the decibel voltage gain decreases by 20 dB.

BODE PLOT OF VOLTAGE GAIN

Whenever we can reduce a circuit to a lag network, we can calculate the cutoff frequency. Then, we know that the decibel voltage gain is down 20 dB when the input frequency is one decade above the cutoff frequency, down 40 dB when the input frequency is two decades above the cutoff frequency, down 60 dB when the input frequency is three decades above the cutoff frequency, and so forth.

Figure 14-25*a* shows the decibel voltage gain of a lag network. This is called a *Bode plot* of voltage gain. In order to see the graph over several decades of frequency, we have compressed the horizontal scale by showing equal distances between each decade in frequency. (Semilogarithmic graph paper does the same thing.) The advantage of marking off the horizontal scale in decades rather than units is this: Well above the cutoff frequency, the decibel voltage gain decreases 20 dB for each decade increase in frequency; therefore, the graph of voltage gain will be a straight line with a slope of -20 dB per decade.

Figure 14-25*b* shows the *ideal* Bode plot of voltage gain. To a first approximation, we neglect the -3 dB at the cutoff frequency, and draw a straight line with slope of -20 dB per decade. Ideal Bode plots are used for preliminary analysis because they are easy to draw.

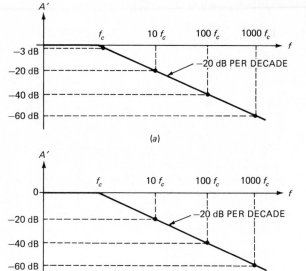

Fig. 14-25
*Bode plot of
voltage gain for
lag network.* (a)
Exact. (b) *Ideal.*

6 dB PER OCTAVE

An *octave* is a factor of 2 in frequency. For instance, if the frequency changes from 100 to 200 Hz, it has changed one octave. Similarly, 100 to 400 Hz is two octaves, 100 to 800 Hz is three octaves, and so on.

Above the cutoff frequency, the decibel voltage gain of a lag network decreases 6 dB per octave. This is easy to prove. When $f/f_c = 10$, $A' = -20$ dB. When $f/f_c = 20$ (an octave change),

$$A' = 20 \log \frac{1}{\sqrt{1 + (20)^2}} = -26 \text{ dB}$$

As you can see, the decibel voltage gain has decreased 6 dB.

In other words, you can describe the frequency response of a lag network above the cutoff frequency in either of two ways. You can say that the decibel voltage gain decreases at a rate of 20 dB per decade, or you can say that it decreases at a rate of 6 dB per octave. Both roll-off rates are used in industry.

PHASE ANGLE

Equation (14-46) can be written as

$$\phi = -\arctan \frac{f}{f_c} \tag{14-48}$$

where ϕ = phase angle in degrees
f_c = cutoff frequency of lag network
f = frequency of input signal

For instance, when $f/f_c = 0.1$, Eq. (14-48) gives

$$\phi = -\arctan 0.1 \cong -6°$$

When $f/f_c = 1$,

$$\phi = -\arctan 1 = -45°$$

When $f/f_c = 10$,

$$\phi = -\arctan 10 \cong -84°$$

Figure 14-26a shows how the phase angle of a lag network varies with frequency. At very low frequencies the phase angle is zero. When $f = 0.1f_c$, the phase angle is $-6°$. When the input frequency equals the cutoff frequency, the phase angle equals $-45°$. For an input frequency of 10 times the critical frequency, the phase angle is $-84°$. Further increases in frequency produce little change because the limiting value is $-90°$. As you can see, the phase angle of a lag network is between 0 and $-90°$. This means that the output voltage lags the input voltage.

BODE PLOT OF PHASE ANGLE

A graph like Fig. 14-26a is called a Bode plot of phase angle. Knowing that the phase angle is $-6°$ at $0.1f_c$ and $-84°$ at $10f_c$ is of little value except to indicate how close the phase angle is to its limiting value. The ideal Bode plot of Fig. 14-26b is much more useful. This is the one to remember because it emphasizes these ideas:

1. When $f = 0.1f_c$, the phase angle is approximately zero.
2. When $f = f_c$, the phase angle is $-45°$.
3. When $f = 10f_c$, the phase angle is approximately $-90°$.

Fig. 14-26
Bode plot of phase angle for lag network. (a) *Exact.* (b) *Ideal.*

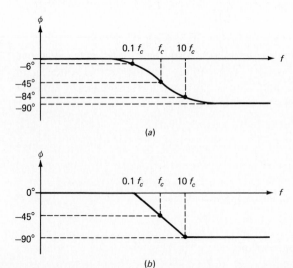

Another way to summarize the Bode plot of phase angle is this: At the cutoff frequency, the phase angle equals −45°. A decade below the cutoff frequency, the phase angle is approximately 0°; a decade above the cutoff frequency, the phase angle is approximately −90°.

LEAD NETWORK

The analysis of a lead network is similar to that given for a lag network. The decibel voltage gain is

$$A' = 20 \log \frac{1}{\sqrt{1 + (f_c/f)^2}} \qquad (14\text{-}49)$$

where A' = decibel voltage gain of a lead network
f_c = cutoff frequency
f = input frequency

The phase angle is

$$\phi = \arctan \frac{f_c}{f} \qquad (14\text{-}50)$$

If you compare these equations with those of a lag network, you will see an inverse symmetry. This is why the ideal Bode plots of decibel voltage gain and phase angle appear as shown in Fig. 14-27. Below the cutoff frequency, the decibel voltage gain rolls off at a rate of 20 dB per decade, equivalent to 6 dB per octave. The phase angle is between 0 and 90°, which means that the output leads the input.

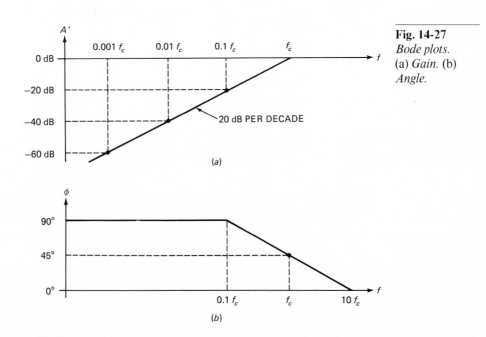

Fig. 14-27
Bode plots.
(a) *Gain.* (b) *Angle.*

14-9 AMPLIFIER RESPONSE

In the midband of an amplifier, the voltage gain is maximum. Below the midband, coupling and bypass capacitors cause the voltage gain to decrease. Above the midband, internal and stray capacitances cause the voltage gain to decrease.

AC AMPLIFIER

Figure 14-28 shows the frequency response of an ac amplifier. The lower cutoff frequency is f_1 and the upper cutoff frequency is f_2. The bandwidth of any amplifier equals

$$B = f_2 - f_1 \tag{14-51}$$

For instance, if an ac amplifier has a lower cutoff frequency of 1 MHz and an upper cutoff frequency of 3 MHz, then its bandwidth is

$$B = 3 \text{ MHz} - 1 \text{ MHz} = 2 \text{ MHz}$$

A tuned amplifier (one with resonant circuits) is often called a *narrowband* amplifier because f_1 and f_2 are close in value. For example, if $f_1 = 450$ kHz and $f_2 = 460$ kHz, then the bandwidth is

$$B = 460 \text{ kHz} - 450 \text{ kHz} = 10 \text{ kHz}$$

This is a narrowband amplifier because f_2 is only slightly larger than f_1.

On the other hand, an untuned amplifier is often called a *wideband* amplifier because f_2 is much greater than f_1. As an example, if $f_1 = 200$ Hz and $f_2 = 10$ MHz, then

$$B = 10 \text{ MHz} - 200 \text{ Hz} \cong 10 \text{ MHz}$$

This is a wideband amplifier because f_2 is much greater than f_1.

DC AMPLIFIER

As mentioned earlier, we can design an amplifier with no coupling or bypass capacitors. An amplifier like this is called a *dc amplifier* (dc amp). Figure 14-29*a* shows the ideal Bode plot of decibel voltage gain for a dc amp. As you can see, the dc amp has a decibel voltage gain of A_{mid} up to the cutoff frequency f_2. Beyond this, the decibel voltage gain drops 20 dB per decade.

The Bode plot of Fig. 14-29*a* assumes one dominant lag network. This is the reason the slope is -20 dB per decade. If another lag network had a cutoff frequency

Fig. 14-28
Frequency response of ac amplifier.

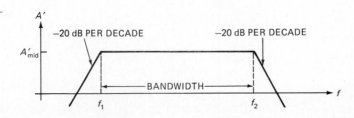

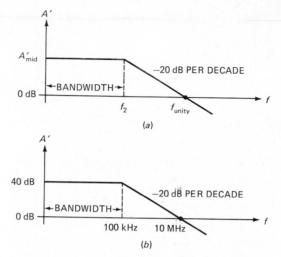

Fig. 14-29
Frequency response of dc amplifier.

near the first, the Bode plot would break again, and the gain would roll off at a rate of 40 dB per decade.

To avoid *oscillations* (discussed in Chap. 20), the voltage gain of a typical dc amplifier is designed to roll off at a rate of 20 dB per decade until the graph crosses the horizontal axis at f_{unity} (Fig. 14-29a). The subscript *unity* means that the ordinary voltage gain equals 1 at this frequency, equivalent to a decibel voltage gain of 0 dB. For instance, Fig. 14-29b shows the Bode plot of a dc amp whose decibel voltage gain drops 20 dB per decade until it crosses the horizontal axis at 10 MHz. Beyond this frequency, other lag networks will reach their cutoff frequencies, but we are not interested in them.

Since there is no lower cutoff frequency in a dc amplifier, the bandwidth is

$$B = f_2 \tag{14-52}$$

In Fig. 14-29b, the bandwidth equals 100 kHz.

BANDWIDTH OF CASCADED STAGES

The overall bandwidth of cascaded stages is less than the bandwidth of any stage. For instance, suppose we have two dc-coupled stages, each with a bandwidth of 10 kHz. Then at 10 kHz, the response of each stage is down 3 dB and the overall response is down 6 dB. This implies that the overall bandwidth is less than 10 kHz.

Appendix 1 proves that the overall bandwidth of n identical cascaded stages is given by

$$B_n = B\sqrt{2^{1/n} - 1} \tag{14-52a}$$

where B_n = overall bandwidth
B = bandwidth of one stage

For instance, if there are two identical stages, each with a bandwidth of 10 kHz, the overall bandwidth is

$$B_2 = (10 \text{ kHz})\sqrt{2^{1/2} - 1} = 6.44 \text{ kHz}$$

Table 14-7. FCC Frequency Ranges

Frequency Range	Abbreviation	Description
30 to 300 kHz	LF	Low frequencies
0.3 to 3 MHz	MF	Medium frequencies
3 to 30 MHz	HF	High frequencies
30 to 300 MHz	VHF	Very high frequencies
0.3 to 3 GHz	UHF	Ultra high frequencies

Three cascaded stages, each with a bandwidth of 10 kHz, produce an overall bandwidth of

$$B_3 = (10 \text{ kHz})\sqrt{2^{1/3} - 1} = 5.1 \text{ kHz}$$

Try to remember this idea because you often have to cascade stages. If the stages are not identical, you will still get bandwidth shrinkage because of the cumulative effects of the stages. In other words, the overall bandwidth is always smaller than the smallest bandwidth of any stage.

FREQUENCY RANGES

Table 14-7 shows the FCC (Federal Communications Commission) frequency allocations from 30 kHz to 3 GHz. These are useful in describing amplifiers and other electronic equipment.

Here are some examples. The RF circuits inside an AM radio receiver operate in the MF range because the received signals have frequencies between 535 and 1605 kHz. FM receivers operate in the VHF range because the RF signals are between 88 and 108 MHz. A TV receiver can operate in either the VHF or the UHF range, depending on the channel being received. Channels 2 through 6 (54 to 88 MHz) and channels 7 through 13 (174 to 216 MHz) are VHF channels. On the other hand, channels 14 through 83 (470 to 890 MHz) are UHF channels.

14-10 RISETIME-BANDWIDTH RELATION

Sine-wave testing of an amplifier means that we drive it with a sinusoidal input and measure the sinusoidal output voltage. To find the upper cutoff frequency, we have to vary the input frequency until the voltage gain drops 3 dB from the midband value. Sine-wave testing is a common approach. But there is a faster and simpler way to test an amplifier, using a *square-wave input* instead of a sine-wave input.

RISETIME

Given a lag circuit like Fig. 14-30a, basic circuit theory tells us what happens after the switch is closed. If the capacitor is initially uncharged, the voltage will rise exponentially toward the supply voltage V. The *risetime* T_R is the amount of time it takes the capacitor voltage to go from $0.1V$ (the 10 percent point) to $0.9V$ (the 90 percent point). If it takes 10 μs for the exponential waveform to go from the 10 percent point to the 90 percent point, the waveform has a risetime of 10 μs.

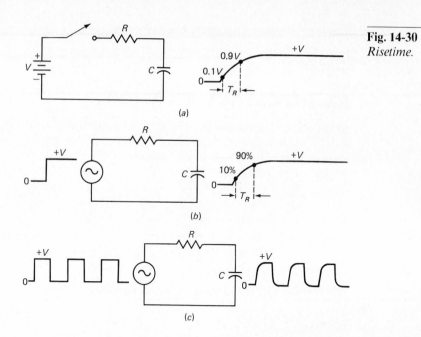

Fig. 14-30
Risetime.

Instead of using a switch to apply the sudden step in voltage, we can use a square-wave generator. For instance, Fig. 14-30*b* shows the leading edge of a square wave driving the same *RC* network as before. The risetime is still the time it takes for the voltage to go from the 10 percent point to the 90 percent point.

Figure 14-30*c* shows how several cycles would look. The input voltage changes suddenly from one voltage level to another. The output voltage takes longer to make its transitions. It cannot suddenly step because the capacitor has to charge and discharge through the resistance.

RELATION BETWEEN T_R AND RC

Basic courses prove that

$$v_C = V(1 - e^{-t/RC})$$

where v_C = capacitor voltage
V = total change in input voltage
t = time after input transition
RC = time constant of lag network

At the 10 percent point, $v_{out} = 0.1V$. At the 90 percent point, $v_{out} = 0.9V$. Substituting these voltages and solving for time difference gives the risetime:

$$T_R = 2.2RC \tag{14-53}$$

where T_R = risetime of capacitor voltage
RC = time constant of lag network

For instance, if R equals 10 kΩ and C is 50 pF,

$$RC = (10 \text{ k}\Omega)(50 \text{ pF}) = 0.5 \ \mu\text{s}$$

and the risetime of the output waveform equals

$$T_R = 2.2RC = 2.2(0.5\ \mu s) = 1.1\ \mu s$$

AN IMPORTANT RELATION

As mentioned earlier, a dc amplifier typically has one dominant lag network that rolls off the voltage gain at a rate of 20 dB per decade until f_{unity} is reached. The cutoff frequency of this lag network is given by

$$f_2 = \frac{1}{2\pi RC}$$

By rearranging,

$$RC = \frac{1}{2\pi f_2}$$

Substituting this into Eq. (14-53) and rearranging gives

$$f_2 = \frac{0.35}{T_R} \tag{14-54}$$

where f_2 = upper cutoff frequency of amplifier
T_R = risetime of amplifier output voltage

This is an important result because it relates sinusoidal and square-wave operation of an amplifier. It means that we can test an amplifier with either a sine wave or a square wave. For instance, Fig. 14-31a shows the leading edge of a square wave driving a dc amplifier. If we measure a risetime of 1 μs, we can calculate a sinusoidal cutoff frequency of

$$f_2 = \frac{0.35}{1\ \mu s} = 350\ \text{kHz}$$

Alternatively, we can measure the cutoff frequency directly by driving the dc amplifier with a sine wave, as shown in Fig. 14-31b. If we vary the frequency until the output is down 3 dB, we will measure a cutoff frequency of 350 kHz.

Fig. 14-31
Relation between risetime and cutoff frequency.

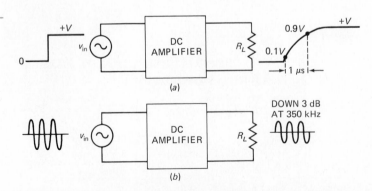

14-11 STRAY EFFECTS

Stray wiring capacitance can seriously degrade the high-frequency response of an amplifier. This is why you have to keep leads as short as possible when building circuits to operate above 100 kHz. In addition to stray wiring capacitance, other unwanted effects can also degrade the high-frequency response.

EQUIVALENT CIRCUITS

Every resistor has a small amount of inductance and capacitance. At lower frequencies the unwanted L and C have negligible effects. But as the frequency increases, the resistor no longer acts as pure resistance. Figure 14-32a shows the equivalent circuit of a resistor with its inductance and capacitance. At lower frequencies, the inductive reactance approaches zero and the capacitive reactance approaches infinity. In other words, the inductor appears shorted and the capacitor appears open. In this case, the resistor acts like a pure resistance.

We will refer to the inductance as the *lead inductance* because it is produced by the leads going into the resistor. And we will refer to the capacitance as the stray capacitance because it is the stray wiring capacitance between the ends of the resistor. For frequencies less than 100 MHz, either the inductive or the capacitive effect dominates. This means that a resistor has an equivalent circuit like Fig. 14-32b or 14-32c.

STRAY CAPACITANCE

The stray capacitance of a typical resistor ($\frac{1}{8}$ to 2 W) is in the vicinity of 1 pF, with the exact value determined by the length of the leads, the size of the resistor body, and other factors. In most applications, you can neglect stray capacitance when

$$\frac{X_C}{R} > 10$$

For instance, if a 10-kΩ resistor has 1 pF of stray capacitance, the X_C at 1 MHz equals

$$X_C = \frac{1}{2\pi(1 \text{ MHz})(1 \text{ pF})} = 159 \text{ k}\Omega$$

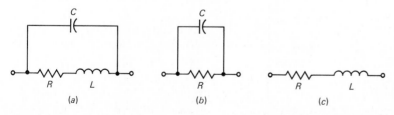

Fig. 14-32   *Resistor equivalent circuits.* (a) *Complete.* (b) *Resistor with stray capacitance.* (c) *Resistor with lead inductance.*

The ratio of reactance to resistance is

$$\frac{X_C}{R} = \frac{159 \text{ k}\Omega}{10 \text{ k}\Omega} = 15.9$$

This is greater than 10; therefore, we can neglect the stray capacitance of a 10-kΩ resistor operating at 1 MHz.

LEAD INDUCTANCE

The lead inductance of a typical resistor is approximately 0.02 μH per in. You can neglect lead inductance when

$$\frac{R}{X_L} > 10$$

For example, suppose the leads of a 1-kΩ resistor are cut to $\frac{1}{2}$ in on each end. Then the total lead length is 1 in, equivalent to approximately 0.02 μH. At 300 MHz, the reactance is

$$X_L = 2 \pi(300 \text{ MHz})(0.02 \text{ μH}) = 37.7 \text{ }\Omega$$

and the ratio of resistance to reactance is

$$\frac{R}{X_L} = \frac{1000}{37.7} = 26.5$$

Therefore, even at 300 MHz we can neglect the lead inductance of a 1-kΩ resistor.

A USEFUL GRAPH

By setting $X_C/R = 10$ and $R/X_L = 10$, we can plot frequency versus resistance, as shown in Fig. 14-33. This graph gives the dividing lines between the resistive, inductive, and capacitive approximations, assuming a stray capacitance of 1 pF and a lead inductance of 0.02 μH.

Here is how to use the graph. For any point below the two lines, you can idealize the resistor, that is, you can neglect its capacitance and inductance. On the other hand, if a point falls above either line, you may need to include the lead inductance or stray capacitance. For instance, a 10-kΩ resistor operating at 1 MHz can be idealized because it falls in the ideal region of Fig. 14-33. But if this 10-kΩ resistor is operating at 5 MHz, you must include stray capacitance in precise calculations. Similarly, a 20-Ω resistor acts ideally up to 16 MHz, but beyond this frequency, the lead inductance becomes important.

Don't lean too heavily on Fig. 14-33; it is only a guide to help you determine when to include stray capacitance or lead inductance in your calculations. If you are working at high frequencies where precise analysis is required, you may have to measure the stray capacitance and the lead inductance with a high-frequency *RLC* bridge or *Q* meter.

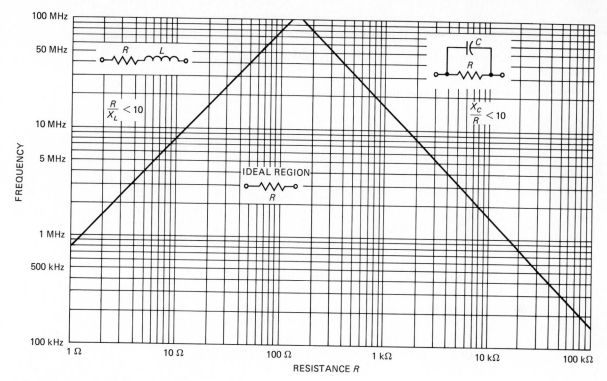

Fig. 14-33 *Approximation guide for resistors.*

PROBLEMS

Straightforward

14-1. If $\beta = 175$ in Fig. 14-34, what are the cutoff frequencies for the input and output lead networks? (Ignore the bypass capacitor.)

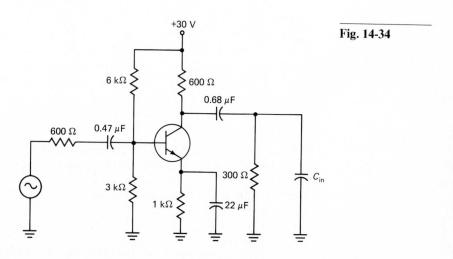

Fig. 14-34

14-2. What is the cutoff frequency of the output lead network in Fig. 14-35? (Ignore the bypass capacitor.)

Fig. 14-35

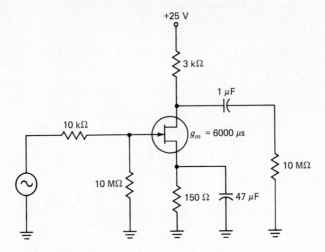

14-3. In Fig. 14-5, the generator resistance changes from 3.6 kΩ to 1 kΩ. If $\beta = 120$, what are the cutoff frequencies of the input and output lead networks?

14-4. If $\beta = 85$ in Fig. 14-34, what is the cutoff frequency of the emitter bypass network?

14-5. What is the cutoff frequency of the source bypass network in Fig. 14-35?

14-6. The FET of Fig. 14-36a has a g_m of 8000 μS. What is the voltage gain of the stage in the midband? At what frequency is the voltage gain down to 0.707 of the midband value?

Fig. 14-36

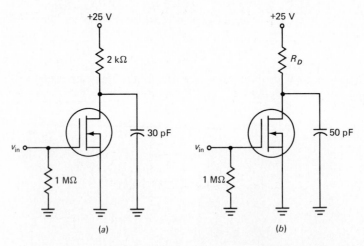

(a)   (b)

14-7. The g_m for the FET of Fig. 14-36b is 5000 μS. If R_D is 5 kΩ, what is the voltage gain when the input frequency is 10 kHz? What is the cutoff frequency?

14-8. The FET of Fig. 14-36b has a g_m of 5000 μS. Calculate the midband voltage gain and the cutoff frequency for each of these:
   a. $R_D = 1$ kΩ
   b. $R_D = 2$ kΩ
   c. $R_D = 4$ kΩ
Describe the relation between voltage gain and cutoff frequency.

14-9. The FET of Fig. 14-35 has these capacitances: $C_{gs} = 6$ pF, $C_{gd} = 4$ pF, and $C_{ds} = 1$ pF. Calculate the cutoff frequencies of the gate and drain lag networks.

14-10. For the same data as in the preceding problem, but with a stray wiring capacitance of 4 pF across the output, what is the cutoff frequency of the gate and drain lag networks?

14-11. A 2N3300 has an f_T of 250 MHz for an I_E of 50 mA. Calculate the value of C_e' for this emitter current.

14-12. The bipolar transistor of Fig. 14-34 has these values: $r_b' = 50\ \Omega$, $\beta = 80$, $r_e' = 2.5\ \Omega$, $f_T = 200$ MHz, $C_c' = 3$ pF, and $C_{in} = 50$ pF. What is the cutoff frequency of the base lag network? Of the collector lag network?

14-13. In Fig. 14-34, the transistor has $r_b' = 75\ \Omega$, $\beta = 20$, $r_e' = 1\ \Omega$, $f_T = 500$ MHz, $C_c' = 2$ pF, and $C_{in} = 100$ pF. Calculate the cutoff frequencies of the base and collector lag networks.

14-14. An amplifier has an input of 15 mW and an output of 2.4 W. What is the decibel power gain?

14-15. How much power does 54 dBm represent?

14-16. An amplifier has a voltage gain of 185. Calculate its decibel voltage gain.

14-17. Sketch the ideal Bode plot of decibel voltage gain for the amplifier of Fig. 14-36a, using a g_m of 10,000 μS. Also show the ideal Bode plot of phase angle.

14-18. In Fig. 14-37a, what is the decibel voltage gain in the midband? What is the bandwidth? What is the decibel voltage gain when $f = 10$ kHz? What is the phase angle at 1 MHz?

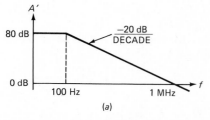

(a)

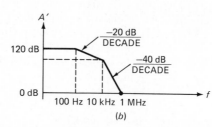

(b)

Fig. 14-37

14-19. In Fig. 14-37b, the voltage gain breaks a second time and rolls off at -40 dB per decade above 10 kHz. What is the decibel voltage gain at 100 kHz? At 1 MHz?

14-20. The amplifier of Fig. 14-38a has a midband voltage gain of 100. If V_{in} equals 20 mV, what is the output voltage at the 90 percent point? What is the upper cutoff frequency of the amplifier?

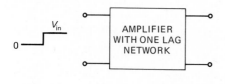

(a)

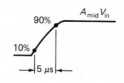

Fig. 14-38

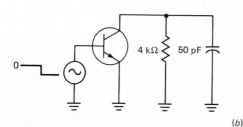

(b)

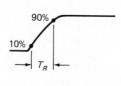

14-21. The negative input step voltage produces a positive-going output in the ac equivalent circuit of Fig. 14-38*b*. What risetime does the output waveform have?

14-22. A dc amplifier has a decibel voltage gain of 60 dB and a cutoff frequency of 10 kHz. If a square wave drives the amplifier, what is the risetime of the output?

14-23. You have the data sheets for two different dc amplifiers. The first shows a cutoff frequency of 1 MHz. The second has a risetime of 1 μs. Which amplifier has the greater bandwidth?

14-24. Use Fig. 14-33 to determine the ideal frequency ranges for each of these resistors:
   a. 10 Ω
   b. 100 Ω
   c. 1 kΩ
   d. 10 kΩ
   e. 100 kΩ

Troubleshooting

14-25. In production, someone mistakenly uses a capacitor of 0.0047 μF for the input coupling capacitor of Fig. 14-34. Which of the following are possible outcomes:
   a. Upper cutoff frequency is reduced
   b. Lower cutoff frequency is reduced
   c. Less bandwidth
   d. Shorter risetime

14-26. In Fig. 14-34, the total collector capacitance to ground is 25 pF; this includes internal transistor capacitance, stray wiring capacitance, and the input capacitance C_{in} of the next stage. If somebody uses an oscilloscope with an input capacitance of 47 pF to look at the collector signal, what happens to the collector cutoff frequency:
   a. Nothing
   b. Decreases it
   c. Increases it
   Explain your answer.

14-27. You are square-wave-testing the amplifier of Fig. 14-35. A peak-to-peak input voltage of 10 mV produces a peak-to-peak output of 180 mV with a risetime of 0.35 μs. Suddenly, the peak-to-peak output drops to approximately 95 mV; the risetime remains at 0.35 μs. Which of the following is the trouble:
   a. Output coupling capacitor opened
   b. Bypass capacitor opened
   c. Supply voltage decreased to 20 V
   d. FET open

Design

14-28. In Fig. 14-34, $\beta = 125$. Redesign the amplifier to have these cutoff frequencies: $f_{in} = 1$ Hz, $f_{out} = 1$ Hz, $f_E = 10$ Hz.

14-29. By shunting a capacitor from collector to ground in Fig. 14-34, we can produce an upper cutoff frequency of 20 kHz. What value should the capacitor have?

14-30. The transistor of Fig. 14-34 has these parameters: $r'_b = 50$ Ω, $\beta = 80$, $r'_e = 2.5$ Ω, $f_T = 200$ MHz, $C'_c = 3$ pF, and $C_{in} = 50$ pF. Redesign the amplifier to get an upper cutoff frequency of 4 MHz.

14-31. You are designing a dc amplifier with a bandwidth of 4 MHz. From Fig. 14-33, what are the minimum and maximum resistance you should use if you want to avoid the effects of lead inductance and stray capacitance?

Challenging

14-32. A noninverting amplifier has a voltage gain of 50. If a feedback capacitor of 10 pF is used, what is the input Miller capacitance? What does this unusual answer mean?

14-33. The label dBV is used to specify voltage with respect to a reference of 1 V rms. Construct three tables similar to Tables 14-4 through 14-6 showing the dBV values for 1 V, 2 V, 4 V, and so on.

14-34. Draw the Bode plots of voltage gain and phase angle for the amplifier shown in Fig. 14-35. Use these data: $C_{gs} = 10$ pF, $C_{gd} = 5$ pF, $C_{ds} = 2$ pF.

Computer

14-35. Here is a program:

```
10 PRINT "ENTER RS": INPUT RS
20 PRINT "ENTER RIN": INPUT RIN
30 PRINT "ENTER CIN": INPUT CIN
40 DEN = 2*3.1416*(RS + RIN)*CIN: FIN = 1/DEN
50 PRINT "INPUT CUTOFF FREQUENCY IS "; FIN
60 GOTO 10
```

What does this program do?

14-36. Here is a program:

```
10 PRINT "ENTER RG": INPUT RG
20 PRINT "ENTER CG": INPUT CG
30 PRINT "ENTER RD": INPUT RD
40 PRINT "ENTER CD": INPUT CD
50 DEN = 2*3.1416*RG*CG: FG = 1/DEN
60 DEN = 2*3.1416*RD*CD: FD = 1/DEN
70 PRINT "GATE CUTOFF FREQUENCY IS ";FG;" HZ"
80 PRINT "DRAIN CUTOFF FREQUENCY IS ";FD;" HZ"
```

What happens in line 10? In line 50? What does the program do?

14-37. Write a program that prints out the cutoff frequencies of the gate and drain lag networks in Fig. 14-15. The inputs are C_{gs}, C_{gd}, C_{ds}, C_{in}, R_S, g_m, and r_D.

14-38. Write a program that prints out the cutoff frequencies of the base and collector lag networks in Fig. 14-20c. The inputs are β, C'_e, C'_c, C_{in}, R_S, r'_b, r_C, and r'_e.

15

Op-Amp Theory

About a third of all linear ICs are *operational amplifiers* (op amps). The typical op amp is a high-gain dc amplifier usable from 0 to over 1 MHz. By connecting external resistors, you can adjust the voltage gain and bandwidth of an op amp to your requirements. There are over 2000 types of commercially available op amps. Most are low-power devices because their power dissipations are less than a watt. Whenever you need voltage gain for a low-power application, check available op amps. You can almost always find an op amp to do the job.

15-1 MAKING AN IC

At one time, op amps were made from discrete components. Nowadays, most op amps are fabricated on chips. Before we discuss op-amp characteristics, circuits, and other topics, it will help to have a basic idea of how integrated circuits are made.

THE p SUBSTRATE

First, the manufacturer produces a p crystal several inches long (Fig. 15-1a). This is sliced into many thin *wafers* like Fig. 15-1b. One side of the wafer is lapped and polished to get rid of surface imperfections. This wafer is called the *p substrate*; it will be used as a chassis for the integrated components.

THE EPITAXIAL n LAYER

Next, the wafers are put in a furnace. A gas mixture of silicon atoms and pentavalent atoms passes over the wafers. This forms a thin layer of n-type semiconductor on the heated surface of the substrate (see Fig. 15-1c). We call this thin layer an *epitaxial* layer. As shown in Fig. 15-1c, the epitaxial layer is about 0.1 to 1 mil thick.

THE INSULATING LAYER

To prevent contamination of the epitaxial layer, pure oxygen is blown over the surface. The oxygen atoms combine with the silicon atoms to form a layer of silicon

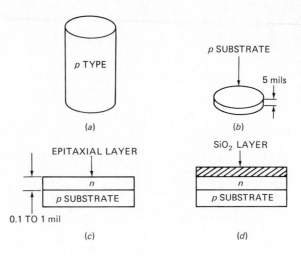

Fig. 15-1
(a) *p crystal.* (b) *Wafer.* (c) *Epitaxial layer.* (d) *Insulating layer.*

dioxide (SiO_2) on the surface, as shown in Fig. 15-1*d*. This glasslike layer of SiO_2 seals off the surface and prevents further chemical reactions; sealing off the surface like this is known as *passivation*.

CHIPS

Visualize the wafer subdivided into areas, as shown in Fig. 15-2. Each of these areas will be a separate chip after the wafer is cut. But before the wafer is cut, the manufacturer produces hundreds of circuits on the wafer, one on each area of Fig. 15-2. This simultaneous mass production is the reason for the low cost of integrated circuits.

FORMING A TRANSISTOR

Here is how an integrated transistor is formed. Part of the SiO_2 is etched off, exposing the epitaxial layer (see Fig. 15-3*a*). The wafer is then put into a furnace and trivalent atoms are diffused into the epitaxial layer. The concentration of trivalent atoms is enough to change the exposed epitaxial layer from *n* material to *p* material. Therefore, we get an island of *n* material under the SiO_2 layer (Fig. 15-3*b*).

Oxygen is again blown over the surface to form the complete SiO_2 layer shown in Fig. 15-3*c*. A hole is now etched in the center of the SiO_2 layer. This exposes the *n* epitaxial layer (Fig. 15-3*d*). The hole in the SiO_2 layer is called a *window*. We are now looking down at what will be the collector of the transistor.

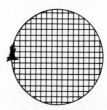

Fig. 15-2
Cutting the wafer into chips.

Fig. 15-3
Steps in making a transistor.

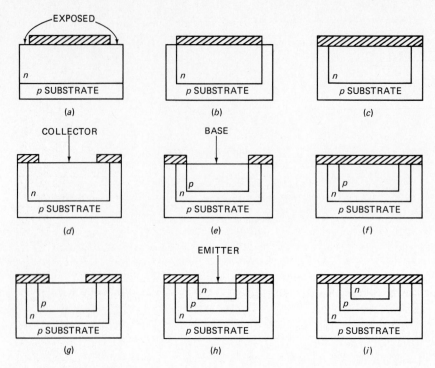

To get the base, we pass trivalent atoms through this window; these impurities diffuse into the epitaxial layer and form an island of *p*-type material (Fig. 15-3*e*). Then the SiO_2 layer is re-formed by passing oxygen over the wafer (Fig. 15-3*f*).

To form the emitter, we etch a window in the SiO_2 layer and expose the *p* island, (Fig. 15-3*g*). By diffusing pentavalent atoms into the *p* island, we can form the small *n* island shown in Fig. 15-3*h*. We then passivate the structure by blowing oxygen over the wafer (Fig. 15-3*i*).

IC COMPONENTS

By etching windows in the SiO_2 layer, we can deposit metal to make electrical contact with the emitter, base, and collector. This gives us the integrated transistor of Fig. 15-4*a*.

To get a diode, we follow the same steps up to the point at which the *p* island has been formed and sealed off (Fig. 15-3*f*). Then we etch windows to expose the *p* and *n* islands. By depositing metal through these windows, we make electrical contact with the cathode and anode of the integrated diode (Fig. 15-4*b*).

By etching two windows above the *p* island of Fig. 15-3*f*, we can make metallic contact with this *p* island; this gives us an integrated resistor (Fig. 15-4*c*).

Transistors, diodes, and resistors are easy to fabricate on a chip. For this reason, almost all integrated circuits use these components. Inductors and large capacitors are not practical to integrate on the surface of a chip.

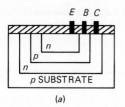
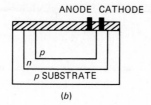
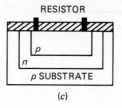

Fig. 15-4
Integrated components. (a) Transistor. (b) Diode. (c) Resistor.

A SIMPLE EXAMPLE

To give you an idea of how a circuit is produced, look at the simple three-component circuit of Fig. 15-5*a*. To fabricate this circuit, we would simultaneously produce hundreds of circuits like this on a wafer. Each chip area would resemble Fig. 15-5*b*. The diode and resistor would be formed at the point mentioned earlier. At a later step, the emitter of the transistor would be formed. Then we would etch windows and deposit metal to connect the diode, transistor, and resistor, as shown in Fig. 15-5*b*.

Regardless of how complicated a circuit may be, producing it is mainly a process of etching windows, forming *p* and *n* islands, and connecting the integrated components.

The *p* substrate isolates the integrated components from one another. In Fig. 15-5*b*, there are depletion layers between the *p* substrate and the three *n* islands that touch it. Because the depletion layers have essentially no current carriers, the integrated components are insulated from one another. This kind of insulation is known as *depletion-layer* or *diode isolation*.

MONOLITHIC ICs

The integrated circuits we have described are called *monolithic* ICs. The word "monolithic" is from Greek and means "one stone." The word is apt because the components are part of one chip.

Monolithic ICs are by far the most common, but there are many other kinds. *Thin-film* and *thick-film* ICs are larger than monolithic ICs but smaller than discrete circuits. For a thin- or thick-film IC, the passive components like resistors and

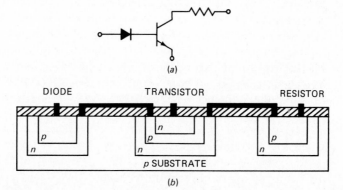

Fig. 15-5
Simple integrated circuit.

capacitors are fabricated simultaneously on a substrate. Then discrete active components like transistors and diodes are connected to form a complete circuit. Therefore, commercially available thin- and thick-film circuits are combinations of integrated and discrete components.

Hybrid ICs either combine two or more monolithic ICs in one package or combine monolithic ICs with thin- or thick-film circuits.

SSI, MSI, AND LSI

Figure 15-5*b* is an example of *small-scale integration* (SSI); only a few components have been integrated to form a complete circuit. As a guide, SSI refers to ICs with less than 12 integrated components.

Medium-scale integration (MSI) refers to ICs that have from 12 to 100 integrated components per chip. *Large-scale integration* (LSI) refers to ICs with more than a hundred components. As mentioned earlier, it takes fewer steps to make an integrated MOSFET. Furthermore, a manufacturer can produce more MOSFETs on a chip than bipolar transistors. For this reason, MOS LSI has become the largest segment of the LSI market.

15-2 THE DIFFERENTIAL AMPLIFIER

Transistors, diodes, and resistors are the only practical components in a monolithic IC. Capacitors have been fabricated on a chip, but these are usually less than 50 pF. Therefore, IC designers cannot use coupling and bypass capacitors as a discrete-circuit designer can. Instead, the stages of a monolithic IC have to be direct-coupled. One of the best direct-coupled stages is the *differential amplifier* (diff amp). This amplifier is widely used as the input stage of an op amp. In this section, we focus on the diff amp because it determines the input characteristics of the typical op amp.

DOUBLE-ENDED INPUT AND OUTPUT

Figure 15-6*a* shows the most general form of a diff amp. It has two inputs, v_1 and v_2. Because of the direct coupling, the input signals can have frequencies all the way down to zero, equivalent to dc. The output voltage v_{out} is the voltage between the collectors. Ideally, the circuit is symmetrical, with identical transistors and collector resistors. As a result, the output voltage is zero when the two inputs are equal. When v_1 is greater than v_2, an output voltage with the polarity shown appears. When v_1 is less than v_2, the output voltage has the opposite polarity.

The diff amp of Fig. 15-6*a* has a *double-ended* input. Input v_1 is called the *noninverting* input because the output voltage is in phase with v_1. On the other hand, v_2 is the *inverting* input because the output is 180° out of phase with v_2. A diff amp amplifies the difference between the two input voltages, producing an output of

$$v_{out} = A(v_1 - v_2) \tag{15-1}$$

where v_{out} = voltage between collectors
$A = R_C/r'_e$
v_1 = noninverting input voltage
v_2 = inverting input voltage

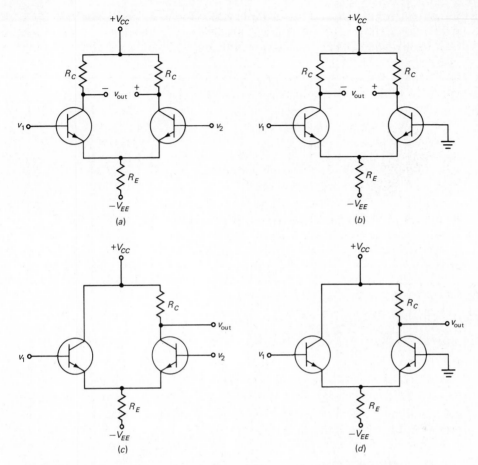

Fig. 15-6 *Differential amplifiers.* (a) *Double-ended input, double-ended output.*
(b) *Single-ended input, double-ended output.* (c) *Double-ended input, single-ended output.*
(d) *Single-ended input, single-ended output.*

SINGLE-ENDED INPUT AND DOUBLE-ENDED OUTPUT

In some applications, only one of the inputs is used, with the other grounded as shown in Fig. 15-6b. This type of input is called *single-ended*. The output remains double-ended and is given by Eq. (15-1). With v_2 equal to zero, $v_{out} = Av_1$.

A double-ended output has few applications because it requires a floating load. In other words, you have to connect both ends of the load to the collectors. This is inconvenient in most applications because loads are usually single-ended, meaning that one end of the load is connected to ground.

DOUBLE-ENDED INPUT AND SINGLE-ENDED OUTPUT

Figure 15-6c shows the most practical and widely used form of diff amp. This has many applications because it can drive single-ended loads like CE amplifiers, emitter followers, and other circuits discussed in earlier chapters. The diff amp of Fig. 15-6c is the type of diff amp used for the input stage of most op amps. For this reason, the remainder of the chapter emphasizes this form of the diff amp.

As will be derived later, the output voltage is given by

$$v_{\text{out}} = A(v_1 - v_2) \tag{15-2}$$

where v_{out} = ac voltage from collector to ground
$A = R_C/2r'_e$
v_1 = noninverting input voltage
v_2 = inverting input voltage

Notice that the voltage gain A is half the value in Eq. (15-1), a direct consequence of using only a single collector resistance R_C.

SINGLE-ENDED INPUT AND OUTPUT

Figure 15-6*d* shows the final form of a diff amp: It has a single-ended input and single-ended output. The output voltage is given by Eq. (15-2). Since v_2 is zero, v_{out} equals Av_1. A diff amp of this form is useful for direct-coupled stages where you are interested in amplifying only one input.

15-3 DC ANALYSIS OF A DIFF AMP

Figure 15-7*a* shows the dc equivalent circuit of a diff amp with a double-ended input and single-ended output. The bases are returned to ground through resistances R_B. These may be actual resistors, or they may represent the Thevenin resistances of the circuits driving the diff amp. Either way, there must be a dc path from each base to ground; otherwise, the transistors go into cutoff.

TAIL CURRENT

A diff amp is sometimes called a *long-tail pair* because it consists of two transistors connected to a single emitter resistor (the tail). The current through this resistor is called the *tail current*. When the transistors are identical, the tail current splits equally between the Q_1 and Q_2 emitters. Because of this, we can draw the equivalent circuit of Fig. 15-7*b*. Notice that each emitter is biased through a resistance of $2R_E$. This circuit produces the same emitter currents as the original circuit.

EMITTER BIAS

In Fig. 15-7*b*, we can recognize emitter bias. Each transistor is emitter-biased by the V_{EE} supply and the $2R_E$ resistor. Therefore,

$$I_E = \frac{V_{EE} - V_{BE}}{2R_E} \tag{15-3}$$

This equation assumes a stiff design, meaning that R_B is less than $\frac{1}{100}$ of $2\,\beta_{\text{dc}}R_E$. In symbols,

$$R_B < 0.02\,\beta_{\text{dc}}R_E \tag{15-4}$$

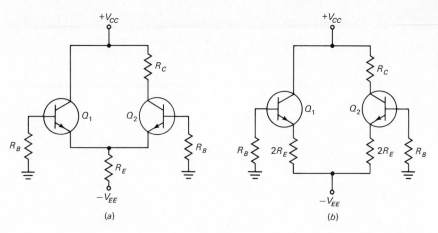

Fig. 15-7
(a) *Diff amp with base resistors.*
(b) *Dc equivalent circuit.*

When this condition is satisfied, the dc voltage from each base to ground is approximately zero.

In Fig. 15-7a, the tail current is the sum of the two emitter currents and is given by

$$I_T = 2I_E$$

or

$$I_T = \frac{V_{EE} - V_{BE}}{R_E} \tag{15-5}$$

When calculating the emitter current in either transistor, you have two choices. You can use Eq. (15-5) to calculate the tail current; dividing by 2 gives the emitter current. Alternatively, you can use Eq. (15-3) to calculate the emitter current.

INPUT OFFSET CURRENT

Base currents I_{B1} and I_{B2} flow to ground through the base resistors in Fig. 15-7a. The *input offset current* is defined as the difference between the base currents. In symbols,

$$I_{\text{in(off)}} = I_{B1} - I_{B2} \tag{15-6}$$

This current indicates how closely matched the transistors are. If the transistors are identical, the input offset current is zero.

As an example, suppose $I_{B1} = 85\ \mu A$ and $I_{B2} = 75\ \mu A$. Then,

$$I_{\text{in(off)}} = 85\ \mu A - 75\ \mu A = 10\ \mu A$$

The Q_1 transistor has 10 μA more base current than the Q_2 transistor. This may cause a problem, depending on how large the base return resistors are. More will be said about this later.

INPUT BIAS CURRENT

The *input bias current* is the average of the two base currents:

$$I_{\text{in(bias)}} = \frac{I_{B1} + I_{B2}}{2} \tag{15-7}$$

For instance, if $I_{B1} = 85 \ \mu\text{A}$ and $I_{B2} = 75 \ \mu\text{A}$, then the input bias current is

$$I_{\text{in(bias)}} = \frac{85 \ \mu\text{A} + 75 \ \mu\text{A}}{2} = 80 \ \mu\text{A}$$

In our later analysis of op amps, we will need these two relations:

$$I_{B1} = I_{\text{in(bias)}} \pm \frac{I_{\text{in(off)}}}{2} \tag{15-8}$$

$$I_{B2} = I_{\text{in(bias)}} \mp \frac{I_{\text{in(off)}}}{2} \tag{15-9}$$

You can easily derive these by a simultaneous solution of Eqs. (15-6) and (15-7). As an example, if a data sheet lists $I_{\text{in(bias)}} = 80$ nA and $I_{\text{in(off)}} = 20$ nA, then $I_{B1} = 90$ nA or 70 nA, and $I_{B2} = 70$ nA or 90 nA, depending on which base current is the larger of the two.

INPUT OFFSET VOLTAGE

Assume that we ground both bases in Fig. 15-7a. If the transistors are identical, the quiescent dc voltage at the output is

$$V_C = V_{CC} - I_C R_C \tag{15-10}$$

where I_C is approximately equal to the I_E of Eq. (15-3). Any deviation from this quiescent value is called the *output offset voltage*. If the transistors are not identical, for instance, the dc emitter currents are not equal and there is an output offset voltage.

The *input offset voltage* is defined as the input voltage needed to zero or null the output offset voltage. For instance, if a data sheet lists an input offset voltage of ±5 mV, then we need to apply a voltage of ±5 mV to one of the inputs to reduce the output offset voltage to zero. In general, the smaller the input offset voltage, the better the diff amp because its transistors are more closely matched.

15-4 AC ANALYSIS OF A DIFF AMP

Figure 15-8a shows a diff amp with a noninverting input v_1 and an inverting input v_2. One way to derive the voltage gain is to apply the superposition theorem. This means working out the voltage gain for each input separately, then combining the two results to get the total gain.

NONINVERTING INPUT

Let us start by applying v_1 while grounding v_2, as shown in Fig. 15-8b. The circuit has been redrawn to emphasize the following: The input signal drives Q_1, which acts like an emitter follower. The output of the emitter follower then drives Q_2, which is a common-base amplifier. Because there is no phase inversion, the final output is in phase with v_1. This is why v_1 is called the noninverting input.

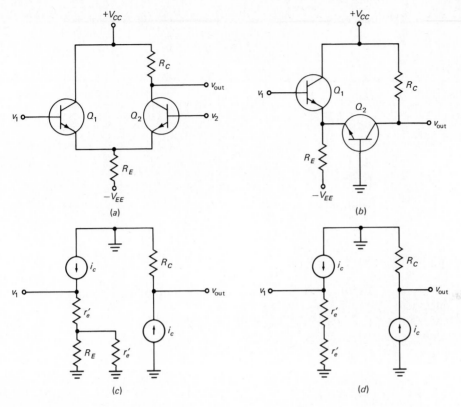

Fig. 15-8 (a) *Diff amp.* (b) *CE stage drives CB stage.* (c) *Ac equivalent circuit.* (d) *Two ac emitter resistances are in series.*

Figure 15-8c shows the ac equivalent circuit. Notice that the upper r'_e is the ac emitter resistance of Q_1, and the lower r'_e is the ac input resistance of the CB amplifier. In any practical circuit, R_E is much greater than r'_e, so that the circuit simplifies to Fig. 15-8d. This means that approximately half of the input voltage reaches the input of the CB amplifier. Stated another way, the ac emitter current is

$$i_e = \frac{v_1}{2r'_e}$$

Since the ac collector current approximately equals i_e, the ac output voltage is

$$v_{\text{out}} = i_c R_C = \frac{v_1}{2r'_e} R_C$$

or

$$\frac{v_{\text{out}}}{v_1} = \frac{R_C}{2r'_e} \tag{15-11}$$

This is the voltage gain for the noninverting input.

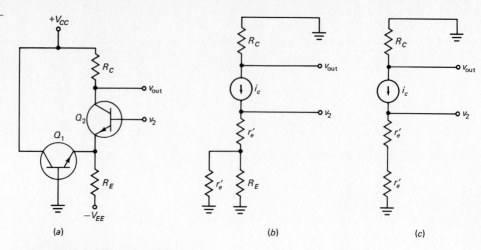

Fig. 15-9
*Equivalent
circuits for invert-
ing input.*

(a)   (b)   (c)

INVERTING INPUT

Next, let us find the voltage gain for the inverting input. This means that we can ground v_1 and redraw the circuit as shown in Fig. 15-9a. Now Q_2 drives Q_1, which has an input resistance of r'_e. Figure 15-9b is the ac equivalent circuit. Again, R_E is always much greater than r'_e, so the circuit simplifies to Fig. 15-9c. The ac emitter current is

$$i_e = \frac{v_2}{2r'_e}$$

and the ac output voltage is

$$v_{\text{out}} = -i_c R_C = -\frac{v_2}{2r'_e} R_C$$

or

$$\frac{v_{\text{out}}}{v_2} = \frac{-R_C}{2r'_e} \tag{15-12}$$

This is the voltage gain for the inverting input. The minus sign indicates phase inversion.

DIFFERENTIAL GAIN

Compare Eqs. (15-11) and (15-12). As you can see, the magnitude of the voltage gain is the same; only the phase differs. Here is how to find the total voltage gain with both inputs active at the same time. Using Eq. (15-11), we can write

$$v_{\text{out(1)}} = \frac{R_C}{2r'_e} v_1$$

From Eq. (15-12),

$$v_{\text{out(2)}} = \frac{-R_C}{2r'_e} v_2$$

The superposition theorem says that we can add these individual outputs to get the total output with both signals present. After factoring and rearranging, we get

$$v_{\text{out}} = v_{\text{out}(1)} + v_{\text{out}(2)} = \frac{R_C}{2r'_e}(v_1 - v_2)$$

This can be written as

$$v_{\text{out}} = A(v_1 - v_2) \qquad (15\text{-}13)$$

where

$$A = \frac{R_C}{2r'_e} \qquad (15\text{-}14)$$

The quantity A is called the *differential voltage gain* because it tells us how much the difference $v_1 - v_2$ is amplified.

INPUT IMPEDANCE

In the midband of a diff amp, the input impedance looking into either input is

$$r_{\text{in}} = 2\,\beta r'_e \qquad (15\text{-}15)$$

This input impedance is twice that of an ordinary CE amplifier. The factor of 2 arises because the r'_e of each transistor is in series (see Figs. 15-8d and 15-9c).

One way to get a higher input impedance with a diff amp is to use Darlington transistors. Another way is to use JFETs instead of bipolars. This is the approach taken with *BIFET* op amps; they use JFETs for the input diff-amp stage. With BIFET op amps the input impedance approaches infinity.

COMMON-MODE GAIN

A *common-mode signal* is one that drives both inputs of a diff amp equally. Most interference, static, and other kinds of undesirable pickup are common-mode. What happens is this. The connecting wires on the input bases act like small antennas. If the diff amp is operating in an environment with a lot of electromagnetic interference, each base picks up an unwanted interference voltage. One of the reasons the diff amp is so popular is because it discriminates against common-mode signals. In other words, a diff amp refuses to amplify common-mode signals. Because of this, you don't get a lot of unwanted interference at the output.

Let us now find out why the diff amp does not amplify common-mode signals. Figure 15-10a shows a common-mode signal driving a diff amp. As you can see, an equal voltage $v_{\text{in(CM)}}$ drives both inputs simultaneously. Assuming that the transistors are identical, the equal inputs imply equal emitter currents. Therefore, we can split R_E as shown in Fig. 15-10b. This equivalent circuit has exactly the same emitter currents as the original circuit.

Figure 15-10c shows the ac equivalent circuit. Can you see what this means? When a common-mode signal drives a diff amp, a large unbypassed emitter resistance appears in the ac equivalent circuit. Therefore, the voltage gain for a common-

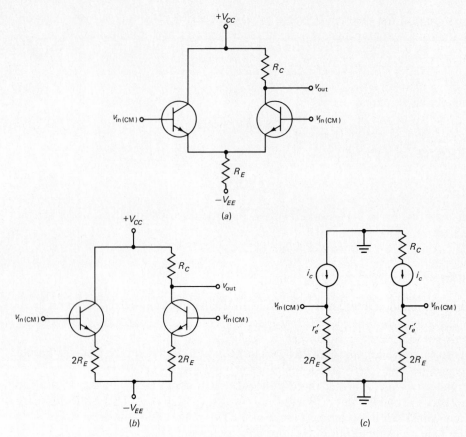

Fig. 15-10 (a) *Common-mode signal drives diff amp.* (b) *Splitting the emitter resistor.* (c) *Ac equivalent circuit.*

mode signal is

$$\frac{v_{\text{out}}}{v_{\text{in(CM)}}} = \frac{-R_C}{r'_e + 2R_E}$$

Since R_E is always much greater than r'_e, we can approximate the *common-mode voltage gain* as $-R_C/2R_E$. In symbols,

$$A_{\text{CM}} = \frac{-R_C}{2R_E} \qquad (15\text{-}16)$$

where A_{CM} = common-mode voltage gain
$\qquad R_C$ = collector resistance
$\qquad R_E$ = emitter resistance

For instance, if $R_C = 10$ kΩ and $R_E = 10$ kΩ, then

$$A_{\text{CM}} = \frac{-10 \text{ k}\Omega}{20 \text{ k}\Omega} = -0.5$$

This means that the diff amp attenuates a common-mode signal, because the voltage gain is less than 1.

COMMON-MODE REJECTION RATIO

Data sheets list the *common-mode rejection ratio* (CMRR). It is defined as the ratio of differential voltage gain to common-mode voltage gain. In symbols,

$$\text{CMRR} = \frac{A}{-A_{\text{CM}}} \tag{15-17}$$

where the minus sign is included to get a positive ratio. For instance, if $A = 200$ and $A_{\text{CM}} = -0.5$, then

$$\text{CMRR} = \frac{200}{0.5} = 400$$

Data sheets almost always specify CMRR in decibels, using the following formula for the conversion:

$$\text{CMRR}' = 20 \log \text{CMRR} \tag{15-18}$$

If CMRR $= 400$, then

$$\text{CMRR}' = 20 \log 400 = 52 \text{ dB}$$

CURRENT-MIRROR BIAS

With an emitter-biased diff amp, the differential voltage gain is $R_C/2r'_e$ and the common-mode voltage gain is $-R_C/2R_E$. From Eq. (15-17),

$$\text{CMRR} = \frac{R_C/2r'_e}{R_C/2R_E} = \frac{R_E}{r'_e}$$

From this it is clear that the higher we can make R_E, the better the CMRR.

One way to get a very high equivalent R_E is to use *current-mirror bias,* as shown in Fig. 15-11. This is typical for the first stage of an integrated op amp. Here you see a current mirror driving the emitters of a diff amp. With integrated circuits, the compensating diode Q_3 is actually a transistor connected as a diode (the base and

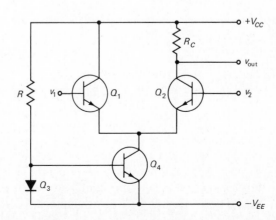

Fig. 15-11
Current mirror sources tail current for diff amp.

Fig. 15-12
Current mirror is active load on diff amp.

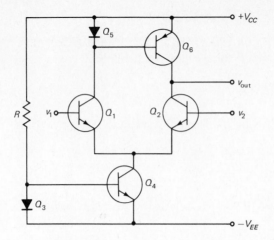

collector are tied together). The current through Q_3 is given by

$$I = \frac{V_{CC} + V_{EE} - V_{BE}}{R}$$

This is the value of mirror current supplied by Q_4. Since Q_4 acts like a current source, it has a very high output resistance, much higher than we can get with conventional emitter bias. This means that the equivalent R_E of the diff amp is hundreds of kilohms and the CMRR is dramatically improved.

CURRENT-MIRROR LOAD

Earlier, we derived a differential voltage gain of $R_C/2r'_e$. The larger we can make R_C, the greater the differential voltage gain. But we have to be careful. Too large an R_C will saturate the output transistor. As a rule, the designer selects an R_C to get a quiescent voltage that is half of V_{CC}. For example, if the collector supply voltage is +15 V, then R_C is selected to get a V_C of +7.5 V. This places a limit on the size of R_C and the voltage gain.

One way around the problem is to use an *active load*. Figure 15-12 shows a current-mirror load. Since Q_5 acts like a diode, it has a very low impedance, and the load on Q_1 still appears almost like an ac short. Q_6, on the other hand, acts like a *pnp* current source. Therefore, Q_2 sees an equivalent R_C that is hundreds of kilohms. As a result, the differential voltage gain is much higher with a current-mirror load. Active loading like this is typical of monolithic op amps.

15-5 THE OPERATIONAL AMPLIFIER

In 1965, Fairchild Semiconductor introduced the μA709, the first widely used monolithic op amp. Although highly successful, this first-generation op amp had many disadvantages. This led to an improved op amp known as the μA741. Because it is inexpensive and easy to use, the μA741 has been an enormous success. Many other 741 designs have appeared from various manufacturers. For instance, Motorola produces the MC1741, National Semiconductor the LM741, and Texas Instruments the SN72741. All these monolithic op amps are equivalent because they have the same specifications on their data sheets. For convenience, most people drop the prefixes and refer to this widely used op amp simply as the 741.

THE INDUSTRY STANDARD

The 741 has become the *industry standard*. As a rule, you first try to use it in your designs. In those cases where you cannot meet a design specification with a 741, you upgrade to a better op amp. Because of its great importance, we will analyze the 741 in detail. Once you understand it, you can easily branch out to other op amps.

Incidentally, the 741 has different versions, numbered 741, 741A, 741C, 741E, 741N, and so on. These differ in their voltage gain, temperature range, noise level, and other such characteristics. The 741C (the C stands for commercial grade) is the least expensive and most widely used. It has a typical input impedance of 2 MΩ, a voltage gain of 100,000, and an output impedance of 75 Ω.

SCHEMATIC DIAGRAM OF THE 741

Figure 15-13 is a simplified schematic diagram of the 741. This circuit is equivalent to the 741 and many later-generation op amps. The input stage is a *diff amp* using *pnp* transistors (Q_1 and Q_2). To get a high CMRR, a current mirror (Q_{13} and Q_{14}) sources tail current to the diff amp. Also, to get as high a voltage gain as possible for the diff amp, a current-mirror load is used (Q_3 and Q_4).

The output of the diff amp (collector of Q_2) drives an emitter follower (Q_5). This stage steps up the impedance level to avoid loading down the diff amp. The signal out of Q_5 goes to Q_6, which is a class B driver. Incidentally, the plus sign on the collector of Q_5 means that it is connected to the positive V_{CC} supply. Similarly, the minus signs on the bottoms of R_2 and R_3 mean that these are connected to the negative V_{EE} supply.

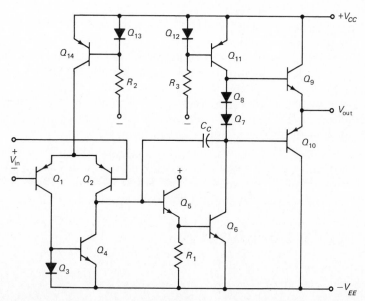

Fig. 15-13
Simplified schematic diagram for 741 and similar op amps.

The last stage is a *class B push-pull emitter follower* (Q_9 and Q_{10}). Because of the *split supply* (equal positive and negative voltages), the quiescent output is ideally 0 V when the input voltages are zero; any deviation from 0 V is called the *output offset voltage*. Q_{11} and Q_{12} are a current-mirror load for the class B driver.

Notice that direct coupling is used between all stages. For this reason, there is no lower cutoff frequency. In other words, an op amp is a dc amplifier because it operates all the way down to zero frequency, equivalent to dc.

ACTIVE LOADING

In Fig. 15-13, we have two examples of *active loading* (using transistors instead of resistors for loads). First, there is a current-mirror load (Q_3 and Q_4) on the diff amp. Second, there is a current-mirror load (Q_{11} and Q_{12}) on the driver stage. Because current sources have high impedances, active loads produce much higher voltage gain than is possible with resistors. These active loads produce a typical voltage gain of 200,000 for the 741.

Active loading is very popular in integrated circuits because it is easier and less expensive to fabricate transistors on a chip than it is to fabricate resistors. MOS digital integrated circuits use active loading almost exclusively; in these ICs, one MOSFET is the active load for another.

COMPENSATING CAPACITOR

In Fig. 15-13, C_C is called a *compensating capacitor*. Because of the Miller effect, this small capacitor (typically 30 pF) has a pronounced effect on the frequency response. C_C is part of a base lag network that rolls off the decibel voltage gain at a rate of 20 dB per decade (equivalent to 6 dB per octave). This is necessary to prevent oscillations (an unwanted signal produced by the amplifier). Later sections tell more about the compensating capacitor and oscillations.

FLOATING INPUTS

In Fig. 15-13, note that the inputs are floating. The op amp cannot possibly work unless each input has a *dc return* path to ground. As mentioned before, these return paths are usually provided by the direct-coupled circuits that drive the op amp. If the driving circuits are capacitively coupled, you have to insert separate base return resistors. The key thing to remember is the dc path to ground. Each base must have such a path; otherwise, the input transistors go into cutoff.

Assuming that there is a dc path to ground for each base, we still have an offset problem to worry about. Because the input transistors are not quite identical, an unwanted offset voltage will exist at the output of the op amp. A later section tells you how to eliminate the output offset voltage.

INPUT IMPEDANCE

Recall that the input impedance of a diff amp is

$$r_{in} = 2\,\beta r'_e$$

With a small tail current in the input diff amp, an op amp can have a fairly high input impedance. For instance, the input diff amp of a 741 has a tail current of approximately 15 μA. Since each emitter gets half of this,

$$r'_e = \frac{25 \text{ mV}}{I_E} = \frac{25 \text{ mV}}{7.5 \ \mu\text{A}} = 3.33 \text{ k}\Omega$$

Each input transistor has a typical β of 300; therefore,

$$r_{\text{in}} = 2 \ \beta r'_e = 2(300)(3.33 \text{ k}\Omega) = 2 \text{ M}\Omega$$

This agrees with the data-sheet value for a 741.

BIFET OP AMPS

If an extremely high input impedance is required, you can use a *BIFET* op amp. BIFET means that bipolar transistors and JFETs are fabricated on the same chip. A BIFET op amp uses JFETs for its input stage, followed by bipolar stages. This combination produces the high input impedance associated with JFETs and the high voltage gain of bipolar transistors. For instance, the 355 is a popular all-purpose BIFET op amp with a typical input bias current of 0.03 nA. This is much lower than 80 nA, the typical input bias current of a 741. Therefore, if you have to upgrade a design because of input impedance, the BIFET op amp is a natural choice.

SCHEMATIC SYMBOL

Given an amplifier like Fig. 15-13, we can save time by using a schematic symbol. Figure 15-14a shows a simple way to represent an amplifier with two inputs and one output. A is the unloaded voltage gain. This is the gain we get when no load resistor is connected to the output. The input and output voltages are with respect to the ground line.

Most of the time, we don't bother drawing the ground line, as shown in Fig. 15-14b. In this abbreviated symbol, you have to remember that the voltages are with respect to ground. Figure 15-14c shows the most widely used symbol for an op amp. The inverting input has a minus sign, a reminder of the phase inversion that

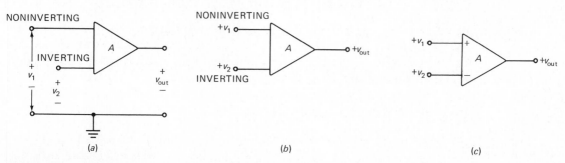

Fig. 15-14 *Symbols for op amp.* (a) *With ground line.* (b) *Ground line implied.* (c) *Common symbol.*

Fig. 15-15
*Input impedance
and Thevenin
output circuit.*

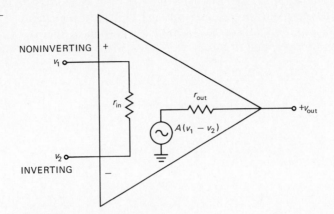

takes place with this input. On the other hand, the noninverting input has a plus sign because no phase inversion occurs with this input.

INPUT IMPEDANCE AND THEVENIN OUTPUT CIRCUIT

In Fig. 15-15, r_{in} is the impedance between the input terminals. For a 741, $r_{in} = 2$ MΩ. A BIFET op amp has a much higher input impedance. In any case, regardless of what kind of op amp we are using, we can visualize an impedance r_{in} between the noninverting and inverting input terminals.

As long as the op amp is operating in its *linear region* (unsaturated transistors), we can thevenize the output, as shown in Fig. 15-15. The Thevenin output voltage is

$$v_{TH} = A(v_1 - v_2)$$

and the Thevenin output resistance is r_{out}. For a 741C, A is 100,000 and r_{out} is 75 Ω.

Whenever you analyze an op-amp circuit, remember that there is an input impedance r_{in} between the input terminals, an unloaded voltage gain A, and an output impedance r_{out}, as shown in Fig. 15-15. The typical op amp has a high r_{in}, a high A, and a low r_{out}. If a manufacturer could build the ideal op amp, it would have infinite input impedance, infinite voltage gain, and zero output impedance.

15-6 OP-AMP CHARACTERISTICS

Because an op amp is a dc amplifier, you have to consider both dc and ac characteristics in troubleshooting or designing op-amp circuits. In this section, we take a closer look at the offset problem, as well as discussing other characteristics that affect op-amp performance.

INPUT OFFSET VOLTAGE

When the inputs of an op amp are grounded, there almost always is an *output offset voltage,* as shown in Fig. 15-16a, because the input transistors have different V_{BE} values. The *input offset voltage* is equal to this difference in V_{BE} values. For instance, a typical 741C has an input offset voltage of ±2 mV, which means that the V_{BE} of one

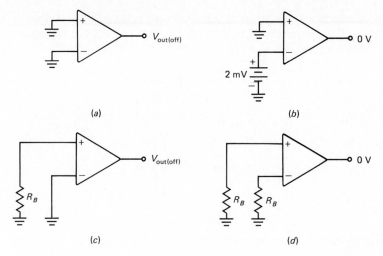

Fig. 15-16 (a) *Output offset voltage.* (b) *Nulling the output offset voltage.* (c) *Return resistor may produce output offset voltage.* (d) *Equal return resistors reduce output offset voltage.*

input transistor differs from the other by 2 mV. This 2 mV is amplified to produce an output offset voltage.

Theoretically, we can apply a voltage of 2 mV to the inverting input, as shown in Fig. 15-16b; then the output offset voltage goes to zero for the typical 741C. (Note: The offset can have either polarity, so it might be necessary to reverse the polarity of the 2 mV.)

INPUT BIAS CURRENT

Suppose we get lucky and happen to use an op amp whose input transistors have equal V_{BE} values. Then the input offset voltage is zero. But a problem can still arise because of the bias currents. If either input to the op amp has a resistance in its return path, an output offset voltage can exist. For instance, Fig. 15-16c shows a resistance R_B between the noninverting input and ground. Since there is a base current I_{B1} through R_B, a voltage appears at the noninverting input, given by

$$v_1 = I_{B1} R_B$$

This unwanted input voltage is amplified to produce an output offset voltage. If R_B is small enough, the resulting output offset voltage may be small enough to ignore.

If the input base currents are equal (which almost never happens), we can eliminate the output offset voltage by inserting an equal base resistor on the inverting input, as shown in Fig. 15-16d. Now the equal base currents produce equal voltage drops across the base resistors to null the output offset voltage, as shown in Fig. 15-16d.

INPUT OFFSET CURRENT

The base currents of the input transistors are almost never equal because the β values are usually different. As mentioned earlier, the *input offset current* equals the difference in the base currents. Therefore, even though we use equal resistors in Fig. 15-16*d*, the input offset current can produce an unwanted difference voltage as follows. The input to the noninverting input is

$$v_1 = I_{B1}R_B$$

The input to the inverting input is

$$v_2 = I_{B2}R_B$$

The differential input is

$$v_1 - v_2 = I_{B1}R_B - I_{B2}R_B = (I_{B1} - I_{B2})R_B$$

or

$$v_1 - v_2 = I_{\text{in(off)}}R_B$$

When amplified, this produces an output offset voltage.

DIFFERENT BASE RESISTANCES

In the op-amp circuits to be discussed later, the base resistors may be equal or unequal, as shown in Fig. 15-17. Because the input offset voltage can have either polarity and because either input current may be larger than the other, the differential input voltage is given by

$$v_1 - v_2 = \pm v_{\text{in(off)}} \pm I_{B1}R_{B1} \mp I_{B2}R_{B2} \qquad (15\text{-}19)$$

where $v_1 - v_2$ = total input offset voltage
$v_{\text{in(off)}}$ = input offset voltage, V_{BE} differences
I_{B1} = base current to noninverting input
I_{B2} = base current to inverting input
R_{B1} = dc return resistance, noninverting input
R_{B2} = dc return resistance, inverting input

This differential input voltage is amplified to produce an output offset voltage. How large the output offset voltage is depends on the voltage gain. Ordinarily, an op amp is used with external resistors to produce negative feedback. This reduces the output offset voltage. In Chap. 16, we will discuss the use of negative feedback with op amps. At that time, we will again examine the size of the output offset voltage.

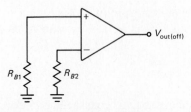

Fig. 15-17   *Offsets.*

CMRR

The *common-mode rejection ratio* was defined earlier. For a 741C, CMRR' = 90 dB at lower frequencies. Given equal signals, one a difference input (desired) and the other a common-mode input (undesired), the difference signal will be 90 dB larger at the output than the common-mode signal. In ordinary numbers, this means that the desired signal will be 30,000 times larger at the output than the common-mode signal. At higher frequencies, reactive effects degrade CMRR', as shown in Fig. 15-18*a*. Notice that CMRR' is approximately 75 dB at 1 kHz, 56 dB at 10 kHz, and so on.

AC OUTPUT COMPLIANCE

The *ac output compliance PP* is the maximum unclipped peak-to-peak output voltage that an op amp can produce. Since the quiescent output is ideally zero, the ac output voltage can swing positive or negative. How far it swings depends on the load resistance. For large load resistances, each peak can swing to within 1 or 2 V of the supply voltages. For instance, if $V_{CC} = +15$ V and $V_{EE} = -15$ V, the maximum unclipped peak-to-peak voltage with a load resistance of 10 kΩ is approximately 27 V.

As the load resistance decreases, the slope of the ac load line changes and the ac output compliance decreases. Figure 15-18*b* shows the typical variation of ac output compliance with load resistance. Notice that PP is approximately 27 V for an R_L of 10 kΩ, 25 V for 1 kΩ, and 7 V for 100 Ω.

SHORT-CIRCUIT OUTPUT CURRENT

In some applications, an op amp may drive a load resistance that is approximately zero. Since the 741C has an output impedance of only 75 Ω, you might think it can supply large output currents. Not so. A monolithic op amp is a low-power device, and so its output current is limited. For instance, the 741C can supply a maximum *short-circuit output current* of only 25 mA. If you are using small load resistors (less than 75 Ω), don't expect to get a large output voltage because the voltage cannot be greater than 25 mA times the load resistance.

FREQUENCY RESPONSE

Negative feedback means sacrificing some voltage gain in exchange for a rock-solid voltage gain, less distortion, and other improvements in amplifier performance. When an op amp uses negative feedback, the operation is called *closed-loop*. If the op amp is running wide open without negative feedback, the operation is known as *open-loop*.

Figure 15-18*c* shows the small-signal frequency response of a 741C. In the midband, the open-loop voltage gain is 100,000. The 741C has an open-loop cutoff frequency f_{OL} of 10 Hz. As indicated, the voltage gain is 70,700 (down 3 dB) at this frequency. Beyond cutoff, the voltage gain rolls off at an ideal rate of 20 dB per decade; this is caused by C_C, the compensating capacitor on the 741 chip.

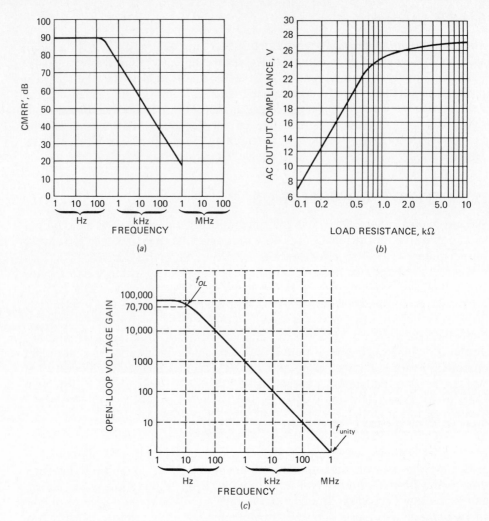

Fig. 15-18 (a) *Common-mode rejection ratio.* (b) *AC output compliance.* (c) *Open-loop voltage gain.*

The *unity-gain* frequency, designated f_{unity}, is the frequency at which the voltage gain equals 1. In Fig. 15-18c, f_{unity} is 1 MHz. Data sheets specify the value of f_{unity} because it represents the upper limit on the useful gain of an op amp. For instance, the data sheet of a 318 lists an f_{unity} of 15 MHz. This means that the 318 can amplify signals at much higher frequencies than can the 741C. Naturally, you have to pay considerably more for a 318.

As shown in Fig. 15-18c, the 741C has a midband voltage gain of 100,000 and a cutoff frequency of 10 Hz for open-loop operation. As a rule, you never run an op amp without negative feedback because it is too unstable. In Chap. 16, you will learn how to connect a few external resistors to an op amp to get closed-loop operation. This results in less voltage gain but greater bandwidth. In fact, you will learn how to trade off voltage gain for increased bandwidth. More about this later.

SLEW RATE

Of all the specifications affecting the ac operation of an op amp, *slew rate* is one of the most important because it limits the size of the output voltage at higher frequencies. To understand slew rate, we have to discuss some basic circuit theory. The charging current in a capacitor is given by

$$i = C\frac{dv}{dt}$$

where i = current into capacitor
C = capacitance
dv/dt = rate of capacitor voltage change

We can rearrange this basic equation to get

$$\frac{dv}{dt} = \frac{i}{C}$$

This says that the rate of voltage change equals the charging current divided by the capacitance.

The greater the charging current, the faster the capacitor charges. If for any reason the charging current is limited to a maximum value, the rate of voltage change is also limited to a maximum value.

Figure 15-19*a* brings out the idea of current limiting and its effect on output voltage. A current of I_{max} charges the capacitor. Because this current is constant, the capacitor voltage increases linearly, as shown in Fig. 15-19*b*. The rate of voltage change with respect to time is

$$\frac{dv_{out}}{dt} = \frac{I_{max}}{C_C} \tag{15-20}$$

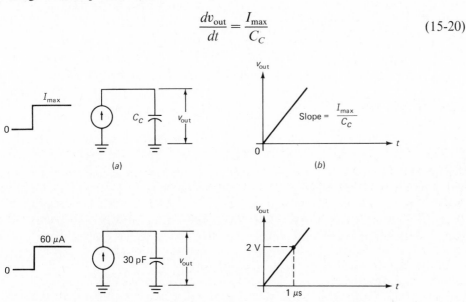

Fig. 15-19 *Slew rate equals maximum charging current divided by capacitance.*

As an example, if $I_{max} = 60 \ \mu A$ and $C_C = 30$ pF (see Fig. 15-19c), the maximum rate of voltage change is

$$\frac{dv_{out}}{dt} = \frac{60 \ \mu A}{30 \ pF} = 2 \ V/\mu s$$

This says that the output voltage across the capacitor changes at a maximum rate of 2 V/μs (Fig. 15-19d). The voltage cannot change faster than this unless we can increase I_{max} or decrease C_C.

Slew rate S_R is defined as the maximum rate of output voltage change. Because of this, we can rewrite Eq. (15-20) as

$$S_R = \frac{I_{max}}{C_C} \qquad (15\text{-}21)$$

In Fig. 15-19a, slew rate pins down the rate at which the output voltage can change. If $I_{max} = 60 \ \mu A$ and $C_C = 30$ pF, the circuit can slew no faster than 2 V/μs.

SLEW-RATE DISTORTION

When a large positive step of input voltage drives the op amp of Fig. 15-13, Q_1 saturates and Q_2 cuts off. Therefore, all the tail current I_T passes through Q_1 and Q_3. Because of the current mirror, the current through Q_4 equals I_T. Since Q_2 is cut off, all Q_4 current passes on to the next stage. Initially, all of it goes to C_C (Q_5 base current is negligible).

As C_C charges, the output voltage rises. Assuming unity voltage gain for the output stage, the rate of output voltage change is equal to the rate of voltage change across C_C. From Eq. (15-21), the maximum rate of output voltage change is

$$S_R = \frac{I_T}{C_C}$$

This says that the output voltage can change no faster than the ratio of I_T to C_C.

Here's an example. In Fig. 15-13, $I_T = 15 \ \mu A$ and $C_C = 30$ pF. Therefore, the slew rate of a 741C is

$$S_R = \frac{15 \ \mu A}{30 \ pF} = 0.5 \ V/\mu s$$

This is the ultimate speed of a 741C; its output voltage can change no faster than 0.5 V/μs. Figure 15-20 illustrates the idea. If we overdrive a 741C with a large step input (Fig. 15-20a), the output slews as shown in Fig. 15-20b. It takes 20 μs for the output voltage to change from 0 to 10 V (nominal output voltage swing with 12-V supplies). It is impossible for the output of 741C to change faster than this.

Fig. 15-20
Overdriving an op amp produces slew-rate limiting.

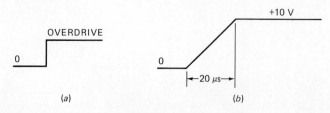

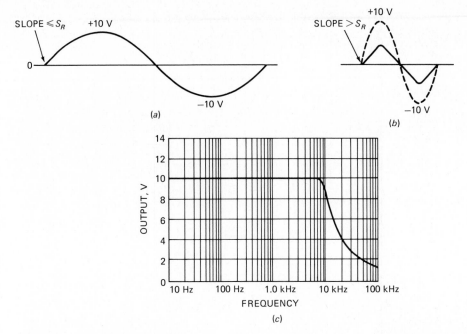

Fig. 15-21 *(a) Initial slope of sine wave cannot exceed slew rate. (b) Slew-rate distortion.*
(c) Large-signal response of 741C.

We can also get slew-rate limiting with a sinusoidal signal. Figure 15-21*a* shows the maximum sinusoidal output from a 741C when the peak voltage is 10 V. As long as the initial slope of the sine wave is less than or equal to S_R, there is no slew-rate limiting. But when the initial slope of the sine wave is greater than S_R, we get the *slew-rate distortion* shown in Fig. 15-21*b*. The output begins to look triangular; the higher the frequency, the smaller the swing and the more triangular the waveform.

POWER BANDWIDTH

Slew-rate distortion of a sine wave starts at the point at which the initial slope of the sine wave equals the slew rate of the op amp. In Appendix 1 we derive this useful equation:

$$f_{\text{max}} = \frac{S_R}{2\pi V_P} \tag{15-22}$$

where f_{max} = highest undistorted frequency
S_R = slew rate of op amp
V_P = peak of output sine wave

As an example, if a 741C has $V_P = 10$ V and $S_R = 0.5$ V/μs, the maximum undistorted frequency for the large-signal operation is

$$f_{\text{max}} = \frac{0.5 \text{ V/}\mu\text{s}}{2\pi(10 \text{ V})} = 7.96 \text{ kHz}$$

Fig. 15-22
*Graphs to trade
off peak amplitude
for power
bandwidth.*

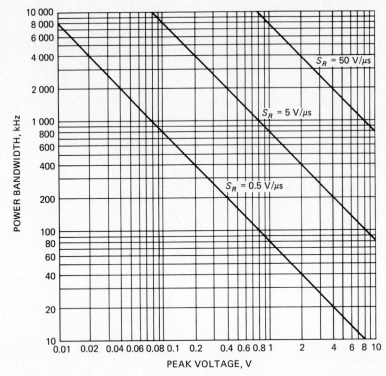

Above this frequency, you will begin to notice slew-rate distortion on an oscilloscope.

Frequency f_{max} is called the *power bandwidth* (also the large-signal bandwidth) of an op amp. We have just found that the 10-V power bandwidth of a 741C is approximately 8 kHz. This means that the undistorted bandwidth for large-signal operation is 8 kHz. If we try to amplify higher frequencies of the same peak value, the output voltage drops off, as shown in Fig. 15-21c.

TRADEOFF

One way to increase the power bandwidth is to accept less output voltage. Figure 15-22 is a graph of Eq. (15-22) for three different slew rates. By *trading off amplitude for frequency,* we can improve the power bandwidth. For instance, if peak amplitudes of 1 V are acceptable in an application, the power bandwidth of a 741C increases to 80 kHz (the bottom curve). If a peak amplitude of 0.1 V is all right, the power bandwidth increases to 800 kHz.

On the other hand, if you really want a peak amplitude of 10 V, you need to use a better op amp than a 741C. In Fig. 15-22, notice that the 10-V power bandwidth increases to 80 kHz for an S_R of 5 V/μs and to 800 kHz for an S_R of 50 V/μs.

POPULAR OP AMPS

Table 15-1 lists some popular op amps. The LF351 through LF356 and the TL071 through TL074 are BIFET op amps. Notice how low the input bias and offset

Table 15-1. Typical Parameters of Popular Op Amps

Number	$V_{in(off)}$, mV	$I_{in(bias)}$, nA	$I_{in(off)}$, nA	$I_{out(max)}$, mA	f_{unity}, MHz	Slew rate, V/μs
LF351	5	0.05	0.025	20	4	13
LF353	5	0.05	0.025	20	4	13
LF355	3	0.03	0.003	20	2.5	5
LF356	3	0.03	0.003	20	5	12
LM10C	0.5	12	0.4	20	0.1	0.12
LM11C	0.1	0.025	0.0005	2	0.5	0.3
LM301C	2	70	3	10	1	0.5
LM307	2	70	3	10	1	0.5
LM308	2	1.5	0.2	5	0.3	0.15
LM312	2	1.5	0.2	6	1	0.1
LM318	4	150	30	21	15	70
LM324	2	45	5	20	1	0.5
LM348	1	30	4	25	1	0.5
LM358	2	45	5	40	1	0.5
LM709	2	300	100	42	*	0.25
LM739	1	300	50	1.5	6	1
LM741C	2	80	20	25	1	0.5
LM747C	2	80	20	25	1	0.5
LM748	2	80	20	27	*	*
LM1458	1	200	80	20	1	0.5
LM4250	3–5	*	*	*	*	*
LM13080	3	*	*	250	1	*
NE531	2	400	50	20	1	35
TL071	3	0.03	0.005	10	3	13
TL072	3	0.03	0.005	10	3	13
TL074	5	0.05	0.025	17	4	13

* Externally controlled by resistors or capacitors.

currents are for these devices. The LM10C through NE531 are bipolar op amps. As indicated, the LM741C has a typical input offset voltage of 2 mV, an input bias current of 80 nA, an input offset current of 20 nA, and so on. If the dc return resistances on the noninverting and inverting inputs must be high, a 741C may produce too much output offset voltage. In a case like this, you can upgrade to an op amp such as the LF355, a general-purpose BIFET op amp.

The table also includes maximum output current, unity-gain frequency, and slew rate. Sometimes a 741C may not slew fast enough to produce adequate power bandwidth for your application. In this case, you can upgrade to a fast-slew-rate device like the TL071, an inexpensive BIFET op amp. The ultimate weapon for slew-rate problems is the LM318; it has a slew rate of 70 V/μs.

All data are typical. For worst-case values and other specifications, you will have to refer to manufacturers' data sheets. For the devices listed in Table 15-1, CMRR is from 80 to 100 dB, and voltage gain from 100,000 to 300,000. Some of the devices are rather unusual. The LM4250, for instance, has a string of asterisks. An asterisk means that the quantity can be varied by the user. In other words, the LM4250 is programmable by a single external resistor that allows you to vary input bias and offset currents, slew rate, unity-gain frequency, and so on.

15-7 OTHER LINEAR ICs

Although the op amp is the most important linear IC, you will encounter other linear ICs in various applications. This section briefly examines some of these ICs. Our survey covers only the main types.

AUDIO AMPLIFIERS

Preamplifiers (preamps) are audio amplifiers with less than 50 mW of output power. Preamps are optimized for low noise because they are used at the front end of audio systems, where they amplify weak signals from phonograph cartridges, magnetic tape heads, microphones, and so on.

An example of an IC preamp is the LM381, a low-noise dual preamplifier. Each amplifier is completely independent of the other. The LM381 has a voltage gain of 112 dB and a 10-V power bandwidth of 75 kHz. It operates from a positive supply of 9 to 40 V. Its input impedance is 100 kΩ, and its output impedance is 150 Ω. The LM381's input stage is a diff amp, which allows differential or single-ended input.

Medium-level audio amplifiers have output powers from 50 to 500 mW. These are useful near the output end of small audio systems like transistor radios or signal generators. An example is the MHC4000P, which has an output power of 250 mW.

Audio power amplifiers deliver more than 500 mW of output power. They are used in phonograph amplifiers, intercoms, AM-FM radios, and other such applications. The LM380 is an example. It has a voltage gain of 34 dB, a bandwidth of 100 kHz, and an output power of 2 W. As another example, the LM2002 has a voltage gain of 40 dB, a bandwidth of 100 kHz, and an output power of 8 W.

Figure 15-23 shows a simplified schematic diagram of the LM380. The input diff amp uses *pnp* ground-referenced inputs (discussed in Sec. 8-5). Because of this, the signal can be directly coupled, which is an advantage with transducers. The diff amp drives a current-mirror load (Q_5 and Q_6). The output of the mirror goes to an emitter follower (Q_7) and a CE driver (Q_8). The output stage is a class B push-pull emitter follower (Q_{13} and Q_{14}).

There is an internal compensating capacitor of 10 pF that rolls off the decibel voltage gain at a rate of 20 dB per decade. This capacitor produces a slew rate of approximately 5 V/μs.

VIDEO AMPLIFIERS

A *video* or *wideband* amplifier has a flat response (constant decibel voltage gain) over a very broad range of frequencies. Typical bandwidths are well into the megahertz region. Video amps are not necessarily dc amps, but they often do have a response that extends down to zero frequency. They are used in applications in which the range of input frequencies is very large. For instance, many oscilloscopes handle frequencies from 0 to over 10 MHz; instruments like these use video amps to

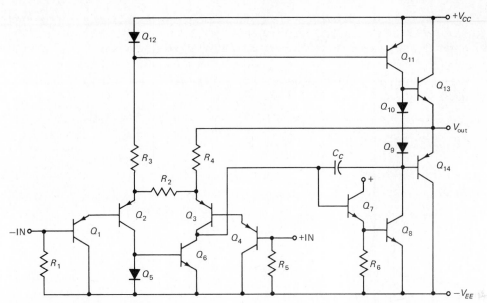

Fig. 15-23
Simplified schematic diagram for LM380 and similar audio IC's.

increase the signal strength before applying it to the cathode-ray tube. As another example, a television receiver uses a video amp to handle frequencies from near zero to about 4 MHz.

IC video amps have voltage gains and bandwidths that you can adjust by connecting different external resistors. For instance, the μA702 has a decibel voltage gain of 40 dB and a cutoff frequency of 5 MHz; by changing external components, you can get useful gain to 30 MHz. The MC1553 has a decibel voltage gain of 52 dB and a bandwidth of 20 MHz; these are adjustable by changing external components. The LM733 has a very wide bandwidth; it can be set up to give 20 dB gain and a bandwidth of 120 MHz.

RF AND IF AMPLIFIERS

A *radio-frequency* (RF) amplifier is usually the first stage in an AM, FM, or TV receiver. *Intermediate-frequency* (IF) amplifiers typically are the middle stages. ICs like the LM703 include RF and IF amplifiers on the same chip. The amplifiers are tuned (resonant) so that they amplify only a narrow band of frequencies. This allows the receiver to tune into a desired signal from a particular radio or television station. As mentioned earlier, it is impractical to integrate inductors and large capacitors on a chip. For this reason, you have to connect external L's and C's to the chip to get tuned amplifiers.

VOLTAGE REGULATORS

Chapter 3 discussed rectifiers and power supplies. After filtering, we have a dc voltage with ripple. This dc voltage is proportional to the line voltage, that is, it will change 10 percent if the line voltage changes 10 percent. In most applications, a 10 percent change in dc voltage is too much, and so *voltage regulation* is necessary. Typical of the new IC voltage regulators is the LM340 series. Chips of this type can hold the output dc voltage to within 0.01 percent for normal changes in line voltage and load resistance. Other features include positive or negative output, adjustable output voltage, and short-circuit protection. Chapter 19 will discuss IC voltage regulators in detail.

PROBLEMS

Straightforward

15-1. Inputs v_1 and v_2 in Fig. 15-24*a* are grounded. What is the dc emitter current in each transistor? The tail current? The dc voltage at the output?

Fig. 15-24

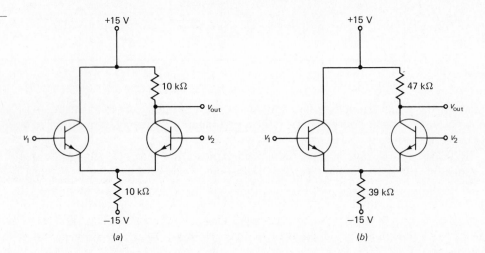

(a)   (b)

15-2. The base currents in a diff amp are 20 μA and 24 μA. What is the value of input offset current? Input bias current?

15-3. In Fig. 15-24*a*, the left transistor has a β_{dc} of 100 and the right transistor has a β_{dc} of 120. If inputs v_1 and v_2 are grounded, what are the dc base currents in each transistor? Calculate the input offset current and the input bias current.

15-4. Repeat Prob. 15-1 for Fig. 15-24*b*.

15-5. Repeat Prob. 15-3 for Fig. 15-24*b*.

15-6. A data sheet lists $I_{in(bias)} = 300$ nA and $I_{in(off)} = 100$ nA. What are the values of I_{B1} and I_{B2}?

15-7. In Fig. 15-24*a*, calculate the differential voltage gain and the input impedance for a β of 150.

15-8. What is the common-mode voltage gain of Fig. 15-24*a*? The CMRR expressed in decibels?

15-9. In Fig. 15-24*b*, calculate the following quantities: differential voltage gain, common-mode voltage gain, and CMRR in decibels.

15-10. Inputs v_1 and v_2 in Fig. 15-25 are grounded. What is the current through Q_3? The tail current? The r'_e of Q_1 and Q_2?

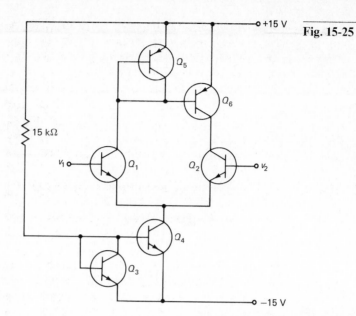

Fig. 15-25

15-11. In Fig. 15-25, the equivalent R_E looking into the collector of Q_4 is 100 kΩ, and the equivalent R_C looking into the collector of Q_6 is 200 kΩ. Calculate the differential voltage gain, the common-mode voltage gain, and the CMRR in decibels.

15-12. If a diff amp has a common-mode rejection ratio of 80 dB and a differential voltage gain of 200, how much output voltage do you get with a common-mode input voltage of 10 mV?

15-13. The amplifier of Fig. 15-26a has an r_{in} of 2 MΩ, an r_{out} of 75 Ω, and an A of 100,000. Calculate the approximate output voltage.

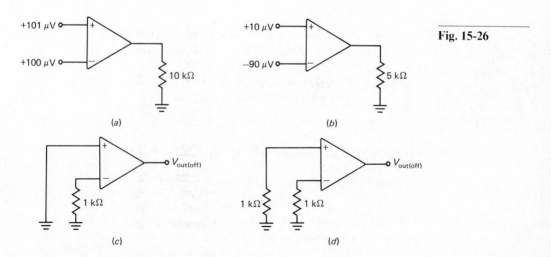

Fig. 15-26

15-14. In Fig. 15-26a, A' equals 92 dB and r_{out} is 75 Ω. How much output voltage is there?

15-15. If A is 100,000 and r_{out} is 75 Ω in Fig. 15-26b, what is the output voltage?

15-16. In Fig. 15-26c, the input base currents are 90 nA (noninverting) and 70 nA (inverting). If the V_{BE} values are the same, what does the input bias current equal? The input offset current? How much voltage is there at the inverting input? If $A = 100,000$, what does the output offset voltage equal?

15-17. In Fig. 15-26c, $v_{in(off)} = 0$, $I_{in(bias)} = 80$ nA, and $I_{in(off)} = 20$ nA. If $A = 100,000$, what is the maximum output offset voltage?

15-18. $I_{in(off)} = 20$ nA in Fig. 15-26d. If $v_{in(off)} = 0$, what is the differential input voltage? If $A = 100,000$, what does the output offset voltage equal?

15-19. Refer to Fig. 15-18 to answer these questions.
   a. What is the CMRR′ of a 741C at 100 kHz?
   b. In Fig. 15-18b, what is the ac output compliance when the load resistance is 500 Ω?
   c. What is the open-loop voltage gain of a 741C at 1 kHz?

15-20. A capacitor has a constant charging current of 1 mA. If the capacitance is 50 pF, what is the rate of voltage change with respect to time?

15-21. A 100-pF capacitor has a maximum charging current of 150 μA. What is the slew rate?

15-22. An op amp has a slew rate of 35 V/μs. How long will it take the output to change from 0 to 15 V?

15-23. The input stage of an op amp like Fig. 15-13 has $I_T = 100$ μA. If $C_C = 30$ pF, what is the slew rate?

15-24. An op amp has a slew rate of 2 V/μs. If the peak output is 12 V, what is the power bandwidth?

15-25. An op-amp data sheet gives a 15-V power bandwidth of 25 kHz. What is the slew rate?

15-26. If $V_{CC} = 15$ V, $V_{EE} = 15$ V, and V and $R_2 = 1$ MΩ in Fig. 15-13, what does I_T equal in the diff-amp stage?

15-27. A 318 has a slew rate of 50 V/μs. Refer to Fig. 15-22 to determine the power bandwidth for
   a. $V_P = 10$ V
   b. $V_P = 4$ V
   c. $V_P = 2$ V

Troubleshooting

15-28. Somebody builds the diff amp of Fig. 15-24a with floating or unconnected inputs. What does the output voltage equal? What does the circuit need in order to work properly?

15-29. If in the circuit of Fig. 15-24b, 3.9 kΩ is mistakenly used instead of 39 kΩ, what does the output voltage equal?

15-30. In Fig. 15-26c, the output is saturated (approximately equal to either supply voltage). The input bias current is 80 nA. Which of the following is the trouble?
   a. 1 kΩ shorted
   b. 100 kΩ mistakenly used for 1 kΩ
   c. 1 kΩ connected noninverting input instead of inverting input
   d. Both inputs accidentally shorted together

Design

15-31. Select an emitter resistor for Fig. 15-24a to get a differential voltage gain of approximately 200.

15-32. Design a diff amp like Fig. 15-24a to meet these specifications: $V_{CC} = 10$ V, $V_{EE} = -10$ V, $A = 100$ (minimum), and CMRR = 150 (minimum).

15-33. Redesign the diff amp of Fig. 15-25 to get a tail current of 100 μA.

15-34. The op amp of Fig. 15-26*d* has an input bias current of 10 nA and an input offset current of 1 nA. Redesign the circuit with the largest possible base return resistors to meet these specifications: V_{BE} values identical, $A = 100{,}000$, and output offset voltage to be less than ± 1 V.

Challenging

15-35. The dc resistance of each signal source in Fig. 15-27*a* is zero. What is r'_e of each transistor? If the output voltage is taken between the collectors, what is the differential voltage gain?

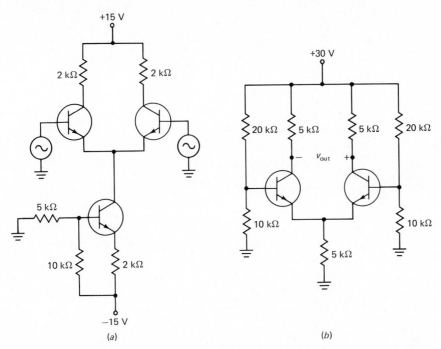

Fig. 15-27

(a)

(b)

15-36. Equation (15-1) applies to a diff amp with a double-ended input and double-ended output. Prove that the equation is true.

15-37. What is the dc emitter current in each transistor of Fig. 15-27*b*? The differential voltage gain?

Computer

15-38. A string variable has the symbol $ as a suffix. Examples are A$, B4$, and NAME$. Instead of representing a number, a string variable represents a group of letters, numbers, or other symbols. For instance, here is a program:

```
10 A$ = "JOE BLOW"
20 B$ = "1342 WEBSTER STREET"
30 C$ = "PALO ALTO, CA 94300"
40 PRINT A$: PRINT B$: PRINT C$
```

Line 10 sets A$ (pronounced A string) equal to JOE BLOW. Notice that quotation marks enclose JOE BLOW; this tells the computer that a string variable is involved.

Line 20 sets B$ equal to the address, and line 30 sets C$ equal to the city, state, and Zip code. Line 40 then prints the following on the screen:

JOE BLOW
1342 WEBSTER STREET
PALO ALTO, CA 94300

Rewrite the foregoing program to print your name, address, city, state, and Zip code.

15-39. Here is a program:

```
 10 A$ = "ENTER SUPPLY VOLTAGE "
 20 B$ = "VEE"
 30 C$ = "VCC"
 40 D$ = "ENTER RESISTANCE "
 50 E$ = "RE"
 60 F$ = "RC"
 70 PRINT A$;B$: INPUT VEE
 80 PRINT A$;C$: INPUT VCC
 90 PRINT D$;E$: INPUT RE
100 PRINT D$;F$: INPUT RC
110 IT = (VEE − 0.7)/RE
120 PRINT "THE TAIL CURRENT IS ";IT
130 V = VCC − (0.5*IT)*RC
140 PRINT "THE QUIESCENT OUTPUT VOLTAGE IS ";V
```

What does the computer screen show after the execution of line 70? Line 100? What kind of circuit does this program analyze?

15-40. Strings can be added. If A$ = "JOE " and B$ = "BLOW", then A$ + B$ = "JOE BLOW". Here is a program:

```
 10 A$ = "ENTER RESISTANCE "
 20 B$ = "RC"
 30 C$ = "RE"
 40 D$ = "R PRIME E"
 50 PRINT A$ + B$: INPUT RC
 60 PRINT A$ + C$: INPUT RE
 70 PRINT A$ + D$: INPUT RPE
 80 A = RC/2*RPE
 90 ACM = −RC/2*RE
100 CMRR = −A/ACM
110 PRINT A, ACM, CMRR
```

What does the computer screen show after the execution of line 50? Line 70? In line 110, what does the computer print out on the screen?

15-41. Write a program that prints out the power bandwidth for peak values of 1, 2, 3, . . . , 10 V. Use an INPUT statement to enter the slew rate and a FOR . . . NEXT statement to generate the required peak voltages.

Negative Feedback

Sometimes you have a great idea that others don't understand and some even ridicule. It happened to the Wright brothers, it happened to Marconi, and it happened to H. S. Black. When he tried to patent a negative-feedback amplifier in 1928, his idea was classified as another perpetual-motion folly. As it has turned out, negative feedback is one of the most valuable ideas ever discovered in electronics.

In a feedback amplifier, the output is sampled and returned to the input. This feedback signal can produce remarkable changes in circuit performance. Negative feedback means that the returning signal has a phase that opposes the input signal. The advantages of negative feedback are stable gain, less distortion, and more bandwidth.

16-1 NONINVERTING VOLTAGE FEEDBACK

The most basic type of negative feedback is *noninverting voltage feedback.* With this type of feedback, the input signal drives the noninverting input of an amplifier; a fraction of the output voltage is then sampled and fed back to the inverting input. An amplifier with noninverting voltage feedback tends to act like a perfect voltage amplifier, one with infinite input impedance, zero output impedance, and constant voltage gain.

ERROR VOLTAGE

Figure 16-1 shows a double-ended input amplifier. This amplifier is usually an op amp, but it may be a discrete amplifier with one or more stages. Notice that the output voltage is being sampled by a voltage divider. Therefore, the feedback voltage to the inverting input is proportional to the output voltage.

In a feedback amplifier, the difference between the noninverting and inverting input voltages is called the *error voltage.* In symbols,

$$v_{error} = v_1 - v_2$$

This error voltage is amplified to get an output voltage of

$$v_{out} = Av_{error}$$

Fig. 16-1
Noninverting
voltage feedback.

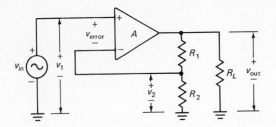

Typically, A is very large. To avoid saturation of the output transistors, therefore, v_{error} is kept very small. For instance, if the differential voltage gain is 100,000, an error voltage of only 0.1 mV produces an output voltage of 10 V.

STABLE VOLTAGE GAIN

In Fig. 16-1, the overall voltage gain is approximately constant, even though the differential voltage gain may change. Why? Suppose A increases for some reason (temperature change, op-amp replacement, or some other reason). Then the output voltage will try to increase. This means that more voltage is fed back to the inverting input, causing the error voltage to decrease. This almost completely offsets the attempted increase in output voltage.

A similar argument applies to a decrease in differential voltage gain. If A decreases, the output voltage tries to decrease. In turn, the feedback voltage decreases, causing v_{error} to increase. This almost completely offsets the attempted decrease in A.

Remember the key idea. When the input voltage is constant, an attempted change in output voltage is fed back to the input, producing an error voltage that automatically compensates for the attempted change.

MATHEMATICAL ANALYSIS

Most op amps have an extremely large voltage gain A, a very high input impedance r_{in}, and a very low output impedance r_{out}. For instance, the 741C has typical values of $A = 100,000$, $r_{in} = 2$ MΩ, and $r_{out} = 75$ Ω.

In Fig. 16-1, the voltage divider returns a sample of output voltage to the inverting input. If the voltage divider appears stiff to the inverting input, the feedback voltage is

$$v_2 = \frac{R_2}{R_1 + R_2} v_{out}$$

This is usually written as

$$v_2 = B v_{out}$$

where B is the fraction of output voltage fed back to the input. In symbols,

$$B \cong \frac{R_2}{R_1 + R_2} \tag{16-1}$$

This assumes that r_{in} is much greater than R_2, a condition usually satisfied in an op-amp circuit. The exact expression is

$$B = \frac{R_2 \parallel r_{in}}{R_1 + R_2 \parallel r_{in}}$$

The error voltage to the amplifier is

$$v_{error} = v_1 - v_2 \cong v_{in} - Bv_{out}$$

This is amplified to get an output voltage of approximately

$$v_{out} = Av_{error} = A(v_{in} - Bv_{out})$$

By rearranging,

$$\frac{v_{out}}{v_{in}} = \frac{A}{1 + AB} \tag{16-2}$$

where v_{out} = output voltage
$\quad v_{in}$ = input voltage
$\quad A$ = differential voltage gain
$\quad B$ = fraction of output voltage fed back to input

APPROXIMATE VOLTAGE GAIN

The product AB is called the *loop gain*. For noninverting voltage feedback to be effective, the designer must deliberately make the loop gain AB much greater than 1, so that Eq. (16-2) reduces to

$$\frac{v_{out}}{v_{in}} \cong \frac{1}{B} \tag{16-3}$$

Why is this result so important? Because it says that the overall voltage gain of the circuit equals the reciprocal of B, the feedback fraction. In other words, the gain is no longer dependent on the gain of the op amp, but rather depends on the feedback of the voltage divider. Since we can use precision resistors with tolerances of 1 percent for the voltage divider, B is a precise and stable value independent of the amplifier. Even with tolerances of 5 percent for the feedback resistors, the voltage gain is predictable to within 5 percent. Therefore, the voltage gain of a feedback amplifier becomes a rock-solid value approximately equal to the reciprocal of B.

SIMPLIFIED VIEWPOINT

Here is a simple way to remember Eq. (16-3). Because of the high differential voltage gain, the inverting input voltage is *bootstrapped* to within microvolts of the noninverting input. This means that

$$v_1 \cong v_2$$

equivalent to

$$v_{in} \cong Bv_{out}$$

or

$$\frac{v_{\text{out}}}{v_{\text{in}}} \cong \frac{1}{B}$$

This is the same result as before, derived with a lot less work.

OPEN-LOOP VOLTAGE GAIN

The *open-loop* voltage gain A_{OL} is the ratio $v_{\text{out}}/v_{\text{in}}$ with the feedback path opened, as shown in Fig. 16-2. When you open the feedback loop, the impedances on each terminal must not be disturbed. This is why the inverting input terminal is returned to ground through an equivalent resistance of

$$R_B = R_1 \parallel R_2$$

and why the output terminal is loaded by an equivalent resistance of

$$r_L = (R_1 + R_2) \parallel R_L$$

Usually, r_L is much greater than the output impedance of the amplifier, so that the open-loop voltage gain A_{OL} is approximately equal to the differential voltage gain A. With a 741C, this means that the open-loop voltage gain is typically 100,000.

CLOSED-LOOP VOLTAGE GAIN

The *closed-loop* voltage gain is the overall voltage gain when the feedback path is closed. Equation (16-2) is sometimes written as

$$A_{CL} = \frac{A_{OL}}{1 + A_{OL}B} \tag{16-4}$$

where A_{CL} = closed-loop voltage gain
A_{OL} = open-loop voltage gain $\cong A$
B = feedback fraction

In most feedback amplifiers, the loop gain $A_{OL}B$ is much greater than 1, and Eq. (16-4) simplifies to

$$A_{CL} \cong \frac{1}{B} \tag{16-5}$$

Fig. 16-2
Open loop.

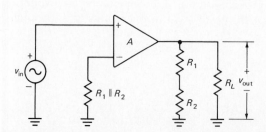

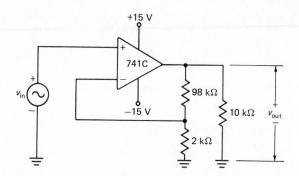

Fig. 16-3

Since $B \cong R_2/(R_1 + R_2)$, an alternative form is

$$A_{CL} \cong \frac{R_1 + R_2}{R_2}$$

which is often written as

$$A_{CL} \cong \frac{R_1}{R_2} + 1 \qquad\qquad (16\text{-}6)$$

EXAMPLE 16-1

If the 741C of Fig. 16-3 has an open-loop gain of 100,000, what is the closed-loop gain? If $v_{in} = 1$ mV, what do the output and error voltages equal?

SOLUTION

The gain of the voltage divider is

$$B \cong \frac{2 \text{ k}\Omega}{100 \text{ k}\Omega} = 0.02$$

The closed-loop gain is approximately

$$A_{CL} \cong \frac{1}{0.02} = 50$$

A more accurate gain is given by Eq. (16-4):

$$A_{CL} = \frac{100,000}{1 + 100,000(0.02)} = 49.975$$

Look how close this is to 50. The point is that $1/B$ is an accurate approximation for the closed-loop voltage of an amplifier that uses noninverting voltage feedback.

If $v_{in} = 1$ mV, the output voltage is

$$v_{out} = A_{CL}v_{in} \cong 50(1 \text{ mV}) = 50 \text{ mV}$$

The error voltage is

$$v_{error} \cong \frac{50 \text{ mV}}{100,000} = 0.5 \ \mu V$$

Notice how small the error voltage is. This is typical of op-amp feedback amplifiers because the open-loop voltage gain is quite high.

EXAMPLE 16-2

Suppose the 741C of Fig. 16-3 is replaced by another that has a voltage gain of only 20,000 (worst-case value on the data sheet). Recalculate the value of A_{CL}. Also, what are the new values of v_{out} and v_{error} for $v_{in} = 1$ mV?

SOLUTION

With Eq. (16-4),

$$A_{CL} = \frac{20,000}{1 + 20,000(0.02)} = 49.875$$

The closed-loop gain is still extremely close to 50, despite the huge drop in open-loop gain. Can you see why negative feedback is useful? Without negative feedback, the overall voltage gain would drop from 100,000 to 20,000, a decrease of 80 percent. Using negative feedback, we have less voltage gain, but in return we get a fabulously stable closed-loop voltage gain. In this example, the closed-loop voltage gain decreases from 49.975 to 49.875, a decrease of only 0.2 percent. Therefore, the closed-loop voltage gain is nearly independent of the op-amp voltage gain.

Since A_{CL} is approximately 50, v_{out} is approximately 50 mV, but the error voltage changes to

$$v_{error} = \frac{50 \text{ mV}}{20,000} = 2.5 \ \mu V$$

Compared with the preceding example, the error voltage has increased from 0.5 to 2.5 μV.

Do you understand what has happened? When the open-loop gain drops by a factor of 5, the error voltage increases by a factor of 5. Therefore the output voltage remains at approximately 50 mV. This echoes our earlier explanation of negative feedback. Attempted changes in output voltage are fed back to the input, producing an error voltage that automatically compensates for the output change.

16-2 OTHER EFFECTS OF NONINVERTING VOLTAGE FEEDBACK

Stable voltage gain is not the only benefit of noninverting voltage feedback. It also improves input impedance, output impedance, nonlinear distortion, and output offset voltage. This section discusses all these improvements.

INPUT IMPEDANCE

Figure 16-4 shows an amplifier with noninverting voltage feedback. The op amp has an open-loop input impedance of approximately r_{in}. The overall amplifier has a closed-loop input impedance of $r_{in(CL)}$. The closed-loop impedance $r_{in(CL)}$ is larger than the open-loop impedance r_{in}.

How much larger is the closed-loop input impedance? To find out, we have to derive an expression for v_{in}/i_{in}. In Fig. 16-4,

$$v_{in} = v_{error} + Bv_{out}$$

or

$$v_{in} = v_{error} + ABv_{error} = (1 + AB)v_{error}$$

Because $v_{error} = i_{in}r_{in}$,

$$v_{in} = (1 + AB)i_{in}r_{in}$$

or

$$\frac{v_{in}}{i_{in}} = (1 + AB)r_{in}$$

The ratio v_{in}/i_{in} is the input impedance seen by the source. Therefore, we can write

$$r_{in(CL)} = (1 + AB)r_{in} \qquad (16\text{-}7)$$

where $r_{in(CL)}$ = closed-loop input impedance
   r_{in} = open-loop input impedance
   AB = loop gain

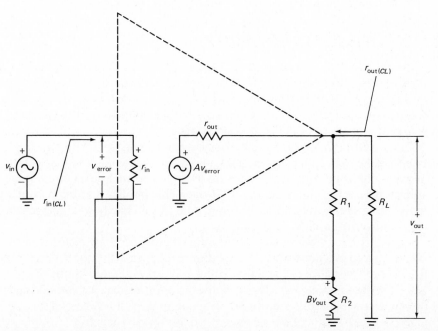

Fig. 16-4
Deriving input and output impedances.

In most feedback amplifiers, AB is much larger than 1, which means that $r_{in(CL)}$ is much larger than r_{in}.

The use of noninverting voltage feedback with op amps leads to input impedances that approach infinity. This means that an op amp using noninverting voltage feedback approximates an ideal voltage amplifier.

OUTPUT IMPEDANCE

The op amp of Fig. 16-4 has an open-loop output impedance of r_{out}. The overall amplifier, however, has a closed-loop impedance of $r_{out(CL)}$. The closed-loop output impedance is lower than the open-loop output impedance. Why? In Fig. 16-4, the op-amp output is equivalent to a Thevenin voltage of Av_{error} and an output impedance of r_{out}. If R_L decreases, more output current flows, producing a larger internal voltage drop across r_{out}. This implies that v_{out} will try to decrease. Since less voltage is fed back to the input, v_{error} increases. This produces a larger Thevenin output voltage, which almost completely offsets the additional voltage drop across r_{out}. The effect is equivalent to decreasing the output impedance of the feedback amplifier.

Appendix 1 proves the following:

$$r_{out(CL)} = \frac{r_{out}}{1 + AB} \tag{16-8}$$

where $r_{out(CL)}$ = closed-loop output impedance
r_{out} = open-loop output impedance
AB = loop gain

When the loop gain is much greater than unity, $r_{out(CL)}$ is much smaller than r_{out}. In fact, noninverting voltage feedback with op amps results in output impedances that approach zero, the ideal case for a voltage amplifier.

NONLINEAR DISTORTION

The final stage of an op amp has nonlinear distortion when the signal swings over most of the ac load line. Large swings in current cause the r'_e of a transistor to change during the cycle. To state this another way, the open-loop voltage gain varies throughout the cycle when a large signal is being amplified. It is this changing voltage gain that is a source of the nonlinear distortion.

Noninverting voltage feedback reduces nonlinear distortion because the feedback stabilizes the closed-loop voltage gain, making it almost independent of the changes in open-loop voltage gain. As long as the loop gain is much greater than 1, the output voltage equals $1/B$ times the input voltage. This implies that the output will be a more faithful reproduction of the input. And this is exactly what happens when we use noninverting voltage feedback.

Figure 16-5 can give you a better understanding of how distortion is reduced. Under large-signal conditions, the op amp produces a distortion voltage, designated v_{dist}. We can visualize v_{dist} as a new voltage source in series with the original source Av_{error}. Without negative feedback, all the distortion voltage v_{dist} would appear at the output. But with negative feedback, a fraction of the distortion voltage is fed back to

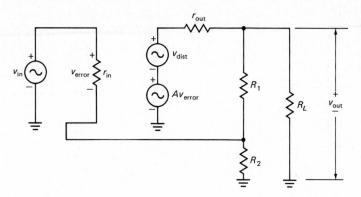

Fig. 16-5
Internally generated nonlinear distortion is reduced.

the inverting input. This is amplified and arrives at the output with inverted phase, almost completely canceling the original distortion produced by the output stage.

How much improvement is there? In Fig. 16-5, the output voltage is

$$v_{out} = Av_{error} + v_{dist} = A(v_{in} - Bv_{out}) + v_{dist}$$

Solving for v_{out} gives

$$v_{out} = \frac{A}{1 + AB} \, v_{in} + \frac{v_{dist}}{1 + AB}$$

The first term is what we want because it represents the amplified input voltage. The second term is the distortion that appears at the final output. This can be written as

$$v_{dist(CL)} = \frac{v_{dist}}{1 + AB} \qquad (16\text{-}9)$$

where $v_{dist(CL)}$ = closed-loop distortion voltage
$\quad\;\; v_{dist}$ = open-loop distortion voltage
$\quad\;\; AB$ = loop gain

When the loop gain is much greater than 1, the closed-loop distortion is much smaller than the open-loop distortion. Again, noninverting voltage feedback produces a real improvement in the quality of the amplifier.

REDUCED OUTPUT OFFSET VOLTAGE

In Chap. 15, we saw that an output offset voltage can exist even though the input voltage is zero. There are three causes of this unwanted output offset voltage: input offset voltage, input bias current, and input offset current. Figure 16-6 shows a feedback amplifier with an output offset voltage in series with the original source Av_{error}. This new voltage source represents the output offset voltage without feedback. The actual output offset voltage with negative feedback is smaller.

Why? The reasoning is similar to that given for distortion. Some of the output offset voltage is fed back to the inverting input. After amplification, an out-of-phase voltage arrives at the output, canceling most of the original output offset voltage. By a

Fig. 16-6
*Output offset
voltage is reduced.*

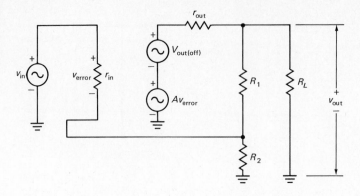

derivation identical to that given for distortion,

$$V_{oo(CL)} = \frac{V_{\text{out(off)}}}{1 + AB} \qquad (16\text{-}10)$$

where $V_{oo(CL)}$ = closed-loop output offset voltage
$V_{\text{out(off)}}$ = open-loop output offset voltage
AB = loop gain

When the loop gain is much greater than 1, the closed-loop output offset voltage is much smaller than the open-loop output offset voltage.

DESENSITIVITY

The closed-loop voltage gain with noninverting voltage feedback is

$$A_{CL} = \frac{A}{1 + AB} \qquad (16\text{-}11)$$

The quantity $1 + AB$ is called the *desensitivity* of a feedback amplifier because it indicates how much the voltage gain is reduced by the negative feedback. For instance, if $A = 100,000$ and $B = 0.02$, then

$$1 + AB = 1 + 100,000(0.02) = 2001$$

The desensitivity is 2001, meaning the voltage gain is reduced by a factor of 2001:

$$A_{CL} = \frac{100,000}{2001} = 50$$

We can rearrange Eq. (16-11) to get

$$1 + AB = \frac{A}{A_{CL}} \qquad (16\text{-}12)$$

This says that desensitivity equals the ratio of open-loop voltage gain to closed-loop voltage gain. For instance, if $A = 100,000$ and $A_{CL} = 250$, the desensitivity is

$$1 + AB = \frac{100,000}{250} = 400$$

Table 16-1. Noninverting Voltage Feedback

Quantity	Symbol	Effect	Formula
Voltage gain	A_{CL}	Decreases	$1/B$
Input impedance	$r_{in(CL)}$	Increases	$(1 + AB)r_{in}$
Output impedance	$r_{out(CL)}$	Decreases	$r_{out}/(1 + AB)$
Distortion	$v_{dist(CL)}$	Decreases	$v_{dist}/(1 + AB)$
Output offset	$V_{oo(CL)}$	Decreases	$V_{out(off)}/(1 + AB)$

Equation (16-12) is convenient to use when you know the values of A and A_{CL}, but not B.

Table 16-1 summarizes the effects of noninverting voltage feedback. As you can see, the desensitivity appears in most of the formulas. This is why it is important to remember how to calculate desensitivity. You can use either $1 + AB$ or A/A_{CL}.

EXAMPLE 16-3

Figure 16-7 shows a 741C with pin numbers. If $A = 100,000$, $r_{in} = 2$ MΩ, and $r_{out} = 75$ Ω, what are the closed-loop input and output impedances?

SOLUTION

The feedback fraction is

$$B = \frac{100}{100,100} = 0.000999$$

and the desensitivity is

$$1 + AB = 1 + 100,000(0.000999) = 101$$

The closed-loop input impedance is

$$r_{in(CL)} = 101(2 \text{ M}\Omega) = 202 \text{ M}\Omega$$

and the closed-loop output impedance is

$$r_{out(CL)} = \frac{75 \text{ }\Omega}{101} = 0.743 \text{ }\Omega$$

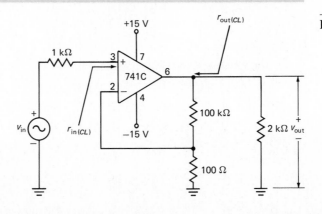

Fig. 16-7

EXAMPLE 16-4

The 741C of Fig. 16-7 has $I_{in(bias)} = 80$ nA, $I_{in(off)} = 20$ nA, and $v_{in(off)} = 2$ mV. What is the output offset voltage?

SOLUTION

The noninverting input sees a Thevenin resistance of 1 kΩ, and the inverting input sees a Thevenin resistance of approximately 100 Ω. (The exact resistance is 100 Ω in parallel with 100 kΩ.) We are given an input bias current of 80 nA and an input offset current of 20 nA. From Eqs. (15-8) and (15-9), the two possible solutions for the input bias currents are

$$I_{B1} = 90 \text{ nA}, I_{B2} = 70 \text{ nA}$$
$$I_{B1} = 70 \text{ nA}, I_{B2} = 90 \text{ nA}$$

The worst combination is 90 nA through the 1 kΩ and 70 nA through the 100 Ω, because this produces more output offset voltage.

From Eq. (15-19), the maximum input offset is

$$v_1 - v_2 = 2 \text{ mV} + (90 \text{ nA})(1 \text{ k}\Omega) - (70 \text{ nA})(100 \text{ }\Omega)$$
$$= 2 \text{ mV} + 90 \text{ }\mu\text{V} - 7 \text{ }\mu\text{V} = 2.08 \text{ mV}$$

From the preceding example, we know that the desensitivity is 101. With Eq. (16-10), we can calculate the closed-loop output offset voltage:

$$V_{oo(CL)} = \frac{100,000(2.08 \text{ mV})}{101} = 2.06 \text{ V}$$

This means that we have lost 2.06 V of the output range because of offset voltages and currents.

We can reduce the closed-loop output offset voltage in three ways. First, we can decrease the closed-loop voltage gain to 100 (done by changing feedback resistors). Then the desensitivity increases to

$$1 + AB = \frac{A}{A_{CL}} = \frac{100,000}{100} = 1000$$

and the closed-loop output offset voltage drops to approximately

$$V_{oo(CL)} = \frac{100,000 \ (2.08 \text{ mV})}{1000} = 0.208 \text{ V}$$

The second option is to upgrade to a better op amp, like the LM11C (see Table 15-1). It has a typical input offset voltage of 0.1 mV, an input bias current of 25 pA, and an input offset current of 0.5 pA. Because the LM11C has much smaller input offsets, the output offset voltage would be lower.

The third alternative is described on the data sheet of a 741C. It involves connecting a 10-kΩ potentiometer between pins 1 and 5 with the wiper tied to the negative supply voltage, as shown in Fig. 16-8. By adjusting this potentiometer, we can null or zero the output offset voltage.

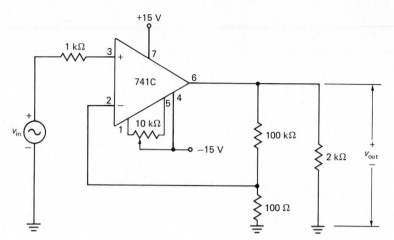

Fig. 16-8
Nulling the output offset voltage.

Without negative feedback the typical op amp saturates immediately because the input offset voltage multiplied by the open-loop gain drives the output stage into saturation. This is important to remember. Monolithic op amps are intended to be used with some form of feedback. Running wide open, they have too much voltage gain to be of any use because they hang up in the saturation region.

A final point: The basic equation

$$V_{oo(CL)} = \frac{V_{out(off)}}{1 + AB}$$

can be rewritten as

$$V_{oo(CL)} = A_{CL} V_{in(off)}$$

where $V_{in(off)}$ includes all input offset voltages, including those caused by input bias and offset currents.

EXAMPLE 16-5

Figure 16-9 shows a circuit called a *voltage follower*. What is its closed-loop voltage gain? Its closed-loop input and output impedances? The output offset voltage? Use typical 741C parameters: $A_{OL} = 100{,}000$, $r_{in} = 2$ MΩ, $r_{out} = 75$ Ω, $V_{in(off)} = 2$ mV, $I_{in(bias)} = 80$ nA, and $I_{in(off)} = 20$ nA.

SOLUTION

All the output voltage is fed back to the inverting input because R_1 is zero and R_2 is infinite. Therefore, $B = 1$ and $A_{CL} = 1$. This is massive negative feedback, the most you can have. In this case, the closed-loop voltage gain equals 1 to a very close approximation.

Fig. 16-9
Voltage follower.

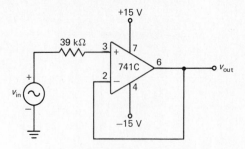

The closed-loop impedance of a voltage follower is

$$r_{in(CL)} = (1 + AB)r_{in} = (1 + A)r_{in}$$
$$\cong Ar_{in} = 100{,}000(2 \text{ M}\Omega) = 2(10^{11}) \text{ }\Omega$$

Likewise,

$$r_{out(CL)} = \frac{r_{out}}{1 + AB} = \frac{r_{out}}{1 + A}$$
$$\cong \frac{r_{out}}{A} = \frac{75 \text{ }\Omega}{100{,}000} = 0.00075 \text{ }\Omega$$

As you can see, $r_{in(CL)}$ approaches infinity and $r_{out(CL)}$ approaches zero. A voltage follower is a great buffer amplifier because of its high input impedance, low output impedance, and unity voltage gain.

Since the closed-loop gain is 1, the desensitivity equals

$$1 + AB = \frac{A}{A_{CL}} = A = 100{,}000$$

In the worst case, the maximum input offset voltage is

$$v_1 - v_2 = 2 \text{ mV} + (90 \text{ nA})(39 \text{ k}\Omega) = 5.51 \text{ mV}$$

So the output offset voltage is

$$V_{oo(CL)} = \frac{100{,}000(5.51 \text{ mV})}{100{,}000} = 5.51 \text{ mV}$$

In other words, the voltage follower is almost immune to offset problems; because it has only unity voltage gain, the output offset voltage can be no more than the input offset voltage.

16-3 NONINVERTING CURRENT FEEDBACK

With *noninverting current feedback,* an input voltage drives the noninverting input of an amplifier, and the output current is sampled to get the feedback voltage. An amplifier with noninverting current feedback tends to act like a perfect voltage-to-current converter, one with infinite input impedance, infinite output impedance, and stable *transconductance.*

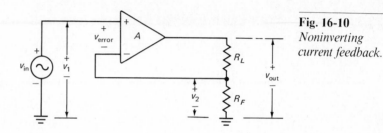

Fig. 16-10
Noninverting current feedback.

AC EQUIVALENT CIRCUIT

Figure 16-10 shows the ac equivalent circuit for a feedback amplifier with noninverting current feedback. The load resistor and feedback resistor are in series. Because of this, the load current passes through the feedback resistor. The feedback voltage is proportional to load current because

$$v_2 = i_{out}R_F$$

Whenever the feedback voltage is proportional to output current, the circuit has current feedback.

STABLE OUTPUT CURRENT

Current feedback stabilizes the output current. This means that a constant input voltage produces an almost constant output current, despite changes in open-loop gain and load resistance. For instance, suppose the open-loop voltage gain decreases. Then the output current tries to decrease. This results in less feedback voltage and more error voltage. The increased error voltage almost completely offsets the decrease in open-loop voltage gain, so that the output current remains almost constant.

A similar argument applies to an increase in open-loop gain. Attempted increases in output current are almost eliminated by the negative feedback.

MATHEMATICAL ANALYSIS

The feedback amplifier of Fig. 16-10 has a closed-loop voltage gain of

$$A_{CL} = \frac{A}{1 + AB}$$

where

$$B = \frac{R_F}{R_L + R_F}$$

The output current is

$$i_{out} = \frac{v_{out}}{R_L + R_F} = \frac{A_{CL}v_{in}}{R_L + R_F}$$

which can be rearranged as

$$\frac{i_{\text{out}}}{v_{\text{in}}} = \frac{A_{CL}}{R_L + R_F} \tag{16-13}$$

When the loop gain is high, A_{CL} approximately equals $1/B$ and Eq. (16-13) gives

$$\frac{i_{\text{out}}}{v_{\text{in}}} \cong \frac{(R_L + R_F)/R_F}{R_L + R_F}$$

or

$$\frac{i_{\text{out}}}{v_{\text{in}}} \cong \frac{1}{R_F} \tag{16-14}$$

This says that the ratio of output current to input voltage equals the reciprocal of R_F. Since R_F is an external resistor, $i_{\text{out}}/v_{\text{in}}$ has a constant value independent of the open-loop voltage gain and load resistance.

TRANSCONDUCTANCE

A feedback amplifier with noninverting current feedback is often called a *transconductance* amplifier, and Eq. (16-14) is written as

$$g_m \cong \frac{1}{R_F} \tag{16-15}$$

where g_m = transconductance, $i_{\text{out}}/v_{\text{in}}$
R_F = current-feedback resistor

The circuit of Fig. 16-10 is also called a *voltage-to-current converter* because an input voltage controls an output current. If we rearrange Eq. (16-14), the output current is

$$i_{\text{out}} \cong \frac{v_{\text{in}}}{R_F} \tag{16-16}$$

OTHER BENEFITS

Compare the circuit of Fig. 16-10 with that of Fig. 16-1. The only difference is the location of the load resistor. In Fig. 16-10, the load resistor is floating to allow output current to pass through the feedback resistor. In Fig. 16-1, the load resistor is connected between the output and ground. Because the circuits are similar except for the location of the load resistor, the negative feedback again reduces distortion and output offset voltage. Also, the input impedance approaches infinity.

The one quantity that is different with noninverting current feedback is the closed-loop output impedance. Since the load is no longer grounded but is part of the feedback circuit, it sees a different Thevenin output impedance from before. It can be shown that

$$r_{\text{out(CL)}} = (1 + A)R_F \tag{16-17}$$

where $r_{\text{out(CL)}}$ = closed-loop output impedance
A = open-loop voltage gain
R_F = feedback resistor

Table 16-2. Noninverting Current Feedback

Quantity	Symbol	Effect	Formula
Transconductance	i_{out}/v_{in}	Stabilizes	$1/R_F$
Input impedance	$r_{in(CL)}$	Increases	$(1 + AB)r_{in}$
Output impedance	$r_{out(CL)}$	Increases	$(1 + A)R_F$
Distortion	$v_{dist(CL)}$	Decreases	$v_{dist}/(1 + AB)$
Output offset	$V_{oo(CL)}$	Decreases	$V_{out(off)}/(1 + AB)$

Since A is very large, $r_{out(CL)}$ approaches infinity. From now on, remember that voltage feedback produces a low output impedance, while current feedback produces a high output impedance.

Table 16-2 summarizes noninverting current feedback. As indicated, transconductance is stabilized, input impedance increases, output impedance increases, and so on.

SIMPLIFIED VIEWPOINT

There is a simple way of looking at the circuit of Fig. 16-10. Since the inverting input voltage is bootstrapped to within microvolts of the noninverting input voltage,

$$v_2 \cong v_{in}$$

Therefore,

$$i_{out} = \frac{v_2}{R_F} \cong \frac{v_{in}}{R_F}$$

This is the same result we got earlier with formal mathematics.

EXAMPLE 16-6

Figure 16-11 shows a sensitive dc voltmeter. The LF355 is a BIFET op amp with the following typical data taken from Table 15-1: $V_{in(off)} = 3$ mV, $I_{in(bias)} = 0.03$ nA, and $I_{in(off)} = 0.003$ nA. How much input voltage does it take to get full-scale deflection of the ammeter?

SOLUTION

Equation (16-16) can be rewritten as

$$v_{in} \cong i_{out}R_F$$

For the switch position shown, $R_F = 10\ \Omega$. The ammeter has a full-scale deflection current of 100 μA. Therefore, the input voltage that produces full-scale deflection is

$$v_{in} = (100\ \mu A)(10\ \Omega) = 1\ \text{mV}$$

Fig. 16-11
*A BIFET volt-
meter.*

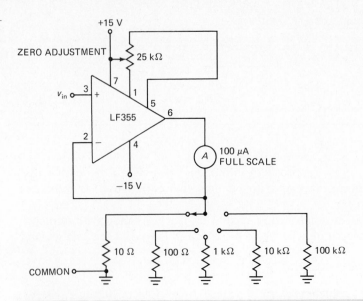

If we switch to a feedback resistor of 100 Ω, the input voltage that produces full-scale deflection is

$$v_{in} = (100\ \mu A)(100\ \Omega) = 10\ mV$$

Now it takes 10 mV to get a full-scale reading on the ammeter.

Similarly, 1 kΩ results in an input voltage of 100 mV, 10 kΩ in 1 V, and 100 kΩ in 10 V. With precision resistors, the circuit is an accurate dc voltmeter with ranges of 1 mV, 10 mV, 100 mV, 1 V, and 10 V. In production, the meter face would be labeled in volts rather than microamperes.

The data sheet of an LF355 indicates that a 25-kΩ potentiometer can be connected as shown for offset nulling. Before taking a reading, you short the input leads and zero the output reading. This eliminates the output offset voltage.

With a BIFET op amp, the input bias and offset currents are so small that they produce negligible input offsets with a circuit resistance up to approximately 1 MΩ. In other words, when you connect the voltmeter to the circuit being tested, the Thevenin resistance of that circuit provides the return path to ground for the noninverting bias current. If this Thevenin resistance is less than 1 MΩ, the 0.03 nA of input bias current produces an input offset voltage of less than 0.03 mV. This introduces almost no error in the output reading.

16-4 INVERTING VOLTAGE FEEDBACK

Figure 16-12*a* shows an amplifier with the noninverting input grounded. The input signal drives the inverting input, and the output voltage is sampled. This produces *inverting voltage feedback*. An amplifier with inverting voltage feedback tends to act like a perfect current-to-voltage converter, a device with zero input impedance, zero output impedance, and a constant v_{out}/i_{in} ratio.

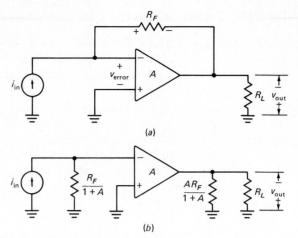

Fig. 16-12
Inverting voltage feedback.

MATHEMATICAL ANALYSIS

With the input signal driving the inverting input, the polarity of output voltage is reversed, as shown in Fig. 16-12a. This output voltage is given by

$$v_{\text{out}} = Av_{\text{error}}$$

To avoid excessive output offset voltage, feedback resistor R_F is usually less than 100 kΩ. Since the input resistance of the typical op amp is in megohms, almost all the input current flows through R_F. Because of this, we can sum voltages around the circuit to get

$$v_{\text{out}} - i_{\text{in}}R_F + v_{\text{error}} = 0$$

or

$$v_{\text{out}} - i_{\text{in}}R_F + \frac{v_{\text{out}}}{A} = 0$$

Now we can rearrange this as

$$\frac{v_{\text{out}}}{i_{\text{in}}} = \frac{AR_F}{1 + A} \tag{16-18}$$

When the open-loop gain is much greater than 1, the foregoing simplifies to

$$\frac{v_{\text{out}}}{i_{\text{in}}} = R_F \tag{16-19}$$

or

$$v_{\text{out}} = i_{\text{in}}R_F \tag{16-20}$$

In Eq. (16-19), the ratio of v_{out} to i_{in} is sometimes referred to as the *transresistance* because it involves a resistance connected between the input and the output. A feedback amplifier with inverting voltage feedback is sometimes called a transresistance amplifier. A circuit like this is also called a *current-to-voltage converter* because an input current controls an output voltage.

INPUT IMPEDANCE

The closed-loop impedance in Fig. 16-12a is the ratio of v_{error} to i_{in}. Because the error voltage approaches zero, the closed-loop input impedance also approaches zero. Appendix 1 derives this formula for input impedance:

$$r_{in(CL)} = \frac{R_F}{1 + A} \tag{16-21}$$

where $r_{in(CL)}$ = closed-loop input impedance
$\quad\quad R_F$ = feedback resistance
$\quad\quad A$ = open-loop voltage gain

In most current-to-voltage converters, R_F is less than 100 kΩ and A is more than 100,000. Therefore, $r_{in(CL)}$ approaches zero.

MILLER'S THEOREM FOR FEEDBACK RESISTOR

Chapter 14 discussed Miller's theorem. The idea was to split the feedback capacitor of an inverting amplifier into an input Miller capacitance and an output Miller capacitance. There is also a Miller's theorem for a feedback resistor. The input Miller resistance is $R_F/(1 + A)$ and the output Miller resistance is $AR_F/(1 + A)$. Whenever you see a feedback resistor between the input and output of an inverting amplifier (Fig. 16-12a), you can split it into an input Miller resistance and an output Miller resistance (Fig. 16-12b). Because A is extremely high, the input Miller resistance approaches zero and the output Miller resistance approaches R_F.

IDEAS TO REMEMBER

We have seen enough feedback amplifiers to state the following ideas. Driving the noninverting input results in a high input impedance, while driving the inverting input produces low input impedance. Furthermore, voltage feedback stabilizes the output voltage (low output impedance), while current feedback stabilizes the output current (high output impedance). Finally, all types of negative feedback reduce nonlinear distortion and output offset voltage. If you remember these ideas, it will help you keep things straight in your mind when you are troubleshooting and analyzing negative-feedback amplifiers.

OTHER BENEFITS

We can redraw the circuit as shown in Fig. 16-13. Now it resembles the noninverting-voltage-feedback amplifier, except that a current source is driving the inverting terminal. Since the current source ideally has infinite impedance, the feedback fraction B is approximately equal to unity, and the circuit has maximum negative feedback. With derivations identical to those given earlier, we can get formulas for closed-loop output impedance, nonlinear distortion, and output offset voltage. Table 16-3 summarizes inverting voltage feedback.

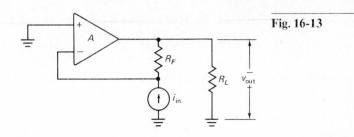

Fig. 16-13

VIRTUAL GROUND

There is a shortcut that you can use when you are analyzing a current-to-voltage converter like that of Fig. 16-14. Ideally, inverting voltage feedback implies that

1. The error voltage is zero.
2. The current into the op amp is zero.

The high open-loop voltage gain is the basis for the first point. The high input resistance of an op amp is the reason for the second point.

These two key ideas are summarized by the concept of *virtual ground,* which is any point in a circuit that has zero voltage and draws no current. The inverting input in Fig. 16-14 is a virtual ground because it acts like ground as far as voltage, but not current, is concerned. An ordinary ground has zero voltage and can sink infinite current. A virtual ground has zero voltage and zero current.

With this in mind, look at Fig. 16-14 and notice the following. Because the virtual ground draws no current, all the i_{in} has to pass through R_F. The voltage across R_F therefore equals $i_{in}R_F$. Since the virtual ground has approximately zero voltage to ground, Kirchhoff's voltage law tells us that the output voltage equals the voltage across R_F:

$$v_{out} = i_{in}R_F$$

This is the same formula we derived earlier, but look at how much easier it was this time around.

VIRTUAL GROUND MAY HAVE DC POTENTIAL

In many circuits the noninverting input is grounded for dc as well as ac signals. Therefore, the virtual ground has a dc potential of zero with respect to ground. But

Table 16-3. Inverting Voltage Feedback

Quantity	Symbol	Effect	Formula
Transresistance	v_{out}/i_{in}	Stabilizes	R_F
Input impedance	$r_{in(CL)}$	Decreases	$R_F/(1 + A)$
Output impedance	$r_{out(CL)}$	Decreases	$r_{out}/(1 + A)$
Distortion	$v_{dist(CL)}$	Decreases	$v_{dist}/(1 + A)$
Output offset	$V_{oo(CL)}$	Decreases	$V_{out(off)}/(1 + A)$

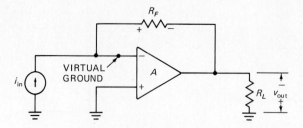

Fig. 16-14 *Virtual ground means inverting input of op amp has zero signal voltage and current.*

don't get the idea that a virtual ground is always at dc ground potential. There are circuits in which the noninverting input is biased at a positive or negative dc level. A bypass capacitor is then used to ac ground the noninverting input. In this case, the virtual ground has a dc potential. But it is still a virtual ground to ac signals, meaning that it has zero ac voltage and draws no ac current.

AN ELECTRONIC AMMETER

With inverting voltage feedback, we can build an electronic ammeter whose input resistance approaches zero. In Fig. 16-15, the typical open-loop voltage gain of a 741C is 100,000. From Eq. (16-21),

$$r_{in(CL)} = \frac{100 \text{ k}\Omega}{100,000} = 1 \ \Omega$$

and

$$v_{out} = (100 \text{ k}\Omega)i_{in}$$

The first equation tells us that we will add only 1 Ω of resistance to the branch whose current is being measured. The second equation tells us how sensitive the ammeter is. If i_{in} equals 50 μA, v_{out} equals

$$v_{out} = (100 \text{ k}\Omega)(50 \ \mu\text{A}) = 5 \text{ V}$$

This is enough voltage to measure easily with an inexpensive voltmeter. Therefore, even though it uses inexpensive parts, the electronic ammeter of Fig. 16-15 is superior to a moving-coil ammeter as far as input impedance is concerned.

Fig. 16-15 *Electronic ammeter.*

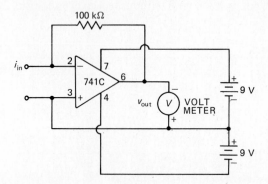

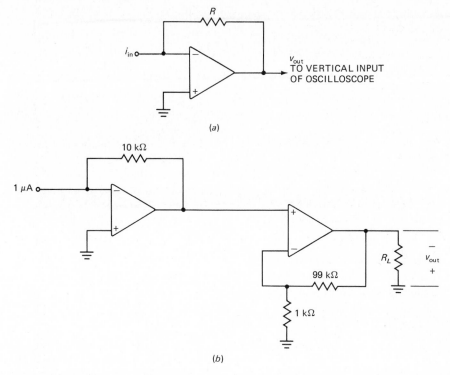

Fig. 16-16 (a) *Using oscilloscope to measure current.* (b) *Current-to-voltage converter drives voltage amplifier.*

EXAMPLE 16-7

Suppose we have an oscilloscope with an input sensitivity of 10 mV/cm. By connecting the current-to-voltage converter of Fig. 16-16a to the vertical input, we can measure current. If we want 1 μA of input current to produce 1 cm of vertical deflection, what value should R have?

SOLUTION

To get 1 cm of deflection, 1 μA must produce 10 mV. From Eq. (16-19),

$$R_F = \frac{10 \text{ mV}}{1 \text{ } \mu\text{A}} = 10 \text{ k}\Omega$$

Therefore, a 10-kΩ resistor in Fig. 16-16a allows us to measure microampere currents with an oscilloscope.

EXAMPLE 16-8

If 1 μA of input current drives the system of Fig. 16-16b, what is the value of the output voltage?

SOLUTION

The first stage is a current-to-voltage converter with an output voltage of

$$v_{out(1)} = (10 \text{ k}\Omega)(1 \text{ }\mu\text{A}) = 10 \text{ mV}$$

The second stage is a voltage amplifier with a gain of 100. Therefore, the final output voltage is

$$v_{out} = 100(10 \text{ mV}) = 1 \text{ V}$$

16-5 INVERTING CURRENT FEEDBACK

In Fig. 16-17, the input signal drives the inverting input, and the output current is sampled. This produces *inverting current feedback*. An amplifier with inverting current feedback tends to act like a perfect current amplifier, one that has zero input impedance, infinite output impedance, and a constant current gain.

With the input signal driving the inverting input, the polarity of output voltage is reversed, as shown in Fig. 16-17. As a result, conventional flow is upward through R_L (electron flow is downward). Appendix 1 proves that the current gain is:

$$\frac{i_{out}}{i_{in}} = \frac{R_1}{R_2} + 1 \tag{16-22}$$

Equation (16-22) says that the current gain ideally depends on the ratio of two resistances, both of which can be of precision tolerance. An amplifier with inverting current feedback is called a *current amplifier* because the current gain has been stabilized.

Inverting current feedback decreases the input impedance, increases the output impedance, decreases distortion, and decreases output offset voltage. With this type of negative feedback,

$$B = \frac{R_2}{R_1 + R_2} \tag{16-23}$$

By derivations similar to those given earlier, we can derive the formulas listed in Table 16-4.

Fig. 16-17
Inverting current feedback.

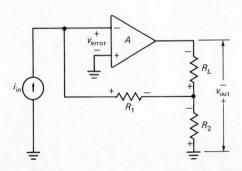

Table 16-4. Inverting Current Feedback

Quantity	Symbol	Effect	Formula
Current gain	i_{out}/i_{in}	Stabilizes	$1/B$
Input impedance	$r_{in(CL)}$	Decreases	$R_1/(1 + AB)$
Output impedance	$r_{out(CL)}$	Increases	$(1 + A)R_2$
Distortion	$v_{dist(CL)}$	Decreases	$v_{dist}/(1 + AB)$
Output offset	$V_{oo(CL)}$	Decreases	$V_{out(off)}/(1 + AB)$

EXAMPLE 16-9

Figure 16-18 shows a sensitive dc ammeter using inverting current feedback. How much input current do we need to get full-scale deflection of the output meter?

SOLUTION

With Eq. (16-22), the current gain is

$$\frac{i_{out}}{i_{in}} = \frac{990\ \Omega}{10\ \Omega} + 1 = 100$$

Since the ammeter has a full-scale deflection of 100 μA, it takes only 1 μA of input current to produce full-scale deflection.

16-6 BANDWIDTH

Noninverting voltage feedback *increases* the bandwidth. Why? When the frequency increases, we can find a frequency f_2 at which the open-loop voltage gain is down 3 dB. In a typical feedback amplifier, the loop gain at f_2 is still very large. Therefore, the closed-loop voltage gain still equals $1/B$ (approximately). This means that the frequency can increase well above f_2 before the output voltage drops 3 dB, which is equivalent to saying that the closed-loop cutoff frequency is greater than the open-loop cutoff frequency.

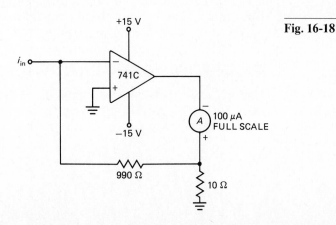

Fig. 16-18

MATHEMATICAL ANALYSIS

The mathematical derivation is too complicated to go into here, but Appendix 1 shows that

$$f_{2(CL)} = (1 + A_{mid}B)f_2 \qquad (16\text{-}24)$$

where $f_{2(CL)}$ = closed-loop upper cutoff frequency
   f_2 = open-loop upper cutoff frequency
   A_{mid} = open-loop voltage gain in midband
   B = feedback fraction

When the midband loop gain is much greater than 1, the closed-loop cutoff frequency is much greater than the open-loop cutoff freqency.

LOWER CUTOFF FREQUENCY

With an op amp, there is no lower cutoff frequency because the stages are direct-coupled. But with discrete amplifiers, ac coupling is typically used between stages. It can be shown that

$$f_{1(CL)} = \frac{f_1}{1 + A_{mid}B} \qquad (16\text{-}25)$$

where $f_{1(CL)}$ = closed-loop lower cutoff frequency
   f_1 = open-loop lower cutoff frequency
   A_{mid} = open-loop voltage gain in midband
   B = feedback fraction

This says that the lower cutoff frequency is reduced by the desensitivity in the midband of the amplifier.

OTHER TYPES OF NEGATIVE FEEDBACK

Negative feedback increases the bandwidth of all four feedback amplifiers discussed earlier. By derivations similar to those for noninverting voltage feedback, we can derive the formulas given in Table 16-5. With inverting inputs, the source impedance approaches infinity because a current source drives the inverting input. In this case, B approximately equals unity. This is the reason the desensitivity is $1 + A_{mid}$ for the inverting types of negative feedback.

OPEN-LOOP GAIN-BANDWIDTH PRODUCT

With Eq. (16-12), the desensitivity in the midband of the amplifier equals

$$1 + A_{mid}B = \frac{A_{mid}}{A_{CL}}$$

Therefore, Eq. (16-24) can be written as

$$f_{2(CL)} = \frac{A_{mid}}{A_{CL}}f_2 \qquad (16\text{-}26)$$

Table 16-5. Bandwidth with Negative Feedback

Type	Lower cutoff	Upper cutoff
Noninverting voltage feedback	$f_1/(1 + A_{mid}B)$	$(1 + A_{mid}B)f_2$
Noninverting current feedback	$f_1/(1 + A_{mid}B)$	$(1 + A_{mid}B)f_2$
Inverting voltage feedback	$f_1/(1 + A_{mid})$	$(1 + A_{mid})f_2$
Inverting current feedback	$f_1/(1 + A_{mid})$	$(1 + A_{mid})f_2$

or

$$A_{CL}f_{2(CL)} = A_{mid}f_2 \qquad (16\text{-}27)$$

The right-hand side of this equation is called the open-loop *gain-bandwidth product* because it is the product of open-loop gain and bandwidth. For a 741C, the typical voltage gain is 100,000 and the cutoff frequency is 10 Hz. Therefore,

$$A_{mid}f_2 = 100{,}000(10 \text{ Hz}) = 1 \text{ MHz}$$

This says that a 741C has an open-loop gain-bandwidth product of 1 MHz.

CLOSED-LOOP GAIN-BANDWIDTH PRODUCT

The left-hand side of Eq. (16-27) is called the closed-loop gain-bandwidth product because it is a product of closed-loop gain and bandwidth. Because of Eq. (16-27), the gain-bandwidth product is the same for open-loop and closed-loop conditions. Given a typical 741C, the product of A_{CL} and $f_{2(CL)}$ always equals 1 MHz, regardless of the values of R_1 and R_2.

Equation (16-27) implies that the gain-bandwidth product is a *constant*. Therefore, even though A_{CL} and $f_{2(CL)}$ change when we change external resistors R_1 and R_2, the product of these two quantities remains constant and equal to $A_{mid}f_2$.

Figure 16-19 summarizes the idea for a 741C. In the midband of the top graph, the open-loop voltage gain is 100,000. The gain decreases 3 dB at 10 Hz, the open-loop cutoff frequency. The open-loop response rolls off at a rate of 20 dB per decade until it reaches an f_{unity} of 1 MHz.

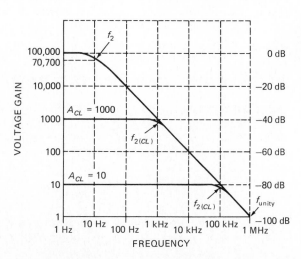

Fig. 16-19
Open-loop and closed-loop responses.

The closed-loop voltage gain is different. If we select resistors to get a closed-loop midband voltage gain of 1000 (middle graph), then the voltage gain of Fig. 16-19 is down 3 dB at 1 kHz, the closed-loop cutoff frequency. Beyond this frequency, the closed-loop response superimposes the open-loop response, rolling off at a rate of 20 dB per decade.

As you reduce the closed-loop voltage gain, you can get more bandwidth. With a midband voltage gain of 10 (bottom graph), the gain drops 3 dB at 100 kHz, the closed-loop cutoff frequency. By looking at the three graphs, you can see that bandwidth increases when midband voltage gain decreases.

UNITY-GAIN FREQUENCY

If $A_{CL} = 1$, then Eq. (16-27) reduces to

$$f_{\text{unity}} = A_{\text{mid}}f_2 \qquad (16\text{-}28)$$

This says that the unity-gain frequency equals the gain-bandwidth product. Data sheets usually list the value of f_{unity} because it equals the gain-bandwidth product. The higher the f_{unity}, the larger the gain-bandwidth product of the op amp. For instance, the 741C has an f_{unity} of 1 MHz. The LM318 has an f_{unity} of 15 MHz. Although it costs more, the LM318 may be a better choice if you need the gain-bandwidth product. With an LM318 you get 15 times more voltage gain than with a 741C that has the same bandwidth.

The gain-bandwidth product gives us a fast way of comparing amplifiers. The greater the gain-bandwidth product, the higher we can go in frequency and still have usable gain. With Eqs. (16-27) and (16-28), we get this useful formula for calculating the closed-loop cutoff frequency:

$$f_{2(CL)} = \frac{f_{\text{unity}}}{A_{CL}} \qquad (16\text{-}29)$$

Equation (16-29) is valid only when the open-loop voltage gain rolls off at a rate of 20 dB per decade (the same as 6 dB per octave). This 20-dB per decade roll-off must continue to the cutoff frequency $f_{2(CL)}$. The 741 has a compensating capacitor fabricated on the chip to produce a 20-dB per decade roll-off above 10 Hz. This single lag network dominates to 1 MHz. Therefore, Eq. (16-29) can be used with the 741 and similar op amps.

SLEW RATE AND POWER BANDWIDTH

Negative feedback has *no effect* on slew rate or power bandwidth. Until the output voltage has changed, there is no feedback signal and no benefits of negative feedback. In other words, slew rate and power bandwidth remain the same with or without negative feedback. By rearranging Eq. (15-22),

$$V_{P(\text{max})} = \frac{S_R}{2\pi f_{\text{max}}}$$

This gives us the peak value of the maximum undistorted output of an op amp.

In our earlier discussion of gain-bandwidth product, we assumed small-signal operation, meaning peak values less than $V_{P(\text{max})}$. Small-signal operation becomes large-signal operation at the point at which the small-signal bandwidth $f_{2(CL)}$ equals the power bandwidth f_{max}. Therefore, the largest undistorted output you can get with op-amp feedback amplifiers is

$$V_{P(\text{max})} = \frac{S_R}{2\pi f_{2(CL)}} \qquad (16\text{-}30)$$

CONCLUSION

We have discussed all four types of negative feedback. The two types of voltage feedback, noninverting and inverting, are the most important because the load resistance is single-ended (grounded on one side). Both types of current feedback, noninverting and inverting, are used less often because the load resistance has to float (not grounded on either end).

In the next two chapters, you will see practical circuits that use the various types of negative feedback. As an aid, Table 16-6 summarizes the effects of negative feedback. Notice the following relations: With noninverting types of feedback, voltage is the input variable. With inverting types of feedback, current is the input variable. This implies that we would use voltage sources to drive noninverting circuits and current sources to drive inverting circuits.

Also notice the following: With voltage feedback, voltage is the stabilized output variable. With current feedback, current is the stabilized output variable. This implies that the output of a negative-feedback circuit appears like a voltage source with voltage feedback, and like a current source with current feedback.

If you memorize these associations:

$$\text{Noninverting input} \rightarrow v_{\text{in}}$$
$$\text{Inverting input} \rightarrow i_{\text{in}}$$
$$\text{Voltage feedback} \rightarrow v_{\text{out}}$$
$$\text{Current feedback} \rightarrow i_{\text{out}}$$

you can better remember the different types of negative feedback and the effects they have on circuit performance.

EXAMPLE 16-10

The data sheet of a 741C shows a decibel graph like Fig. 16-20a. In Fig. 16-20c, calculate the closed-loop cutoff frequency. Draw the decibel closed-loop response.

Table 16-6. Effects of Negative Feedback

Feedback	Stabilized	$r_{\text{in}(CL)}$	$r_{\text{out}(CL)}$	$v_{\text{dist}(CL)}$	$V_{oo(CL)}$	Bandwidth
Noninverting voltage	$v_{\text{out}}/v_{\text{in}}$	Higher	Lower	Lower	Lower	Larger
Noninverting current	$i_{\text{out}}/v_{\text{in}}$	Higher	Higher	Lower	Lower	Larger
Inverting voltage	$v_{\text{out}}/i_{\text{in}}$	Lower	Lower	Lower	Lower	Larger
Inverting current	$i_{\text{out}}/i_{\text{in}}$	Lower	Higher	Lower	Lower	Larger

Fig. 16-20

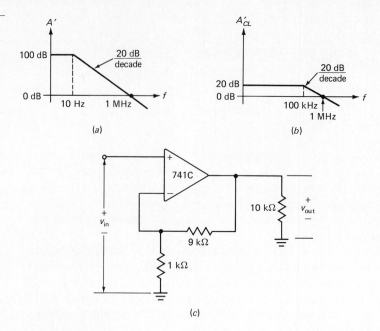

(a)

(b)

(c)

SOLUTION

In Fig. 16-20*a*, the midband decibel voltage gain of the 741C is 100 dB, equivalent to a voltage gain of

$$A_{\text{mid}} = 100{,}000$$

As shown, the open-loop cutoff frequency of the 741C is

$$f_2 = 10 \text{ Hz}$$

One way to get the gain-bandwidth product is

$$A_{\text{mid}}f_2 = 100{,}000(10 \text{ Hz}) = 1 \text{ MHz}$$

Alternatively, you can read the unity-gain frequency, which is

$$f_{\text{unity}} = 1 \text{ MHz}$$

Either way, the 741C has a gain-bandwidth product of 1 MHz.

In the feedback amplifier of Fig. 16-20*c*, the closed-loop voltage gain equals

$$A_{CL} = \frac{9 \text{ k}\Omega}{1 \text{ k}\Omega} + 1 = 10$$

With Eq. (16-29),

$$f_{2(CL)} = \frac{1 \text{ MHz}}{10} = 100 \text{ kHz}$$

Figure 16-20*b* shows the decibel voltage gain of the feedback amplifier. In the midband of the feedback amplifier, the decibel gain is 20 dB. The closed-loop cutoff frequency is 100 kHz, the roll-off rate is 20 dB per decade, and the unity-gain frequency is 1 MHz.

EXAMPLE 16-11

By changing the resistors in Fig. 16-20*c*, we can get different values of A_{CL}. Calculate the $f_{2(CL)}$ for each of the following values: A_{CL} = 1000, 100, 10, and 1.

SOLUTION

With Eq. (16-29), an A_{CL} of 1000 gives

$$f_{2(CL)} = \frac{1 \text{ MHz}}{1000} = 1 \text{ kHz}$$

When A_{CL} equals 100,

$$f_{2(CL)} = \frac{1 \text{ MHz}}{100} = 10 \text{ kHz}$$

When A_{CL} is 10,

$$f_{2(CL)} = \frac{1 \text{ MHz}}{10} = 100 \text{ kHz}$$

When A_{CL} equals 1,

$$f_{2(CL)} = \frac{1 \text{ MHz}}{1} = 1 \text{ MHz}$$

The point is this: We can trade off closed-loop gain for bandwidth. By changing the external resistors in Fig. 16-20*c*, we can tailor a voltage amplifier for a particular application. The values in this example give these choices of gain and bandwidth:

$$A_{CL} = 1000, f_{2(CL)} = 1 \text{ kHz}$$
$$A_{CL} = 100, f_{2(CL)} = 10 \text{ kHz}$$
$$A_{CL} = 10, f_{2(CL)} = 100 \text{ kHz}$$
$$A_{CL} = 1, f_{2(CL)} = 1 \text{ MHz}$$

Notice that the product of gain and cutoff frequency is a constant value of 1 MHz.

EXAMPLE 16-12

If an op amp with the decibel graph of Fig. 16-21*a* is used in the ac equivalent circuit of Fig. 16-21*b*, what is the cutoff frequency of the feedback amplifier?

Fig. 16-21

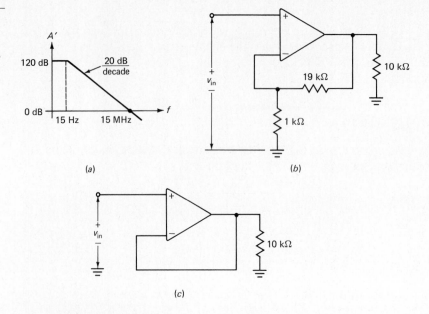

(a)

(b)

(c)

SOLUTION

The closed-loop voltage gain is

$$A_{CL} = \frac{19 \text{ k}\Omega}{1 \text{ k}\Omega} + 1 = 19 + 1 = 20$$

In Fig. 16-21a, f_{unity} is 15 MHz. With Eq. (16-29),

$$f_{2(CL)} = \frac{15 \text{ MHz}}{20} = 750 \text{ kHz}$$

EXAMPLE 16-13

Suppose an op amp with the decibel graph of Fig. 16-21a is used in the voltage follower of Fig. 16-21c. What is the closed-loop cutoff frequency?

SOLUTION

From Eq. (16-29),

$$f_{2(CL)} = \frac{f_{unity}}{A_{CL}} = \frac{15 \text{ MHz}}{1} = 15 \text{ MHz}$$

Because its voltage gain is only 1, the voltage follower has maximum bandwidth.

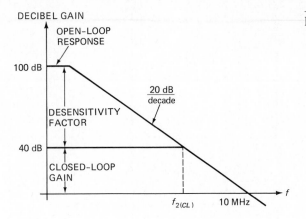

Fig. 16-22

EXAMPLE 16-14

Figure 16-22 is the decibel response of a noninverting voltage-feedback amplifier. What does the midband desensitivity equal? The closed-loop cutoff frequency? The open-loop cutoff frequency?

SOLUTION

The desensitivity is

$$1 + AB = \frac{A}{A_{CL}}$$

which can be written as

$$20 \log (1 + AB) = 20 \log A - 20 \log A_{CL}$$

This says that the desensitivity in decibels equals the difference between the open-loop gain and the closed-loop gain in decibels. In the midband,

$$20 \log (1 + AB) = 100 \text{ dB} - 40 \text{ dB} = 60 \text{ dB}$$

Taking the antilog gives

$$1 + AB = 1000$$

This tells us that the midband desensitivity is 1000.

Since the roll-off rate is 20 dB per decade, the closed-loop cutoff frequency is two decades below f_{unity}:

$$f_{2(CL)} = 100 \text{ kHz}$$

Figure 16-22 also makes it clear that the open-loop cutoff frequency is five decades below f_{unity}, or three decades below $f_{2(CL)}$; therefore,

$$f_2 = 100 \text{ Hz}$$

Fig. 16-23

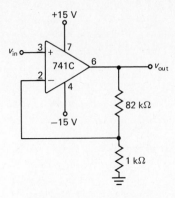

EXAMPLE 16-15

In Fig. 16-23, what is the closed-loop voltage gain in the midband of the feedback amplifier? The small-signal bandwidth? The largest output peak value without slew-rate distortion?

SOLUTION

The closed-loop gain is

$$A_{CL} = \frac{82 \text{ k}\Omega}{1 \text{ k}\Omega} + 1 = 83$$

With Eq. (16-29), the small-signal bandwidth is

$$f_{2(CL)} = \frac{1 \text{ MHz}}{83} = 12 \text{ kHz}$$

With Eq. (16-30), we can calculate the largest output peak value without slew-rate distortion:

$$V_{P(\max)} = \frac{0.5 \text{ V/}\mu s}{2\pi(12 \text{ kHz})} = 6.63 \text{ V}$$

As long as you keep the peak output less than 6.63 V, you will have a closed-loop voltage gain of 83, a closed-loop cutoff frequency of 12 kHz, and no slew-rate distortion.

EXAMPLE 16-16

If R_1 is changed from 82 kΩ to 15 kΩ in Fig. 16-23, what are the new values of A_{CL}, $f_{2(CL)}$, and $V_{P(\max)}$?

SOLUTION

The closed-loop gain decreases to

$$A_{CL} = \frac{15 \text{ k}\Omega}{1 \text{ k}\Omega} + 1 = 16$$

The closed-loop cutoff frequency increases to

$$f_{2(CL)} = \frac{1 \text{ MHz}}{16} = 62.5 \text{ kHz}$$

The largest peak value without slew-rate distortion is

$$V_{P(\text{max})} = \frac{0.5 \text{ V}/\mu\text{s}}{2\pi(62.5 \text{ kHz})} = 1.27 \text{ V}$$

Because the small-signal bandwidth has increased from 12 kHz to 62.5 kHz, the maximum allowable peak output decreases from 6.63 V to 1.27 V.

16-7 NEGATIVE FEEDBACK WITH DISCRETE AMPLIFIERS

The monolithic op amp is ideally suited for use in negative-feedback amplifiers because it has a high open-loop voltage gain, high input impedance, and low output impedance. Before monolithic op amps were invented, discrete amplifiers were used to provide the open-loop voltage gain.

NONINVERTING VOLTAGE FEEDBACK

Figure 16-24 is an example of noninverting voltage feedback. The two CE stages produce an open-loop voltage gain of A. The output voltage drives a voltage divider formed by R_1 and R_2. Because the bottom of R_2 is at ac ground, the feedback fraction

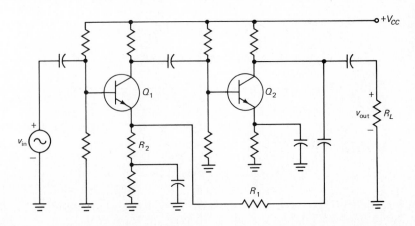

Fig. 16-24
Discrete noninverting voltage feedback.

is approximately

$$B \cong \frac{R_2}{R_1 + R_2}$$

(This ignores the loading effect of the Q_1 emitter.)

The input voltage drives the Q_1 base, while the feedback voltage drives the Q_1 emitter. The error voltage appears across the base-emitter diode. The mathematical analysis is similar to that given earlier. The closed-loop voltage gain approximately equals $1/B$, the input impedance is $(1 + AB)r_{in}$, the output impedance is $r_{out}/(1 + AB)$, the distortion is $v_{dist}/(1 + AB)$, the lower cutoff frequency is $f_1/(1 + AB)$, and the upper cutoff frequency is $(1 + AB)f_2$.

NONINVERTING CURRENT FEEDBACK

Figure 16-24 can be modified as follows to get noninverting current feedback. The load resistor R_L can be moved to the position of R_1. In this case, the output current flows through R_2, and we have current feedback. This means that the output current is stabilized, and the circuit is a voltage-to-current converter.

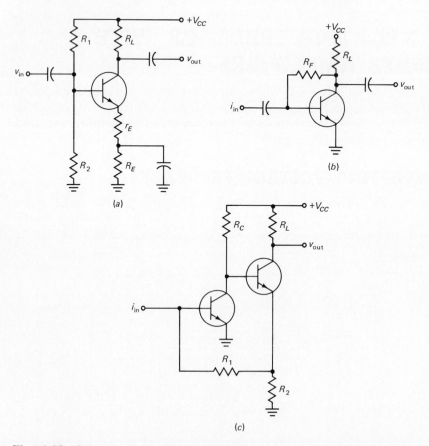

Fig. 16-25 *Discrete negative feedback. (a) Noninverting current feedback. (b) Inverting voltage feedback. (c) Inverting current feedback.*

Figure 16-25a is another example of noninverting current feedback. Since the α_{dc} of the transistor is close to unity, approximately the same value of output current flows through the feedback resistor r_E. Therefore, the output current is being sampled and stabilized. This output current approximately equals

$$i_{out} = \frac{v_{in}}{r_E}$$

INVERTING VOLTAGE FEEDBACK

Figure 16-25b is an example of inverting voltage feedback. The circuit acts like a current-to-voltage converter with a transresistance of approximately

$$\frac{v_{out}}{i_{in}} = R_F$$

Therefore, the output voltage is

$$v_{out} = i_{in} R_F$$

INVERTING CURRENT FEEDBACK

Figure 16-25c is an example of inverting current feedback. The output current is sampled to get the feedback voltage. The circuit acts like a current amplifier with a gain of

$$\frac{i_{out}}{i_{in}} = \frac{R_1}{R_2} + 1$$

PROBLEMS

Straightforward

16-1. An amplifier with noninverting voltage feedback has an input voltage of 20 mV, an output voltage of 1 V, a feedback voltage of 20 mV, and an error voltage of 1 μV. Calculate the open-loop voltage gain, the feedback fraction B, and the closed-loop voltage gain.

16-2. How much output voltage is there in Fig. 16-26a? If the op amp is a 741C with a typical gain of 100,000, what is the value of desensitivity? The closed-loop voltage gain?

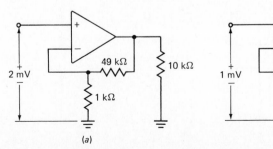

Fig. 16-26

(a)   (b)

16-3. If the op amp of Fig. 16-26*b* has a voltage gain of 1,000,000, what are the approximate values of v_{error} and v_2? If the gain decreases to 100,000, what are the approximate values of v_{error} and v_2? What is the approximate closed-loop voltage gain in Fig. 16-26*b*?

16-4. Calculate the output voltage in Fig. 16-26*a*. If the r_{in} = 10 kΩ, r_{out} = 100 Ω, and A = 50,000, what are the values of $r_{in(CL)}$ and $r_{out(CL)}$?

16-5. Suppose the data sheet of an op amp lists an open-loop voltage gain of 100,000, an r_{in} of 500 kΩ, and an r_{out} of 200 Ω. If this op amp is used in Fig. 16-26*b*, what do $r_{in(CL)}$ and $r_{out(CL)}$ equal?

16-6. An amplifier has A = 100,000 and B = 0.01. If the open-loop distortion voltage is 1.5 V, what is the closed-loop distortion voltage?

16-7. The source resistance driving the noninverting input of Fig. 16-26*a* is 100 Ω. If $v_{in(off)}$ = 2 mV, $I_{in(bias)}$ = 100 nA, and $I_{in(off)}$ = 15 nA, what is closed-loop output offset voltage in the worst case?

16-8. Figure 16-27*a* shows a sensitive dc voltmeter. Calculate the input dc voltages that produce full-scale deflection of the ammeter in each switch position.

Fig. 16-27

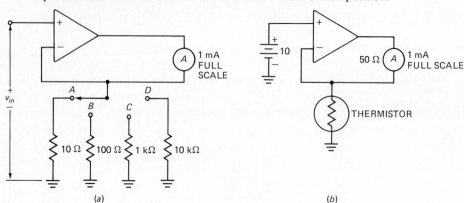

(a)   (b)

16-9. The ammeter of Fig. 16-27*a* has a resistance of 40 Ω. If the op amp has an open-loop voltage gain of 1,000,000 and an r_{in} of 100 kΩ, what is the closed-loop input impedance?

16-10. Figure 16-27*b* shows an electronic thermometer. At 0°C, the thermistor has a resistance of 20 kΩ. The resistance decreases 200 Ω for each degree rise, so that $R_{thermistor}$ equals 19.8 kΩ, 19.6 kΩ, 19.4 kΩ, and so on for T = 1°C, 2°C, 3°C, and so on. What does the ammeter read at 0°C? At 25°C? At 50°C?

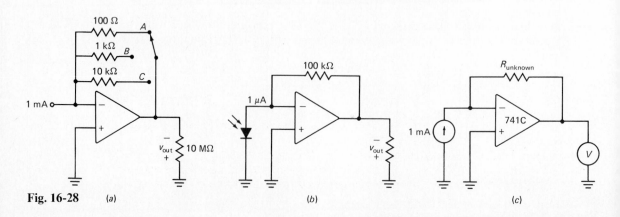

Fig. 16-28   (a)   (b)   (c)

16-11. An input current of 1 mA drives the current-to-voltage converter of Fig. 16-28a. What is the output voltage for each switch position?

16-12. Figure 16-28b shows a photodiode driving a current-to-voltage converter. If $A = 100,000$ what does $r_{in(CL)}$ equal? With 1 μA coming out of the photodiode, how much output voltage is there?

16-13. In Fig. 16-28c, the voltmeter across the output has ranges of 1 mV, 10 mV, and 100 mV (full scale). We can use the circuit as an electronic ohmmeter. What is the value of $R_{unknown}$ that produces full-scale deflection for each given voltage range? If we change the current source to 1 μA, what is the value of $R_{unknown}$ that produces full-scale deflection for each range?

16-14. Many transducers are resistive types; a nonelectrical input quantity changes their resistance. A carbon microphone is an example of a resistive transducer; the sound-wave input produces changes in resistance. Other examples are strain gauges, thermistors, and photoresistors. Suppose we use a resistive transducer in the circuit of Fig. 16-29; we can convert the changes in resistance to changes in output voltage.
   a. What is the value of i_{in}?
   b. If the quiescent value of the transducer resistance is 1 kΩ, how much output voltage is there?

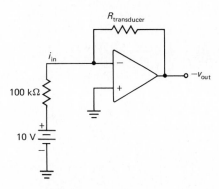

Fig. 16-29

16-15. The meter resistance in Fig. 16-30 is 50 Ω. The amplifier has an open-loop voltage gain of 1,000,000. Calculate the value of i_{out}.

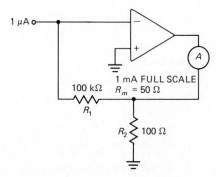

Fig. 16-30

16-16. To get a current gain of 200 in Fig. 16-30, what value should we use for R_1 if R_2 is kept at 100 kΩ?

16-17. A noninverting voltage-feedback amplifier has a desensitivity of 1000. If the amplifier has an open-loop cutoff frequency f_2 of 10 Hz, what is the closed-loop cutoff frequency $f_{2(CL)}$? If the slew rate is 1 V/μs, what is the largest output peak value without slew-rate distortion?

16-18. The op amp of Fig. 16-31 has an open-loop voltage gain of 1,000,000 and a cutoff frequency of 15 Hz. Work out the value of $f_{2(CL)}$ for each position of the switch.

Fig. 16-31

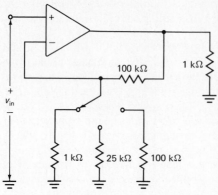

16-19. In Fig. 16-31, the op amp is a 741C with an f_{unity} of 1 MHz. What are the closed-loop voltage gain and bandwidth for each switch position?

16-20. What is the value of f_{unity} for each of these:
a. $A = 50,000$ and $f_2 = 100$ Hz
b. $A' = 100$ dB and $f_2 = 20$ Hz
c. $A' = 120$ dB and $f_2 = 15$ Hz

16-21. If an amplifier has the decibel voltage gain shown in Fig. 16-32a, what does its gain-bandwidth product equal? What is the value of f_{unity}?

Fig. 16-32

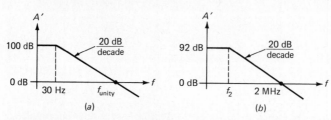

16-22. For an amplifier whose decibel voltage gain looks like Fig. 16-32b, what is the value of its open-loop cutoff frequency f_2? The value of its gain-bandwidth product?

16-23. If a voltage follower uses an op amp with the decibel voltage gain shown in Fig. 16-32b, what will the $f_{2(CL)}$ equal?

Troubleshooting

16-24. Somebody mistakenly reverses the battery of Fig. 16-27b. What happens?

16-25. In Fig. 16-28c, the output voltmeter reads five times too high for all values of $R_{unknown}$. Which of the following is a possible trouble:
a. Inverting input grounded by solder bridge
b. 741C has an open-loop gain of 500,000 instead of 100,000
c. Current source produces 5 mA instead of 1 mA
d. Noninverting input is open

16-26. In Fig. 16-9, a large output offset voltage exists. Which of the following is a possible trouble:
a. 741C defective
b. 39 kΩ shorted
c. Supplies are 10 V instead of 15 V
d. Noninverting input grounded by solder bridge

16-27. In Fig. 16-11, the zero adjust doesn't work. Name some possible troubles.

Design

16-28. Redesign the feedback amplifier of Fig. 16-3 to get a closed-loop voltage gain of approximately 75.

16-29. Design a feedback amplifier like Fig. 16-7 with a closed-loop voltage gain that meets these specifications: typical closed-loop output offset voltage less than 0.5 V, op amp is 741C, and source resistance is 1 kΩ.

16-30. Use a 741C with $A = 100,000$ and $f_{unity} = 1$ MHz. Design a circuit like Fig. 16-7 that has a closed-loop bandwidth of approximately 20 kHz.

Challenging

16-31. In Fig. 16-24, the feedback voltage divider is loaded by the input impedance looking into the emitter. Derive a formula for B that takes this loading into account.

16-32. In Fig. 16-33a, what does the output voltage equal?

16-33. What does v_{out} equal in Fig. 16-33b?

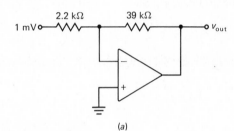 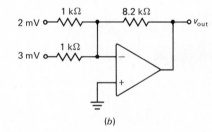

Fig. 16-33

(a)   (b)

Computer

16-34. Here is a program:

```
10 PRINT "ENTER R1": INPUT R1
20 PRINT "ENTER R2": INPUT R2
30 PRINT "ENTER A": INPUT A
40 B = R2/(R1 + R2)
50 ACL = A/(1 + A*B)
60 PRINT "ACL EQUALS ":ACL
```

What does the program calculate in line 40? What does the program print out in line 60?

16-35. One reason for spacing the lines 10 apart is to allow new lines to be entered later if necessary. For instance, we can insert lines 11, 12, . . . , 19 between lines 10 and 20 of an existing program. Insert new lines into the program of Prob. 16-34 so that the new program will print out the closed-loop distortion and the closed-loop output offset voltage.

16-36. Write a program that calculates A_{CL}, g_m, v_{out}/i_{in}, or i_{out}/i_{in}. The program should start with a menu of the four types of negative feedback.

16-37. Write a program that calculates the closed-loop lower and upper cutoff frequencies of a noninverting voltage-feedback amplifier.

17

Linear Op-Amp Circuits

This chapter is about *linear* op-amp circuits, the kind that preserve the shape of the input signal. For instance, if the input signal is sinusoidal, the output signal is also sinusoidal. The chapter begins with noninverting and inverting amplifiers, followed by op-amp current sources, differential amplifiers, and a variety of other linear op-amp circuits.

17-1 NONINVERTING VOLTAGE AMPLIFIERS

A noninverting voltage-feedback amplifier is approximately an ideal *voltage amplifier* because of its high input impedance, low output impedance, and stable voltage gain. Let's take a look at different circuits that utilize noninverting voltage feedback.

BASIC CIRCUIT

Figure 17-1 shows the basic circuit for a noninverting voltage-feedback amplifier. As derived in Chap. 16, it has a closed-loop voltage gain of

$$A_{CL} = \frac{R_1}{R_2} + 1$$

and a bandwidth of

$$f_{2(CL)} = \frac{f_{unity}}{A_{CL}}$$

Since the gain-bandwidth product is a constant, you have to decrease the closed-loop voltage gain if you want more bandwidth.

AC AMPLIFIER

In some applications, you don't need a response that extends down to zero frequency because only ac signals drive the input. In this case, you can insert *coupling*

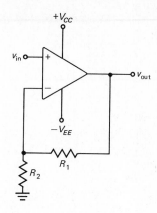

Fig. 17-1
Amplifier with noninverting voltage feedback.

capacitors on the input and output sides, as shown in Fig 17-2. While you're at it, you can minimize the output offset voltage by inserting a *bypass capacitor* in the feedback loop, as shown. In the midband of the amplifier, the bypass capacitor appears shorted, and the closed-loop ac voltage gain is $R_1/R_2 + 1$.

But at zero frequency the capacitor appears open and the feedback fraction increases to 1. Therefore, the desensitivity for dc signals is $1 + A$, the maximum value it can have. This reduces the output offset voltage to a minimum, allowing maximum swing of the ac output signal. In other words, you get maximum ac output compliance by using a bypass capacitor.

The bypass capacitor produces a cutoff frequency of

$$f_{BY} = \frac{1}{2\pi R_2 C_{BY}} \tag{17-1}$$

At this frequency the closed-loop voltage gain is down approximately 3 dB from the midband value. Ten times above this frequency, the closed-loop voltage gain is within half a percent of the midband voltage gain.

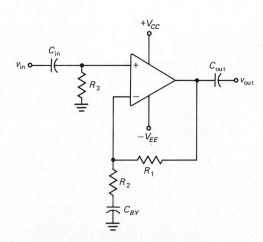

Fig. 17-2
Ac-coupled voltage amplifier.

Fig. 17-3
*Single-supply
ac-coupled voltage
amplifier.*

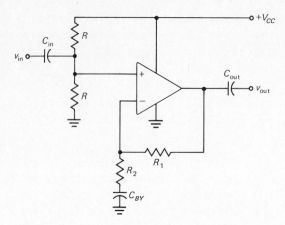

SINGLE-SUPPLY OPERATION

Most op-amp circuits use dual or split supplies, such as $V_{CC} = +15$ V and $V_{EE} = -15$ V. But sometimes you will see an op-amp circuit running off a *single supply*, as shown in Fig. 17-3. Notice that the V_{EE} input is grounded. To get maximum ac output compliance, you need to bias the noninverting input at half the supply voltage, which is conveniently done with an equal-resistor voltage divider. This produces a dc input of $+0.5V_{CC}$ at the noninverting input. Because of the bootstrap effect, the inverting input automatically has a quiescent value of $+0.5V_{CC}$, and so does the output.

The ac operation is the same as with Fig. 17-2, except that the ac output compliance is limited to 1 or 2 V less than V_{CC}. For $V_{CC} = +15$ V, this means a maximum unclipped output of 13 or 14 V.

AUDIO AMPLIFIER

Figure 17-4 shows another single-supply design. The collector of the bipolar stage typically has a quiescent voltage of approximately half of V_{CC}. Therefore, we can

Fig. 17-4
*Bipolar CE stage
is direct-coupled to
op-amp stage.*

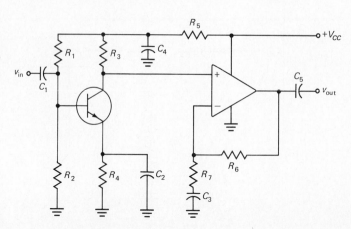

direct-couple to the noninverting input. This neatly eliminates the coupling capacitor and the voltage divider shown earlier, while providing additional voltage gain.

Most of the components for the bipolar stage should appear familiar from earlier discussions. For instance, R_1 and R_2 provide voltage-divider bias, with C_2 bypassing the emitter to ground for maximum voltage gain. The only new components are R_5 and C_4. This lag circuit is called a *decoupling* network. It has a low cutoff frequency to prevent oscillations caused by unwanted feedback between stages (discussed in Chap. 20).

An audio amplifier covers frequencies from 20 Hz to 20 kHz. If a 741C is used for the op amp, a closed-loop voltage gain of 50 produces an upper cutoff frequency of 20 kHz. With a 15-V supply, the bipolar stage will have a voltage gain somewhere in the vicinity of 200, with the exact value depending on the design. Therefore, the audio amplifier has an overall voltage gain of around 10,000, equivalent to 80 dB.

JFET-SWITCHED VOLTAGE GAIN

Some applications require a change in closed-loop voltage gain. Figure 17-5 shows a *JFET-controlled* amplifier. The control voltage for the JFET switch comes from another circuit that produces a two-level output, either 0 V or a voltage that is equal to $V_{GS(off)}$. When the control voltage equals $V_{GS(off)}$, the JFET switch is open and the closed-loop voltage gain is $R_1/R_2 + 1$. When the control voltage is zero, the JFET switch is closed and the closed-loop voltage gain is

$$A_{CL} = \frac{R_1}{R_2 \parallel R_3} + 1$$

A typical JFET for an application like this is the 2N4860, which has a maximum $r_{ds(on)}$ of 40 Ω. In most designs, R_3 is made much larger than $r_{ds(on)}$ to prevent $r_{ds(on)}$ from affecting the closed-loop voltage gain. Often, you will see several JFET switches and resistors connected in parallel with R_2 to provide a selection of closed-loop voltage gains. (TTL control is typically used. TTL is a type of digital integrated circuit with a two-level output: high and low.)

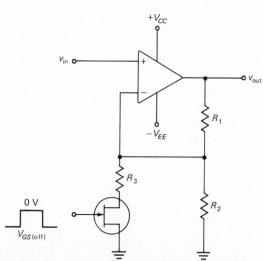

Fig. 17-5
JFET switch controls voltage gain of op-amp circuit.

AUDIO AGC

AGC stands for *automatic gain control*. Figure 17-6 shows an audio AGC circuit. Q_1 is a JFET used as a *voltage-variable resistance*. As discussed in Sec. 12-11, for small-signal operation with drain voltages near zero, the JFET operates in the ohmic region and has a resistance of $r_{ds(on)}$ to ac signals. The $r_{ds(on)}$ of a JFET can be controlled by the gate voltage. The more negative V_{GS} is, the larger $r_{ds(on)}$ becomes. With a JFET like the 2N4861, $r_{ds(on)}$ can vary from 100 Ω to more than 10 MΩ. If R_3 is around 100 kΩ, the R_3-Q_1 combination acts like a voltage divider whose output varies between $0.001v_{in}$ and v_{in}. Therefore, the noninverting input voltage is between $0.001v_{in}$ and v_{in}, a 60-dB range. The amplified output voltage is $R_1/R_2 + 1$ times this.

In Fig. 17-6, the output voltage is coupled to the base of Q_2. For peak-to-peak outputs less than 1.4 V, Q_2 is cut off because there is no bias on it. In this case, capacitor C_2 is uncharged and the gate of Q_1 is at $-V_{EE}$, enough to cut off the JFET. This means that almost all the input voltage reaches the noninverting input.

When the output has a peak-to-peak voltage greater than 1.4 V, Q_2 conducts during part of the negative half cycle. This charges capacitor C_2 and raises the gate voltage above the quiescent level of $-V_{EE}$. When this happens, $r_{ds(on)}$ decreases. As it does, the output of the R_3-Q_1 voltage divider decreases, and so less input voltage reaches the noninverting input. Stated another way, the overall voltage gain of the circuit decreases when the peak-to-peak output voltage gets above 1.4 V.

The whole purpose of the AGC circuit is to change the voltage gain when the input signal changes, so that the output voltage stays approximately *constant*. One reason for doing this is to prevent sudden increases in signal level from overdriving a loudspeaker. If you're listening to a radio, for instance, you don't want an unexpected increase in signal to bombard your hearing. Likewise, you don't want the signal to become inaudible because of a decrease in signal strength.

Fig. 17-6
JFET used as voltage-variable resistance in AGC circuit.

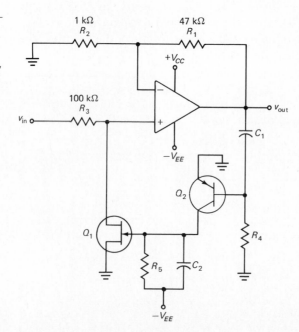

In summary, even though the input voltage of Fig. 17-6 varies over a 60-dB range, the peak-to-peak output voltage is restricted to slightly more than 1.4 V. Later, you will see other examples of AGC circuits.

17-2 THE INVERTING VOLTAGE AMPLIFIER

The noninverting voltage amplifier produces an output voltage that is in phase with the input voltage, which is desirable in many applications. But there are many other applications in which we prefer an inverted output signal. Figure 17-7 shows a very popular *inverting voltage amplifier*. The circuit is a current-to-voltage converter driven by a voltage source instead of a current source. Because of the source resistor R_S, the feedback fraction changes and some feedback properties change.

SIMPLIFIED ANALYSIS

The inverting input of a current-to-voltage converter is a virtual ground. Since the right end of R_S is grounded, the input current equals

$$i_{in} \cong \frac{v_{in}}{R_S} \qquad (17\text{-}2)$$

Because the virtual ground can sink no current, all the input current flows through R_F, producing an output voltage of

$$v_{out} \cong -i_{in}R_F$$

or

$$v_{out} \cong \frac{-v_{in}R_F}{R_S} \qquad (17\text{-}3)$$

By rearranging,

$$\frac{v_{out}}{v_{in}} \cong \frac{-R_F}{R_S} \qquad (17\text{-}4)$$

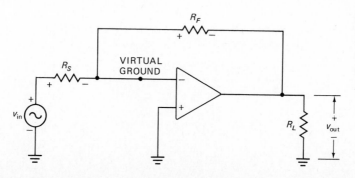

Fig. 17-7
Inverting voltage amplifier using inverting voltage feedback.

or

$$A_{CL} \cong \frac{-R_F}{R_S} \tag{17-5}$$

where the minus sign indicates phase inversion. This says that the closed-loop voltage gain equals minus the ratio of the feedback resistor to the source resistor.

IMPEDANCES

Because of the virtual ground, the right end of R_S appears grounded, and the source sees a closed-loop input impedance of

$$r_{in(CL)} \cong R_S \tag{17-6}$$

This tells us that the input impedance of an inverting voltage amplifier is controlled by the designer because it equals R_S. One of the reasons the inverting voltage amplifier is popular is that it allows the designer to set up a precise value of *input impedance* as well as voltage gain.

The circuit can be redrawn as shown in Fig. 17-8. The feedback fraction is

$$B \cong \frac{R_S}{R_F + R_S} \tag{17-7}$$

By a derivation similar to that given in Sec. 16-2, the closed-loop output impedance is

$$r_{out(CL)} \cong \frac{r_{out}}{1 + AB} \tag{17-8}$$

(Note: When R_S approaches infinity, B approaches 1, and the circuit becomes the current-to-voltage converter analyzed in Sec. 16-4.)

CLOSED-LOOP GAIN-BANDWIDTH PRODUCT

Because of the negative feedback, the closed-loop cutoff frequency of an inverting amplifier is greater than the open-loop cutoff frequency. Appendix 1 proves that

$$f_{2(CL)} = (1 + A_{mid}B)f_2 \tag{17-9}$$

Fig. 17-8

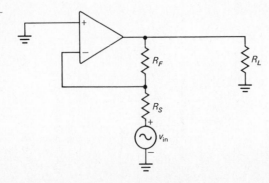

This says that the closed-loop cutoff frequency is increased by a factor of $1 + A_{mid}B$.

The closed-loop gain-bandwidth product is $A_{CL}f_{2(CL)}$. Appendix 1 shows that

$$A_{CL}f_{2(CL)} = \frac{-R_F}{R_F + R_S} f_{unity} \qquad (17\text{-}10)$$

This says that the closed-loop gain-bandwidth product equals a fraction of f_{unity}. When R_F is much larger than R_S, $A_{CL}f_{2(CL)}$ approximately equals f_{unity}. But when R_F is not large enough to swamp out R_S, the closed-loop gain-bandwidth product may be significantly less than f_{unity}.

Equation (17-10) can be rearranged as

$$A_{CL}f_{2(CL)} = \frac{A_{CL}}{1 - A_{CL}} f_{unity} \qquad (17\text{-}10a)$$

In this form, you can see the effect that closed-loop voltage gain has. When A_{CL} is large, the closed-loop gain-bandwidth product approximately equals $-f_{unity}$. When A_{CL} is not large, the closed-loop gain-bandwidth product shrinks. For instance, if $A_{CL} = -1$, then

$$A_{CL}f_{2(CL)} = \frac{-1}{1 + 1} f_{unity} = -0.5 f_{unity}$$

Dividing both sides of Eq. (17-10) by A_{CL} and rearranging gives

$$f_{2(CL)} = B f_{unity} \qquad (17\text{-}11)$$

This is useful because it relates the closed-loop cutoff frequency to the feedback fraction and the open-loop gain bandwidth product.

In summary, as the voltage gain decreases, an inverting voltage amplifier has *less available bandwidth* than a noninverting voltage amplifier.

OFFSET CAUSED BY INPUT BIAS CURRENT

In Fig. 17-9, there is an input offset voltage caused by the input bias current:

$$v_2 = I_{B2}(R_S \parallel R_F)$$

In some designs, a *resistor is added* between the noninverting input and ground, as shown in Fig. 17-10. This cancels out most of the bias-current offset because the input now is

$$v_1 - v_2 = I_{B1}(R_S \parallel R_F) - I_{B2}(R_S \parallel R_F)$$

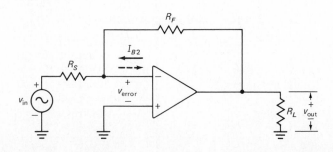

Fig. 17-9
Inverting voltage amplifier.

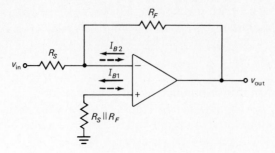

Fig. 17-10 *Reducing output offset voltage by using resistor on noninverting input.*

or

$$v_1 - v_2 = I_{in(off)}(R_S \parallel R_F)$$

Since $I_{in(off)}$ is usually much smaller than $I_{in(bias)}$, the bias-current offset is minimized. The added resistor has no effect on the closed-loop voltage gain because there is no ac voltage across it.

PROPERTIES OF AN INVERTING VOLTAGE AMPLIFIER

Table 17-1 summarizes the properties of an inverting voltage amplifier. The negative feedback stabilizes the voltage gain and input impedance. Because of the voltage feedback, the amplifier has a very low output impedance. As always, negative feedback decreases distortion and output offset voltage.

EXAMPLE 17-1

Calculate the output voltage, the input impedance, and the cutoff frequency in Fig. 17-11. The 741C has an f_{unity} of 1 MHz.

SOLUTION

The closed-loop voltage gain is

$$A_{CL} = \frac{-2 \text{ k}\Omega}{1 \text{ k}\Omega} = -2$$

Table 17-1. Inverting Voltage Amplifier

Quantity	Effect	Formula
v_{out}/v_{in}	Stabilizes	$-R_F/R_S$
Input impedance	Stabilizes	R_S
Output impedance	Decreases	$r_{out}/(1 + AB)$
Distortion	Decreases	$v_{dist}/(1 + AB)$
Output offset	Decreases	$V_{out(off)}/(1 + AB)$
Bandwidth	Increases	Bf_{unity}

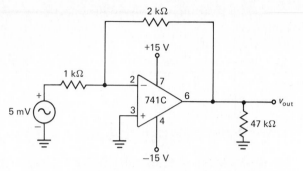

Fig. 17-11

So the output voltage is

$$v_{\text{out}} = -2(5 \text{ mV}) = -10 \text{ mV}$$

The closed-loop input impedance seen by the ac source is

$$r_{\text{in}(CL)} = 1 \text{ k}\Omega$$

The feedback fraction is

$$B = \frac{1 \text{ k}\Omega}{2 \text{ k}\Omega + 1 \text{ k}\Omega} = 0.333$$

With Eq. (17-11), the closed-loop cutoff frequency is

$$f_{2(CL)} = 0.333(1 \text{ MHz}) = 333 \text{ kHz}$$

The bandwidth of an inverting voltage amplifier is less than that of a noninverting voltage amplifier. The loss of bandwidth becomes noticeable for closed-loop voltage gains less than approximately 10. In the worst case, when $A_{CL} = -1$, the inverting voltage amplifier has a bandwidth of 500 kHz, while the noninverting voltage amplifier has a bandwidth of 1 MHz.

17-3 OP-AMP INVERTING CIRCUITS

The inverting voltage amplifier has a stable voltage gain and input impedance. These properties allow designers to come up with a variety of inverting op-amp circuits for different applications. In analyzing these circuits, it helps to remember that the inverting input is *bootstrapped* to within microvolts of the noninverting input because the open-loop voltage gain is extremely high.

SWITCHABLE INVERTER

Figure 17-12*a* shows an op amp that can function as either an inverter or a noninverter. With the switch in the lower position, the noninverting input is grounded. Since the feedback and source resistors are equal, we have an inverting voltage amplifier with a closed-loop voltage gain of

$$A_{CL} = -1$$

When the switch is moved to the upper position, the input signal also drives the noninverting input. Since the inverting input is bootstrapped to the noninverting input, there is approximately zero current through the R on the left. Furthermore, with the output voltage fed back to the inverting input, we conclude that the output voltage equals the input voltage. This means that the closed-loop voltage gain is

$$A_{CL} = 1$$

JFET-CONTROLLED SWITCHABLE INVERTER

Figure 17-12b is a modification of Fig. 17-12a. This time, a JFET is used as a switch, a voltage-variable resistance that equals either a very low or a very high resistance. Recall that the drain curves of a JFET extend on both sides of the origin. For this reason, no dc supply voltage is needed. The ac signal voltage on the drain is sufficient.

When the gate of the JFET is at 0 V, the JFET switch is closed and the circuit is an inverter with a voltage gain of -1. On the other hand, when the gate voltage is at $V_{GS(\text{off})}$, the JFET switch is open and the circuit is a noninverter with a voltage gain of 1. For proper operation, R should be at least 100 times greater than $r_{ds(\text{on})}$ when the JFET switch is closed.

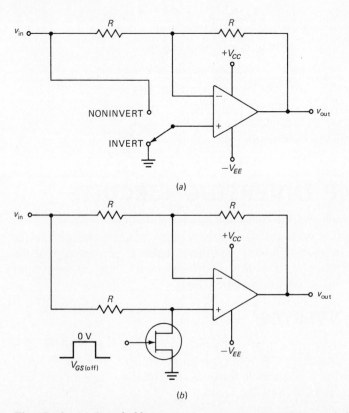

Fig. 17-12 (a) *Switchable inverter/noninverter.* (b) *JFET switch for inverter/noninverter.*

ADJUSTABLE BANDWIDTH

Sometimes we would like to change the closed-loop bandwidth of an inverting voltage amplifier without changing the closed-loop voltage gain. Sound impossible? Not at all. Figure 17-13*a* shows an adjustable resistor R connected between the inverting input and ground. Figure 17-13*b* shows an equivalent circuit with the input side thevenized. The effective source resistance driving the inverting input is now R_S in parallel with R. For this reason, the feedback fraction is

$$B = \frac{R_S \parallel R}{R_S \parallel R + R_F}$$

The closed-loop bandwidth is

$$f_{2(CL)} = Bf_{unity}$$

Since R is adjustable, we can change B and therefore control $f_{2(CL)}$.

The output voltage is

$$v_{out} = \frac{-R_F}{R_S \parallel R} \frac{R}{R_S + R} v_{in}$$

which reduces to

$$v_{out} = \frac{-R_F}{R_S} v_{in}$$

This means that the output voltage is constant even though the bandwidth changes.

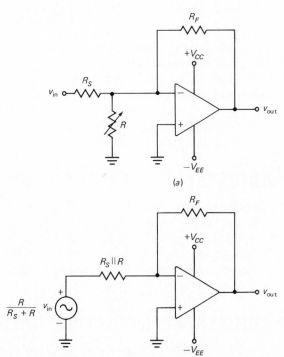

Fig. 17-13
Circuit with constant voltage gain but adjustable bandwidth.

Fig. 17-14
*Single-supply
inverting ampli-
fier.*

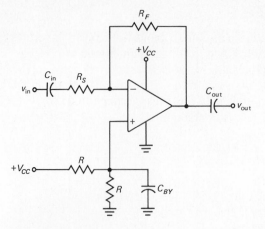

SINGLE-SUPPLY OPERATION

Figure 17-14 shows a single-supply inverting voltage amplifier that can be used with ac signals. The V_{EE} supply is grounded, and half the V_{CC} supply is applied to the noninverting input. Because of the bootstrap effect, the inverting input rides at a quiescent level of approximately $+V_{CC}/2$. Since the input coupling capacitor is open at zero frequency, the closed-loop voltage gain is -1 for dc signals, which means minimum output offset voltage.

The input coupling capacitor produces a lower cutoff frequency of

$$f_{in} = \frac{1}{2\pi R_S C_{in}} \qquad (17\text{-}12)$$

Ten times above this frequency the closed-loop voltage gain is within a half percent of the midband value, $-R_F/R_S$.

Incidentally, a bypass capacitor is typically used on the noninverting input, as shown in Fig. 17-14. This reduces the amount of power supply ripple and noise appearing at the noninverting input. To be effective, the cutoff frequency of this bypass circuit (a lag network with an equivalent resistance of $R/2$) should be much lower than the ripple frequency.

ADJUSTABLE INVERTED GAIN

When the adjustable resistor of Fig. 17-15a is reduced to zero, the noninverting input is grounded, and we get a maximum voltage gain of $-R_F/R_S$. When the adjustable resistor is increased to R_F, equal voltages drive the noninverting and inverting inputs. Because of the common-mode rejection, the output voltage is approximately zero. Therefore, the circuit of Fig. 17-15a has an adjustable closed-loop voltage gain from approximately zero to $-R_F/R_S$.

ADJUSTABLE INVERTER/NONINVERTER

Figure 17-15b shows a circuit that allows us to adjust the closed-loop voltage gain between $-n$ and $+n$. When the adjustable resistor is at zero, the noninverting input is

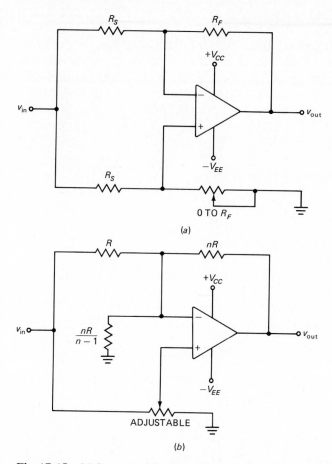

Fig. 17-15 (a) *Inverter with adjustable gain.* (b) *Circuit with gain adjustable from +1 to −1.*

grounded and the closed-loop voltage gain is derived as for Fig. 17-13:

$$A_{CL} \cong \frac{-nR}{R} = -n$$

When the adjustable resistor is at the other extreme, the input voltage is applied directly to the noninverting input. Because of the bootstrap effect, the inverting input is at the same approximate voltage, and resistor R has no effect because v_{in} appears on both ends of R. The feedback fraction is

$$B \cong \frac{nR/(n-1)}{nR/(n-1) + nR} = \frac{1}{n}$$

Therefore, the closed-loop gain is

$$A_{CL} \cong \frac{1}{1/n} = n$$

When the adjustable resistor is turned between its extremes, the closed-loop voltage gain varies from $-n$ to $+n$.

LOW-LEVEL VIDEO AGC

Figure 17-16a shows a standard technique for *video* AGC that has been used for frequencies up to 10 MHz. In this circuit the JFET acts like a voltage-variable resistance. When the AGC voltage is zero, the JFET is cut off by the negative bias, and its $r_{ds(on)}$ is maximum. As the AGC voltage increases, the $r_{ds(on)}$ of the JFET decreases. The signal driving the inverting voltage amplifier is

$$v_A = \frac{R_2 + r_{ds(on)}}{R_1 + R_2 + r_{ds(on)}} v_{in}$$

The output voltage from the inverting amplifier is

$$v_{out} = \frac{-R_F}{R_S} v_A$$

In this circuit the JFET acts like a voltage-variable resistance controlled by $+V_{AGC}$. The more positive the AGC voltage, the smaller the value of $r_{ds(on)}$ and the lower the voltage to the inverting amplifier. This means that the AGC voltage controls the overall voltage gain of the circuit.

With a wideband op amp, the circuit works well for input signals up to approximately 100 mV. Beyond this level, the JFET resistance becomes a function of the

Fig. 17-16
(a) *Low-level video AGC.* (b) *High-level video AGC.*

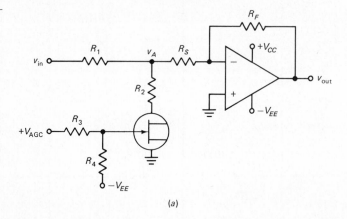

(a)

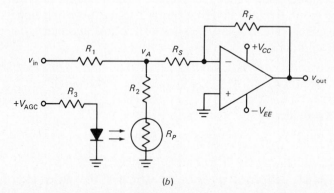

(b)

signal level in addition to the AGC voltage. This is undesirable because only the AGC voltage should control the overall voltage gain.

HIGH-LEVEL VIDEO AGC

For high-level video signals, we can replace the JFET by a LED-photoresistor combination like Fig. 17-16b. The resistance R_P of the photoresistor decreases as the amount of light increases. Therefore, the larger the AGC voltage, the lower the value of R_P. As before, the input voltage divider controls the amount of voltage driving the inverting voltage amplifier. This voltage is given by

$$v_A = \frac{R_2 + R_P}{R_1 + R_2 + R_P}\, v_{\text{in}}$$

The circuit can handle high-level input voltages up to 10 V because the photocell resistance is unaffected by larger voltages and is only a function of V_{AGC}. Also note that there is almost total isolation between the AGC voltage and the input voltage v_{in}.

17-4 THE SUMMING AMPLIFIER

Another advantage of the inverting voltage amplifier is its ability to handle more than one input at a time. To understand why, look at Fig. 17-17a. Because of the virtual ground, both input resistors are effectively grounded on the right side. The input current through R_1 is

$$i_1 = \frac{v_1}{R_1}$$

and the input current through R_2 is

$$i_2 = \frac{v_2}{R_2}$$

The sum, $i_1 + i_2$, flows through R_3. So the output voltage is

$$v_{\text{out}} = -(i_1 + i_2)R_3$$

or

$$v_{\text{out}} = -\frac{R_3}{R_1}\, v_1 - \frac{R_3}{R_2}\, v_2 \qquad (17\text{-}13)$$

This means that we can have a different voltage gain for each input; the output is the *sum* of the amplified inputs. The same idea applies to any number of inputs; add another resistor for each new input signal.

SUMMER AND MIXER

Often, we need a circuit that adds two or more inputs. In this case, we can use a *summer,* an inverting amplifier with several inputs, each with unity voltage gain.

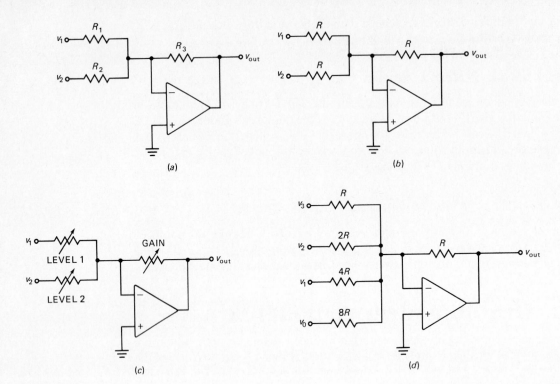

Fig. 17-17
Summing circuits.

Figure 17-17b shows a summer with two inputs. Because all resistors are equal, each input has unity voltage gain and the output is given by

$$v_{\text{out}} = - (v_1 + v_2)$$

Figure 17-17c shows a convenient way to mix two audio signals. The adjustable resistors allow us to set the level of each input, and the gain control allows us to adjust the output volume.

D/A CONVERTER

If you study digital electronics, you will encounter the *digital-to-analog* (D/A) converter. Figure 17-17d shows how an op amp can be used to build a D/A converter. There are four inputs representing a binary number. Because of the input resistors, the output is

$$v_{\text{out}} = - (v_3 + 0.5v_2 + 0.25v_1 + 0.125v_0)$$

The input voltages are digital, meaning that they have only two values: low or high. The output voltage is the analog equivalent of the input voltage.

17-5 CURRENT BOOSTERS FOR VOLTAGE AMPLIFIERS

The maximum output current of a typical op amp is limited. For instance, the 741C has a maximum output current of 25 mA. If the load requires more than this, you can add a *current booster* to the output.

UNIDIRECTIONAL LOAD CURRENT

If a *unidirectional* load current is all right, then add an emitter follower to the output of an op amp, as shown in Fig. 17-18. With the transistor inside the feedback loop, the negative feedback automatically adjusts V_{BE} to the required value. Since the circuit is a noninverting voltage-feedback amplifier, the closed-loop voltage gain is

$$A_{CL} = \frac{R_1}{R_2} + 1$$

and the output impedance is

$$r_{out(CL)} = \frac{r_{out}}{1 + AB}$$

where r_{out} is the open-loop output impedance looking back into the emitter. Unlike previous voltage amplifiers, the op amp no longer has to supply the load current. Instead, it only has to supply base current to the transistor. As a result, the current booster allows us to use smaller load resistances than before.

BIDIRECTIONAL LOAD CURRENT

The major disadvantage of Fig. 17-18 is its unidirectional load current (conventional flow is down through the load, and electron flow is up). One way to get *bidirectional* load current is with a class B *push-pull* emitter follower, as shown in Fig. 17-19. In

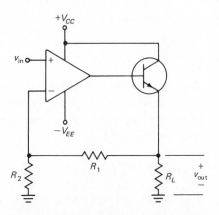

Fig. 17-18
Bipolar transistor is current booster for op amp.

Fig. 17-19
Class B current booster for op amp.

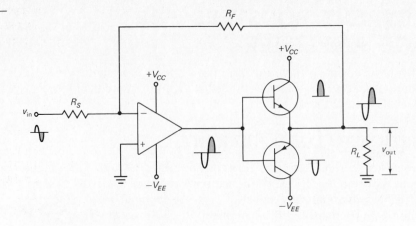

this case, the closed-loop voltage gain is

$$A_{CL} = \frac{-R_F}{R_S}$$

The output impedance is still $r_{out}/(1 + AB)$ because of the voltage feedback.

Again, the negative feedback automatically adjusts the V_{BE} values to whatever they need to be. Furthermore, there's no need to provide bias to eliminate crossover distortion because the negative feedback reduces it by a factor of $1 + AB$. When the input voltage goes positive, the lower transistor is conducting and the load voltage is negative. On the other hand, when the input goes negative, the upper transistor conducts and the output voltage is positive.

17-6 VOLTAGE-CONTROLLED CURRENT SOURCES

Figure 17-20a shows noninverting current feedback. As you recall, the load current is stabilized against changes in load resistance and open-loop gain. Because the inverting input is bootstrapped to the noninverting input, voltage v_{in} appears across R, and the load current is

$$i_{out} \cong \frac{v_{in}}{R} \tag{17-14}$$

The value of load resistance does not appear in this equation. Therefore, the output current is independent of the value of load resistance. Stated another way, the output impedance facing the load is $(1 + A)R$. Therefore, the load appears to be driven by a very stiff current source.

GROUNDED LOAD

If a floating load is all right, then a circuit like Fig. 17-20a works quite well. If the load has to be grounded on one end (the usual case), we can modify the basic circuit as

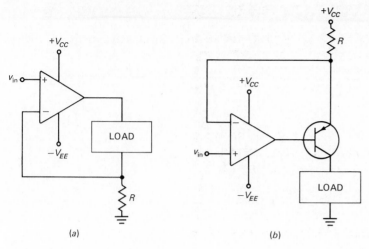

Fig. 17-20 *Voltage-controlled current sources. (a) Floating load. (b) Load grounded on one end.*

shown in Fig. 17-20*b*. Since the collector and emitter currents are equal to a close approximation, we can say that load current flows through the feedback resistor *R*. On account of this, we still have current feedback, which means that the load current is stabilized.

Because of the bootstrap effect, the inverting input voltage equals v_{in}. This means that the current through *R* equals

$$i_{out} \cong \frac{V_{CC} - v_{in}}{R} \tag{17-15}$$

If v_{in} comes from a zener diode or other stiff dc voltage source, then the transistor appears like a stiff direct current source to the load.

There is a limit to the output current that the circuit can supply. The base current in the transistor equals i_{out}/β_{dc}. Since the op amp has to supply this base current, i_{out}/β_{dc} must be less than the $I_{out(max)}$ of the op amp, typically 10 to 25 mA. Therefore, if you design a circuit like this, make sure that the i_{out} given by Eq. (17-15) satisfies the condition: i_{out}/β_{dc} less than the $I_{out(max)}$ of the op amp.

There is also a limit on the output voltage in Fig. 17-20*b*. As the load resistance increases, the load voltage increases. Eventually, the transistor runs out of compliance because it goes into saturation. Since the emitter is at v_{in} with respect to ground, the maximum load voltage is slightly less than v_{in} (transistor saturated). Therefore, the second thing to check with this kind of circuit is that the output current multiplied by the largest load resistance does not exceed v_{in}.

Incidentally, the nice thing about negative feedback is that it automatically adjusts the V_{BE} in Fig. 17-20*b* to the exact value needed to hold the load current constant. For instance, if the load resistance decreases, the load current tries to increase. This means that more voltage is fed back to the inverting input, which decreases V_{BE} just enough to almost completely nullify the attempted increase in load current.

GROUNDED VOLTAGE-TO-CURRENT CONVERTER

In Eq. (17-15), the load current decreases when the input voltage increases. Figure 17-21 shows a circuit in which the load current increases when the input voltage increases. Because of the bootstrap effect, the inverting input of the first op amp has a • voltage of v_{in}. The current through the first transistor is

$$i \cong \frac{v_{\text{in}}}{R} \tag{17-16}$$

This current produces a collector voltage of

$$V_C \cong V_{CC} - v_{\text{in}} \tag{17-17}$$

Since this voltage drives the noninverting input of the second op amp, the inverting voltage is $V_{CC} - v_{\text{in}}$ to a close approximation. This implies that the voltage across the final R is

$$V_{CC} - (V_{CC} - v_{\text{in}}) = v_{\text{in}}$$

and the output current is

$$i_{\text{out}} \cong \frac{v_{\text{in}}}{R} \tag{17-18}$$

As before, this output current must satisfy the condition: $i_{\text{out}}/\beta_{\text{dc}}$ must be less than the $I_{\text{out(max)}}$ of the op amp. Furthermore, the load voltage cannot exceed $V_{CC} - v_{\text{in}}$ because of transistor saturation; therefore, $i_{\text{out}}R_L$ must be less than $V_{CC} - v_{\text{in}}$.

The circuit of Fig. 17-21 is rather nice, since it is a voltage-to-current converter that can drive a grounded load. A grounded load is used a lot more than a floating

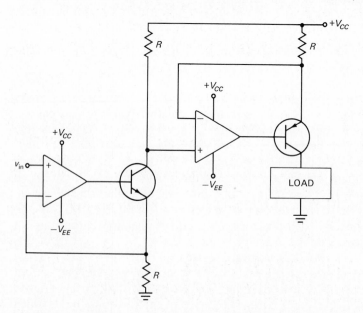

Fig. 17-21 *Voltage-controlled current source with output current proportional to input voltage.*

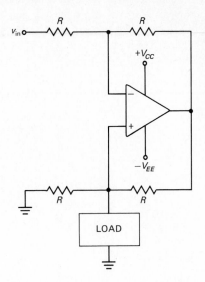

Fig. 17-22
Howland current source can supply bidirectional current.

load, so try to remember this circuit, because it is very useful and practical whenever you want to convert an input voltage to an output current. The limitation of this circuit is that only positive voltages can be generated across the load because of the unidirectional current in the output transistor.

HOWLAND CURRENT SOURCE

The current source of Fig. 17-21 produces a unidirectional load current (conventional flow down and electron flow up). Figure 17-22 shows a *Howland* current source; it can produce a bidirectional load current. The circuit is difficult to analyze mathematically. By writing four loop equations and rearranging as needed, it is possible to prove that

$$i_{\text{out}} \cong \frac{v_{\text{in}}}{R} \tag{17-19}$$

The maximum load current is approximately V_{CC}/R. One way to see this is to short the load. Then, the noninverting input is grounded, and we have a standard inverting voltage amplifier with a closed-loop voltage gain of -1. This means that the output voltage equals $-v_{\text{in}}$, where v_{in} can be positive or negative. Since the maximum output voltage of an op amp is within 1 or 2 V of V_{CC}, the maximum current through the lower right R is roughly V_{CC}/R. A circuit like this typically has a load resistance that is much smaller than R.

17-7 DIFFERENTIAL AND INSTRUMENTATION AMPLIFIERS

Figure 17-23a shows an op amp connected as a *differential* amplifier. It amplifies the difference of v_1 and v_2. The voltage gain of the inverting input is

Fig. 17-23
(a) *Differential amplifier.* (b) *Instrumentation amplifier.*

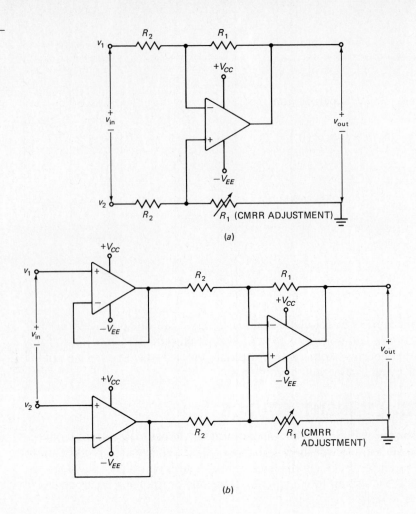

$$\frac{v_{\text{out}}}{v_1} \cong \frac{-R_1}{R_2}$$

The voltage gain of the noninverting input is

$$\frac{v_{\text{out}}}{v_2} \cong \left(\frac{R_1}{R_1 + R_2}\right)\left(\frac{R_1}{R_2} + 1\right)$$

which reduces to

$$\frac{v_{\text{out}}}{v_2} \cong \frac{R_1}{R_2}$$

Therefore, the voltage gains are equal in magnitude for each single-ended input.

Usually, the circuit is driven with a differential input; that is, a single input voltage v_{in} is applied between the v_1 and v_2 points. In this case, $v_{\text{in}} = v_1 - v_2$ and the output voltage is given by

$$v_{\text{out}} \cong - \frac{R_1}{R_2} v_{\text{in}} \tag{17-20}$$

(You have to apply the superposition theorem to derive this formula.) Note the adjustable resistor in Fig. 17-23a. This allows us to balance out common-mode signals. In this way, we can maximize the common-mode rejection ratio.

Figure 17-23b is an example of an *instrumentation* amplifier, a differential amplifier optimized for high input impedance and high CMRR. An instrumentation amplifier is typically used in applications in which a small differential voltage and a large common-mode voltage are the inputs. In this example of an instrumentation amplifier, voltage followers on each input produce a very high input impedance. Again, a CMRR adjustment is included to balance out the common-mode signals.

Manufacturers can put voltage followers and differential amplifiers on a single chip to get an IC instrumentation amplifier. Examples include the LH0036, LF352, and AD521. The LF352 is an example of a BIFET device, with JFETs used for the input voltage followers and bipolar transistors for the diff amp. This results in an input impedance of approximately $2(10^{12})$ Ω and input bias currents of only 3 pA. The JFETs have extremely low noise, an essential characteristic of a good instrumentation amplifier. The LF352 has other outstanding features, like a CMRR of at least 110 dB, a supply current of only 1 mA, and a single external resistor to control gain.

17-8 ACTIVE FILTERS

At lower frequencies, inductors are bulky and expensive. By using op amps, it is possible to build *active RC* filters that produce the sharp roll-off associated with passive *LC* filters. There are many possible filter designs. They are known as Butterworth, Chebyshev, Bessel, and others. You can find entire books on the subject of filter design. To keep our discussion to a reasonable length, we will discuss only the most popular active filters, known as the Butterworth or maximally flat filters.

LOW-PASS FILTER

In Fig. 17-24, a lag network is tacked on to the input side of a noninverting voltage amplifier. In the midband of the amplifier, the closed-loop voltage gain is

$$A_{CL} = \frac{R_1}{R_2} + 1$$

This is the gain from the noninverting input to the output. If the cutoff frequency f_c of the lag network is much lower than $f_{2(CL)}$, the overall voltage gain $v_{\text{out}}/v_{\text{in}}$ will be down 3 dB at

$$f_c = \frac{1}{2\pi RC} \tag{17-21}$$

This is the cutoff frequency of the lag network.

Fig. 17-24
One-pole low-pass filter.

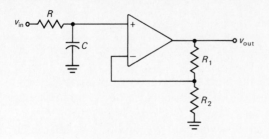

Above the cutoff frequency, the voltage gain rolls off at a rate of 20 dB per decade, equivalent to 6 dB per octave. The mathematical expression for voltage gain is

$$\frac{v_{\text{out}}}{v_{\text{in}}} = \frac{A_{CL}}{1 + jf/f_c} \qquad (17\text{-}22)$$

where v_{out} = output of filter
$\qquad v_{\text{in}}$ = input to filter
$\qquad A_{CL}$ = noninverting closed-loop voltage gain
$\qquad f$ = input frequency
$\qquad f_c$ = cutoff frequency of lag network

The active filter of Fig. 17-24 passes all frequencies up to the cutoff frequency, above which the frequency response rolls off. A filter like this is called a *low-pass* filter. You can recognize a low-pass filter because it usually has one or more lag networks. Another way to recognize a low-pass filter is the presence of $1 + jf/f_c$ factors in the denominator of the transfer function (the formula for voltage gain).

POLES

You often see the word "pole" used with filters. For instance, a circuit can be described as a one-pole low-pass filter, a two-pole low-pass filter, and so on. The number of poles in a circuit equals the number of j factors in the denominator of the transfer function. In Eq. (17-22), one j factor appears in the denominator, and so the circuit is a one-pole filter. Another way to get the number of poles is to count lag networks in the circuit. Since there is one lag network in Fig. 17-24, the circuit is a one-pole low-pass filter.

TWO-POLE LOW-PASS FILTER

Figure 17-25 is a two-pole low-pass filter because it has two lag networks. The capacitor for the first lag network receives a feedback voltage. This modifies the cutoff frequency and the response of the active filter. A mathematical analysis reveals that a closed-loop voltage gain A_{CL} of 1.586 is a critical value. For gains less than 1.586, the filter response approaches a linear phase shift with frequency (Bessel response); for gains greater than 1.586, you get ripples in the midband (Chebyshev response). When the gain is 1.586, you get the flattest possible response in the midband; this response is called the *Butterworth* or *maximally flat* response and is the one that is most popular.

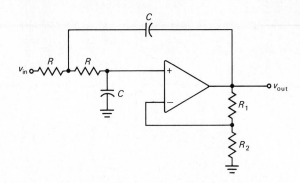

Fig. 17-25
Two-pole low-pass filter.

Since the closed-loop voltage gain must be 1.586 for a Butterworth response,

$$1.586 = \frac{R_1}{R_2} + 1$$

or

$$R_1 = 0.586 R_2$$

If $R_1 = 1$ kΩ, then $R_2 = 0.586$ kΩ. By using the nearest standard value, 560 Ω, we get approximately a maximally flat response.

When $A_{CL} = 1.586$, the cutoff frequency is

$$f_c = \frac{1}{2\pi RC} \qquad (17\text{-}23)$$

where f_c = cutoff frequency of two-pole Butterworth
    R = resistance of each resistor
    C = capacitance of each capacitor

A two-pole Butterworth filter like Fig. 17-25 has the advantage of using equal-value components. (Not all designs are like this.)

At the cutoff frequency, the overall voltage gain is down 3 dB. Above the cutoff frequency, the voltage gain decreases 40 dB per decade, equivalent to 12 dB per octave. This roll-off is twice as fast as before. The reason is that we have a two-pole filter, with each lag network producing a 20-dB decrease per decade. In general, a three-pole filter produces 60 dB per decade, a four-pole produces 80 dB per decade, and so on.

THREE-POLE LOW-PASS FILTER

The simplest way to build a three-pole low-pass filter is to cascade a one-pole filter (first section) with a two-pole (second section), as shown in Fig. 17-26. The voltage gain of the first section is optional; you can set it at whatever you want.

The voltage gain of the second section, however, affects the flatness of the overall response. If we keep the closed-loop gain at 1.586, then the overall gain will be down 6 dB (3 dB for each section) at the cutoff frequency, given by Eq. (17-23). By increasing the voltage gain of the second section slightly, we can offset this cumulative loss of voltage gain. Using an advanced mathematical derivation, we can prove

Fig. 17-26
Three-pole low-pass filter.

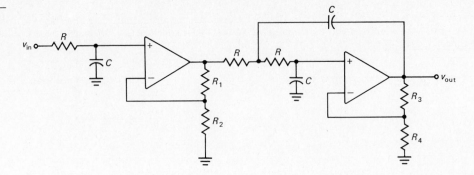

that an A_{CL} of 2 is the critical value needed for a maximally flat response. In this case,

$$R_3 = R_4$$

So, if $R_4 = 1$ kΩ, then $R_3 = 1$ kΩ.

When $A_{CL} = 2$, the cutoff frequency is

$$f_c = \frac{1}{2\pi RC} \tag{17-24}$$

where f_c = cutoff frequency of three-pole filter
$\quad\quad R$ = resistance in all sections
$\quad\quad C$ = capacitance in all sections

At the cutoff frequency, the overall voltage gain is down 3 dB. Above the cutoff frequency, the voltage gain decreases at the rate of 60 dB per decade, equivalent to 18 dB per octave.

MORE POLES

Figure 17-27 shows a four-pole low-pass filter, a cascade of a two-pole and a two-pole. If we try to use an A_{CL} of 1.586 for both sections, the voltage gain will be down 6 dB at the frequency given by Eq. (17-23). But by using different gains for each section, we can strike a compromise that produces a maximally flat response.

Fig. 17-27
Four-pole low-pass filter.

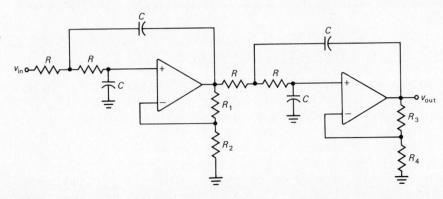

An advanced derivation shows that you need to use $A_{CL} = 1.152$ for the first section and $A_{CL} = 2.235$ for the second section. In all our Butterworth designs, the cutoff frequency is given by $1/2\pi RC$.

BUTTERWORTH TABLE

Table 17-2 lists the voltage gains you need to build low-pass Butterworth filters. As indicated, a one-pole filter has an optional A_{CL}. A two-pole filter needs an A_{CL} of 1.586, as previously discussed. A three-pole filter requires two sections, the first a one-pole with an optional A_{CL}, and the second a two-pole with an A_{CL} of 2.

A four-pole filter has two sections; the first section is a two-pole with a gain of 1.152, and the second section is a two-pole with a gain of 2.235. A five-pole filter has three sections, as shown in Fig. 17-28a. As indicated in Table 17-2, the first section is a one-pole with an optional A_{CL}, the second section is a two-pole with a gain of 1.382, and the third section is a two-pole with a gain of 2.382.

A six-pole filter is a cascade of three two-pole sections, as shown in Fig. 17-28b. From Table 17-2, the first section needs an A_{CL} of 1.068, the second a gain of 1.586, and the third a gain of 2.482. This Butterworth filter produces a roll-off rate of 120 dB per decade.

In all filters, the same resistance and capacitance values are used in the lag networks, a definite convenience in selection of components and ease of construction. Furthermore, the 3-dB cutoff frequency is always the same, given by

$$\frac{1}{2\pi RC}$$

HIGH-PASS FILTERS

You can change a low-pass Butterworth filter into a *high-pass* Butterworth filter by using lead networks instead of lag networks. The cutoff frequency is still given by $1/2\pi RC$, and the voltage gains are the same as those listed in Table 17-2. For instance, Fig. 17-29 shows a four-pole high-pass filter. Instead of lag networks, we use lead networks with resistances of R and capacitances of C. From Table 17-2, the first section needs an A_{CL} of 1.152 and the second section an A_{CL} of 2.235. With a filter like this, the overall voltage gain is down 3 dB at the cutoff frequency. Below the cutoff frequency, the voltage gain rolls off at a rate of 80 dB per decade.

Table 17-2. Gains for Butterworth Filters

Poles	Roll-off (decade)	1st section (1 or 2 poles)	2nd section (2 poles)	3rd section (2 poles)
1	20 dB	Optional		
2	40 dB	1.586		
3	60 dB	Optional	2	
4	80 dB	1.152	2.235	
5	100 dB	Optional	1.382	2.382
6	120 dB	1.068	1.586	2.482

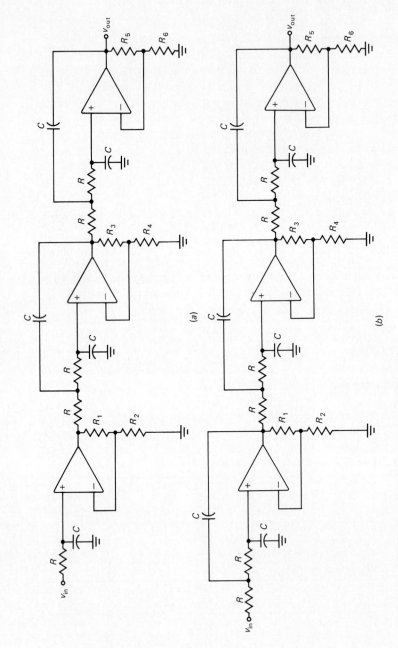

Fig. 17-28 *Low-pass filters.* (a) *Five-pole.* (b) *Six-pole.*

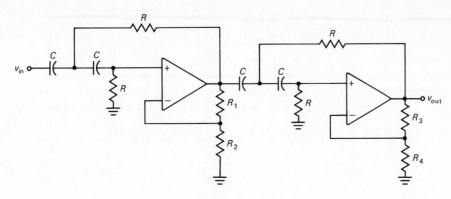

Fig. 17-29
Four-pole high-pass filter.

BANDPASS FILTERS

A *bandpass* filter has a lower cutoff frequency and an upper cutoff frequency. If the upper cutoff frequency is at least 10 times the lower cutoff frequency, you can cascade a low-pass filter and a high-pass filter. The low-pass filter can be designed using Table 17-2, and so it can have from one to six poles. Likewise, the high-pass filter can have from one to six poles.

PROBLEMS

Straightforward

17-1. In Fig. 17-1, $R_1 = 18$ kΩ and $R_2 = 2$ kΩ. What does the voltage gain equal?

17-2. In Fig. 17-2, $R_1 = 18$ kΩ, $R_2 = 2$ kΩ, $R_3 = 18$ kΩ, $C_{in} = 1$ μF, $C_{out} = 0.1$ μF, and $C_{BY} = 10$ μF. What is the midband voltage gain of the amplifier? What is the input cutoff frequency? The bypass cutoff frequency? If the load resistance is 10 kΩ, what is the output cutoff frequency?

17-3. What is the voltage gain in Fig. 17-30?

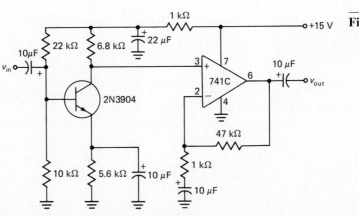

Fig. 17-30

17-4. Assume $R_S = 0$, $R_L = 10$ kΩ, and $\beta = 150$. Calculate the input, output, and bypass cutoff frequencies in Fig. 17-30 (use an f_{unity} of 1 MHz for the 741C). What is the bandwidth?

17-5. In Fig. 17-5, $R_1 = 18$ kΩ, $R_2 = 3$ kΩ, $R_3 = 6$ kΩ, and $r_{ds(on)} = 40$ Ω. What is the voltage gain when the gate voltage equals $V_{GS(off)}$? When the gate voltage is zero?

17-6. $R_1 = 47$ kΩ, $R_2 = 1$ kΩ, and $R_3 = 100$ kΩ in Fig. 17-6. Calculate the voltage gain for each of the $r_{ds(on)}$ values: 100 Ω, 1 kΩ, 10 kΩ, 100 kΩ, 1 MΩ, and 10 MΩ. Express the answers in decibels.

17-7. In Fig. 17-31, what are the voltage gain and bandwidth?

Fig. 17-31

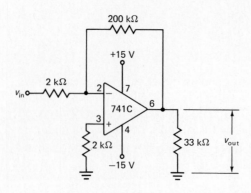

17-8. A 741C with an f_{unity} of 1 MHz is used in Fig. 17-12a. What is the bandwidth in each switch position?

17-9. In Fig. 17-17d, calculate v_{out} for $v_3 = 1$ V, $v_2 = 1$ V, $v_1 = 0$, and $v_0 = 0$.

17-10. The transistors of Fig. 17-32 have $\beta_{dc} = 75$. If v_{in} is 0.5 V, what is the base current?

Fig. 17-32

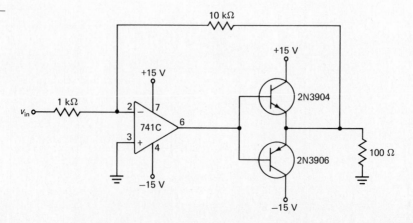

17-11. What does the output current equal in Fig. 17-33*a*? For normal operation, what is the largest load resistance that can be used?

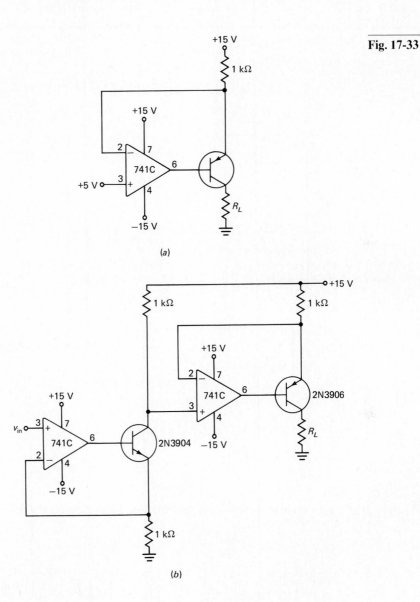

Fig. 17-33

(*a*)

(*b*)

17-12. If v_{in} is 5 V in Fig. 17-33*b*, what does the output current equal? If R_L is 470 Ω, what is the load voltage? If the load resistance is reduced to zero, what does the output current equal?

17-13. In Fig. 17-34, what does the output current equal if v_{in} is 1 V? The load voltage if R_L is 500 Ω?

Fig. 17-34

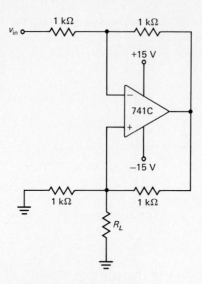

17-14. If $v_{in} = 2$ V in Fig. 17-34, what is the load voltage for an R_L of 1 kΩ? The voltage at the op-amp output?

17-15. What is the cutoff frequency in Fig. 17-35? At what rate does the voltage decrease beyond cutoff?

Fig. 17-35

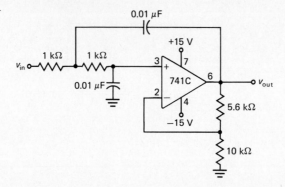

17-16. Calculate the cutoff frequency in Fig. 17-36. What is the roll-off rate below cutoff?

Fig. 17-36

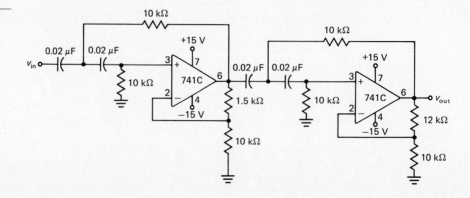

17-17. In Fig. 17-37, what is the lower cutoff frequency? The upper cutoff frequency? At what rate does the voltage gain decrease outside the passband of the amplifier?

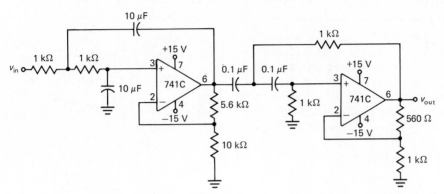

Fig. 17-37

Troubleshooting

17-18. If you measure a voltage gain of about 46 dB for Fig. 17-30, which of the following is a possible cause?
   a. No supply voltage
   b. Emitter bypass capacitor open
   c. Feedback bypass capacitor open
   d. Transistor open
17-19. What will typically happen to the output voltage of Fig. 17-32 if the 10 kΩ opens?
17-20. For a v_{in} of 5 V in Fig. 17-33b, there is no current through the load. Name at least three possible causes.

Design

17-21. Redesign the output stage of Fig. 17-30 so that the amplifier has an upper cutoff frequency of approximately 50 kHz. Assume an f_{unity} of 1 MHz.
17-22. Design a circuit like Fig. 17-15a to meet these specifications: voltage gain adjustable from 0 to 40 dB, input impedance at least 1 kΩ, the smallest resistor must be less than 10 kΩ to avoid excessive offsets.
17-23. Design a current source similar to Fig. 17-21 with these specifications: 15-V supplies, maximum v_{in} is 10 V, maximum output current is 10 mA.
17-24. Design a six-pole low-pass Butterworth filter with a cutoff frequency of 20 kHz.

Challenging

17-25. Derive the following closed-loop bandwidth for an inverting voltage amplifier like Fig. 17-7:

$$f_{2(CL)} = \frac{1}{1 - A_{CL}} f_{unity}$$

17-26. The four input voltages of Fig. 17-17d can be either 0 or 2 V. Because of this, there are 16 possible input conditions. Calculate the 16 possible output voltages.

17-27. Prove that Eq. (17-19) is true.

17-28. In Fig. 17-33, v_{in} is 2 V. Calculate the approximate value of all node voltages and all branch currents.

Computer

17-29. We want to know the bandwidth for an inverting amplifier with closed-loop voltage gains of $-1, -2, -3, \ldots, -10$. Write a program that prints out the 10 different combinations of gain and bandwidth.

17-30. Write a program that inputs the values of v_{in} and R for the circuit of Fig. 17-21. The program should print out the value of i_{out}.

17-31. Write a program that designs Butterworth active filters, using the data of Table 17-2. Your program should let you choose the number of poles.

Nonlinear Op-Amp Circuits

Monolithic op amps are inexpensive, versatile, and reliable. For this reason, they can be used not only for voltage amplifiers, current sources, and active filters, but also for active diode circuits, comparators, and waveshaping. This chapter is about *nonlinear* op-amp circuits, the kind in which the shape of the output signal is different from that of the input.

18-1 ACTIVE DIODE CIRCUITS

Op amps can enhance the performance of diode circuits. For one thing, an op amp with negative feedback reduces the effect of the *diode offset voltage,* allowing us to rectify, peak-detect, clip, and clamp *low-level* signals (those with amplitudes less than the diode offset voltage). And because of their buffering action, op amps can eliminate the effects of source and load on diode circuits. What follows is a discussion of active diode circuits.

HALF-WAVE RECTIFIER

Figure 18-1 is an *active half-wave rectifier*. When the input signal goes positive, the output goes positive and turns on the diode. The circuit then acts like a voltage follower, and the positive half cycle appears across the load resistor. On the other hand, when the input goes negative, the op-amp output goes negative and turns off the diode. Since the diode is open, no voltage appears across the load resistor. This is why the final output is almost a perfect half-wave signal.

The high gain of the op-amp virtually eliminates the effect of diode offset voltage. For instance, if the diode drop is 0.7 V and A is 100,000, the input that just turns on the diode is

$$v_{\text{in}} = \frac{0.7 \text{ V}}{100,000} = 7 \ \mu\text{V}$$

When the input is greater than 7 μV, the diode turns on, and the circuit acts like a

Fig. 18-1
Active half-wave rectifier.

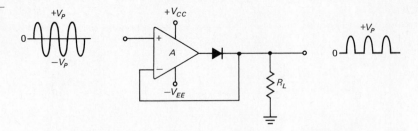

voltage follower. The effect is equivalent to reducing the offset potential by a factor of A. In symbols,

$$\phi' = \frac{\phi}{A} \tag{18-1}$$

where ϕ' = offset potential seen by input signal
ϕ = offset potential of diode

Because ϕ' is so small, the active half-wave rectifier is useful with low-level signals in the millivolt region.

ACTIVE PEAK DETECTOR

To peak-detect small signals, we can use an *active peak detector* like Fig. 18-2a. Again, the input offset potential ϕ' is in the microvolt region, which means that we can peak-detect low-level signals. When the diode is on, the heavy noninverting voltage feedback produces a Thevenin output impedance that approaches zero. This means that the charging time constant is very low, so that the capacitor can quickly charge to the positive peak value. On the other hand, when the diode is off, the capacitor has to discharge through R_L. Because the discharging time constant R_LC can be made much longer than the period of the input signal, we can get almost perfect peak detection of low-level signals.

If the peak-detected signal has to drive a small load, we can avoid loading effects by using an op-amp buffer. For instance, if we connect point A of Fig. 18-2a to point B of Fig 18-2b, the voltage follower isolates the small load resistor from the peak detector. This prevents the small load resistor from discharging the capacitor prematurely.

If possible, the R_LC time constant should be at least 100 times longer than the period of the lowest input frequency. If this condition is satisfied, the output voltage will be within 1 percent of the peak input. For instance, if the lowest frequency is 1 kHz, the period is 1 ms. In this case, the R_LC time constant should be at least 100 ms if you want an error of less than 1 percent.

Often, a *reset* function is included with an active positive peak detector, as shown in Fig. 18-2c. When the reset input is low, the transistor switch is off. This allows the circuit to work as previously described. When the reset input is high, the transistor switch is closed; this rapidly discharges the capacitor. The reason you may need a reset is because the long discharge time constant means that the capacitor will hold its charge for a long time, even though the input signal is removed. By using a high

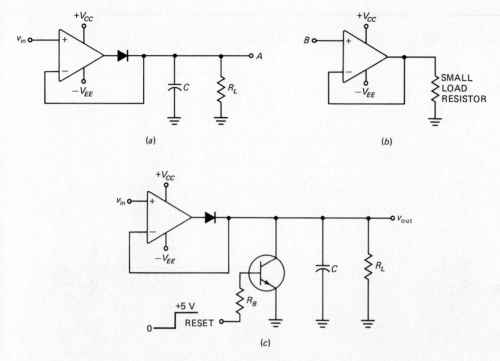

Fig. 18-2   (a) *Active peak detector.* (b) *Buffer amplifier.* (c) *Peak detector with reset.*

reset, we can quickly discharge the capacitor in preparation for another input signal with a different peak value.

ACTIVE POSITIVE LIMITER

Figure 18-3a is an *active positive limiter*. With the wiper all the way to the left, v_{ref} is zero and the noninverting input is grounded. When v_{in} goes positive, the error voltage drives the op-amp output negative and turns the diode on. This provides feedback, so that the final output v_{out} is at virtual ground for any positive value of v_{in}.

When v_{in} goes negative, the op-amp output is positive, which turns off the diode and opens the loop. As this happens, the virtual ground is lost, and the final output v_{out} is free to follow the negative half cycle of input voltage. This is why the negative half cycle appears at the output. The final output is

$$v_{out} = \frac{R_L}{R + R_L} v_{in}$$

To change the clipping level, all we do is adjust v_{ref} as needed. In this case, clipping occurs at v_{ref}, as shown in Fig. 18-3a. As usual, the offset voltage is reduced to ϕ' at the input, which means that the circuit is suitable for low-level inputs.

Figure 18-3b shows an active circuit that clips on both half cycles. Notice the back-to-back zener diodes in the feedback loop. Below the breakdown voltage, the circuit has a closed-loop gain of $-R_2/R_1$. When the output tries to exceed the zener

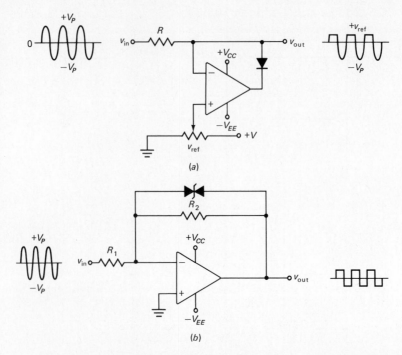

Fig. 18-3 (a) *Active positive limiter with adjustable reference voltage.* (b) *Zener diodes used in limiter to produce rectangular wave.*

voltage plus one forward diode drop, the zener diode breaks down and the output is clipped as shown.

ACTIVE POSITIVE CLAMPER

Figure 18-4*a* is an *active positive clamper*. The first negative input half cycle produces a positive op-amp output which turns on the diode. This allows the capacitor to charge to the peak value of the input with the polarity shown in Fig. 18-4*a*. Just beyond the negative peak, the diode turns off, the loop opens, and the virtual ground is lost. From Kirchhoff's voltage law,

$$v_{out} = v_{in} + V_P$$

Since V_P is being added to a sinusoidal input voltage, the final output waveform is shifted positively through V_P volts. In other words, we get the positive clamped waveform of Fig. 18-4*a*; it swings from 0 to $2V_P$. Again, the reduction of the diode offset voltage allows excellent clamping with low-level inputs.

Figure 18-4*b* shows the op-amp output. During most of the cycle, the op amp operates in negative saturation. Right at the negative input peak, however, the op amp produces a sharp positive-going pulse that replaces any charge lost by the clamping capacitor between negative input peaks.

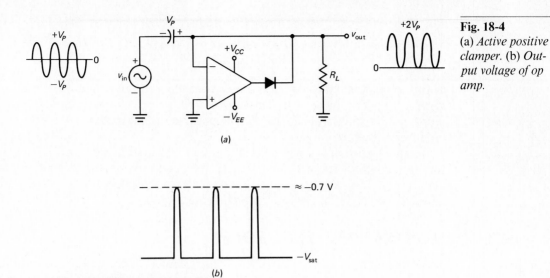

Fig. 18-4
(a) *Active positive clamper.* (b) *Output voltage of op amp.*

18-2 COMPARATORS

Often we want to compare one voltage with another to see which is larger. All we need is a yes/no answer. A *comparator* is a circuit with two input voltages (noninverting and inverting) and one output voltage. When the noninverting voltage is larger than the inverting voltage, the comparator produces a high output voltage. When the noninverting input is less than the inverting input, the output is low. The high output symbolizes the "yes" answer, and the low output stands for the "no" answer.

BASIC CIRCUIT

The simplest way to build a comparator is to connect an op amp without feedback resistors as shown in Fig. 18-5a. When the inverting input is grounded, the slightest input voltage (in fractions of a millivolt) is enough to saturate the op amp. For instance, if the supplies are 15 V, then the output compliance is from approximately -13 V to $+13$ V. With a 741C, the open-loop voltage gain is typically 100,000.

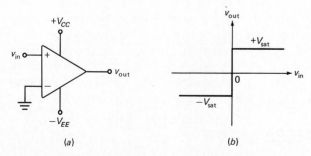

Fig. 18-5   (a) *Op amp used as comparator.* (b) *Transfer characteristic of comparator.*

Therefore, the input voltage needed to produce positive saturation is

$$v_{in} = \frac{13 \text{ V}}{100,000} = 0.13 \text{ mV}$$

This is so small that the transfer characteristic of Fig. 18-5b has what appears to be a vertical transition at $v_{in} = 0$. This is not actually vertical. With a 741C it takes $+0.13$ mV of input voltage to produce positive saturation and -0.13 mV to get negative saturation.

Because the input voltages needed to produce saturation are so small, the transition of Fig. 18-5b appears to be vertical. As an approximation, we will treat it as vertical. This means that a positive input voltage produces positive saturation, while a negative input voltage produces negative saturation.

MOVING THE TRIP POINT

The *trip point* (also called the threshold, the reference, and so on) of a comparator is the value of input voltage at which the output switches states (low to high, or vice versa). In Fig. 18-5a, the trip point is zero, because this is the value of input voltage at which the output switches states. When v_{in} is greater than the trip point, the output is high. When v_{in} is less than the trip point, the output is low. A circuit like Fig. 18-5a is often called a *zero-crossing detector*.

In Fig. 18-6a, a reference voltage is applied to the inverting input:

$$v_{ref} = \frac{R_2}{R_1 + R_2} V_{CC} \tag{18-2}$$

When v_{in} is less than v_{ref}, the error voltage is negative and the output is low. When v_{in} is greater than v_{ref}, the error voltage is positive and the output is high.

Incidentally, a bypass capacitor is typically used on the inverting input, as shown in Fig. 18-6a. This reduces the amount of power supply ripple and noise appearing at the inverting input. To be effective, the cutoff frequency of this bypass circuit (a lag network with an equivalent resistance of $R_1 \| R_2$) should be much lower than the ripple frequency.

Figure 18-6b shows the transfer characteristic. The trip point is now equal to v_{ref}. When v_{in} is slightly more than v_{ref}, the output of the comparator goes into positive saturation. When v_{in} is less than v_{ref}, the output goes into negative saturation. A comparator like this is sometimes called a *limit detector* because a positive output indicates that the input voltage exceeds a specific limit. With different values of R_1 and R_2, we can set the positive trip point anywhere between zero and V_{CC}.

If a negative trip point is preferred, then connect the $-V_{EE}$ to the voltage divider, as shown in Fig. 18-6c. Now a negative reference voltage is applied to the inverting input. When v_{in} is more positive than v_{ref}, the error voltage is positive and the output is high, as shown in Fig. 18-6d. When v_{in} is more negative than v_{ref}, the output is low.

SINGLE-SUPPLY COMPARATOR

As you know, a typical op amp like the 741C can be run from a single positive supply be grounding the $-V_{EE}$ pin, as shown in Fig. 18-7a. Now the output voltage has only

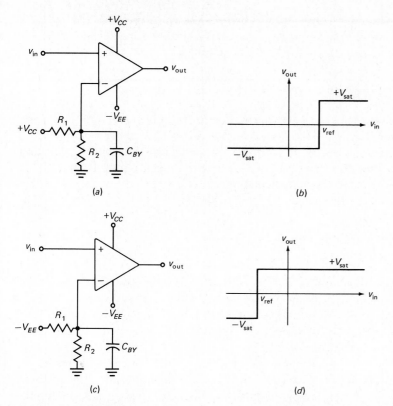

(a)

(b)

(c)

(d)

Fig. 18-6 (a) *Comparator with adjustable positive trip point.* (b) *Transfer characteristic.*
(c) *Comparator with negative trip point.* (d) *Transfer characteristic with negative trip point.*

one polarity, either a low or a high positive voltage. For instance, with V_{CC} equal to
+15 V, the output compliance is from approximately 1 or 2 V (low state) to around
13 or 14 V (high state).

The reference voltage applied to the inverting input is positive and equal to

$$v_{ref} = \frac{R_2}{R_1 + R_2} V_{CC}$$

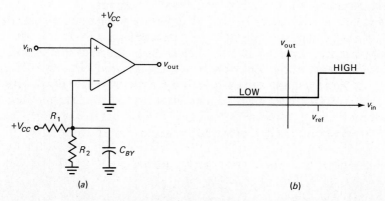

(a)

(b)

Fig. 18-7
(a) *Single-supply comparator.* (b) *Transfer characteristic.*

When v_{in} is greater than v_{ref}, the output is high, as shown in Fig. 18-7b. When v_{in} is less than v_{ref}, the output is low. In either case, the output has a positive polarity. For most digital applications, this is the kind of comparator output that is preferred.

SPEED PROBLEMS

An op amp like a 741C can be used as a comparator, but it has speed limitations. As you know, the slew rate limits the rate of output voltage change. With a 741C, the output can change no faster than 0.5 V/μs. Because of this, a 741C takes more than 50 μs to switch between a low output of -13 V and a high output of $+13$ V. One approach to speeding up the switching action is to use faster-slew-rate op amp, like the 318. Since it has a slew rate of 70 V/μs, it can switch from -13 V to $+13$ V in approximately 0.3 μs.

IC COMPARATORS

The compensating capacitor found in a typical op amp is the source of the slew-rate problem. For linear op-amp circuits, this capacitor is essential because it rolls off the open-loop voltage gain at a rate of 20 dB per decade and prevents oscillations. Furthermore, the typical op amp has a class B push-pull output stage that ultimately determines the output compliance.

A comparator is a nonlinear circuit, so there is really no need to include a compensating capacitor. Furthermore, in most comparator applications it's better to let the user determine the output compliance. For these two reasons, a manufacturer can redesign the typical op amp by deleting the compensating capacitor and changing the output stage. When an IC has been optimized for use as a comparator, the device is listed in a separate section of the manufacturer's catalog. In other words, you will find op amps in one place and comparators in another.

Figure 18-8a is a simplified schematic diagram for an IC comparator. The input stage is a diff amp (Q_1 and Q_2). A current mirror, Q_6 and Q_7, supplies the tail current. As before, a current mirror, Q_3 and Q_4, is an active load. The output stage is a single transistor Q_5 with an open collector. The manufacturer leaves this collector open on purpose. This allows the user to connect any desired load resistor and positive supply voltage.

For the circuit to work properly, you have to connect the open collector to an external resistor and supply voltage, as shown in Fig. 18-8b. The resistor is called a *pullup resistor* because it literally pulls the output voltage up to the supply voltage when the output transistor is cut off. When the transistor is saturated, the output voltage is low. Basically, the output stage is a transistor switch. This is why the comparator produces a two-state output, either a low or a high voltage.

When the noninverting input is more positive than the inverting input, the base voltage of Q_5 decreases, and the transistor cuts off. This means that the output voltage is high and equal to $+V$. On the other hand, when the noninverting input is less positive than the inverting input, the base voltage of Q_5 increases, and the transistor goes into saturation. Therefore, the output voltage is low, only a few tenths of a volt.

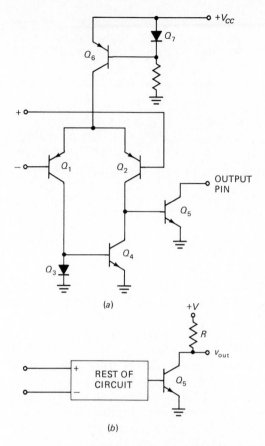

(a)

(b)

Fig. 18-8 (a) *Simplified schematic diagram of IC comparator.* (b) *Open-collector transistor needs external pullup resistor to work properly.*

With no compensating capacitor, the output in Fig. 18-8*a* can slew very rapidly because only small stray capacitances remain in the circuit. One limitation on the switching speed is the amount of capacitance across Q_5. This output capacitance is the sum of collector capacitance and stray wiring capacitance. The output time constant is the product of the pullup resistance and the output capacitance. For this reason, the smaller the pullup resistance in Fig. 18-8*b*, the faster the output voltage can change. Typically, R is from a couple of hundred to a couple of thousand ohms.

Examples of IC comparators are the LM311, LM339, and NE529. These all have an open-collector output stage, and so you have to connect the output pin to a pullup resistor and positive supply voltage. Because of their high slew rates, these IC comparators can switch output states in a microsecond or less. The LM339 is actually a quad comparator, four comparators in a single IC package. Because it is inexpensive and easy to use, the LM339 has become a popular comparator for general-purpose applications.

Fig. 18-9
(a) *LM339
comparator.* (b)
*Transfer charac-
teristic.*

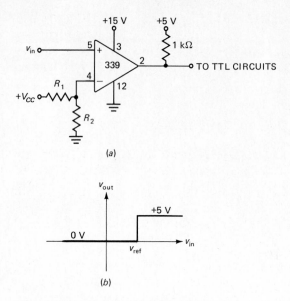

(a)

(b)

DRIVING TTL DEVICES

Often, the output of a comparator drives TTL (transistor-transistor logic) devices; these are digital integrated circuits used in computers, digital systems, and other switching applications. Typical input voltages for TTL devices are between 0 and $+5$ V. Figure 18-9a shows how an LM339 can be connected to *interface* with TTL devices. Notice that the open-collector output is connected to a supply of $+5$ V through a pullup resistor of 1 kΩ. Because of this, the output can be either 0 V or $+5$ V, as shown in Fig. 18-9b. This drive is ideal for TTL devices.

PART OF AN A/D CONVERTER

Figure 18-10 is part of an *analog-to-digital* (A/D) converter used in digital voltmeters and many other applications. The input voltage being measured or converted is applied to the noninverting input. A staircase voltage drives the inverting input. As the inverting voltage increases, the error voltage becomes less positive. Somewhere along the staircase, the inverting input becomes more positive than the noninverting input. When this happens, the output of the comparator switches to the low state.

Fig. 18-10
*Comparator as
part of analog-to-
digital converter.*

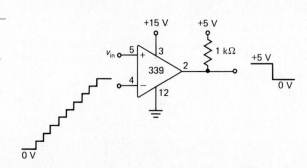

The amount of time it takes the staircase voltage to exceed v_{in} is the key to how the circuit works. The larger v_{in} is, the more time it takes for the staircase voltage to exceed v_{in}. In other words, the time is directly proportional to v_{in}. With other circuits not shown, we can measure this time and display the voltage with a seven-segment indicator.

18-3 THE WINDOW COMPARATOR

An ordinary comparator indicates when the input voltage exceeds a certain limit or threshold. A *window comparator* (also called a double-ended limit detector) detects when the input voltage is between two limits. This section will discuss two examples of window comparators.

OP-AMP EXAMPLE

Figure 18-11*a* is an example of a window comparator that uses an op amp. The noninverting input is referenced by a Thevenin voltage $+V_{CC}/3$, and the inverting input by a Thevenin voltage $+V_{CC}/4$. Since V_{CC} is 12 V, the Thevenin references are $+4$ V for the noninverting input and $+3$ V for the inverting input.

When the input voltage is zero, the upper diode is on and the lower diode is off. Since the noninverting input is clamped one diode drop above the input voltage, the noninverting input is $+0.7$ V. The inverting input, on the other hand, is at $+3$ V. Therefore, the error voltage is negative and the comparator output is low.

As the input voltage increases, the noninverting input also increases, remaining 0.7 V higher than v_{in}. When v_{in} reaches $+2.3$ V, the noninverting input is clamped at $+3$ V. Since the inverting input is still at $+3$ V, the error voltage is now zero. If the input voltage v_{in} rises above $+2.3$ V, the output of the comparator goes high. An input of $+2.3$ V is a critical value because the output of the comparator is on the verge of switching from low to high. This input voltage is called the *lower trip point* (LTP). When v_{in} is greater than the LTP, the output voltage switches into the high state, as shown in Fig. 18-11*b*.

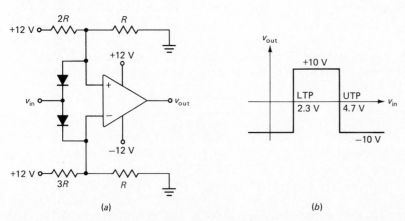

(a)   (b)

Fig. 18-11 (a) *Window comparator with two diodes and one op amp.* (b) *Transfer characteristic.*

As the input voltage increases, the comparator output stays high until v_{in} equals $+4.7$ V. At this value of input voltage, the lower diode is on and the inverting input is at $+4$ V; therefore, the error voltage is again at zero. Once again, the comparator is on the verge of switching its output. When v_{in} is greater than $+4.7$ V, the error voltage goes negative, driving the output into the low state. An input of $+4.7$ V is called the *upper trip point* (UTP) because slightly above this level, the output switches back to the low state.

The transfer characteristic of Fig. 18-11*b* is called a *window* because the output is high only when the input is between the LTP and the UTP. With a V_{CC} of $+12$ V, the window comparator of Fig. 18-11*a* has an LTP of $+2.3$ V and a UTP of $+4.7$ V. By changing the voltage dividers, we can change the width of the window. The window comparator is a useful circuit whenever we are trying to check whether the input is between two limits.

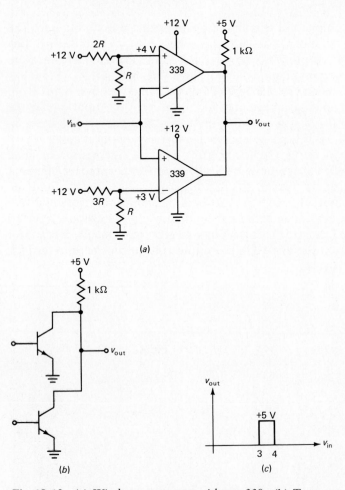

Fig. 18-12 (a) *Window comparator with two 339s.* (b) *Two open-collector outputs are connected to pullup resistor.* (c) *Transfer characteristic.*

USING THE LM339

Figure 18-12a shows how you can connect two comparators ($\frac{1}{4}$ of an LM339) to get a window comparator. With a positive supply of +12 V, the reference voltages are +4 V for the upper comparator and +3 V for the lower comparator. When v_{in} is between +3 and +4 V, both comparators have a positive error voltage, and their output transistors are open (Fig. 18-12b). Because of this, the final output is high. When v_{in} is less than +3 V or more than +4 V, one of the comparators will have a saturated transistor and the other will have a cut-off transistor. The saturated transistor pulls the output voltage down to a low level. Figure 18-12c shows the transfer characteristic. The LTP is +3 V and the UTP is +4 V.

18-4 THE SCHMITT TRIGGER

If the input to a comparator contains noise, the output may be erratic when v_{in} is near a trip point. For instance, with a zero-crossing detector, the output is high when v_{in} is positive and low when v_{in} is negative. If the input contains a noise voltage with a peak of 1 mV or more, then the comparator will detect the zero crossings produced by the noise. A similar thing happens when the input is near the trip points of a limit detector or a window comparator; the noise causes the output to jump back and forth between its low and high states. We can avoid this noise triggering by using a *Schmitt trigger,* a comparator with positive feedback.

BASIC CIRCUIT

Figure 18-13a shows an op-amp Schmitt trigger. Because of the voltage divider, we have positive voltage feedback. When the output voltage is positively saturated, a positive voltage is fed back to the noninverting input; this positive input holds the output in the high state. On the other hand, when the output voltage is negatively saturated, a negative voltage is fed back to the noninverting input, holding the output in the low state. In either case, the positive feedback reinforces the existing output state.

The feedback fraction is

$$B = \frac{R_2}{R_1 + R_2} \tag{18-3}$$

When the output is positively saturated, the reference voltage applied to the noninverting input is

$$v_{ref} = +BV_{sat} \tag{18-4}$$

When the output is negatively saturated, the reference voltage is

$$v_{ref} = -BV_{sat} \tag{18-5}$$

As will be shown, these reference voltages are the same as the trip points of the circuit: UTP = $+BV_{sat}$, and LTP = $-BV_{sat}$.

The output will remain in a given state until the input exceeds the reference voltage for that state. For instance, if the output is positively saturated, the reference

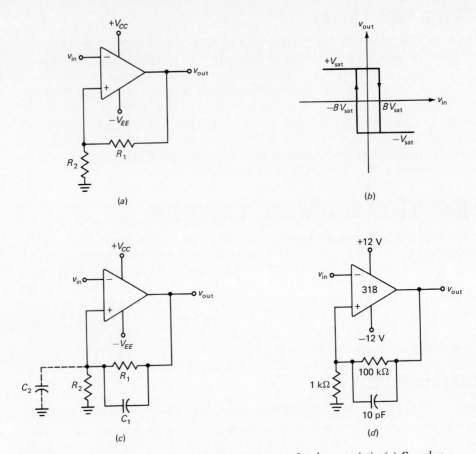

Fig. 18-13 (a) *Schmitt trigger.* (b) *Hysteresis in transfer characteristic.* (c) *Speedup capacitor compensates for stray capacitance.* (d) *Example.*

voltage is $+BV_{sat}$. The input voltage v_{in} must be increased to slightly more than $+BV_{sat}$. Then the error voltage reverses and the output voltage changes to the low state, as shown in Fig. 18-13b. Once the output is in the negative state, it will remain there indefinitely until the input voltage becomes more negative than $-BV_{sat}$. Then the output switches from negative to positive (Fig. 18-13b).

HYSTERESIS

Positive feedback has an unusual effect on the circuit. It forces the reference voltage to have the same polarity as the output voltage; the reference voltage is positive when the output is high and negative when the output is low. This is why we get an upper and a lower trip point. In a Schmitt trigger, the difference between the two trip points is called *hysteresis*. Because of the positive feedback, the transfer characteristic has the hysteresis shown in Fig. 18-13b. If there were no positive feedback, B would equal zero and the hysteresis would disappear, because the trip points would both equal zero. But there is positive feedback, and this spreads those trip points as shown.

Some hysteresis is desirable because it prevents noise from causing false triggering. Imagine a Schmitt trigger with no hysteresis. Then any noise present at the input will cause the Schmitt trigger to randomly jump from the low state to the high state, and vice versa. Next, visualize a Schmitt trigger with hysteresis. If the peak-to-peak noise voltage is less than the hysteresis, there is no way the noise can produce false triggering. A circuit with enough hysteresis is immune to noise triggering. For instance, if the UTP equals $+1$ V and the LTP equals -1 V, peak-to-peak noise less than 2 V cannot trigger the circuit.

SPEEDUP CAPACITOR

Besides suppressing the effects of noise, the positive feedback speeds up the switching of output states. When the output voltage begins to change, this change is fed back to the noninverting input and amplified, forcing the output to change faster. Sometimes a capacitor C_1 is connected in parallel with R_1, as shown in Fig. 18-13c. Known as a *speedup capacitor,* it helps to cancel out the lag network formed by the stray capacitance C_2 across R_2. This stray capacitance has to be charged before the noninverting input voltage can change. The speedup capacitor supplies this charge.

To neutralize the stray capacitance, the capacitive voltage divider formed by C_2 and C_1 must have the same impedance ratio as the resistive voltage divider:

$$\frac{X_{C2}}{X_{C1}} = \frac{R_2}{R_1}$$

or

$$C_1 = \frac{R_2}{R_1} C_2 \tag{18-6}$$

where C_1 = speedup capacitance
R_2 = resistance from noninverting input to ground
R_1 = feedback resistance
C_2 = stray capacitance across R_2

The value of C_1 given by this equation is the minimum value that neutralizes the lag effects of stray capacitance C_2. As long as C_1 is equal to or greater than the value given by Eq. (18-6), the output will switch states at maximum speed. Since you often have to estimate the value of stray capacitance, it's best to make C_1 at least two times larger than the value given by Eq. (18-6). In typical circuits, C_1 is from 10 pF to 100 pF.

As an example, Fig. 18-13d shows a 318 op amp connected as a zero-crossing detector with hysteresis. Because of the 12-V supplies, V_{sat} is approximately 10 V. With B approximately equal to 0.01, the UTP is $+0.1$ V and the LTP is -0.1 V. A speedup capacitor of 10 pF neutralizes any stray capacitance across R_2.

An alternative form of Eq. (18-6) is

$$R_1C_1 \geqslant R_2C_2 \tag{18-6a}$$

This says that the time constant of the speedup section must be equal to or greater than the time constant of the stray-capacitance section.

MOVING THE TRIP POINTS

Figure 18-14*a* shows how to move the trip points. An additional resistor R_3 is connected between the noninverting input and $+V_{CC}$. This sets up the *center* of the hysteresis loop:

$$v_{cen} = \frac{R_2}{R_2 + R_3} V_{CC} \tag{18-7}$$

The positive feedback spreads the trip point on each side of the center voltage. To understand why this happens, apply Thevenin's theorem to get Fig. 18-14*b*. The feedback fraction is

$$B = \frac{R_2 \parallel R_3}{R_1 + R_2 \parallel R_3} \tag{18-8}$$

When the output is positively saturated, the noninverting reference voltage is

$$UTP = v_{cen} + BV_{sat} \tag{18-9}$$

When the output is negatively saturated, the noninverting voltage is

$$LTP = v_{cen} - BV_{sat} \tag{18-10}$$

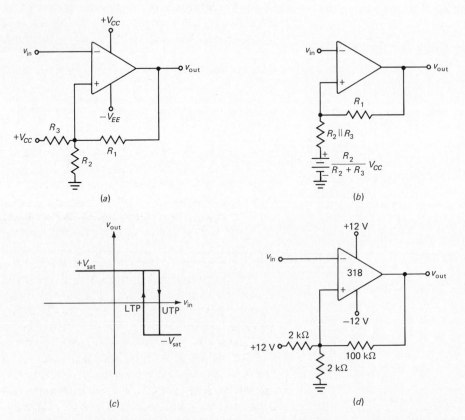

Fig. 18-14 (a) *Schmitt trigger with positive reference voltage on noninverting input.* (b) *Equivalent circuit.* (c) *Center of hysteresis is positive in transfer characteristic.* (d) *Example.*

Figure 18-14c is the transfer characteristic. You can calculate the center voltage using Eq. (18-7). After working out the value of B from Eq. (18-8), you can then calculate the trip points using Eqs. (18-9) and (18-10).

As an example, Fig. 18-14d shows a 318 used as a limit detector with hysteresis. The center voltage is

$$v_{cen} = \frac{2\ k\Omega}{4\ k\Omega}\ 12\ V = 6\ V$$

The feedback fraction is

$$B = \frac{2\ k\Omega\ \|\ 2\ k\Omega}{101\ k\Omega} \cong 0.01$$

The trip points are

$$UTP = 6\ V + 0.01(10\ V) = 6.1\ V$$

and

$$LTP = 6\ V - 0.01(10\ V) = 5.9\ V$$

Incidentally, if a speedup capacitor is used, it would be connected across the 100-kΩ resistor.

NONINVERTING CIRCUIT

Figure 18-15a shows a *noninverting* Schmitt trigger. This is a zero-crossing detector with hysteresis, as shown in Fig. 18-15b. Here is how it works. Assume that the output is negatively saturated. Then the feedback voltage is negative. This feedback voltage will hold the output in negative saturation until the input voltage becomes positive enough to make the error voltage positive. When this happens, the output goes into positive saturation. With the output in positive saturation, the feedback voltage is positive. To switch output states, the input voltage has to become negative enough to make the error voltage negative. When it does, the output can change to the negative state.

Here is how to derive the trip points. The output changes states when v_{error} crosses through zero. When v_{error} is zero,

$$v_{in} = i_{in}R_2$$

Because of the virtual ground, almost all the input current passes through R_1, and

$$v_{out} = -i_{in}R_1$$

or

$$v_{in} = -v_{out}\frac{R_2}{R_1}$$

When the output is in negative saturation, $v_{out} = -V_{sat}$ and

$$UTP = V_{sat}\frac{R_2}{R_1} \tag{18-11}$$

Fig. 18-15
*Noninverting
Schmitt trigger
with zero center
voltage.*

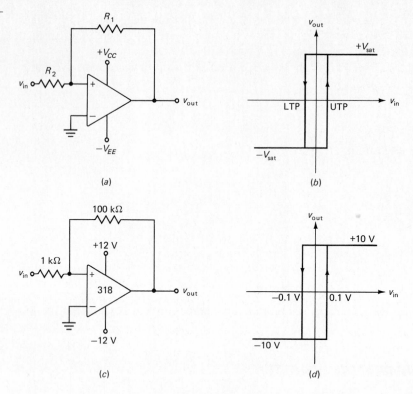

(a)

(b)

(c)

(d)

When the output is in positive saturation, $v_{out} = +V_{sat}$ and

$$\text{LTP} = -V_{sat} \frac{R_2}{R_1} \tag{18-12}$$

For instance, Fig. 18-15c shows a 318 used as a noninverting Schmitt trigger. If V_{sat} is 10 V, the UTP is $+0.1$ V and the LTP is -0.1 V. Figure 18-15d shows the transfer characteristic.

If you want to move the trip points, then apply a reference voltage to the inverting input, as shown in Fig. 18-16a. The reference voltage equals

$$v_{ref} = \frac{R_4}{R_3 + R_4} V_{CC} \tag{18-13}$$

By a derivation similar to that given earlier, we can prove that the center voltage is

$$v_{cen} = \left(1 + \frac{R_2}{R_1}\right) v_{ref} \tag{18-14}$$

The trip points are spread on both sides of this center:

$$\text{UTP} = v_{cen} + \frac{R_2}{R_1} V_{sat} \tag{18-15}$$

and

$$\text{LTP} = v_{cen} - \frac{R_2}{R_1} V_{sat} \tag{18-16}$$

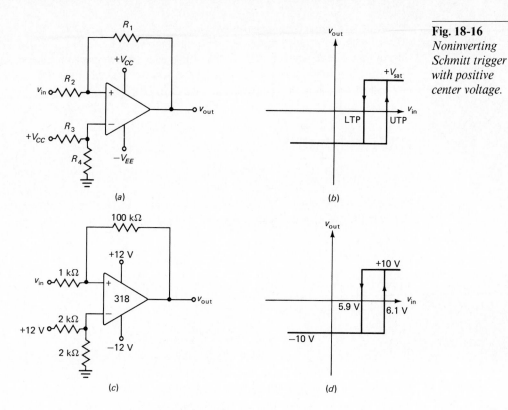

Fig. 18-16
Noninverting Schmitt trigger with positive center voltage.

For example, the noninverting Schmitt trigger of Fig. 18-16c has a reference voltage of 6 V and a center voltage of

$$v_{\text{cen}} = \left(1 + \frac{1 \text{ k}\Omega}{100 \text{ k}\Omega}\right) 6 \text{ V} \cong 6 \text{ V}$$

Notice that the center voltage is approximately the same as the reference voltage because only a small amount of positive feedback is used. Since $R_2/R_1 = 0.01$, the trip points are

$$\text{UTP} = 6 \text{ V} + 0.01(10 \text{ V}) = 6.1 \text{ V}$$

and

$$\text{LTP} = 6 \text{ V} - 0.01(10 \text{ V}) = 5.9 \text{ V}$$

Figure 18-16d summarizes the circuit operation with a transfer characteristic.

18-5 THE INTEGRATOR

An *integrator* is a circuit that performs the mathematical operation of integration because it produces an output voltage that is proportional to the integral of the input. A common application is to use a constant input voltage to produce a *ramp* of output voltage. (A ramp is a linearly increasing or decreasing voltage.) For instance, if you drive a 741C with a voltage step, the output slews at a rate of 0.5 V/μs. This means that the output voltage changes 0.5 V during each microsecond. This is an example of a ramp, a voltage that changes linearly with time. With an op amp, we can build an integrator, a circuit that produces a well-defined ramp output for a rectangular or constant input.

BASIC CIRCUIT

Figure 18-17a is an op-amp integrator. Here is how it works. The typical input to an integrator is a rectangular pulse like Fig. 18-17b. The capitalized V_{in} represents a constant voltage during pulse time T. Visualize V_{in} applied to the left end of R. Because of the virtual ground, the input current is constant and equals

$$I_{in} \cong \frac{V_{in}}{R}$$

Approximately all this current goes to the capacitor. The basic capacitor law says that

$$C = \frac{Q}{V}$$

or

$$V = \frac{Q}{C} \tag{18-17}$$

Since a constant current is flowing into the capacitor, the charge Q increases linearly. This means that the capacitor voltage increases linearly with the polarity shown in Fig. 18-17a. Because of the phase reversal of the op amp, the output voltage is a negative ramp, as shown in Fig. 18-17c. At the end of the pulse period, the input voltage returns to zero, and the charging current stops. Because the capacitor holds its charge, the output voltage remains constant at a negative level.

To get a formula for the output voltage, divide both sides of Eq. (18-17) by T:

$$\frac{V}{T} = \frac{Q/T}{C}$$

Since the charging is constant, we can write

$$\frac{V}{T} = \frac{I}{C}$$

or

$$V = \frac{IT}{C} \tag{18-18}$$

where V = capacitor voltage
I = charging current, V_{in}/R
T = charging time
C = capacitance

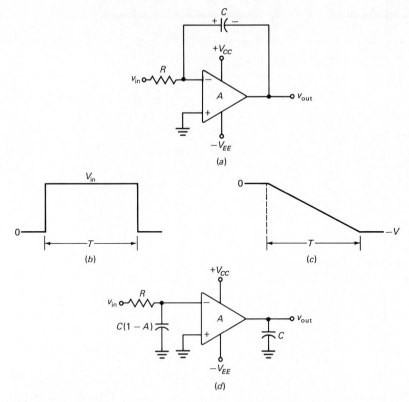

Fig. 18-17 (a) *Integrator.* (b) *Typical input is rectangular pulse.* (c) *Typical output is ramp.* (d) *Equivalent circuit with Miller capacitances.*

This is the voltage across the capacitor. Because of the phase reversal, $v_{\text{out}} = -V$. For instance, if $I = 4$ mA, $T = 2$ ms, and $C = 1$ μF, then the capacitor voltage at the end of the charging period is

$$V = \frac{(4 \text{ mA})(2 \text{ ms})}{1 \ \mu\text{F}} = 8 \text{ V}$$

Because of the phase reversal, the output voltage is -8 V after 2 ms.

A final point: Because of the Miller effect, an integrator can be visualized as shown in Fig. 18-17d. The time constant for the input lag network is

$$\tau = RC(1 - A)$$

For the integrator to work properly, this time constant should be much greater than the width T of the input pulse (at least 10 times greater). In the typical op-amp integrator, the large value of A produces an extremely long time constant, and so you rarely have any problem satisfying the condition that τ be much greater than T.

ROLLING OFF THE GAIN AT ZERO

The circuit of Fig. 18-17a needs a slight modification to make it practical. Because the capacitor acts like an open to dc signals, the closed-loop voltage gain equals the

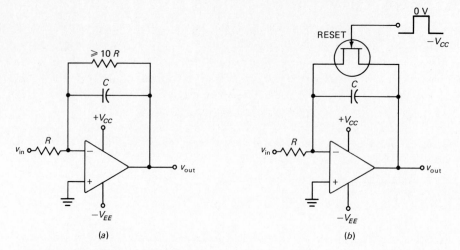

Fig. 18-18 (a) *Resistor across capacitor reduces output offset voltage.* (b) *JFET used to reset integrator.*

open-loop voltage gain at zero frequency. This will produce too much output offset voltage. Without negative feedback at zero frequency, the circuit will treat input offsets the same as a valid input signal. The input offsets will eventually charge the capacitor and drive the output into positive or negative saturation.

One way to reduce the effect of input offsets is to roll off the voltage gain at low frequencies by inserting a resistor in parallel with the capacitor, as shown in Fig. 18-18a. This resistor should be at least 10 times larger than the input resistor. If the added resistance equals $10R$, the closed-loop voltage gain is -10 and the output offset voltage is greatly reduced. The integrator works approximately as previously described because the bulk of the input current still goes to the capacitor.

JFET RESET

Another way to suppress the effect of input offsets is to use a JFET *reset* switch, as shown in Fig. 18-18b. This allows us to discharge the capacitor just before the pulse is applied to the input. For instance, Fig. 18-18b shows a JFET switch that can reset the integrator. When the gate voltage is $-V_{CC}$, the JFET switch is open and the circuit works as previously described. When the gate voltage is changed to 0 V, the JFET switch closes and discharges the capacitor. When the gate voltage again goes negative, the JFET opens, and the capacitor can be recharged by the next input pulse.

18-6 THE DIFFERENTIATOR

A *differentiator* is a circuit that performs the mathematical operation of differentiation. It produces an output voltage proportional to the slope of the input voltage. Common applications of a differentiator are to detect the leading and trailing edges of a rectangular pulse, or to produce a rectangular output from a ramp input.

RC DIFFERENTIATOR

A lead network like Fig. 18-19*a* can be used to differentiate the input signal. When used like this, it is called an *RC* differentiator. Instead of a sinusoidal signal, the typical input is a rectangular pulse, as shown in Figure 18-19*b*. The output of the circuit is positive and negative spikes. The positive spike occurs at the same instant as the leading edge of the input; the negative spike occurs at the same instant as the trailing edge. Other circuits can use these spikes in timing applications.

To understand how the *RC* differentiator works, look at Fig. 18-19*c*. When the input voltage changes from zero to *V*, the capacitor begins to charge exponentially, as shown. After approximately five time constants, the capacitor voltage is within 1 percent of the final voltage *V*. To satisfy Kirchhoff's voltage law, the voltage across the resistor of Fig. 18-19*a* is

$$v_R = v_{in} - v_C$$

This means that the output voltage suddenly jumps from zero to *V*, then decays exponentially, as shown in Fig. 18-19*c*. On the trailing edge of a pulse, the input voltage steps negatively, and, by a similar argument, we get a negative spike. Incidentally, notice that each spike in Fig. 18-19*b* has a peak value of approximately *V*, the size of the voltage step.

If an *RC* differentiator is to produce narrow spikes, the time constant should be at least 10 times smaller than the pulse width *T*. For instance, if the pulse width is 1 ms, then the *RC* time constant should be less than or equal to 0.1 ms. Figure 18-19*d* shows an *RC* differentiator with a time constant of 0.1 ms. The smaller the time constant, the sharper the spikes.

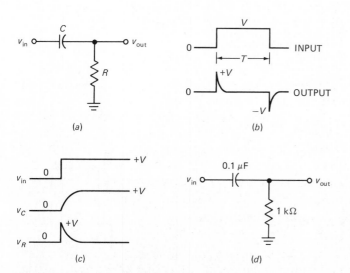

Fig. 18-19   (a) *RC differentiator.* (b) *Rectangular input pulse produces narrow output spikes.* (c) *Voltage waveforms.* (d) *Example.*

OP-AMP DIFFERENTIATOR

Figure 18-20*a* shows an op-amp differentiator. Notice the similarity to the op-amp integrator. The difference is that the resistor and capacitor are interchanged. When the input voltage varies, the capacitor charges or discharges. Because of the virtual ground, the capacitor current passes through the feedback resistor, producing a voltage. This voltage is proportional to the slope of the input voltage.

An input often used with op-amp differentiators is a ramp like the top waveform of Fig. 18-20*b*. Because of the virtual ground, all the input voltage appears across the capacitor. The ramp of voltage implies that the capacitor current is constant. Since all this constant current flows through the feedback resistor, we get an inverted pulse at the output, as shown in Fig. 18-20*b*.

Here is how to derive the current. At the end of the ramp, the capacitor voltage is

$$V = \frac{Q}{C}$$

Dividing both sides by the ramp time gives

$$\frac{V}{T} = \frac{Q/T}{C}$$

or

$$\frac{V}{T} = \frac{I}{C}$$

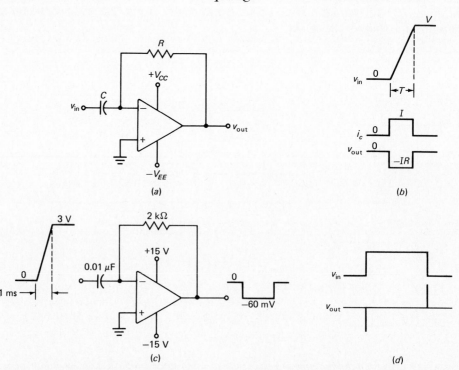

(a)

(b)

(c)

(d)

Fig. 18-20 (a) *Op-amp differentiator.* (b) *Ramp input produces rectangular output.* (c) *Example.* (d) *Rectangular input produces spike output.*

Solving for current, we get

$$I = \frac{CV}{T} \tag{18-19}$$

where I = capacitor current
$\quad C$ = capacitance
$\quad V$ = voltage at end of ramp
$\quad T$ = time between start and end of ramp

This current is the key. Once you have calculated it, you can get the output voltage with

$$v_{\text{out}} = -IR \tag{18-20}$$

As an example, Fig. 18-20c shows a ramp of 3 V driving an op-amp differentiator. The capacitor current is

$$I = \frac{(0.01 \ \mu\text{F})(3 \ \text{V})}{1 \ \text{ms}} = 30 \ \mu\text{A}$$

The output voltage is

$$v_{\text{out}} = (-30 \ \mu\text{A})(2 \ \text{k}\Omega) = -60 \ \text{mV}$$

Therefore, the output waveform is a negative pulse with a peak of -60 mV.

On an oscilloscope, the leading edge of a rectangular pulse may look perfectly vertical. But if you shorten the time base enough, you will see that the leading edge is usually a rising exponential wave. As an approximation, we can treat this rising exponential as a positive ramp.

One common application of the op-amp differentiator is to produce very narrow spikes, as shown in Fig. 18-20d. The leading edge of the pulse is approximately a positive ramp, so that the output will be a negative-going spike of very short duration. Similarly, the trailing edge of the input pulse is approximately a negative ramp, so that the output is a very narrow positive spike. The advantage of an op-amp differentiator over a simple RC differentiator is that the spikes are coming from a low-impedance source, which makes driving typical load resistances easier.

PRACTICAL OP-AMP DIFFERENTIATOR

The op-amp differentiator of Fig. 18-20a has a tendency to *oscillate,* an undesirable condition. To avoid this, a practical op-amp differentiator usually includes some resistance in series with the capacitor, as shown in Fig. 18-21. A typical value for this added resistance is between $0.01R$ and $0.1R$. With this resistor, the closed-loop voltage gain is between -10 and -100. The effect is to limit the closed-loop voltage gain at higher frequencies, where the oscillation problem arises. (Chapter 20 discusses oscillations in more detail.)

Incidentally, the source driving the op-amp differentiator has an output impedance. If this is a resistance between $0.01R$ and $0.1R$, you don't have to include an extra resistor because the source impedance supplies it.

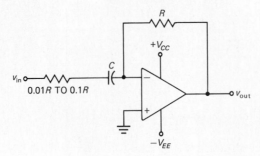

Fig. 18-21   *Resistance in series with capacitor prevents high-frequency oscillations.*

18-7 WAVEFORM CONVERSION

With op amps we can convert sine waves to rectangular waves, rectangular waves to triangular waves, and so on. This section is about some basic circuits that convert an input waveform to an output waveform of a different shape.

SINE TO RECTANGULAR

Figure 18-22*a* shows a Schmitt trigger, and Fig. 18-22*b* is the transfer characteristic. When the input signal is *periodic* (repeating cycles), the Schmitt trigger produces a rectangular-wave output. This assumes that the input signal is large enough to pass through both trip points of Fig. 18-22*c*. When the input voltage exceeds the UTP on the upward swing of the positive half cycle, the output voltage switches to $-V_{sat}$. Half a cycle later, the input voltage becomes more negative than the LTP, and the output switches back to $+V_{sat}$.

A Schmitt trigger always produces a rectangular-wave output, regardless of the shape of the input signal. In other words, the input voltage does not have to be sinusoidal, as it is in Fig. 18-22*a*. As long as the waveform is periodic and has an amplitude large enough to pass through the trip points, we get a rectangular-wave output from the Schmitt trigger. This rectangular wave has the same frequency as the input signal (see Fig. 18-22*c*).

As an example, Fig. 18-22*d* shows a Schmitt trigger with trip points of approximately UTP = $+0.1$ V and LTP = -0.1 V. If the input signal is repetitive and has a peak-to-peak value greater than 0.2 V, then the output is a rectangular wave with a peak-to-peak value of approximately 20 V (two times V_{sat}).

RECTANGULAR TO TRIANGULAR

In Fig. 18-23*a*, a rectangular wave is the input to an integrator. Since the input signal has a dc or average value of zero, the dc or average value of the output is also zero (assuming negligible output offset). As shown in Fig. 18-23*b*, the ramp is decreasing during the positive half cycle of input voltage and increasing during the negative half cycle. Therefore, the output is a periodic triangular wave with the same frequency as the input.

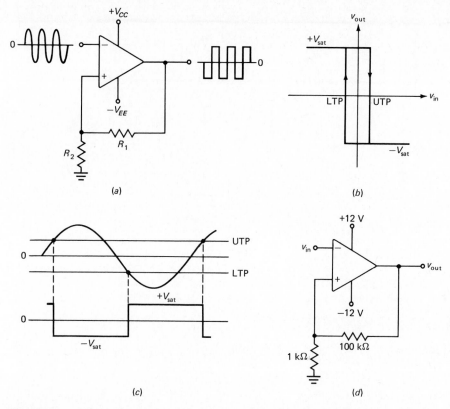

Fig. 18-22 (a) *Periodic input to Schmitt trigger produces rectangular output.* (b) *Inverting transfer characteristic.* (c) *Output transitions occur when input hits trip points.* (d) *Example.*

By analyzing the voltage change in a ramp, Appendix 1 derives this formula for the peak-to-peak output:

$$v_{\text{out(p-p)}} = \frac{v_{\text{in(p-p)}}}{4fRC} \qquad (18\text{-}21)$$

where $v_{\text{out(p-p)}}$ = peak-to-peak output voltage, triangular
$\quad v_{\text{in(p-p)}}$ = peak-to-peak input voltage, rectangular
$\qquad f$ = input frequency
$\qquad R$ = resistance of integrator
$\qquad C$ = capacitance of integrator

For instance, suppose a square wave drives the integrator of Fig. 18-23c. If the frequency is 1 kHz and the peak-to-peak input voltage is 10 V, then the output is a triangular wave with a frequency of 1 kHz and a peak-to-peak voltage of

$$v_{\text{out(p-p)}} = \frac{10 \text{ V}}{4(1 \text{ kHz})(1 \text{ k}\Omega)(10 \text{ } \mu\text{F})} = 0.25 \text{ V}$$

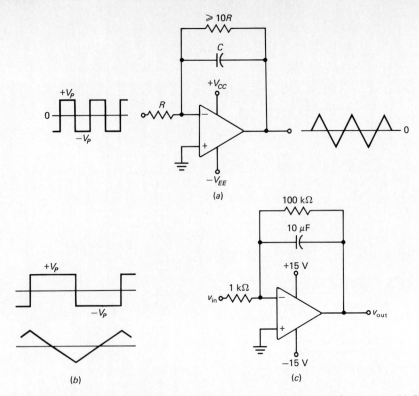

Fig. 18-23 (a) *Rectangular input to integrator produces triangular output.* (b) *Input and output waveforms.* (c) *Example.*

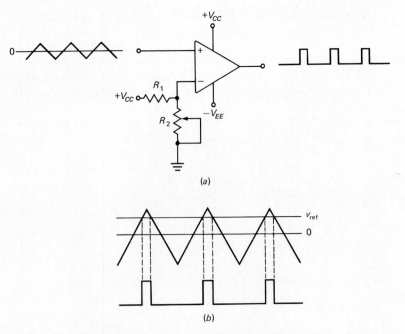

Fig. 18-24 (a) *Output has adjustable duty cycle.* (b) *Waveforms of triangular input and pulse output.*

TRIANGLE TO PULSE

In some applications we would like to produce a variable-duty-cycle pulse. Figure 18-24a shows one way this can be done, with an op-amp limit detector. This is a comparator (with no hysteresis) with a reference voltage that is adjustable. This allows us to move the trip point from zero to some positive level. As long as the input triangular voltage exceeds the reference voltage, the output is high, as shown in Fig. 18-24b. Since the v_{ref} is adjustable, we can vary the width of the output pulse, equivalent to changing the duty cycle. With a circuit like this, we can vary the duty cycle from approximately zero to 50 percent. The output frequency will equal the input frequency.

18-8 WAVEFORM GENERATION

With positive feedback it is also possible to build oscillators, circuits that generate or create an output signal with no external input signal. This section is a brief look at some op-amp circuits that can generate nonsinusoidal signals. Chapter 20 discusses sinusoidal oscillators and more advanced nonsinusoidal oscillators.

RELAXATION OSCILLATOR

In Fig. 18-25a, there is no input signal. Nevertheless, the circuit generates an output rectangular wave. How does it work? Assume that the output is in positive saturation. The capacitor will charge exponentially toward $+V_{sat}$. It never reaches $+V_{sat}$ because its voltage hits the UTP (Fig. 18-25b). The output then switches to $-V_{sat}$. Now a negative voltage is being fed back, and so the capacitor reverses its charging direction. The capacitor voltage decreases as shown. When the capacitor voltage hits the LTP, the output switches back to $+V_{sat}$. Because of the continuous charging and discharging of the capacitor, the output is a rectangular wave.

By analyzing the exponential charge and discharge of the capacitor, Appendix 1 derives this formula for the period of the output rectangular wave:

$$T = 2RC \ln \frac{1 + B}{1 - B} \tag{18-22}$$

where T = period of output signal
R = feedback resistance
C = capacitance
B = feedback fraction, $R_2/(R_1 + R_2)$

Note: ln is the logarithm to the base e. To calculate the frequency of the output rectangular wave, take the reciprocal of the period:

$$f = \frac{1}{T} \tag{18-23}$$

As an example, suppose $R = 1$ kΩ, $C = 0.1$ μF, $R_1 = 2$ kΩ, and $R_2 = 18$ kΩ.

Fig. 18-25
(a) *Relaxation oscillator.* (b) *Capacitor and output waveforms.*

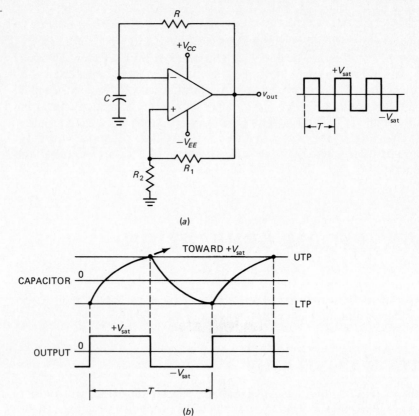

(a)

(b)

Then the feedback fraction B is

$$B = \frac{18 \text{ k}\Omega}{20 \text{ k}\Omega} = 0.9$$

The RC time constant is

$$RC = (1 \text{ k}\Omega)(0.1 \text{ }\mu\text{F}) = 100 \text{ }\mu\text{s}$$

The period of the output signal is

$$T = 2(100 \text{ }\mu\text{s}) \ln \frac{1.9}{0.1} = 589 \text{ }\mu\text{s}$$

and the frequency is

$$f = \frac{1}{589 \text{ }\mu\text{s}} = 1.7 \text{ kHz}$$

Figure 18-25*a* is an example of a *relaxation oscillator,* a circuit that generates an output signal whose frequency depends on the charging or discharging of a capacitor or inductor. If we increase the RC time constant, it takes longer for the capacitor

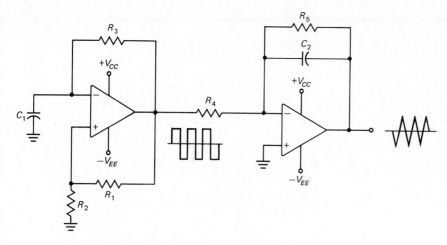

Fig. 18-26 *Relaxation oscillator drives integrator to produce triangular output.*

voltage to reach the trip points; therefore, the frequency is lower. By making R adjustable, we can typically get a 50:1 tuning range.

GENERATING TRIANGULAR WAVES

Cascade a relaxation oscillator and an integrator, and what have you got? A *triangular-waveform* generator. In Fig. 18-26, the rectangular wave out of the relaxation oscillator drives the integrator, which produces a triangular output waveform. The rectangular wave swings between $+V_{sat}$ and $-V_{sat}$; you can calculate its frequency from Eqs. (18-22) and (18-23). The triangular wave has the same frequency; you can calculate its peak-to-peak value from Eq. (18-21).

ANOTHER TRIANGULAR GENERATOR

In Fig. 18-27a, a noninverting Schmitt trigger produces a rectangular wave that drives an integrator. The output of the integrator is a triangular wave. This triangular wave is fed back and used to drive the Schmitt trigger. So we have a very interesting circuit. The first stage drives the second, and the second drives the first.

Figure 18-27b is the transfer characteristic of the Schmitt trigger. When the output is low, the input must increase to the UTP to switch the output to high. Likewise, when the output is high, the input must decrease to the LTP to switch the output to low.

The triangular wave from the integrator is perfect for driving the Schmitt trigger. When the Schmitt trigger output is low in Fig. 18-27a, the integrator produces a positive ramp. This positive ramp increases until it reaches the UTP, as shown in Fig. 18-27c. At this point, the output of the Schmitt trigger switches to the high state and forces the triangular wave to reverse direction. The negative ramp now decreases until it reaches the LTP, where another Schmitt output change takes place.

There is one missing concept in the foregoing explanation: How does the circuit get started in the first place? When you first power up, the Schmitt-trigger output

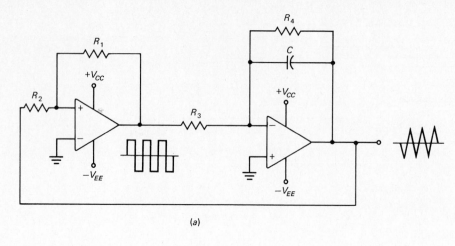

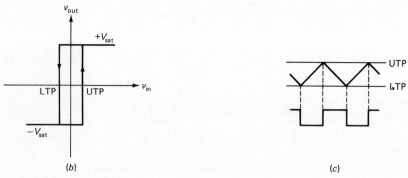

Fig. 18-27 *Feedback circuit with Schmitt trigger and integrator produces triangular output.*

must be either low or high. If it is low, the integrator produces a rising ramp. If it is high, the integrator produces a falling ramp. Either way, the triangular waveform has started, and the positive feedback will keep it going.

PROBLEMS

Straightforward

18-1. A sine wave with a peak of 2 V drives the circuit of Fig. 18-28a. If the op amp has a gain of 100,000, what is the value of input voltage that just turns on the diode? What is the peak current through the load resistor? The average current over the cycle?

Fig. 18-28

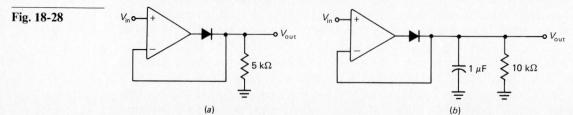

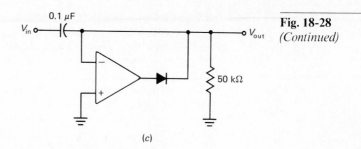

Fig. 18-28
(Continued)

(c)

18-2. If a sine wave with a peak of 500 mV drives the circuit of Fig. 18-28*b*, what is the approximate output voltage? The discharging time constant is 100 times greater than the period of the input signal. What is the input frequency?

18-3. The input signal of Fig. 18-28*c* is a sine wave with a peak of 300 mV. What are the minimum and maximum values of output voltage? If the $R_L C$ time constant should be at least 100 times greater than the input period, what is the lowest frequency allowed?

18-4. In Fig. 18-2*c*, $R_L = 10\ \text{k}\Omega$ and $C = 0.1\ \mu\text{F}$. If the transistor is equivalent to $10\ \Omega$ when saturated, what is the discharging time constant when the reset input is zero? When it is $+5$ V?

18-5. What is the reference voltage in Fig. 18-29*a*? What is the approximate output when v_{in} is 1 V? When v_{in} is 10 V?

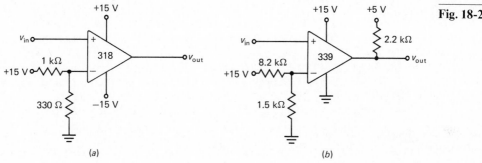

Fig. 18-29

(a)   (b)

18-6. What does v_{out} equal in Fig. 18-29*b* when v_{in} is 1 V? When v_{in} is 4 V? What is the trip point of the circuit?

18-7. What are the trip points for the Schmitt trigger of Fig. 18-30*a*?

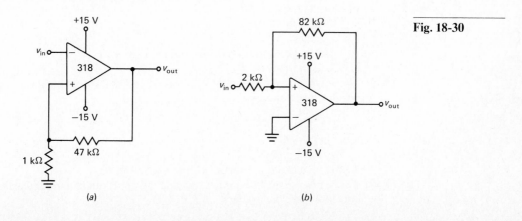

Fig. 18-30

(a)   (b)

18-8. Calculate the trip points in Fig. 18-30*b*.

18-9. In Fig. 18-30*a*, a stray capacitance of 30 pF is across the 1-kΩ resistor. What is the minimum value the speedup capacitor should have?

18-10. If the supplies of Fig. 18-16*c* are changed to +15 V and −15 V, what are the trip points?

18-11. Initially, the output voltage in Fig. 18-31 is zero. What is the approximate output voltage at the end of the input pulse?

Fig. 18-31

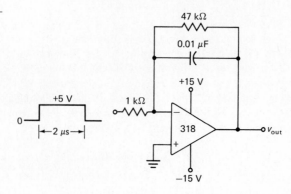

18-12. If the feedback resistor is changed to 10 kΩ in Fig. 18-20*c*, what is the negative output peak?

18-13. A rectangular wave with a frequency of 2.75 kHz drives the integrator of Fig. 18-23*c*. What is the peak-to-peak output voltage if the input has a peak-to-peak voltage of 10 V?

18-14. A relaxation oscillator like Fig. 18-25*a* has $R_1 = 1$ kΩ, $R_2 = 10$ kΩ, $R = 4.7$ kΩ, and $C = 0.022$ μF. What is the frequency of the output rectangular wave?

18-15. In Fig. 18-27*a*, the supply voltages are +15 V and −15 V. Also $R_1 = 2$ kΩ, $R_2 = 1$ kΩ, $R_3 = 3.9$ kΩ, $R_4 = 100$ kΩ, and $C = 0.01$ μF. What are the frequency and peak-to-peak output voltages?

Troubleshooting

18-16. The output of Fig. 18-29*a* remains low until the input is raised to +15 V. Which of these is the trouble:
   a. 330 Ω open
   b. 1 kΩ open
   c. −15 V applied to voltage divider instead of +15 V
   d. 330 Ω shorted

18-17. The Schmitt trigger of Fig. 18-30*a* has no hysteresis. Its trip point is at zero. Name two possible troubles.

18-18. When the input in Fig. 18-31 is high, the output is high. Which of these is the trouble:
   a. 1 kΩ shorted
   b. 47 kΩ shorted
   c. Capacitor open
   d. Inverting and noninverting connections reversed

18-19. The output of Fig. 18-26 is a rectangular wave instead of a triangular wave. Name at least three possible troubles.

Design

18-20. Change the circuit of Fig. 18-29*b* to get a trip point of $+3.5$ V for supplies of $+10$ V and -10 V.

18-21. Design a window comparator like Fig. 18-12*a* with 15-V supplies. The limits should be as close to $+2$ V and $+4$ V as 5 percent resistors allow.

18-22. Redesign the integrator of Fig. 18-31 to produce a ramp with a slope of -1 V/μs for an input of $+5$ V.

18-23. Design a triangular-wave generator like Fig. 18-27*a* that produces a peak-to-peak output of 5 V. Use 15-V supplies.

Challenging

18-24. In Fig. 18-12*a*, the total capacitance between the output and ground is 50 pF. After the input passes through a limit, how long does it take the output voltage to get within 1 percent of its final value?

18-25. The staircase of Fig. 18-10 has 256 steps between 0 and $+5$ V. Each step has a duration of 1 μs. If v_{in} is $+2.75$ V, how long after the staircase starts does the output change states?

18-26. In Fig. 18-12*a*, the outputs of two comparators are connected together. If you try connecting the outputs of two 741Cs together, you probably will damage one of the op amps. Explain why this works with open-collector devices but not with push-pull output devices.

18-27. Derive Eq. (18-22), the period of a relaxation oscillator.

Computer

18-28. Here is a program:

```
10 PRINT "INVERTING SCHMITT TRIGGER"
20 PRINT "ENTER R1": INPUT R1
30 PRINT "ENTER R2": INPUT R2
40 PRINT "ENTER POSITIVE SUPPLY VOLTAGE": INPUT VCC
50 B = R2/(R1 + R2)
60 VREF = B * (VCC − 2)
70 PRINT "UTP = ";VREF: PRINT "LTP = ";−VREF
80 STOP
```

What does line 40 do? Line 60? Line 70?

18-29. Write a program that prints out the limits for a window comparator like Fig. 18-12*a*. The only input is V_{CC}.

18-30. Write a program that prints out the peak-to-peak value of a triangular wave, Eq. (18-21).

CHAPTER 19

Regulated Power Supplies

Chapter 4 introduced the zener diode, a device that has a constant voltage in the breakdown region. With a zener diode, we are able to build simple voltage regulators that can hold the load voltage constant. Chapter 8 cascaded a zener diode and an emitter follower to produce a voltage regulator capable of more load current than a simple zener-diode regulator.

Now we will discuss ways to improve voltage regulation even further. In this chapter, negative feedback will be used to hold the output voltage almost constant despite relatively large changes in line voltage and load current. Although we start with discrete circuits, we end with integrated circuits because these are now commonly used. The chapter will conclude with switching regulators, the type of voltage regulator preferred for large load currents.

19-1 VOLTAGE-FEEDBACK REGULATION

In critical applications, zener voltages near 6 V are used because the *temperature coefficient* approaches zero. The highly stable zener voltage, sometimes called a *reference* voltage, can be amplified with a noninverting voltage-feedback amplifier to get higher output voltage that has the same temperature stability as the reference voltage.

BASIC IDEA

Figure 19-1a shows a discrete *voltage regulator*. Transistor Q_2 is an emitter follower. Therefore, its base voltage is one V_{BE} drop higher than the output voltage across the load. As before, Q_2 is called a pass transistor because all the load current passes through it.

Notice that a voltage divider samples the output voltage V_{out} and delivers a feedback voltage V_F to the base of Q_1. This transistor operates in the active region as a linear amplifier. The feedback voltage V_F controls the Q_1 collector current. Since the output voltage is being sampled, it is stabilized against changes in open-loop gain, load resistance, line voltage, and so on. The larger the feedback voltage, the greater the Q_1 collector current.

The dc input voltage V_{in} comes from an unregulated power supply, such as a bridge rectifier and a capacitor-input filter. Typically, V_{in} has a peak-to-peak ripple of about 10 percent of the dc voltage. The output voltage V_{out} is almost perfectly constant, even though the input voltage and load current may change. Why? Because any change in output voltage is fed back through the voltage divider to the base of Q_1. This produces an error voltage that automatically compensates for the attempted change.

For instance, if V_{out} tries to increase, more V_F is fed back to the base of Q_1, producing a larger Q_1 collector current through R_3 and less base voltage at Q_2. The reduced base voltage to the emitter follower results in less output voltage. Similarly, if the output voltage tries to decrease, there is less base voltage at Q_1, more base voltage at Q_2, and more output voltage. In either case, the attempted change in V_{out} produces an amplified output change in the opposite direction. The overall effect is to cancel attempted changes in output voltage almost completely.

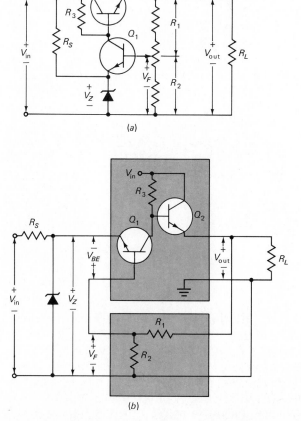

Fig. 19-1 *Discrete voltage regulator. (a) Circuit. (b) Redrawn to emphasize gain and feedback sections.*

OUTPUT VOLTAGE

Figure 19-1*b* shows the circuit redrawn to allow easy recognition of the amplifier and the negative-feedback circuit. The reference voltage V_Z is the input to the emitter of a CB stage, which drives an emitter follower. The output voltage V_{out} is applied to a voltage divider to produce a feedback voltage V_F for the base of Q_1. The feedback fraction is

$$B \cong \frac{R_2}{R_1 + R_2}$$

The closed-loop voltage gain is

$$A_{CL} \cong \frac{1}{B}$$

or

$$A_{CL} \cong \frac{R_1}{R_2} + 1 \qquad (19\text{-}1)$$

Summing voltages around the input loop gives

$$-V_Z - V_{BE} + V_F = 0$$

or

$$V_F = V_Z + V_{BE}$$

If you look at Fig. 19-1*a* or *b*, you can see that V_F equals the sum of V_Z and V_{BE}. Since $V_F = BV_{out}$,

$$BV_{out} = V_Z + V_{BE}$$

or

$$V_{out} = \frac{V_Z + V_{BE}}{B}$$

Since $A_{CL} = 1/B$, we can write

$$V_{out} = A_{CL}(V_Z + V_{BE}) \qquad (19\text{-}2)$$

where V_{out} = regulated output voltage
$\quad A_{CL}$ = closed-loop voltage gain, $R_1/R_2 + 1$
$\quad V_Z$ = zener voltage
$\quad V_{BE}$ = base-emitter voltage of Q_1

For instance, if $R_1 = 2$ kΩ and $R_2 = 1$ kΩ, then $A_{CL} = 3$ and the output voltage is three times the sum of $V_Z + V_{BE}$.

Because of the closed-loop voltage gain, we can use a low zener voltage (5 to 6 V) where the temperature coefficient approaches zero. The amplified output voltage then has the same temperature coefficient. The potentiometer of Fig. 19-1*a* allows us to adjust the output voltage to the exact value required in a particular application. In this way, we can adjust for the tolerance in zener voltages, V_{BE} drops, and feedback resistors. Once the potentiometer is adjusted to the desired output voltage, V_{out} remains almost constant, despite changes in line voltage or load current.

POWER DISSIPATION IN PASS TRANSISTOR

As discussed in Chap. 8, the pass transistor (the emitter follower) is in series with the load. This is why the circuit is called a *series regulator*. The main disadvantage of a series regulator is the power dissipation in the pass transistor:

$$P_D = V_{CE}I_C \qquad (9\text{-}4)$$

where V_{CE} = collector-emitter voltage, $V_{in} - V_{out}$
I_C = load current plus divider current

As long as the load current is not too large, the pass transistor does not get too hot. But when the load current is heavy, the pass transistor has to dissipate a lot of power. This implies heavier heat sinks and a bulkier power supply. In some cases, a fan may be needed to remove the excess heat. Depending on the application, a designer may prefer to use a switching regulator (discussed later) for heavy load currents.

EXAMPLE 19-1

You often see a Darlington pair used for the pass transistor in a series regulator. This allows the regulator to drive small load resistors. What are the minimum and maximum output voltage in Fig. 19-2?

SOLUTION

To begin with, the closed-loop voltage gain is still given by $A_{CL} = R_1/R_2 + 1$. If the wiper is turned all the way up, then $R_1 = 360\ \Omega$ and $R_2 = 720\ \Omega$. The closed-loop voltage gain is

$$A_{CL} = \frac{360}{720} + 1 = 1.5$$

and the regulated output voltage is

$$V_{out} = 1.5(5.6\ \text{V} + 0.7\ \text{V}) = 9.45\ \text{V}$$

On the other hand, when the wiper is moved all the way down, $R_1 = 460\ \Omega$ and $R_2 = 620\ \Omega$. Now,

$$A_{CL} = \frac{460}{620} + 1 = 1.74$$

and

$$V_{out} = 1.74(5.6\ \text{V} + 0.7\ \text{V}) = 11\ \text{V}$$

So, we can adjust the regulated output voltage between 9.45 and 11 V.

EXAMPLE 19-2

In Fig. 19-2, $V_{out} = 10$ V and $R_L = 5\ \Omega$. If the 2N3055 has a β_{dc} of 50 and the 2N3904 has a β_{dc} of 100, what is the current through the zener diode?

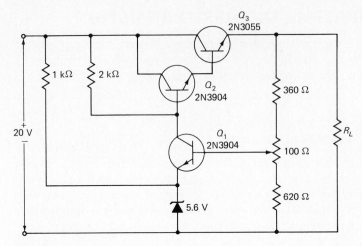

Fig. 19-2 *Darlington pair used for pass transistor to increase maximum load current.*

SOLUTION

The load current is

$$I_L = \frac{10 \text{ V}}{5 \text{ Ω}} = 2 \text{ A}$$

and the current through the voltage divider is

$$I \cong \frac{10 \text{ V}}{1080 \text{ Ω}} = 9.26 \text{ mA}$$

This is negligible compared with 2 A, and so the total emitter current through the 2N3055 is approximately 2 A and the base current is

$$I_B = \frac{2 \text{ A}}{50} = 40 \text{ mA}$$

The 2N3904 has to supply the 40 mA of base current for the 2N3055. This means that the base current of the 2N3904 is

$$I_B = \frac{40 \text{ mA}}{100} = 0.4 \text{ mA}$$

Because of the two V_{BE} drops, the voltage at the bottom of the 2-kΩ resistor is 11.4 V. The current through the 2-kΩ resistor is

$$I = \frac{20 \text{ V} - 11.4 \text{ V}}{2 \text{ kΩ}} = 4.3 \text{ mA}$$

This current splits into the base current for the upper 2N3904 and the collector current for the lower 2N3904. Therefore, the lower 2N3904 has a collector

current of
$$I_C = 4.3 \text{ mA} - 0.4 \text{ mA} = 3.9 \text{ mA}$$

The current through the 1-kΩ resistor is
$$I = \frac{20 \text{ V} - 5.6 \text{ V}}{1 \text{ k}\Omega} = 14.4 \text{ mA}$$

The total current through the zener diode is the sum of the current through the 1-kΩ resistor and the Q_1 emitter current:
$$I_Z = 14.4 \text{ mA} + 3.9 \text{ mA} = 18.3 \text{ mA}$$

Incidentally, the 2N3055 is a workhorse power transistor. This widely used industry standard can handle up to 15 A of continuous current, has breakdown voltages of 60 V or more, and can dissipate 115 W at 25°C.

EXAMPLE 19-3

Calculate the power dissipation of the 2N3055 in Fig. 19-2 for a load voltage of 10 V and a load current of 2 A.

SOLUTION

The power dissipation is
$$P_D = (20 \text{ V} - 10 \text{ V})(2 \text{ A}) = 20 \text{ W}$$

This is a fair amount of power if you think of the nearest standard light bulb (25 W). The 2N3055 would need a heat sink to keep its junction temperature at a safe level.

19-2 CURRENT LIMITING

The series regulator of Fig. 19-2 has no short-circuit protection. If we accidentally short the load terminals, we get an enormous load current that will destroy the pass transistor or a diode in the unregulated supply. To avoid this possibility, regulated supplies usually include *current limiting*.

SIMPLE LIMITING

Figure 19-3 shows one way to limit the load current to safe values even though the output terminals are shorted. R_4 is called a *current-sensing* resistor. For load currents less than 600 mA, the voltage drop across R_4 is less than 0.6 V, and Q_3 is cut off. In this case, the regulator works as previously described. When the load current is between 600 and 700 mA, the voltage across R_4 is between 0.6 and 0.7 V, enough to turn on Q_3. The collector current of Q_3 flows through R_3 and decreases the base voltage to Q_2; this decreases the output voltage as much as is needed to prevent further increases in load current.

Fig. 19-3
*Voltage regulator
with simple
current limiting.*

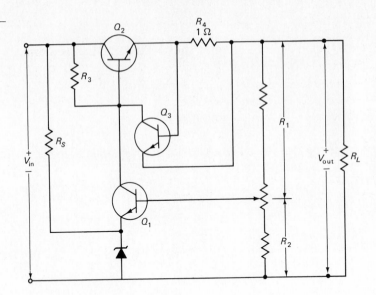

Let us symbolize the load current with shorted-load terminals as I_{SL}. When the load terminals are shorted in Fig. 19-3, the voltage across R_4 is

$$V_{BE} = I_{SL}R_4$$

or

$$I_{SL} = \frac{V_{BE}}{R_4} \qquad (19\text{-}3)$$

where I_{SL} = short-circuit load current
V_{BE} = base-emitter voltage, 0.6 to 0.7 V
R_4 = current-sensing resistance

By selecting other values of R_4, we can change the level of current limiting. For instance, if R_4 is 0.1 Ω, the current limiting is between 6 and 7 A.

DISADVANTAGE OF SIMPLE LIMITING

The simple current limiting just described is a big improvement because it will protect the pass transistor and rectifier diodes in case the load terminals are accidentally shorted. But it has the disadvantage of a relatively large power dissipation in the pass transistor when the load terminals are shorted. With a short across the load, almost all the input voltage appears across the pass transistor. Therefore, it has to dissipate approximately

$$P_D \cong (V_{in} - V_{BE})I_{SL} \qquad (19\text{-}4)$$

For instance, if V_{in} is 20 V, V_{BE} = 0.7 V, and I_{SL} is 6 A, then P_D is

$$P_D = (20 \text{ V} - 0.7 \text{ V})(6 \text{ A}) = 116 \text{ W}$$

(*Note:* V_{BE} is the drop across the current-sensing resistor.)

FOLDBACK LIMITING

One way to reduce the high power dissipation under shorted-load conditions is with the circuit shown in Fig. 19-4a. The load current I_{out} flows through R_4, producing a voltage drop of approximately $I_{out}R_4$. This means that a voltage of $I_{out}R_4 + V_{out}$ is fed to a voltage divider (R_5 and R_6) whose output controls Q_3. The feedback fraction of the voltage divider is approximately

$$K \cong \frac{R_6}{R_5 + R_6} \qquad (19\text{-}5)$$

The analysis of this circuit is too complicated to go into here, but Appendix 1 derives formulas for the shorted-load current and the maximum load current. When

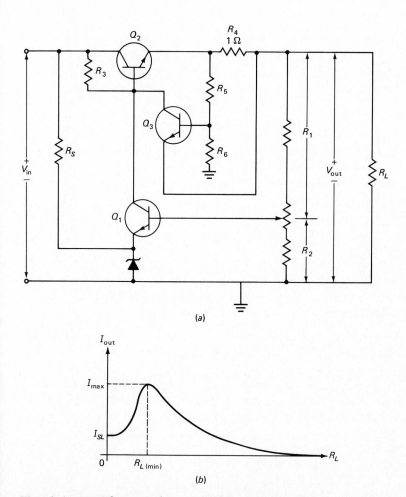

(a)

(b)

Fig. 19-4   (a) *Voltage regulator with foldback current limiting.* (b) *Foldback means that shorted-load current is less than maximum load current.*

the load terminals are shorted, the output current is

$$I_{SL} = \frac{V_{BE}}{KR_4} \tag{19-6}$$

where I_{SL} = shorted-load current
$\quad V_{BE}$ = Q_3 base-emitter voltage, 0.6 to 0.7 V
$\quad\quad K$ = feedback fraction of R_5-R_6 voltage divider
$\quad\quad R_4$ = current-sensing resistance

When the load terminals are not shorted, the maximum output current is

$$I_{max} = I_{SL} + \frac{(1 - K)V_{out}}{KR_4} \tag{19-7}$$

where I_{max} = maximum load current
$\quad V_{out}$ = regulated output voltage

This equation says that the maximum load current is higher than the shorted-load current. Typically, K is selected to produce a maximum load current of two to three times the shorted-load current. The main advantage of foldback current limiting is reduced power dissipation in the pass transistor when the load terminals are accidentally shorted.

Figure 19-4b shows how the output current varies with load resistance. When R_L is large, I_{out} is small. When R_L decreases, I_{out} increases until it reaches a maximum value of I_{max}. The circuit still has a regulated output voltage at this maximum load current. Beyond this point, foldback current limiting takes over. Any further decrease in R_L forces I_{out} to decrease. When R_L is zero, I_{out} equals I_{SL}.

EXAMPLE 19-4

The output voltage of Fig. 19-5 is adjusted to 10 V. If V_{BE} is 0.7 V for the current-limiting transistor, what are the shorted-load current and the maximum load current?

SOLUTION

The feedback fraction of the current-limiting voltage divider is

$$K = \frac{180}{20 + 180} = 0.9$$

From Eq. (19-6),

$$I_{SL} = \frac{0.7 \text{ V}}{0.9(1 \text{ }\Omega)} = 0.778 \text{ A}$$

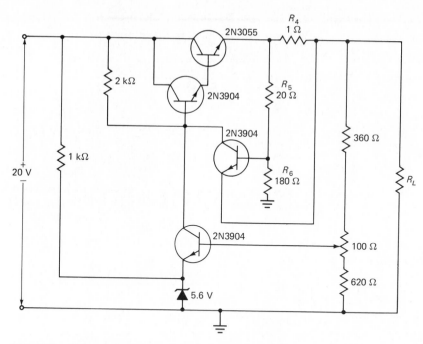

Fig. 19-5 *Voltage regulator uses Darlington pass transistor and foldback current limiting.*

From Eq. (19-7),

$$I_{max} = 0.778 \text{ A} + \frac{(1 - 0.9)(10 \text{ V})}{0.9(1 \text{ }\Omega)} = 1.89 \text{ A}$$

The regulator can supply a maximum load current of 1.89 A. The minimum load resistance is 10 V/1.89 A, or 5.29 Ω. If the load resistance is less than this, the load current is less than 1.89 A. When the load terminals are shorted, the load current drops to 0.778 A.

EXAMPLE 19-5

For the preceding example, calculate the power dissipation of the pass transistor when the load terminals are shorted.

SOLUTION

With a shorted load, the load voltage is zero and the emitter voltage of the 2N3055 is

$$V_E = (0.778 \text{ A})(1 \text{ }\Omega) = 0.778 \text{ V}$$

Since the input voltage is 20 V, the voltage across the 2N3055 is

$$V_{CE} = 20 \text{ V} - 0.778 \text{ V} = 19.2 \text{ V}$$

and the power dissipation is

$$P_D = (19.2 \text{ V})(0.778 \text{ A}) = 14.9 \text{ W}$$

This power dissipation is much lower than it would be with simple current limiting.

19-3 POWER SUPPLY CHARACTERISTICS

The quality of a power supply depends on its load voltage, load current, voltage regulation, and other factors. In this section, we will take a look at some characteristics of regulated power supplies.

LOAD REGULATION

The *load regulation* (also called load effect) is defined as the change in regulated output voltage when the load current changes from minimum to maximum:

$$\text{LR} = V_{NL} - V_{FL} \tag{19-8}$$

where LR = load regulation
V_{NL} = load voltage with no load current
V_{FL} = load voltage with full load current

For instance, if a power supply delivers 10 V at zero load current and 9.9 V at full load current, then its load regulation is

$$\text{LR} = 10 \text{ V} - 9.9 \text{ V} = 0.1 \text{ V}$$

As another example, the Hewlett-Packard 6214A is a regulated power supply with a maximum load voltage of 10 V and a maximum load current of 1 A. Its data sheet lists a load regulation of 4 mV. This means that the load voltage changes only 4 mV when the load current varies from 0 to 1 A.

Load regulation is often expressed as a percent by dividing the change in load voltage by the no-load voltage:

$$\%\text{LR} = \frac{V_{NL} - V_{FL}}{V_{NL}} \times 100\% \tag{19-9}$$

where %LR = percent load regulation
V_{NL} = load voltage with no load current
V_{FL} = load voltage with full load current

For instance, if the no-load voltage is 10 V and the full-load voltage is 9.9 V, then the percent load regulation is

$$\%\text{LR} = \frac{10 \text{ V} - 9.9 \text{ V}}{10 \text{ V}} \times 100\% = 1\%$$

As another example, if the change in load voltage is 4 mV and the no-load voltage is 10 V, then

$$\%LR = \frac{4 \text{ mV}}{10 \text{ V}} \times 100\% = 0.04\%$$

SOURCE REGULATION

Source regulation (also called source effect or line regulation) is the change in regulated load voltage for the specified range of line voltage, typically 115 V ± 10%, a range of approximately 103 V to 127 V. As an example, if the load voltage changes from 10 V to 9.8 V when the line voltage changes from 127 to 103 V, then the source regulation is

$$SR = 10 \text{ V} - 9.8 \text{ V} = 0.2 \text{ V}$$

A quality power supply like the Hewlett-Packard 6214A has a source regulation of 4 mV.

Percent source regulation is

$$\%SR = \frac{SR}{V_{\text{nom}}} \times 100\% \qquad (19\text{-}10)$$

where %SR = percent source regulation
  SR = change in load voltage for full line change
  V_{nom} = nominal load voltage

For instance, if the change in load voltage is 5 mV and the nominal load voltage is 10 V, then

$$\%SR = \frac{5 \text{ mV}}{10 \text{ V}} \times 100\% = 0.05\%$$

OUTPUT IMPEDANCE

A regulated power supply is a very stiff dc voltage source. This implies that its output impedance at low frequencies is very small. In the voltage regulator discussed earlier, an emitter follower supplies the load voltage. The emitter follower already has a low output impedance. The use of voltage feedback further reduces this output impedance because

$$r_{\text{out}(CL)} = \frac{r_{\text{out}}}{1 + AB}$$

Regulated power supplies have typical output impedances in milliohms.

RIPPLE REJECTION

Voltage regulators stabilize the output voltage against changes in input voltage. Ripple is equivalent to a change in input voltage; therefore, a voltage regulator attenuates the ripple that comes in with the input voltage. *Ripple rejection* RR is usually specified in decibels. For example, an RR of 80 dB means that the output ripple is 80 dB less than the input ripple. In ordinary numbers, this means that the output ripple is 10,000 times smaller than the input ripple.

19-4 THREE-TERMINAL IC REGULATORS

In the late 1960s, IC manufacturers began producing a voltage regulator on a chip. First-generation devices like the μA723 and the LM300 included a zener diode, a high-gain amplifier, current limiting, and other useful features. The disadvantage of these early IC regulators was the need for many external components, plus the eight or more pins that had to be connected in various ways to get optimum performance.

The latest generation of IC voltage regulators are devices with only three pins: one for the unregulated input voltage, one for the regulated output voltage, and one for ground. The new devices can supply load current from 100 mA to more than 3 A. Available in plastic or metal packages, these *three-terminal regulators* have become extremely popular because they are so inexpensive and easy to use. Aside from a couple of bypass capacitors, the new three-terminal IC voltage regulators require no external components.

THE LM340 SERIES

The LM340 series is typical of the new breed of three-terminal voltage regulators. Figure 19-6 shows the block diagram. The built-in reference voltage drives the noninverting input of an amplifier. The feedback voltage comes from an internal voltage divider, preset to give any of the following output voltages: 5, 6, 8, 10, 12, 15, 18, and 24 V. For instance, the LM340-5 produces a regulated output voltage of 5 V. The LM340-18 produces an output of 18 V.

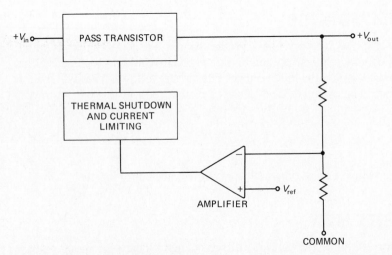

Fig. 19-6 *Functional block diagram of typical three-terminal IC voltage regulator.*

The chip includes a pass transistor that can handle more than 1.5 A of load current, provided that adequate heat sinking is used. Also included are *thermal shutdown* and current limiting. Thermal shutdown means that the chip will automatically turn itself off if the internal temperature becomes dangerously high, around 175°C. This is a precaution against excessive power dissipation, which depends on the ambient temperature, the amount of heat sinking, and other variables. Because of thermal shutdown and current limiting, devices in the LM340 series are almost indestructible.

FIXED REGULATOR

Figure 19-7a shows an LM340-5 connected as a fixed voltage regulator. Pin 1 is the input, pin 2 is the output, and pin 3 is ground. The LM340-5 has an output voltage of +5 V ± 2%, a maximum load current of 1.5 A, a load regulation of 10 mV, a source regulation of 3 mV, and a ripple rejection of 80 dB. With an output impedance of approximately 0.01 Ω, the LM340-5 is a very stiff voltage source to all loads within its maximum current rating.

When the IC is more than a few inches from the filter capacitor of the unregulated power supply, the lead inductance may produce oscillations within the IC. This is why you often see a *bypass* capacitor C_1 on pin 1 (Fig. 19-7b). To improve the transient response of the regulated output voltage, a bypass capacitor C_2 is sometimes used. Typical values for either bypass capacitor are from 0.1 to 1 μF. (The data sheet of the LM340 series suggests 0.22 μF for the input capacitor and 0.1 μF for the output capacitor.)

Any device in the LM340 series needs an input voltage at least 2 to 3 V greater than the regulated output voltage; otherwise, it stops regulating (sometimes called brownout). Furthermore, there is a limit to the input voltage because of excessive power dissipation. For instance, the LM340-5 will regulate over an input range of approximately 7 to 20 V. At the other extreme, the LM340-24 regulates over an input range of approximately 27 to 38 V.

TWO MORE APPLICATIONS

Figure 19-8a shows external components added to an LM340 to get an adjustable output voltage. The common terminal of the LM340 is not grounded, but rather is connected to the top of R_2. This means that the regulated output V_{reg} is across R_1. A

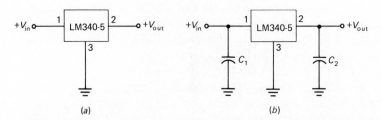

Fig. 19-7 (a) *LM340 connected as voltage regulator.* (b) *Input bypass capacitor prevents oscillations, and output bypass capacitor improves transient response.*

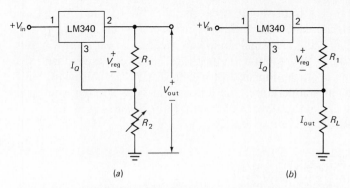

Fig. 19-8 *LM340 circuits.* (a) *Adjustable output voltage.* (b) *Load current is regulated.*

quiescent current I_Q flows through pin 3 and through R_2. Therefore, the output voltage from pin 2 to ground is

$$V_{\text{out}} = V_{\text{reg}} + \left(I_Q + \frac{V_{\text{reg}}}{R_1}\right)R_2 \tag{19-11}$$

Since I_Q shows little variation with line and load changes, the foregoing equation says that V_{out} is regulated and adjustable. (For the LM340 series, I_Q has a maximum value of 8 mA and varies only 1 mA for all line and load changes.)

Figure 19-8*b* is another application, this time a current source (or current regulator). A load resistor R_L takes the place of R_2. As before, the quiescent current I_Q and the R_1 current flow through R_L. Therefore, the load current is

$$I_{\text{out}} = I_Q + \frac{V_{\text{reg}}}{R_1} \tag{19-12}$$

As an example, suppose $V_{\text{reg}} = 5$ V and $R_1 = 10\ \Omega$. Then I_{out} is approximately 500 mA, large enough to swamp out the small variations in I_Q. In other words, I_{out} is essentially constant and independent of R_L. This means that we can change R_L and still have a fixed output current.

THE LM320 SERIES

The LM320 series is a group of negative voltage regulators with preset voltages of -5, $-6, -8, -12, -15, -18$, and -24 V. For instance, an LM320-5 produces a regulated output voltage of -5 V. At the other extreme, an LM320-24 produces an output of -24 V. With the LM320 series, load-current capability is approximately 1.5 A with adequate heat sinking. The LM320 series is similar to the LM340 series, and includes current limiting, thermal shutdown, and excellent ripple rejection.

By combining an LM320 and an LM340, we can regulate the output of a split supply (see Fig. 19-9). The LM340 regulates the positive output, and the LM320 handles the negative output. The input capacitors prevent oscillations, and the output capacitors improve transient response. The manufacturer's data sheet recommends the addition of two diodes, as shown; these diodes ensure that both regulators can turn on under all operating conditions.

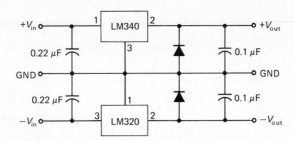

Fig. 19-9
Using LM340 and LM320 to produce split-supply voltages.

ADJUSTABLE REGULATORS

A number of IC regulators, like the LM317, LM338, and LM350, are adjustable. These have maximum load currents from 1.5 to 5 A. For instance, the LM317 is a three-terminal positive voltage regulator that can supply 1.5 A of load current over an adjustable output range of 1.25 to 37 V. The load regulation is 0.1 percent. The line regulation is 0.01 percent per volt; this means that the output voltage changes only 0.01 percent for each volt of input change. The ripple rejection is 80 dB, equivalent to 10,000.

Figure 19-10 shows an unregulated supply driving a typical LM317 circuit. The data sheet of an LM317 gives this formula for output voltage:

$$V_{out} = 1.25\left(\frac{R_2}{R_1} + 1\right) \qquad (19\text{-}13)$$

which is valid from 1.25 to 37 V. Typically, the filter capacitor is selected to get a peak-to-peak ripple of about 10 percent. Since the regulator has about 80 dB of ripple rejection, the final peak-to-peak ripple is around 0.001 percent. In other words, a voltage regulator also filters the input ripple; this eliminates the need for *RC* and *LC* filters in most power supplies.

DUAL-TRACKING REGULATORS

When a split supply is needed, *dual-tracking* regulators like the RC4194 and RC4195 are convenient. These voltage regulators produce equal positive and negative output

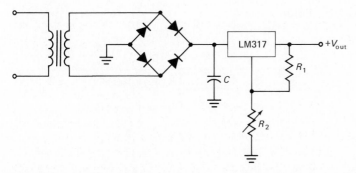

Fig. 19-10 *Bridge rectifier and capacitor-input filter produce unregulated input voltage for LM317.*

Fig. 19-11
Dual-tracking voltage regulator produces split-supply voltages.

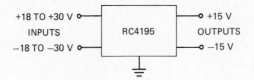

voltages. The RC4194 is adjustable from ±0.05 to ±32 V, while the RC4195 produces fixed output voltages of ±15 V. For instance, Fig. 19-11 shows an RC4195. It needs two unregulated input voltages; the positive input may be from +18 to +30 V, and the negative input from −18 to −30 V. As indicated, the two outputs are ±15 V. The data sheet of an RC4195 lists a maximum output current of 150 mA for each supply, a load regulation of 3 mV, a line regulation of 2 mV, and a ripple rejection of 75 dB.

REGULATOR TABLE

Table 19-1 lists some data for popular IC regulators. For instance, the LM309 is a fixed positive regulator with an output of +5 V, a maximum load current of 1 A, a load regulation of 15 mV, a source regulation of 4 mV, and a ripple rejection of 75 dB. For the adjustable regulators, the LR and SR are given in percent rather than millivolts.

The table also includes dropout voltage, the minimum allowable difference between the input voltage and the output voltage. For example, an LM340-5 has a dropout voltage of 2.3 V. This means that the input voltage must be at least 2.3 V greater than the output. Since the output is 5 V, the input must be at least 7.3 V.

19-5 DC-TO-DC CONVERTER

Sometimes we want to convert a dc voltage to another dc voltage. For instance, if we have a system with a positive supply of +5 V, we can use a *dc-to-dc converter* to produce an output of +15 V. Then we would have two supply voltages for our system: +5 V and +15 V. All kinds of designs are possible for dc-to-dc converters. In

Table 19-1. IC Voltage Regulators

Number	V_{out}, V	I_{max}, A	LR, mV	SR, mV	RR, dB	Dropout, V	Comment
LM309	+5	1	15	4	75	2	Fixed positive
LM317	—	1.5	0.1%	0.2%	65	2.5	Adjustable: 1.2 to 32 V
LM320-5	−5	1.5	50	10	65	2	Fixed negative
LM320-15	−15	1.5	30	5	80	2	Fixed negative
LM338	—	5	0.1%	0.1%	60	2.7	Adjustable: 1.2 to 32 V
LM340-5	+5	1.5	10	3	80	2.3	Fixed positive
LM340-15	+15	1.5	12	4	80	2.5	Fixed positive
LM350	—	3	0.1%	0.1%	65	2.5	Adjustable: 1.2 to 33 V
RC4194	—	0.15	0.2%	0.2%	75	3	Dual-tracking: 0 to 32 V
RC4195	±15	0.15	3	2	70	3	Dual-tracking

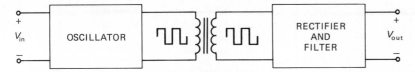

Fig. 19-12 *Output of relaxation oscillator is transformed to a different peak value before rectifying and filtering.*

this section, we discuss a hypothetical design to get an idea of how a dc-to-dc converter works.

BASIC IDEA

In most dc-to-dc converters, the input dc voltage is applied to a square-wave oscillator whose output drives a transformer, as shown in Fig. 19-12. Typically, the frequency is from 1 kHz to 100 kHz. The higher the frequency, the smaller the transformer and filter components. On the other hand, if the frequency is too high, it is difficult to produce a square wave with vertical sides. As a rule, 20 kHz works out as the best compromise, and you see this frequency used a lot.

By selecting different turns ratios, we can get a smaller or larger secondary voltage. To improve the conversion efficiency, the transformer usually has a toroidal core with a rectangular hysteresis loop. This results in a secondary voltage that is a square wave. The secondary voltage can then be rectified and filtered to get a dc output voltage. It is relatively easy to filter the signal because it is a rectified square wave at a high frequency.

One of the most common conversions is +5 to +15 V. In digital systems, +5 V is a standard supply voltage for most ICs. But a few ICs, like op amps, may require +15 V. In a case like this, you commonly find a low-power dc-to-dc converter producing +15 V and −15 V. These voltages are used for the few ICs that require the higher voltages.

AN EXAMPLE

There are many ways to design a dc-to-dc converter, depending on whether the voltage is stepped up or down, the maximum load current needed, and other factors. To give you a basic idea of how it works, Fig. 19-13 shows a hypothetical dc-to-dc converter. This design uses only circuits that have already been discussed, so that you can follow the action. Later, Chaps. 20 and 21 will introduce circuits that are more practical for use in dc-to-dc converters.

Here is how the circuit works. A relaxation oscillator produces a square wave, whose frequency is set by R_3 and C_2. Typically, this frequency is in kilohertz. The square wave drives a phase splitter Q_1, whose outputs are equal and opposite square waves. These square waves are the input to the class B switching transistors Q_2 and Q_3. Transistor Q_2 conducts during one half cycle, and Q_3 during the other half cycle. The square wave out of the secondary winding drives a bridge rectifier and a capacitor-input filter. Because the signal is a rectified square wave in kilohertz, it is easy to filter and get an unregulated dc voltage for the input to the three-terminal regulator. The final output is then a dc voltage at some level different from the input.

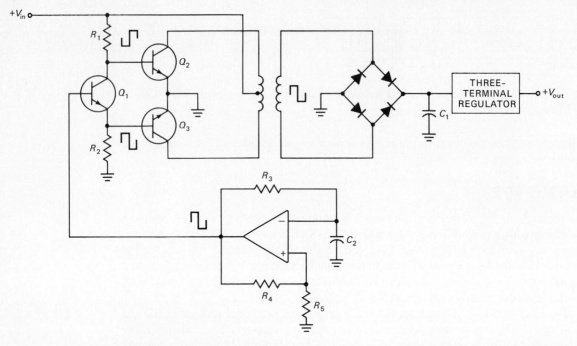

Fig. 19-13   *Dc-to-dc converter.*

19-6 SWITCHING REGULATORS

Series regulators are very popular and fill most of our needs. Their big disadvantage is the power dissipation of the pass transistor. As the load current increases, the pass transistor has to dissipate more power, which implies bigger heat sinks. Because of this, series regulators tend to get bulky. In some cases, a fan may be needed to remove the heat generated by the pass transistor. One way to solve the problem is to use a *switching regulator*. It can produce large load currents with much less power dissipation in the pass transistor.

BASIC IDEA

Figure 19-14 illustrates the key ideas behind a switching regulator. A string of pulses drives the base of the pass transistor. When the base voltage is high, the transistor is saturated. When the base voltage is low, the transistor is cut off. This is class S operation, discussed in Chap. 11. The main idea is that the transistor acts like a switch. Ideally, a switch dissipates no power when it is closed or open. In reality, the transistor switch is not perfect, and so it does dissipate some power, but this power is much less than that dissipated by a series regulator.

A diode is connected from the emitter to ground. This is necessary because of *inductive kickback*. An inductor will try to keep the current through it constant. When the transistor cuts off, the diode continues to provide a path for current

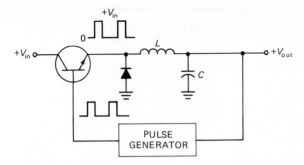

Fig. 19-14　*Transistor acting as a switch controls the duty cycle and final dc output voltage.*

through the inductor. Without the diode, the inductive kickback would produce enough reverse voltage to destroy the transistor.

The duty cycle D is the ratio of the on time W to the period T. By controlling the duty cycle out of the pulse generator, we control the duty cycle of the input voltage to the LC filter. Ideally, this input voltage varies from 0 to V_{in}, as shown. Although it is almost obsolete in ordinary power supplies, the LC filter is very popular in switching regulators because the switching frequency is typically around 20 kHz. This means that a smaller inductor and capacitor can be used. The output of the LC filter is a dc voltage with only a small ripple. This dc output voltage depends on the duty cycle, and is given by

$$V_{out} = DV_{in} \qquad\qquad (19\text{-}14)$$

For instance, if the duty cycle is 0.25 and the dc input voltage is 20 V, the dc output voltage is 5 V.

The output voltage is fed back to the pulse generator. In most switching regulators, the duty cycle is inversely proportional to the output voltage. If the output voltage tries to increase, the duty cycle will decrease. This means that narrower pulses will drive the LC filter, and its output will decrease. In effect, we have negative feedback. Since the output voltage is being sampled and fed back, output voltage is the quantity being stabilized.

AN EXAMPLE

To give you a concrete idea of how a switching regulator works, Fig. 19-15 shows a low-power design using circuits that you already are familiar with. The relaxation oscillator produces a square wave whose frequency is set by R_5 and C_3. The square wave is integrated to get a triangular wave, which drives the noninverting input of a triangular-to-pulse converter. The pulse train out of this circuit then drives the pass transistor, as previously described. The output of the LC filter is sampled by a voltage divider, which returns a feedback voltage to the comparator. This feedback voltage is compared with a reference voltage from a zener diode or other source. The output of the comparator then drives the inverting input of the triangular-to-pulse generator.

Here is how the regulation works. If the regulated output voltage tries to increase, the comparator produces a higher output voltage, which raises the reference voltage

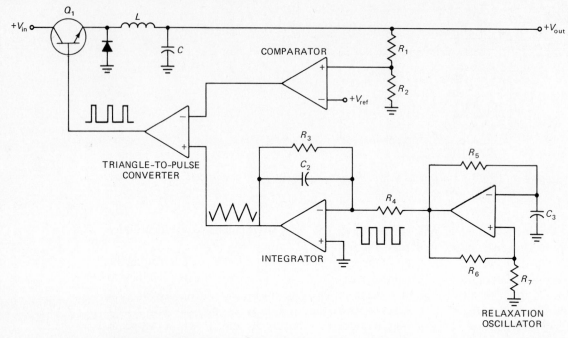

Fig. 19-15 *Switching regulator.*

of the triangular-to-pulse converter. This means that narrower pulses drive the base of the pass transistor. Since the duty cycle is lower, the filtered output is less, which tends to cancel almost all the original increase in output voltage.

Conversely, if the regulated output voltage tries to decrease, the output of the comparator decreases the reference voltage of the triangular-to-pulse converter. Since wider pulses drive the pass transistor, more voltage comes out of the *LC* filter. The net effect is to cancel most of the original decrease in output voltage.

There is enough open-loop gain in the system to ensure a well-regulated output voltage. The bootstrap effect in the comparator implies that

$$V_{\text{ref}} = \frac{R_2}{R_1 + R_2} V_{\text{out}}$$

or

$$V_{\text{out}} = \left(\frac{R_1}{R_2} + 1\right) V_{\text{ref}} \tag{19-15}$$

If $R_1 = 3 \text{ k}\Omega$, $R_2 = 1 \text{ k}\Omega$, and $V_{\text{ref}} = 1.25$ V, then $V_{\text{out}} = 5$ V.

IC SWITCHING REGULATORS

Low-power switching regulators are available on chips. A good example is the Fairchild μA78S40. Known as a universal switching regulator, this chip includes an oscillator, a comparator, a pass transistor, a reference voltage, an op amp, and other circuits. To understand how it it works, you need to have some digital background because the chip includes circuits called AND gates and *RS* flip-flops.

PROBLEMS

Straightforward

19-1. In the voltage-feedback regulator of Fig. 19-1a, $R_1 = 1$ kΩ and $R_2 = 250$ Ω. If $V_Z = 6.2$ V, what does the regulated output voltage equal?

19-2. In Fig. 19-1a, the feedback circuit consists of the following resistors (from top to bottom): 3.9 kΩ, 5-kΩ potentiometer, and 8.2 kΩ. If $V_Z = 7.5$ V, what is the adjustable range of V_{out}?

19-3. Q_3 conducts in Fig. 19-3 when its $V_{BE} = 0.63$ V. What is the maximum load current?

19-4. In Fig. 19-3, $R_1 = 1.5$ kΩ, $R_2 = 700$ Ω, and $V_Z = 5.6$ V. What is the regulated output voltage?

19-5. In the regulator of Fig. 19-3, the feedback resistors have the following values: 4.7 kΩ, 5-kΩ potentiometer, and 6.8 kΩ (from top to bottom). If $V_Z = 5.6$ V, what is the adjustable range of output voltage?

19-6. In Fig. 19-3, $V_{in} = 30$ V, $R_2 = 10$ kΩ, $R_1 = 5$ kΩ, $V_Z = 8.2$ V, and $R_L = 100$ Ω. What is the output voltage?

19-7. R_4 is changed to 0.3 Ω in Fig. 19-5 and the wiper is all the way down. What is the maximum load current? The shorted-load current? The power dissipation of the pass transistor for maximum load current and for shorted-load current?

19-8. The LM340T-12 of Fig. 19-16a has a preset output voltage of 12 V. (The T stands for plastic-tab package, similar to Fig. 10-26b.) If the tolerance on the regulated output voltage is ±2 percent, what are the minimum and maximum values of output voltage?

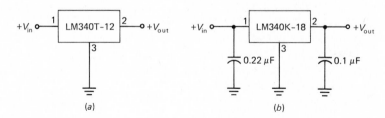

Fig. 19-16

(a)

(b)

19-9. The data sheet of an LM340T-12 indicates that the output voltage is a nominal 12 V. The change in output voltage from no load to full load is 12 mV. What is the percent load regulation?

19-10. The LM340T-12 of Fig. 19-16a has a ripple rejection of 72 dB. If the input ripple is 2 V rms, what is the output ripple?

19-11. In Fig. 19-16b, the LM340K-18 has a preset output voltage of 18 V. (The K stands for metal package, similar to Fig. 10-26c.) If the regulated output changes 38 mV from no load to full load, what is the percent load regulation?

19-12. The ripple rejection of the LM340K-18 shown in Fig. 19-16b is 68 dB. If the input ripple is 3 V p-p, what is the output ripple?

19-13. In Fig. 19-17a, what is the adjustable range of output voltage if I_Q is 8 mA?

19-14. The 50-Ω resistor of Fig. 19-17a is changed to 82 Ω. If $I_Q = 8$ mA, what is the adjustable range of output voltage?

19-15. In Fig. 19-17b, $I_Q = 8$ mA. What is the current through R_L?

19-16. If the 10-Ω resistor of Fig. 19-17b is changed to 15 Ω, what is the load current for an I_Q of 8 mA?

Fig. 19-17

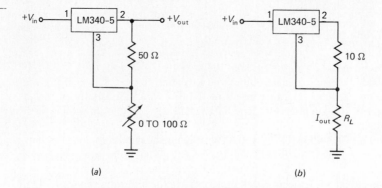

(a)   (b)

19-17. Figure 19-18 shows an LM317 regulator with electronic shutdown. When the shutdown voltage is zero, the transistor is cut off and has no effect on the operation. But when the shutdown voltage is approximately 5 V, the transistor saturates. What is the adjustable range of the output voltage when the shutdown voltage is zero? What does the output voltage equal when the shutdown voltage is 5 V?

Fig. 19-18

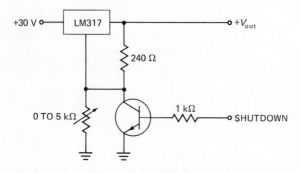

19-18. The transistor of Fig. 19-18 is cut off. To get an output voltage of 15 V, what value should the adjustable resistor have?

19-19. In Fig. 19-14, $V_{in} = 20$ V. If the switching frequency is 20 kHz and the on time is 20 μs, what is the dc output voltage?

19-20. The switching regulator of Fig. 19-15 has $R_1 = 2$ kΩ and $R_2 = 1$ kΩ. If V_{ref} is 5.6 V, what is V_{out}?

Troubleshooting

19-21. In Fig. 19-2, does the output voltage increase, decrease, or remain the same for each of these troubles:

    a. 2N3055 shorted
    b. 2N3055 open
    c. Zener diode shorted
    d. Zener diode open

19-22. In Fig. 19-5, the shorted-load current is approximately 1.89 A, the same as the maximum load current. Which of the following is the trouble:

    a. 20-Ω resistor shorted
    b. 180-Ω resistor shorted
    c. Potentiometer open
    d. Zener diode shorted

19-23. Is the output voltage of Fig. 19-10 likely to increase, decrease, or remain about the same for each of these:
  a. One diode open
  b. Filter capacitor shorted
  c. LM317 defective
  d. R_1 open
19-24. The switching regulator of Fig. 19-15 has the following symptoms: output voltage almost equals input voltage, triangular wave out of integrator is all right, and V_{ref} is normal. Which of these is a possible trouble:
  a. Q_1 open
  b. R_1 open
  c. R_3 shorted
  d. R_7 open
  e. Triangle-to-pulse inputs reversed

Design

19-25. In Fig. 19-2, change the 360-Ω resistor as needed to get an output voltage of approximately 8 V when the wiper is centered.
19-26. Redesign the foldback current limiting of Fig. 19-5 to meet these specifications: $I_{max} = 5$ A, $I_{SL} = 1.25$ A, and wiper is centered.
19-27. Design a power supply that produces a regulated output of +15 V at 1 A. In your design, use a secondary voltage of 17.7 V ac, a bridge rectifier working into a capacitor-input filter, and an LM340 series regulator. Select the capacitor to produce a peak-to-peak ripple of about 10 percent of the dc input voltage.
19-28. The reference voltage of Fig. 19-15 is +1.25 V. Select a pair of values for R_1 and R_2 to get a regulated output voltage of +7.5 V. Keep each resistor less than 10 kΩ to avoid excessive offset.

Challenging

19-29. The output of Fig. 19-2 is adjusted to 10 V for an R_L of 10 Ω. The 2N3904s have $h_{FE} = 100$ and $h_{fe} = 130$. The 2N3055 has $h_{FE} = 60$ and $h_{fe} = 75$. Calculate the open-loop voltage gain, the desensitivity, and the closed-loop output impedance.
19-30. When a bridge rectifier and a capacitor-input filter drive a resistance, the discharge is exponential. But when they drive a voltage regulator, the discharge is almost a perfect ramp. Explain why this is true and what significance this has for Eq. (3-11).
19-31. The 2N3055 has a β_{dc} of 50 in Fig. 19-5. The 2N3904s have a β_{dc} of 150. Calculate the zener current for an output voltage of 10 V and a load current of 1 A.
19-32. Figure 19-19 shows a switching regulator in which the LC filter is collector-driven instead of emitter-driven. The transistor still acts like a switch, and the circuit regulates as previously described. If the output voltage is 15 V, what is the direct current through the inductor? If the duty cycle is 25 percent, what is the direct current through the diode?

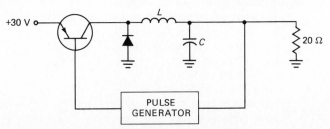

Fig. 19-19

Computer

19-33. Here is a program:

```
10 PRINT "FIGURE 19-4"
20 PRINT "ENTER VBE": INPUT VBE
30 PRINT "ENTER R4": INPUT R4
40 PRINT "ENTER R5": INPUT R5
50 PRINT "ENTER R6": INPUT R6: K = R6/(R5 + R6)
60 ISL = VBE/(K * R4)
70 X = (1 − K) * VOUT/(K * R4)
80 IMAX = ISL + X
90 PRINT "MAXIMUM CURRENT = ";IMAX
100 PRINT "SHORTED-LOAD CURRENT = ";ISL
```

What circuit in the book does this program analyze? What is being calculated in line 60? In line 90?

19-34. Modify the program of Prob. 19-33 so that it also prints out the power dissipation of the pass transistor for maximum load current and for shorted-load current.

19-35. Write a program that calculates and prints the value of V_{out} for Fig. 19-8a.

19-36. Write a program that prints the ten values of V_{out} in Eq. (19-13) that correspond to $R_2/R_1 = 1, 2, 3, \ldots, 9, 10$.

19-37. Here is a program:

```
10 REM MENU FOR AVAILABLE FIGURES
20 PRINT "1. FIG. 19-1"
30 PRINT "2. FIG. 19-2"
40 PRINT "3. FIG. 19-3"
50 PRINT
60 PRINT "ENTER CHOICE"
70 INPUT A
80 REM CHECK CHOICE VALIDITY
90 IF A < 1 THEN GOTO 120
100 IF A > 3 THEN GOTO 120
110 GOTO 1000
120 PRINT
130 PRINT "TRY AGAIN"
140 PRINT
150 GOTO 60
1000 REM MENU FOR ALL FIGURES
1010 PRINT "1. ANALYSIS"
1020 PRINT "2. DESIGN"
1030 PRINT "3. TROUBLESHOOTING"
1040 PRINT
1050 PRINT "ENTER CHOICE"
1060 INPUT B
1070 REM CHECK CHOICE VALIDITY
1080 IF B < 1 THEN GOTO 1110
1090 IF B > 3 THEN GOTO 1110
1100 GOTO 2000
1110 PRINT
1120 PRINT "TRY AGAIN"
1130 PRINT
1140 GOTO 1050
```

```
2000 ON A GOTO 3000, 4000, 5000
2010 END
3000 REM ROUTINE FOR FIG. 19-1
3010 ON B GOTO 3100, 3400, 3700
3100 REM ANALYSIS ROUTINE STARTS HERE
3390 REM ANALYSIS ROUTINE ENDS HERE
3400 REM DESIGN ROUTINE STARTS HERE
3690 REM DESIGN ROUTINE ENDS HERE
3700 REM TROUBLESHOOTING ROUTINE STARTS HERE
3990 REM TROUBLESHOOTING ROUTINE ENDS HERE
4000 REM ROUTINE FOR FIG. 19-2
4010 ON B GOTO 4100, 4400, 4700
4100 REM ANALYSIS ROUTINE STARTS HERE
4390 REM ANALYSIS ROUTINE ENDS HERE
4400 REM DESIGN ROUTINE STARTS HERE
4690 REM DESIGN ROUTINE ENDS HERE
4700 REM TROUBLESHOOTING ROUTINE STARTS HERE
4990 REM TROUBLESHOOTING ROUTINE ENDS HERE
5000 REM ROUTINE FOR FIG. 19-3
5010 ON B GOTO 5100, 5400, 5700
5100 REM ANALYSIS ROUTINE STARTS HERE
5390 REM ANALYSIS ROUTINE ENDS HERE
5400 REM DESIGN ROUTINE STARTS HERE
5690 REM DESIGN ROUTINE ENDS HERE
5700 REM TROUBLESHOOTING ROUTINE STARTS HERE
5990 REM TROUBLESHOOTING ROUTINE ENDS HERE
```

If your choices are Fig. 19-2 and Troubleshooting, what are the values of A and B? What will the screen display if you enter a 4 in line 1060? If the choices are Fig. 19-3 and Design, which line does line 4010 branch to?

CHAPTER 20

Oscillators and Timers

Chapter 18 introduced the relaxation oscillator, a circuit that generates a square-wave output even though it has no input signal. As you recall, positive feedback alternately drives the output into positive and negative saturation. The principle behind a relaxation oscillator is to let the charging and discharging of a capacitor determine the frequency of the output square wave. This chapter discusses a variety of more advanced oscillators with sine-wave outputs. These include *RC* oscillators, *LC* oscillators, and quartz-crystal oscillators. The chapter concludes with a discussion of the NE555, a widely used IC oscillator that has all kinds of applications.

20-1 THEORY OF SINUSOIDAL OSCILLATION

To build a sinusoidal oscillator, we need an amplifier with positive feedback. The idea is to use the feedback signal in place of an input signal. If the *loop gain* and *phase* are correct, there will be an output signal even though there is no external input signal. In other words, an oscillator is an amplifier that has been modified by positive feedback to supply its own input signal. This may sound like perpetual motion, and in a way it is. But remember one thing: The oscillator does not create energy. It only changes dc energy from the power supply into ac energy.

LOOP GAIN AND PHASE

Figure 20-1*a* shows a voltage source v_{in} driving the input terminals of an amplifier. The amplified output voltage is

$$v_{\text{out}} = Av_{\text{in}}$$

This voltage drives a feedback circuit that is usually a *resonant* circuit. Because of this, we get maximum feedback at one frequency. The feedback voltage returning to point x is given by

$$v_f = ABv_{\text{in}}$$

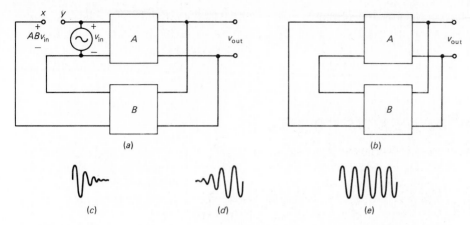

Fig. 20-1 (a) *Positive feedback returns a voltage of ABv_{in} to point x.* (b) *Connecting points x and y.* (c) *Oscillations die out.* (d) *Oscillations increase.* (e) *Oscillations are fixed in amplitude.*

If the phase shift through the amplifier and feedback circuit is $0°$, then ABv_{in} is in phase with the signal v_{in} that drives the input terminals of the amplifier.

Suppose we connect point x to point y and simultaneously remove voltage source v_{in}. Then the feedback voltage ABv_{in} drives the input terminals of the amplifier, as shown in Fig. 20-1b. What happens to the output voltage? If AB is less than 1, ABv_{in} is less than v_{in}, and the output signal will die out, as shown in Fig. 20-1c. On the other hand, if AB is greater than 1, ABv_{in} is greater than v_{in} and the output voltage builds up (Fig. 20-1d). If AB equals 1, then ABv_{in} equals v_{in} and the output voltage is a steady sine wave like Fig. 20-1e. In this case, the circuit supplies its own input signal and produces a sine-wave output.

In an oscillator the value of loop gain AB is greater than 1 when the power is first turned on. A small starting voltage is applied to the input terminals, and the output voltage builds up, as shown in Fig. 20-1d. After the output voltage reaches a desired level, the value of AB automatically decreases to 1, and the output amplitude remains constant (Fig. 20-1e).

THE STARTING VOLTAGE

Where does the *starting voltage* for an oscillator come from? Every resistor contains some free electrons. Because of the ambient temperature, these free electrons move randomly in different directions and generate a noise voltage across the resistor. The motion is so random that it contains frequencies to over 1000 GHz. You can think of each resistor as a small ac voltage source producing all frequencies.

In Fig. 20-1b, here is what happens. When you first turn on the power, the only signals in the system are the noise voltages generated by the resistors. These noise voltages are amplified and appear at the output terminals. The amplified noise drives the resonant feedback circuit. By deliberate design, we can make the phase shift around the loop equal $0°$ at the resonant frequency. In this way, we get oscillations at only one frequency.

In other words, the amplified noise is filtered so that there is only one sinusoidal component with exactly the right phase for positive feedback. When loop gain AB is greater than 1, the oscillations build up at this frequency (Fig. 20-1d). After a suitable level is reached, AB decreases to 1, and we get a constant-amplitude output signal (Fig. 20-1e).

AB DECREASES TO UNITY

There are two ways in which AB can decrease to 1: either A can decrease or B can decrease. In some oscillators, the signal is allowed to build up until clipping occurs because of saturation and cutoff; this is equivalent to reducing the voltage gain A. In other oscillators, the signal builds up and causes B to decrease before clipping occurs. In either case, the product AB decreases until it equals unity.

Here are the key ideas behind any feedback oscillator:

1. Initially, loop gain AB must be greater than 1 at the frequency at which the loop phase shift is 0°.
2. After the desired output level is reached, AB must decrease to 1 through reduction of either A or B.

20-2 THE WIEN-BRIDGE OSCILLATOR

The *Wien-bridge oscillator* is the standard oscillator circuit for low to moderate frequencies, in the range of 5 Hz to about 1 MHz. It is almost always used in commercial audio generators and is usually preferred for other low-frequency applications.

LEAD-LAG NETWORK

The Wien-bridge oscillator uses a feedback circuit called a *lead-lag network* (Fig. 20-2a). At very low frequencies, the series capacitor looks open to the input signal, and there is no output signal. At very high frequencies, the shunt capacitor looks shorted, and there is no output. In between these extremes, the output voltage from the lead-lag network reaches a maximum value (see Fig. 20-2b). The frequency at which the output is maximized is called the *resonant frequency* f_r. At this frequency, the feedback fraction reaches a maximum value of $\frac{1}{3}$.

Figure 20-2c shows the phase angle of the output voltage with respect to the input voltage. At very low frequencies, the phase angle is positive, and the circuit acts like a lead network. On the other hand, at very high frequencies, the phase angle is negative, and the circuit acts like a lag network. In between, there is a resonant frequency f_r at which the phase shift equals 0°.

The lead-lag network of Fig. 20-2a acts like a resonant circuit. At the resonant frequency f_r, the feedback fraction reaches a maximum value of $\frac{1}{3}$ and the phase angle equals 0°. Above and below the resonant frequency, the feedback fraction is less than $\frac{1}{3}$ and the phase angle no longer equals 0°.

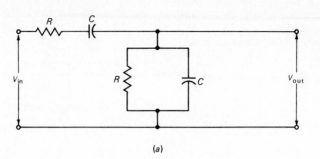

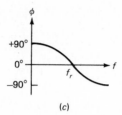

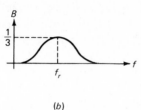

Fig. 20-2
(a) *Lead-lag network.* (b) *Voltage gain.* (c) *Phase shift.*

(a)

(b)   (c)

FORMULA FOR RESONANT FREQUENCY

In Fig. 20-2a, the output of the lead-lag network is

$$\mathbf{V}_{out} = \frac{R \parallel (-jX_C)}{R - jX_C + R \parallel (-jX_C)}\mathbf{V}_{in}$$

By expanding and simplifying, the foregoing equation leads to these two formulas:

$$B = \frac{1}{\sqrt{9 + (X_C/R - R/X_C)^2}} \tag{20-1}$$

and

$$\phi = \arctan \frac{X_C/R - R/X_C}{3} \tag{20-2}$$

Graphing these formulas produces Fig. 20-2b and c.

Equation (20-1) has a maximum when $X_C = R$. For this condition, $B = \frac{1}{3}$ and $\phi = 0°$. This represents the resonant frequency of the lead-lag network. Since $X_C = R$, we can write

$$\frac{1}{2\pi f_r C} = R$$

or

$$f_r = \frac{1}{2\pi RC} \tag{20-3}$$

HOW IT WORKS

Figure 20-3a shows a Wien-bridge oscillator; it uses positive and negative feedback. The positive feedback helps the oscillations to build up when the power is turned on.

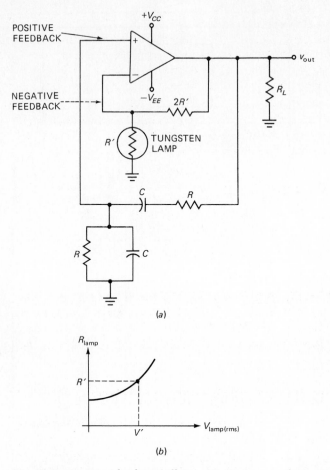

(a)

(b)

Fig. 20-3   (a) *Wien-bridge oscillator.* (b) *Resistance of tungsten lamp increases with voltage.*

After the output signal reaches the desired level, the negative feedback reduces the loop gain to 1. The positive feedback is through the lead-lag network to the noninverting input. Negative feedback is through the voltage divider to the inverting input.

At power-up, the tungsten lamp has a low resistance, and we do not get much negative feedback. For this reason, the loop gain $A_{CL}B$ is greater than 1, and oscillations can build up at the resonant frequency f_r. As the oscillations build up, the tungsten lamp heats slightly, and its resistance increases. (*Note:* In most circuits, the current through the lamp is not enough to make it glow.) At the desired output level, the tungsten lamp has a resistance R'. At this point,

$$A_{CL} = \frac{R_1}{R_2} + 1 = \frac{2R'}{R'} + 1 = 3$$

Since the lead-lag network has a B of $\frac{1}{3}$, the loop gain $A_{CL}B$ equals unity.

INITIAL CONDITIONS

At power-up, the lamp resistance is less than R'; therefore, A_{CL} is greater than 3. Since B equals $\frac{1}{3}$ at the resonant frequency, the loop gain is initially greater than 1. This means that the output voltage will build up as previously described.

As the output voltage increases, the resistance of the lamp increases, as shown in Fig. 20-3b. At some voltage V' the tungsten lamp has a resistance of R'. This means that A_{CL} has a value of 3 and the loop gain becomes 1. When this happens, the output amplitude levels off and becomes constant. (In a practical oscillator, the tungsten lamp does not become luminescent because this would waste signal power.)

AMPLIFIER PHASE SHIFT

In a Wien-bridge oscillator, the phase shift of the lead-lag network equals $0°$ when the oscillations have a frequency of

$$f_r = \frac{1}{2\pi RC}$$

Because of this, we can adjust the frequency by varying the value of R or C. This assumes that the phase shift of the amplifier is negligibly small. Stated another way, the amplifier must have a closed-loop cutoff frequency well above the resonant frequency f_r. Then, the amplifier introduces no additional phase shift. If the amplifier did introduce phase shift, the neat formula $f_r = 1/2\pi RC$ would no longer hold.

WHY CALLED A WIEN-BRIDGE OSCILLATOR

Figure 20-4 shows another way to draw the Wien-bridge oscillator. The lead-lag network is the left side of a bridge, and the voltage divider is the right side. This ac bridge, called a Wien bridge, is used in other applications besides oscillators. The error voltage is the output of the bridge. When the bridge approaches balance, the error voltage approaches zero.

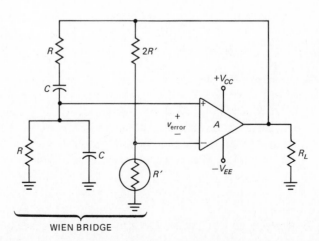

Fig. 20-4
Wien-bridge oscillator.

WIEN BRIDGE

The Wien bridge is an example of a *notch filter,* a circuit with zero output at one particular frequency. For a Wien bridge, the notch frequency equals

$$f_r = \frac{1}{2\pi RC}$$

Because the error voltage to the amplifier is so small, the Wien bridge is approximately balanced and the oscillation frequency approximately equals f_r.

OTHER WAYS TO REDUCE *AB* TO UNITY

A low-power incandescent lamp is the standard method of reducing AB to unity in Wien-bridge oscillators. (Lamps like the #80, 327, 1869, 2158, 7218, and others have been used.) There are alternatives to an incandescent lamp. Figure 20-5a shows a Wien-bridge oscillator that relies on diodes to limit the amplitude of the output signal. On power-up, the diodes are off, and the feedback fraction is less than $\frac{1}{3}$ because the ratio R_1/R_2 is more than 2. This allows the output signal to build up.

Fig. 20-5
Limiting ampli-tude. (a) *Diodes.*
(b) *Zener diode.*

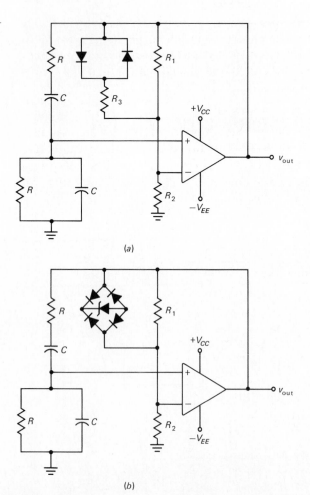

(a)

(b)

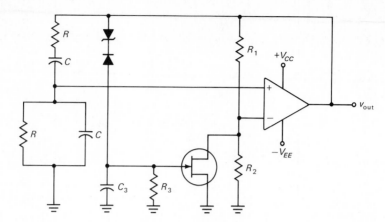

Fig. 20-6
JFET used as voltage-variable resistance to limit output amplitude.

After the desired output level is reached, the diodes conduct on alternate half cycles. This places R_3 in parallel with R_1 and increases the feedback fraction to $\frac{1}{3}$. The output voltage then stabilizes. Sometimes LEDs are used instead of ordinary diodes; LEDs that are lit indicate that the circuit is oscillating.

In Fig. 20-5b, a zener diode is the limiting element. At power-up, the bridge diodes are off, and the feedback fraction is less than $\frac{1}{3}$ because the ratio R_1/R_2 is more than 2. As the output builds up, the bridge diodes are forward-biased, but nothing happens below zener breakdown. At some high output level, the zener diode breaks down and the output level stabilizes.

Figure 20-6 shows another approach. This time, a JFET acting as a voltage-variable resistance limits the output amplitude. At power-up, the JFET has minimum resistance because its gate voltage is zero. By design, the feedback fraction is less than $\frac{1}{3}$, and so the oscillations can start. When the output level exceeds the zener voltage plus one diode drop, we get negative peak detection, and the gate voltage goes negative. When this happens, the $r_{ds(on)}$ of the JFET increases, which increases the feedback fraction until it equals $\frac{1}{3}$. The output then stabilizes.

EXAMPLE 20-1

Calculate the minimum and maximum frequencies in the Wien-bridge oscillator of Fig. 20-7a.

SOLUTION

The ganged rheostats can vary from 0 to 100 kΩ; therefore, the value of R goes from 1 to 101 kΩ. The minimum frequency of oscillation is

$$f_r = \frac{1}{2\pi(101 \text{ k}\Omega)(0.01 \text{ }\mu\text{F})} = 158 \text{ Hz}$$

and the maximum frequency is

$$f_r = \frac{1}{2\pi(1 \text{ k}\Omega)(0.01 \text{ }\mu\text{F})} = 15.9 \text{ kHz}$$

Fig. 20-7

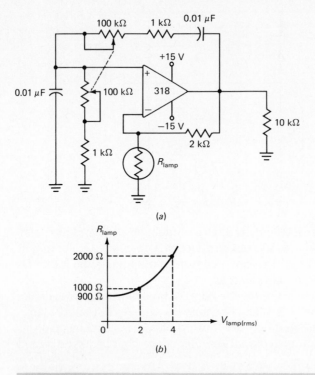

(a)

(b)

EXAMPLE 20-2

Figure 20-7*b* shows the lamp resistance of Fig. 20-7*a*. Calculate the output voltage.

SOLUTION

In Fig. 20-7*a* the output amplitude becomes constant when the lamp resistance equals 1 kΩ. In Fig. 20-7*b* this means that the lamp voltage is 2 V rms. The current flowing through the lamp also flows through the 2-kΩ resistor, which means that there is a 4 V rms signal across this resistor. Therefore, the output voltage equals the sum of 4 V and 2 V, or

$$V_{out} = 6 \text{ V rms}$$

20-3 OTHER *RC* OSCILLATORS

Although the Wien-bridge oscillator is the industry standard for frequencies up to 1 MHz, you occasionally see different *RC* oscillators. This section discusses two other types, called the twin-T oscillator and the phase-shift oscillator.

TWIN-T OSCILLATOR

Figure 20-8*a* is a *twin-T* filter. A mathematical analysis of this circuit shows that it acts like a lead-lag network with a phase angle as shown in Fig. 20-8*b*. Again, there is

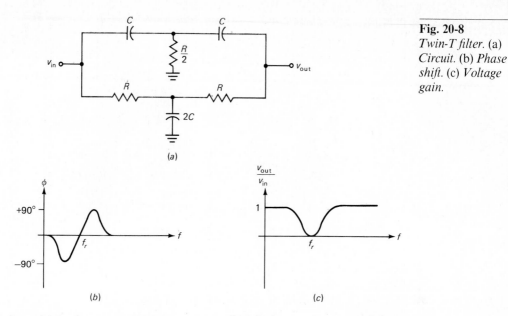

Fig. 20-8
*Twin-T filter. (a)
Circuit. (b) Phase
shift. (c) Voltage
gain.*

a frequency f_r at which the phase shift equals zero. The voltage gain equals unity at low and high frequencies. In between, there is a frequency f_r at which the voltage gain drops to zero (Fig. 20-8c). The twin-T filter is sometimes called a notch filter because it can notch out or attenuate those frequencies near f_r. Frequency f_r, known as the *notch frequency* (also called the resonant frequency), is given by

$$f_r = \frac{1}{2\pi RC} \tag{20-4}$$

Figure 20-9 shows a twin-T oscillator. The positive feedback is through the voltage divider to the noninverting input. The negative feedback is through the twin-T filter. When power is first turned on, the lamp resistance R_1 is low, and the positive feedback is maximum. As the oscillations build up, the lamp resistance increases and the positive feedback decreases. As the feedback decreases, the oscillations level off and become constant. In this way, the lamp stabilizes the level of output voltage.

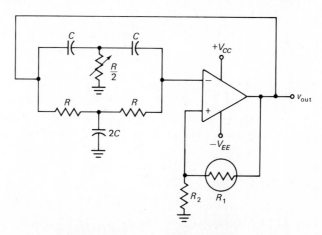

Fig. 20-9
Twin-T oscillator.

Fig. 20-10
*JFET used as
voltage-variable
resistance to limit
output amplitude.*

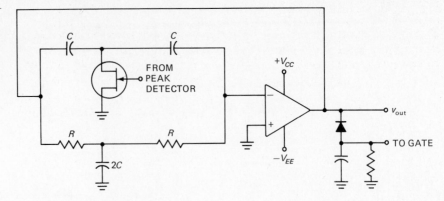

In the twin-T filter, resistance $R/2$ is adjusted. This is necessary because the circuit oscillates at a frequency slightly different from the ideal notch frequency of Eq. (20-4). To ensure that the oscillation frequency is close to the notch frequency, the voltage divider should have R_1 much larger than R_2. As a guide, R_1/R_2 is in the range of 10 to 1000. This forces the oscillator to operate at a frequency near the notch frequency.

Figure 20-10 shows an alternative method of limiting the output level. In this circuit, a JFET is used as a voltage-variable resistance. The gate of the JFET is connected to the output of a negative peak detector. At some output level, the negative voltage out of the peak detector increases the $r_{ds(on)}$ to approximately $R/2$. At this point, the twin-T filter is resonant and the oscillator output stabilizes.

Fig. 20-11
*Phase-shift oscil-
lators.* (a) *Lead
networks.* (b) *Lag
networks.*

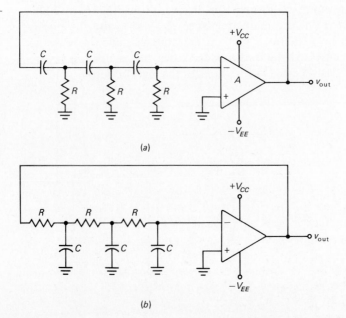

PHASE-SHIFT OSCILLATOR

Figure 20-11*a* is a *phase-shift oscillator* with three lead networks in the feedback path. The amplifier has 180° of phase shift because the signal drives the inverting input. As you recall, a lead network produces a phase shift between 0° and 90°, depending on the frequency. Therefore, at a particular frequency, the total phase shift of the three lead networks equals 180° (approximately 60° each). As a result, the phase shift around the loop will be 360°, equivalent to 0°. If *AB* is greater than unity at this particular frequency, oscillations can start.

Figure 20-10*b* shows an alternative design. It uses three lag networks. The operation is similar. The amplifier produces 180° of phase shift, and the lag networks contribute another 180° at some higher frequency. If *AB* is greater than unity at this frequency, oscillations can start.

Although it is occasionally used, the phase-shift oscillator is not a popular circuit. The main reason for introducing it is because you may accidentally build a phase-shift oscillator when you are trying to build an amplifier. This is discussed later in the chapter under motorboating and parasitic oscillations.

20-4 THE COLPITTS OSCILLATOR

Although it is superb at low frequencies, the Wien-bridge oscillator is not suited to high frequencies (well above 1 MHz). The main problem is the phase shift through the amplifier. One alternative is an *LC* oscillator, a circuit that can be used for frequencies between 1 MHz and 500 MHz. This frequency range is beyond the f_{unity} of most op amps. This is why a bipolar transistor or FET is typically used for the amplifier.

With an amplifier and *LC* tank circuit, we can feed back a signal with the right amplitude and phase to sustain oscillations. The analysis and design of high-frequency oscillators is more of an art than a science. At higher frequencies, the stray capacitances and lead inductances in the transistor and wiring affect the oscillation frequency, feedback fraction, output power, and other ac quantities. For this reason, exact analysis turns into a nightmare. Most people use ball-park approximations for an initial design and adjust the built-up oscillator as needed to get the desired performance. In this section, we examine the *Colpitts oscillator,* one of the most widely used *LC* oscillators.

CE CONNECTION

Figure 20-12*a* shows a Colpitts oscillator. The voltage-divider bias sets up a quiescent operating point. The circuit then has a low-frequency voltage gain of r_C/r'_e, where r_C is the ac resistance seen by the collector. Because of the base and collector lag networks, the high-frequency voltage gain is less than r_C/r'_e.

Figure 20-12*b* is a simplified ac equivalent circuit. The circulating or loop current in the tank flows through C_1 in series with C_2. Notice that v_{out} equals the ac voltage across C_1. Also, the feedback voltage v_f appears across C_2. This feedback voltage drives the base and sustains the oscillations developed across the tank circuit, provided there is enough voltage gain at the oscillation frequency. Since the emitter is at ac ground, the circuit is a CE connection.

Fig. 20-12
(a) *Colpitts oscil-
lator.* (b) *Ac
equivalent circuit.*

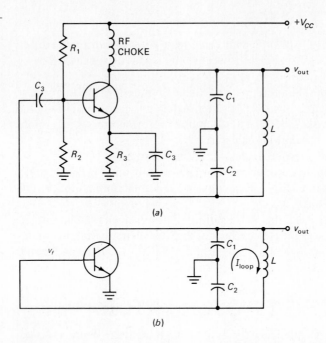

(a)

(b)

You will encounter many variations of the Colpitts oscillator. One way to recognize it is by the capacitive voltage divider formed by C_1 and C_2. This capacitive voltage divider produces the feedback voltage necessary for oscillations. In other kinds of oscillators, the feedback voltage is produced by transformers, inductive voltage dividers, and so on.

RESONANT FREQUENCY

Most LC oscillators use tank circuits with a Q greater than 10. Because of this, we can calculate the approximate *resonant frequency* as

$$f_r \cong \frac{1}{2\pi\sqrt{LC}} \tag{20-5}$$

This is accurate to better than 1 percent when Q is greater than 10.

The capacitance to use in Eq. (20-5) is the equivalent capacitance the circulating current passes through. In the Colpitts tank of Fig. 20-12b, the circulating current flows through C_1 in series with C_2. Therefore, the equivalent capacitance is

$$C = \frac{C_1 C_2}{C_1 + C_2} \tag{20-6}$$

For instance, if C_1 and C_2 are 100 pF each, you would use 50 pF in Eq. (20-5).

STARTING CONDITION

The required starting condition for any oscillator is

$$AB > 1$$

at the resonant frequency of the tank circuit. This is equivalent to

$$A > \frac{1}{B}$$

The voltage gain A in this expression is the voltage gain at the oscillation frequency. In Fig. 20-12b, the output voltage appears across C_1 and the feedback voltage across C_2. Since the circulating current is the same for both capacitors,

$$B = \frac{v_f}{v_{\text{out}}} \cong \frac{X_{C2}}{X_{C1}} = \frac{1/2\pi f C_2}{1/2\pi f C_1}$$

or

$$B \cong \frac{C_1}{C_2}$$

Therefore, the starting condition is

$$A > \frac{C_2}{C_1} \tag{20-7}$$

Remember that this is a crude approximation because it ignores the impedance looking into the base. An exact analysis would take the base impedance into account because it is in parallel with C_2.

What does A equal? This depends on the upper cutoff frequencies of the amplifier. As you recall, there are base and collector lag networks in a bipolar amplifier. If the cutoff frequencies of these lag networks are greater than the oscillation frequency, then A is approximately equal to r_C/r'_e. If the cutoff frequencies are lower than the oscillation frequency, the voltage gain is less than r_C/r'_e, and there is additional phase shift through the amplifier that may prevent oscillations.

OUTPUT VOLTAGE

With light feedback (small B), the value A is only slightly larger than $1/B$, and the operation is approximately class A. When you first turn on the power, the oscillations build up, and the signal swings over more and more of the ac load line. With this increased signal swing, the operation changes from small-signal to large-signal. As this happens, the voltage gain decreases slightly. With light feedback, the value of AB can decrease to 1 without excessive clipping.

With heavy feedback (large B), the large feedback signal drives the base of Fig. 20-12a into saturation and cutoff. This charges capacitor C_3, producing negative dc clamping at the base and changing the operation from class A to class C. The negative clamping automatically adjusts the value of AB to 1. If the feedback is too heavy, you may lose some of the output voltage because of stray power losses.

When you build an oscillator, you can adjust the amount of feedback to maximize the output voltage. The trick is to use enough feedback to start under all conditions (different transistors, temperature, voltage, etc.), but not so much that you lose more output than necessary.

COUPLING TO A LOAD

The exact frequency of oscillation depends on the Q of the circuit and is given by

$$f_r = \frac{1}{2\pi\sqrt{LC}} \left(\frac{Q^2}{1 + Q^2} \right) \tag{20-8}$$

Usually, Q is greater than 10 and this exact equation simplifies to the ideal value given earlier, Eq. (20-5). If Q is less than 10, the frequency is pulled lower than the ideal value. Furthermore, a low Q may prevent the oscillator from starting by lowering the high-frequency gain below $1/B$.

Figure 20-13a shows one way to couple to the load resistance. If the load resistance is large, then it will not load down the resonant circuit too much, and the Q will be greater than 10. On the other hand, if the load resistance is small, the Q drops under 10, and the oscillations may not start. One solution to a small load resistance is to use a small capacitance C_4, one whose X_C is large compared with the load resistance. This prevents excessive loading of the tank circuit.

Figure 20-13b shows *link coupling,* another way of coupling the signal to a small load resistance. Link coupling means using only a few turns on the secondary winding of an RF transformer. This light coupling ensures that the load resistance will not lower the Q of the tank circuit to a point at which the oscillator will not start.

Whether capacitor or link coupling is used, the loading effect should be kept as

Fig. 20-13
Types of output coupling.
(a) *Capacitor.*
(b) *Link.*

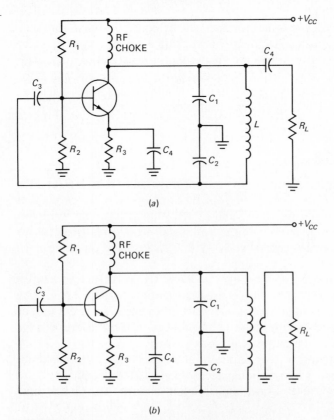

(a)

(b)

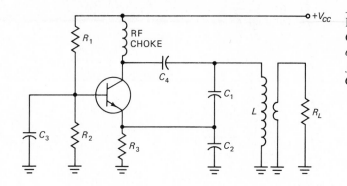

Fig. 20-14
CB oscillator can oscillate at higher frequencies than CE oscillator.

small as possible. In this way, the high Q of the tank ensures an undistorted sinusoidal output with a reliable start for oscillations.

CB CONNECTION

When the feedback signal drives the base, a Miller capacitance appears across the input. This produces a relatively low cutoff frequency and rolls off the gain at a rate of 20 dB per decade. To get a higher cutoff frequency, the feedback signal can be applied to the emitter, as shown in Fig. 20-14. Capacitor C_3 ac grounds the base, and so the transistor acts like a CB amplifier. A circuit like this can oscillate at higher frequencies because its high-frequency gain is larger than that of a comparable CE oscillator. With link coupling on the output, the tank is lightly loaded, and the resonant frequency is still given by Eq. (20-5).

The feedback fraction is slightly different. The output voltage appears across C_1 and C_2 in series, while the feedback voltage appears across C_2. Ideally, the feedback fraction is

$$B = \frac{v_f}{v_{\text{out}}} = \frac{X_{C2}}{X_{C1} + X_{C2}}$$

After expanding and simplifying, this becomes

$$B \cong \frac{C_1}{C_1 + C_2}$$

For the oscillations to start, A must be greater than $1/B$. As an approximation, this means that

$$A > \frac{C_1 + C_2}{C_1} \tag{20-9}$$

This is a rough approximation because it ignores the input impedance of the emitter, which is in parallel with C_2. An exact analysis would include the emitter impedance.

FET COLPITTS

Figure 20-15 is an example of a FET Colpitts oscillator in which the feedback signal is applied to the gate. Since the gate has a high input resistance, the loading effect on

Fig. 20-15
JFET oscillator has less loading effect on tank circuit.

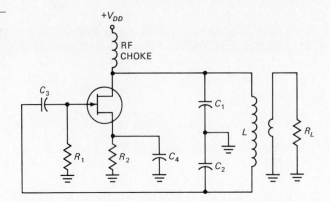

the tank circuit is much less than with a bipolar transistor. In other words, the approximation

$$B \cong \frac{C_1}{C_2}$$

is more accurate with a FET because the impedance looking into the gate is higher. The starting condition for this FET oscillator is

$$A > \frac{C_2}{C_1} \qquad (20\text{-}10)$$

In a FET oscillator, the low-frequency voltage gain is $g_m r_C$. Above the cutoff frequency of the FET amplifier, the voltage gain rolls off. In Eq. (20-10), A is the gain at the oscillation frequency. As a rule, try to keep the oscillation frequency lower than the cutoff frequency of the FET amplifier; otherwise, the additional phase shift through the amplifier may prevent the oscillator from starting. One way to get a higher cutoff frequency for the amplifier is to use a common-gate FET instead of a common-source FET.

EXAMPLE 20-3

What is the frequency of oscillation in Fig. 20-16? The feedback fraction? How much voltage gain does the circuit need to start oscillating?

SOLUTION

The equivalent capacitance of the tank circuit is

$$C = \frac{(0.001\ \mu\text{F})(0.01\ \mu\text{F})}{0.001\ \mu\text{F} + 0.01\ \mu\text{F}} = 909\ \text{pF}$$

The inductance is 15 μH; therefore, the frequency of oscillation is

$$f_r = \frac{1}{2\pi\sqrt{(15\ \mu\text{H})(909\ \text{pF})}} = 1.36\ \text{MHz}$$

Fig. 20-16

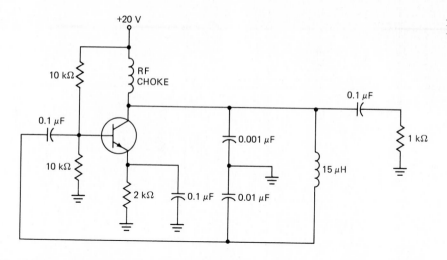

The feedback fraction is

$$B = \frac{0.001 \ \mu F}{0.01 \ \mu F} = 0.1$$

For the oscillator to start, the voltage gain must be greater than 10 at 1.36 MHz. If possible, the cutoff frequency should be higher than 1.36 MHz to avoid additional phase shift around the loop. In other words, the circuit already has a loop phase shift of 0° below the cutoff frequency. If the amplifier starts rolling off, the extra phase shift may prevent oscillations well above cutoff.

20-5 OTHER *LC* OSCILLATORS

The Colpitts is the most widely used *LC* oscillator. The capacitive voltage divider in the resonant circuit is a convenient way to develop the feedback voltage. But there are other kinds of oscillators that are also used. In this section, we will discuss the Armstrong, Hartley, Clapp, and crystal oscillators.

ARMSTRONG OSCILLATOR

Figure 20-17a is an example of an *Armstrong oscillator*. In this circuit, the collector drives an *LC* resonant tank. The feedback signal is taken from a small secondary winding and fed back to the base. There is a phase shift of 180° in the transformer, which means that the phase shift around the loop is zero. Stated another way, the feedback is positive. Ignoring the loading effect of the base, the feedback fraction is

$$B \cong \frac{M}{L} \tag{20-11}$$

where M is the mutual inductance and L is the primary inductance. For the Armstrong oscillator to start, the voltage gain must be greater than $1/B$.

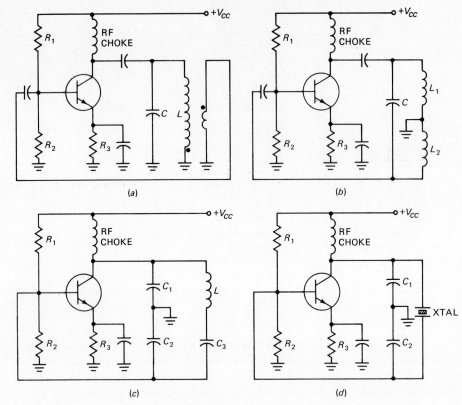

Fig. 20-17
Oscillators.
(a) *Armstrong.*
(b) *Hartley.* (c)
Clapp. (d) *Crystal.*

(a)

(b)

(c)

(d)

An Armstrong oscillator uses transformer coupling for the feedback signal. This is how you can recognize variations on this basic circuit. The secondary winding is sometimes called a *tickler coil* because it feeds back the signal that sustains the oscillations. The resonant frequency is given by Eq. (20-5), using the L and C shown in Fig. 20-17a. As a rule, you don't see the Armstrong oscillator used much because most designers avoid transformers whenever possible.

HARTLEY

Figure 20-17b is an example of a *Hartley oscillator.* When the LC tank is resonant, the circulating current flows through L_1 in series with L_2. So, the equivalent L to use in Eq. (20-5) is

$$L = L_1 + L_2 \tag{20-12}$$

In a Hartley oscillator, the feedback voltage is developed by the inductive voltage divider, L_1 and L_2. Since the output voltage appears across L_1 and the feedback voltage across L_2, the feedback fraction is

$$B = \frac{v_f}{v_\text{out}} \cong \frac{X_{L2}}{X_{L1}}$$

or

$$B \cong \frac{L_2}{L_1} \qquad (20\text{-}13)$$

As usual, this ignores the loading effects of the base. For oscillations to start, the voltage gain must be greater than $1/B$.

Often, a Hartley oscillator uses a single tapped inductor instead of two separate inductors. The action is basically the same either way. Another variation sends the feedback signal to the emitter instead of the base. Also, you may see a FET used instead of a bipolar transistor. The output signal can be either capacitively or link coupled.

CLAPP OSCILLATOR

The *Clapp oscillator* of Fig. 20-17c is a refinement of the Colpitts oscillator. The capacitive voltage divider produces the feedback signal as before. An additional capacitor C_3 is in series with the inductor. Since the circulating tank current flows through C_1, C_2, and C_3 in series, the equivalent capacitance used to calculate the resonant frequency is

$$C = \frac{1}{1/C_1 + 1/C_2 + 1/C_3} \qquad (20\text{-}14)$$

In a Clapp oscillator, C_3 is much smaller than C_1 and C_2. As a result, C is approximately equal to C_3, and the resonant frequency is given by

$$f_r \cong \frac{1}{2\pi\sqrt{LC_3}} \qquad (20\text{-}15)$$

Why is this important? Because C_1 and C_2 are shunted by transistor and stray capacitances. These extra capacitances alter the values of C_1 and C_2 slightly. In a Colpitts oscillator, the resonant frequency depends to some extent on the transistor and stray capacitances. But in a Clapp oscillator, the transistor and stray capacitances have no effect on C_3, which means that the oscillation frequency is more stable and accurate. This is why you occasionally see the Clapp oscillator used instead of a Colpitts oscillator.

CRYSTAL OSCILLATOR

When accuracy and stability of the oscillation frequency are important, a *quartz-crystal oscillator* is used. In Fig. 20-17d, the feedback signal comes from a capacitive tap. As will be discussed in the next section, the crystal (abbreviated XTAL) acts like a large inductor in series with a small capacitor (similar to the Clapp). Because of this, the resonant frequency is almost totally unaffected by transistor and stray capacitances.

EXAMPLE 20-4

If 50 pF is added in series with the 15-μH inductor of Fig. 20-16, the circuit becomes a Clapp oscillator. What is the frequency of oscillation?

SOLUTION

The added capacitor C_3 is only 50 pF; therefore,

$$C = \frac{1}{1/0.001\ \mu F + 1/0.01\ \mu F + 1/50\ pF} \cong 50\ pF$$

The approximate oscillation frequency is

$$f_r \cong \frac{1}{2\pi\sqrt{(15\ \mu H)(50\ pF)}} = 5.81\ MHz$$

20-6 QUARTZ CRYSTALS

Some crystals found in nature exhibit the *piezoelectric effect;* when you apply an ac voltage across them, they vibrate at the frequency of the applied voltage. Conversely, if you mechanically force them to vibrate, they generate an ac voltage. The main substances that produce this piezoelectric effect are quartz, Rochelle salts, and tourmaline.

Rochelle salts have the greatest piezoelectric activity; for a given ac voltage, they vibrate more than quartz or tourmaline. Mechanically, they are the weakest; they break easily. Rochelle salts have been used to make microphones, phonograph pickups, headsets, and loudspeakers.

Tourmaline shows the least piezoelectric activity but is the strongest of the three. It is also the most expensive. It is occasionally used at very high frequencies.

Quartz is a compromise between the piezoelectric activity of Rochelle salts and the strength of tourmaline. Because it is inexpensive and readily available in nature, quartz is widely used for RF oscillators and filters.

CRYSTAL CUTS

The natural shape of a quartz crystal is a hexagonal prism with pyramids at the ends (see Fig. 20-18a). To get a usable crystal out of this, we have to slice a rectangular slab out of the natural crystal. Figure 20-18b shows this slab with thickness *t*. The number of slabs we can get from a natural crystal depends on the size of the slabs and the angle of the cut.

There are a number of different ways to cut the natural crystal; these cuts have names like X cut, Y cut, XY cut, and AT cut. For our purposes, all we need to know is that different cuts have different piezoelectric properties. (Manufacturer's catalogs are usually the best source of information on different cuts and their properties.)

For use in electronic circuits, the slab must be mounted between two metal plates, as shown in Fig. 20-18c. In this circuit the amount of crystal vibration depends on the

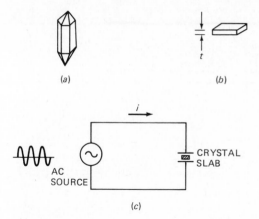

Fig. 20-18 (a) *Natural quartz crystal.* (b) *Slab.* (c) *Input current reaches maximum at crystal resonance.*

frequency of the applied voltage. By changing the frequency, we can find resonant frequencies at which the crystal vibrations reach a maximum. Since the energy for the vibrations must be supplied by the ac source, the ac current is maximized at each resonant frequency.

FUNDAMENTAL AND OVERTONES

Most of the time, the crystal is cut and mounted to vibrate best at one of its resonant frequencies, usually the *fundamental* or lowest frequency. Higher resonant frequencies, called *overtones,* are almost exact multiples of the fundamental frequency. As an example, a crystal with a fundamental frequency of 1 MHz has a first overtone of approximately 2 MHz, a second overtone of approximately 3 MHz, and so on.

The formula for the fundamental frequency of a crystal is

$$f = \frac{K}{t} \tag{20-16}$$

where K is a constant that depends on the cut and other factors, and t is the thickness of the crystal. As we can see, the fundamental frequency is inversely proportional to the thickness. For this reason, there is a practical limit to how high in frequency we can go. The thinner the crystal, the more fragile it becomes and the more likely it is to break because of vibrations.

Quartz crystals work well up to 10 MHz on the fundamental frequency. To reach higher frequencies, we can use a crystal mounted to vibrate on overtones; in this way, we can reach frequencies up to 100 MHz. Occasionally, the more expensive but stronger tourmaline is used at higher frequencies.

AC EQUIVALENT CIRCUIT

What does the crystal look like as far as the ac source is concerned? When the mounted crystal of Fig. 20-19a is not vibrating, it is equivalent to a capacitance C_m

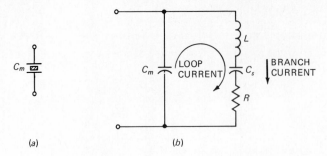

Fig. 20-19   (a) *Mounting capacitance.* (b) *Ac equivalent circuit of vibrating crystal.*

because it has two metal plates separated by a dielectric. C_m is known as the mounting capacitance.

When the crystal is vibrating, it acts like a tuned circuit. Figure 20-19b shows the ac equivalent circuit of a crystal vibrating at or near its fundamental frequency. Typical values are L in henrys, C_s in fractions of a picofarad, R in hundreds of ohms, and C_m in picofarads. As an example, here are the values for one available crystal: $L = 3$ H, $C_s = 0.05$ pF, $R = 2000$ Ω, and $C_m = 10$ pF. Among other things, the cut, thickness, and mounting of the slab affect these values.

The outstanding feature of crystals compared with discrete LC tank circuits is their incredibly high Q. For the values given, we can calculate a Q of over 3000. Q's can easily be over 10,000. On the other hand, a discrete LC tank circuit seldom has a Q of over 100. The extremely high Q of a crystal leads to oscillators with very stable frequency values.

SERIES AND PARALLEL RESONANCE

Besides the Q, L, C_s, R, and C_m of the crystal, there are two other characteristics to know about. The *series* resonant frequency f_s of a crystal is the resonant frequency of the LCR branch in Fig. 20-19b. At this frequency the branch current reaches a maximum value because L resonates with C_s. The formula for this resonant frequency is

$$f_s = \frac{1}{2\pi\sqrt{LC_s}} \tag{20-17}$$

The *parallel* resonant frequency f_p of the crystal is the frequency at which the circulating or loop current of Fig. 20-19b reaches a maximum value. Since this loop current must flow through the series combination of C_s and C_m, the equivalent C_{loop} is

$$C_{\text{loop}} = \frac{C_m C_s}{C_m + C_s} \tag{20-18}$$

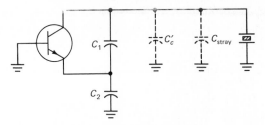

Fig. 20-20 *Stray circuit capacitances are in parallel with mounting capacitance.*

and the parallel resonant frequency is

$$f_p = \frac{1}{2\pi\sqrt{LC_{\text{loop}}}} \qquad (20\text{-}19)$$

Two capacitances in series always produce a capacitance smaller than either; therefore, C_{loop} is less than C_s, and f_p is greater than f_s.

In any crystal, C_s is much smaller than C_m. For instance, with the values given earlier, C_s was 0.05 pF and C_m was 10 pF. Because of this, Eq. (20-18) gives a value of C_{loop} just slightly less than C_s. In turn, this means that f_p is only slightly more than f_s. When you use a crystal in an ac equivalent circuit like Fig. 20-20, the additional circuit capacitances appear in shunt with C_m. Because of this, the oscillation frequency will lie between f_s and f_p. This is the advantage of knowing the values of f_s and f_p; they set lower and upper limits on the frequency of the crystal oscillator.

CRYSTAL STABILITY

The frequency of an oscillator tends to change slightly with time; this *drift* is produced by temperature, aging, and other causes. In a crystal oscillator the frequency drift with time is very small, typically less than 1 part in 10^6 (0.0001 percent) per day. Stability like this is important in electronic wristwatches; they use quartz-crystal oscillators as the basic timing device.

By using crystal oscillators in precision temperature-controlled ovens, crystal oscillators have been built that have frequency drift of less than 1 part in 10^{10} per day. Stability like this is needed in frequency and time standards. To give you an idea of how precise 1 part in 10^{10} is, a clock with this drift will take 300 years to gain or lose 1 s.

CRYSTAL OSCILLATORS

Figure 20-21*a* shows a Colpitts crystal oscillator. The capacitive voltage divider produces the feedback voltage for the base of the transistor. The crystal acts like an inductor that resonates with C_1 and C_2. The oscillation frequency is between the series and parallel resonant values.

Figure 20-21*b* is a variation of the Colpitts crystal oscillator. Figure 20-21*c* is a FET Colpitts oscillator. Figure 20-21*d* is a circuit called a *Pierce* crystal oscillator. It has the advantage of simplicity.

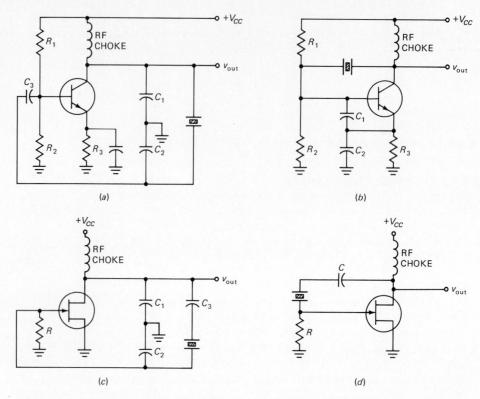

Fig. 20-21 *Crystal oscillators.* (a) *Colpitts.* (b) *Variation of Colpitts.* (c) *Clapp.* (d) *Pierce.*

EXAMPLE 20-5

A crystal has these values: $L = 3$ H, $C_s = 0.05$ pF, $R = 2000\ \Omega$, and $C_m = 10$ pF. Calculate the f_s and f_p of the crystal to three significant digits.

SOLUTION

From Eq. (20-17),

$$f_s = \frac{1}{2\pi\sqrt{(3\text{ H})(0.05\text{ pF})}} = 411\text{ kHz}$$

From Eq. (20-18),

$$C_{\text{loop}} = \frac{(10\text{ pF})(0.05\text{ pF})}{10\text{ pF} + 0.05\text{ pF}} = 0.0498\text{ pF}$$

From Eq. (20-19),

$$f_p = \frac{1}{2\pi\sqrt{(3\text{ H})(0.0498\text{ pF})}} = 412\text{ kHz}$$

If this crystal is used in an oscillator, the frequency of oscillation must lie between 411 and 412 kHz.

20-7 UNWANTED OSCILLATIONS

Oscillators can be frustrating. Sometimes you try to build an oscillator and you wind up with an amplifier because the loop gain is less than 1 at the desired frequency of oscillation. After some tinkering, you finally manage to get the circuit to oscillate. On the other hand, there will be times when you build a high-gain amplifier that turns into an oscillator. You will swear that the circuit cannot oscillate because there is no path for positive feedback. But there are ways to get positive feedback through the power supply, ground loops, and other undesirable paths.

LOW-FREQUENCY OSCILLATIONS

Look at Fig. 20-22a. There is no feedback path from output to input; therefore, the circuit cannot possibly oscillate. Right? Wrong! There are subtle feedback paths that can make any high-gain amplifier produce unwanted oscillations.

Motorboating is a putt-putt sound from a loudspeaker connected to an amplifier like Fig. 20-22a. This sound represents very low-frequency oscillations, like a few hertz. The feedback path is by way of the power supply. Ideally, the power supply looks like a perfect ac short to ground. But to a better approximation the supply has an impedance z, as shown in Fig. 20-22b. Because of the nonzero supply impedance at low frequencies, we can get current feedback through the supply impedance and the voltage-divider bias of the first stage.

Figure 20-22c illustrates this current feedback from the last stage to the first stage. The alternating current out of the last stage flows through R_C and through the supply impedance z. The voltage across z can then drive the R_1-R_2 voltage divider of the first stage. Frequency of the oscillation is determined by the lead networks in the amplifier and the reactance of the power supply. At some frequency below the midband of the amplifier, the loop phase shift is $0°$. If AB is greater than unity at this frequency, motorboating occurs.

What is the cure for motorboating? Use a regulated power supply (Chap. 19). This kind of supply has an internal impedance under 0.1Ω (some as low as 0.0005Ω). The current feedback is then too small to produce oscillations.

HIGH-FREQUENCY OSCILLATIONS

You can get unwanted oscillations above the midband of the amplifier. The electric or magnetic fields around the last stage may induce feedback voltages in an earlier stage with the right phase for oscillations. If AB is greater than unity at the frequency where this happens, you will get oscillations.

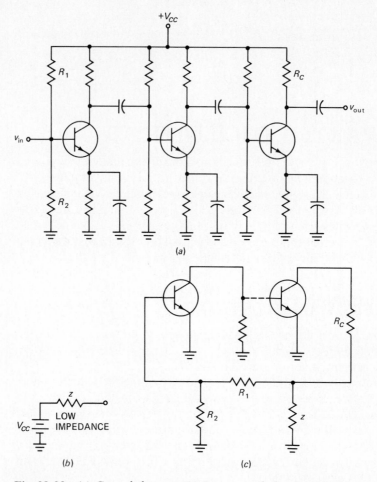

Fig. 20-22 (a) *Cascaded stages.* (b) *Power supply has low impedance.* (c) *Current feedback through power supply impedance.*

Figure 20-23*a* illustrates this idea. The output circuit acts like one plate of a capacitor, the input circuit like the other plate. The capacitance between input and output is small (usually under 1 pF); but at high frequency, the capacitance may feed back enough signal to produce oscillations.

Magnetic coupling is also possible. The output wire labeled "primary" in Fig. 20-23*a* can act like a primary winding of a transformer; the input wire labeled "secondary" can act like a secondary winding. As a result, alternating current in the primary can induce a voltage in the secondary. If the feedback signal is strong enough and the phase correct, we get oscillations.

What is the cure for high-frequency oscillations? One approach is to increase the *distance* between stages; this cuts down the coupling between them. If this is not practical, you can enclose each stage in a *shield* or metallic container (see Fig. 20-23*b*). Shielding like this is common in many high-frequency applications; it

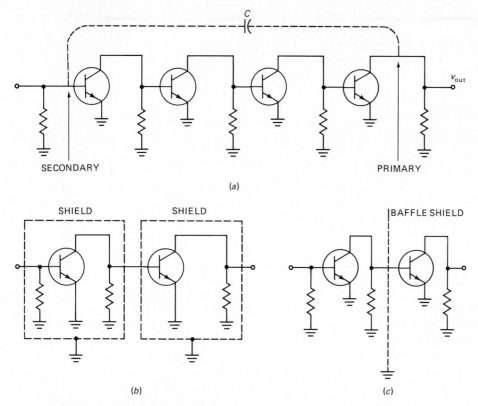

Fig. 20-23 (a) *Capacitive and magnetic coupling from output to input.* (b) *Shields enclose each stage and prevent oscillations.* (c) *Baffle shield prevents capacitive coupling.*

blocks high-frequency electric and magnetic fields. If only capacitive coupling is a problem, a *baffle shield* (a metallic plate) between stages may eliminate high-frequency oscillations (Fig. 20-23c).

GROUND LOOPS

Another subtle cause of high-frequency oscillations is a *ground loop,* a difference of potential between two ground points. In Fig. 20-23a, all ac grounds are ideally at the same potential. But in reality, the chassis or whatever serves as ground has some nonzero impedance that increases with frequency. Therefore, if ac ground currents from the last stage happen to flow through part of the chassis being used by an earlier stage, we can get enough unwanted positive feedback to cause oscillations.

The solution to a ground-loop problem is proper layout of the stages to prevent ac ground currents of later stages from flowing through ground paths of earlier stages. One way to accomplish this is to use a *single* ground point, as shown in Fig. 20-24. If this is done, there can be no difference of potential between two ground points because there is only one ground point.

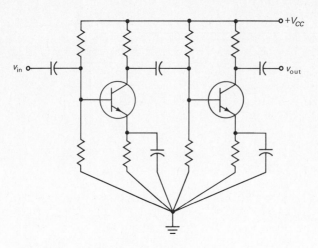

Fig. 20-24
Single ground point prevents ground loops.

SUPPLY BYPASSING

Watch out for lead inductance between the power supply and the circuit (see Fig. 20-25). A long lead may have enough inductance to result in current feedback at high frequencies. The solution is to add a large *bypass* capacitor across the circuit, as shown in Fig. 20-25.

Bypassing the supplies is almost always essential with ICs. Depending on the particular IC, bypass capacitors from about 0.1 to over 1 μF may be needed to prevent oscillations. The bypass capacitors should be located as close as possible to the IC.

PARASITIC OSCILLATIONS

The small transistor capacitances and lead inductances distributed throughout a circuit may form a random Colpitts or Hartley oscillator. The resulting oscillations are called *parasitic* oscillations. Usually, these oscillations are at a very high frequency and have a weak amplitude because the feedback is light. Parasitic oscillations tend to make circuits act erratically: oscillators produce more than one frequency, op amps have too much offset, power supplies have unexplained ripple, amplifiers produce distorted signals, and video displays contain snow. An old troubleshooting trick is to touch low-voltage parts of the circuit that are suspected of

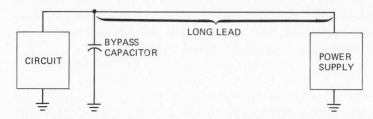

Fig. 20-25 *Bypass capacitor prevents current feedback caused by long lead between power supply and circuit.*

having parasitic oscillations. If the trouble clears, you almost certainly have parasitic oscillations.

What is the cure? Well, you can leave your finger permanently embedded in the circuit. Or you can reduce the feedback for the parasitic oscillations. One way to do this is to add *small resistors* in the base leads of transistors. Another approach that has been used is to slip a *ferrite bead* over each base lead. In either case, the feedback fraction is reduced or the phase shift is changed enough to kill the parasitic oscillations.

NEGATIVE-FEEDBACK AMPLIFIERS

Figure 20-26a shows a three-stage negative-feedback amplifier. In the midband of the internal amplifier the phase shift is 180° because there is an odd number of inverting stages; therefore, the phase shift around the entire loop is 180°, and the feedback is negative.

But outside the midband the internal lag networks of the stages produce additional phase shifts. Because of this, at some high frequency the phase shift around the entire loop is 0°, and the feedback becomes positive. In other words, Fig. 20-26a acts like a phase-shift oscillator using lag networks. At some higher frequency, the three lag networks can produce a phase shift of 180° (60° each if the networks are identical).

The only way to prevent oscillations outside the midband is to make sure that loop gain AB is less than unity when the phase shift reaches 0°. The safest and most widely used method is this: Make one of the lag networks dominant enough to produce a 20-dB roll-off until the loop gain crosses the 0-dB axis. A 20-dB roll-off at the horizontal crossing implies that only one lag network is working beyond its cutoff frequency; all others are still operating below cutoff. This means that the loop phase shift is around 270° at f_{unity}, which makes it impossible to have oscillations.

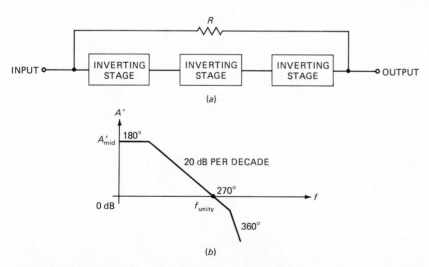

(a)

(b)

Fig. 20-26 (a) *Feedback around three inverting stages.* (b) *One dominant lag network prevents oscillations.*

Figure 20-26b illustrates the idea of one dominant lag network. In the midband the gain is high and the phase shift is 180°. One of the stages has a dominant lag network, so that the gain breaks at a low frequency and rolls off at 20 dB per decade. A decade above this frequency, the phase shift is 270°. It stays at approximately 270° until the gain crosses the horizontal axis at f_{unity}. Beyond this point, oscillations are impossible because the loop gain is less than unity.

With monolithic op amps, the dominant lag network is usually integrated on the chip and automatically provides the 20-dB roll-off until f_{unity} is reached. For instance, the 741 uses a 30-pF compensating capacitor that is part of a Miller lag network; the gain breaks at 10 Hz and rolls off at 20 dB per decade until an f_{unity} of 1 MHz is reached.

With uncompensated op amps like the 709, you have to add external resistors and capacitors to get this 20-dB roll-off; the manufacturer's data sheet tells you the sizes of R and C to use.

20-8 THE 555 TIMER

The 555 *timer* combines a relaxation oscillator, two comparators, an *RS* flip-flop, and a discharge transistor. This versatile IC has so many applications that it has become an industry standard. Once you understand how it works, you can join the many designers who are constantly finding new uses for this amazing IC.

RS FLIP-FLOP

Figure 20-27a shows a pair of cross-coupled transistors. Each collector drives the opposite base through a resistance R_B. In a circuit like this, one transistor is saturated and the other is cut off. For instance, if the right transistor is saturated, its collector voltage is approximately zero. This means no base drive for the left transistor, and so it goes into cutoff and its collector voltage approaches $+V_{CC}$. This high voltage produces enough base current to keep the right transistor in saturation.

On the other hand, if the right transistor is cut off, then its collector voltage drives the left transistor into saturation. The low collector voltage out of this left transistor then keeps the right transistor in cutoff.

Depending on which transistor is saturated, the Q output is either low or high. By adding more components to the circuit, we get an *RS flip-flop*, a circuit that can set

Fig. 20-27
(a) *Part of an RS flip-flop.*
(b) *Symbol for RS flip-flop.*

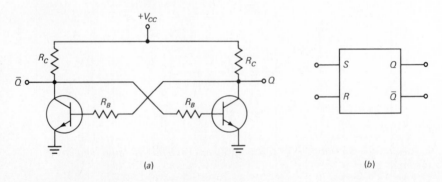

(a)   (b)

the Q output to high or reset it to low. Incidentally, a complementary (opposite) output $\bar{Q}$ is available from the collector of the other transistor.

Figure 20-27b shows the schematic symbol for an RS flip-flop of any design. Whenever you see this symbol, remember the action: The circuit latches in either of two states. A high S input sets Q to high; a high R input resets Q to low. Output Q remains in a given state until it is triggered into the opposite state.

BASIC TIMING CONCEPT

Figure 20-28a illustrates some basic ideas that we will need in our later discussion of the 555 timer. Assume that output Q is high. This saturates the transistor and clamps the capacitor voltage at ground. In other words, the capacitor is shorted and cannot charge.

The noninverting input voltage of the comparator is called the *threshold* voltage, and the inverting input voltage is referred to as the *control* voltage. With the RS flip-flop set, the saturated transistor holds the threshold voltage at 0. The control voltage, on the other hand, is fixed at $+10$ V because of the voltage divider.

Suppose we apply a high voltage to the R input. This resets the RS flip-flop. Output Q goes low and cuts off the transistor. Capacitor C is now free to charge. As the capacitor charges, the threshold voltage increases. Eventually, the threshold voltage becomes slightly greater than the control voltage ($+10$ V). The output of the comparator then goes high, forcing the RS flip-flop to set. The high Q output saturates the transistor, and this quickly discharges the capacitor. Notice the two

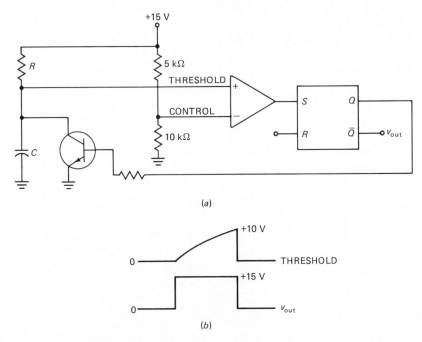

(a)

(b)

Fig. 20-28 (a) *Basic timing circuit.* (b) *Capacitor voltage is exponential and output voltage is rectangular.*

waveforms in Fig. 20-28b. An exponential rise is across the capacitor, and a positive-going pulse appears at the $\overline{Q}$ output.

555 BLOCK DIAGRAM

Figure 20-29 is a simplified block diagram of the NE555 timer, an 8-pin IC timer introduced by Signetics Corporation. Notice that the upper comparator has a threshold input (pin 6) and a control input (pin 5). In most applications, the control input is not used, so that the control voltage equals $+2V_{CC}/3$. As before, whenever the threshold voltage exceeds the control voltage, the high output from the comparator will set the flip-flop.

The collector of the discharge transistor goes to pin 7. When this pin is connected to an external timing capacitor, a high Q output from the flip-flop will saturate the transistor and discharge the capacitor. When Q is low, the transistor opens and the capacitor can charge as previously described.

The complementary signal out of the flip-flop goes to pin 3, the output. When the external reset (pin 4) is grounded, it inhibits the device (prevents it from working). This on/off feature is sometimes useful. In most applications, however, the external reset is not used, and pin 4 is tied directly to the supply voltage.

Notice the lower comparator. Its inverting input is called the *trigger* (pin 2). Because of the voltage divider, the noninverting input has a fixed voltage of $+V_{CC}/3$. When the trigger input voltage is slightly less than $+V_{CC}/3$, the op-amp output goes high and resets the flip-flop.

Finally, pin 1 is the chip ground, while pin 8 is the supply pin. The 555 timer will work with any supply voltage between 4.5 and 16 V.

MONOSTABLE OPERATION

Figure 20-30a shows the 555 timer connected for *monostable* (also called one-shot) operation. The circuit works as follows. When the trigger input is slightly less than $+V_{CC}/3$, the lower comparator has a high output and resets the flip-flop. This cuts off

Fig. 20-29
Simplified schematic diagram of 555 timer.

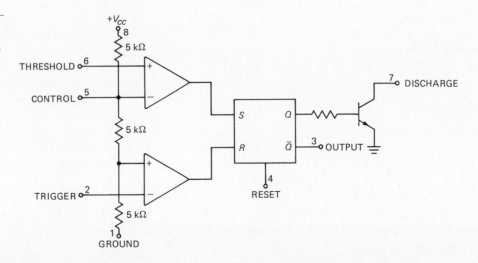

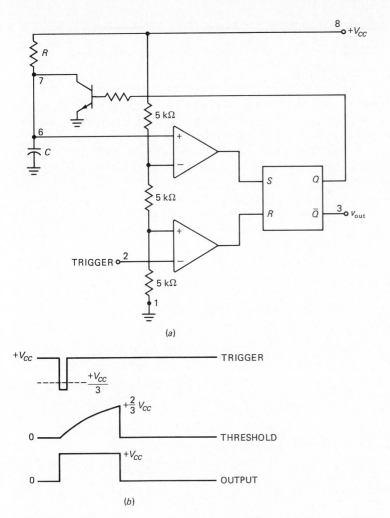

Fig. 20-30 (a) *555 timer connected as monostable multivibrator.* (b) *Trigger, output, and threshold waveforms.*

the transistor, allowing the capacitor to charge. When the capacitor voltage is slightly greater than $+2V_{CC}/3$, the upper comparator has a high output, which sets the flip-flop. As soon as Q goes high, it turns on the transistor; this quickly discharges the capacitor.

Figure 20-30b shows typical waveforms. The trigger input is a narrow pulse with a quiescent value of $+V_{CC}$. The pulse must drop below $+V_{CC}/3$ to reset the flip-flop and allow the capacitor to charge. When the threshold voltage slightly exceeds $+2V_{CC}/3$, the flip-flop sets; this saturates the transistor and discharges the capacitor. As a result, we get one rectangular output pulse.

The capacitor C has to charge through resistance R. The larger the RC time constant, the longer it takes for the capacitor voltage to reach $+2V_{CC}/3$. In other words, the RC time constant controls the width of the output pulse. Appendix 1

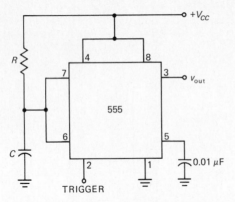

Fig. 20-31
*555 monostable
timer circuit.*

derives this formula for the pulse width:

$$W = 1.1RC \tag{20-20}$$

For instance, if $R = 22$ kΩ and $C = 0.068$ μF, then

$$W = 1.1(22 \text{ k}\Omega)(0.068 \text{ }\mu\text{F}) = 1.65 \text{ ms}$$

Normally, a schematic diagram does not show the comparators, flip-flop, and other components inside the 555 timer. Rather, you will see a schematic diagram like Fig. 20-31 for the monostable 555 circuit. Only the pins and external components are shown. Incidentally, notice that pin 5 (control) is bypassed to ground through a small capacitor, typically 0.01 μF. This provides noise filtering for the control voltage.

Recall that grounding pin 4 inhibits the 555 timer. To avoid accidental reset, pin 4 is usually tied to the supply voltage, as shown in Fig. 20-31.

In summary, the monostable 555 timer produces a single pulse whose width is determined by the external R and C used in Fig. 20-31. The pulse begins with the leading edge of the negative trigger input. One-shot operation like this has a number of applications, as you will see in later studies.

ASTABLE OPERATION

Figure 20-32a shows the 555 timer connected for *astable* (free-running) operation. When Q is low, the transistor is cut off and the capacitor is charging through a total resistance of $R_A + R_B$. Because of this, the charging time constant is $(R_A + R_B)C$. As the capacitor charges, the threshold voltage increases. Eventually, the threshold voltage exceeds $+2V_{CC}/3$; then the upper comparator has a high output, and this sets the flip-flop. With Q high, the transistor saturates and grounds pin 7. Now the capacitor discharges through R_B. Therefore, the discharging time constant is R_BC. When the capacitor voltage drops slightly below $+V_{CC}/3$, the lower comparator has a high output, and this resets the flip-flop.

Figure 20-32b illustrates the waveforms. As you can see, the timing capacitor has an exponentially rising and falling voltage. The output is a rectangular wave. Since the charging time constant is longer than the discharging time constant, the output is not symmetrical; the high output state lasts longer than the low output state.

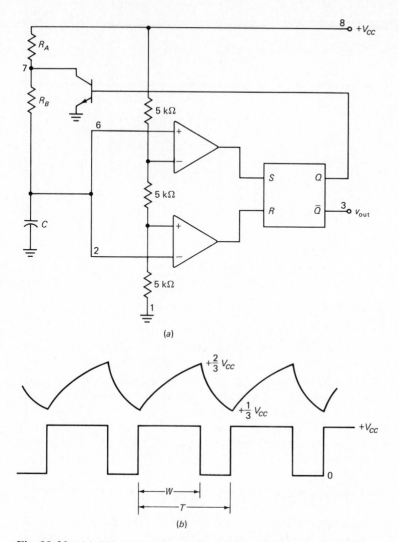

Fig. 20-32 (a) *555 timer connected as astable multivibrator.* (b) *Capacitor and output waveforms.*

To specify how unsymmetrical the output is, we will use the *duty cycle,* defined as

$$D = \frac{W}{T} \times 100\% \tag{20-21}$$

As an example, if $W = 2$ ms and $T = 2.5$ ms, then the duty cycle is

$$D = \frac{2 \text{ ms}}{2.5 \text{ ms}} \times 100\% = 80\%$$

Depending on resistances R_A and R_B, the duty cycle is between 50 and 100 percent. A mathematical solution (Appendix 1) to the charging and discharging equations

Fig. 20-33
555 astable timer circuit.

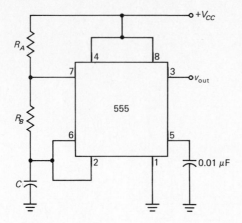

gives the following formulas. The output frequency is

$$f = \frac{1.44}{(R_A + 2R_B)C} \tag{20-22}$$

and the duty cycle is

$$D = \frac{R_A + R_B}{R_A + 2R_B} \times 100\% \tag{20-23}$$

If R_A is much smaller than R_B, the duty cycle approaches 50 percent.

Figure 20-33 shows the astable 555 timer as it usually appears on a schematic diagram. Again notice how pin 4 (reset) is tied to the supply voltage and how pin 5 (control) is bypassed to ground through a 0.01-μF capacitor. An astable 555 timer is often called a *free-running multivibrator* because it produces a continuous train of rectangular pulses.

VOLTAGE-CONTROLLED OSCILLATOR

Figure 20-34*a* shows a *voltage-controlled* oscillator (VCO), one application for a 555 timer. The circuit is sometimes called a voltage-to-frequency converter because an input voltage can change the output frequency. Here is how the circuit works. Recall that pin 5 (control) connects to the inverting input of the upper comparator. Normally, the control voltage is $+2V_{CC}/3$ because of the internal voltage divider. In Fig. 20-34*a*, however, the voltage from an external potentiometer overrides the internal voltage. In other words, by adjusting the potentiometer, we can change the control voltage.

Figure 20-34*b* illustrates the voltage across the timing capacitor. Notice that it varies between $+V_{control}/2$ and $+V_{control}$. If we increase $V_{control}$, it takes the capacitor longer to charge and discharge; therefore, the frequency decreases. As a result, we can change the frequency of the circuit by varying the control voltage.

Incidentally, the control voltage may come from a potentiometer or it may be the output of a transistor circuit, op amp, or some other device. One of the more interesting applications for a VCO is the phase-locked loop (PLL), discussed in Chap. 23.

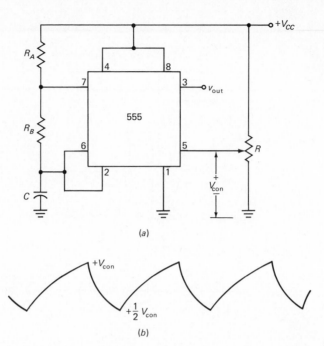

RAMP GENERATOR

Here is another application for the 555 timer. Charging a capacitor through a resistor produces an exponential waveform. If we use a constant current source to charge a capacitor, however, we get a *ramp*. This is the idea behind the circuit of Fig. 20-35a. Here we have replaced the resistor of previous circuits with a *pnp* current source that produces a constant charging current of

$$I_C = \frac{V_{CC} - V_E}{R_E} \tag{20-24}$$

where

$$V_E = \frac{R_2}{R_1 + R_2} V_{CC} + V_{BE} \tag{20-25}$$

For instance, if $V_{CC} = 15$ V, $R_E = 20$ kΩ, $R_1 = 5$ kΩ, $R_2 = 10$ kΩ, and $V_{BE} = 0.7$ V, then

$$V_E = 10 \text{ V} + 0.7 \text{ V} = 10.7 \text{ V}$$

and

$$I_C = \frac{15 \text{ V} - 10.7 \text{ V}}{20 \text{ k}\Omega} = 0.215 \text{ mA}$$

When a trigger starts the monostable 555 timer of Fig. 20-35a, the *pnp* current source forces a constant charging current into the capacitor. Therefore, the voltage across the capacitor is a ramp, as shown in Fig. 20-35b. As derived in Chap. 18, the

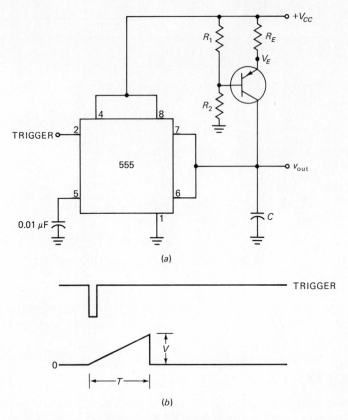

Fig. 20-35   (a) *Using a 555 timer and a bipolar current source to produce a ramp output voltage.* (b) *Trigger and ramp waveforms.*

slope S of the ramp is given by

$$S = \frac{I}{C} \qquad (20\text{-}26)$$

If the charging current is 0.215 mA and the capacitance is 0.022 μF, the ramp will have a slope of

$$S = \frac{0.215 \text{ mA}}{0.022 \text{ }\mu\text{F}} = 9.77 \text{ V/ms}$$

PROBLEMS

Straightforward

20-1. The Wien-bridge oscillator of Fig. 20-36*a* uses a lamp with the characteristics of Fig. 20-36*b*. How much output voltage is there?

Fig. 20-36

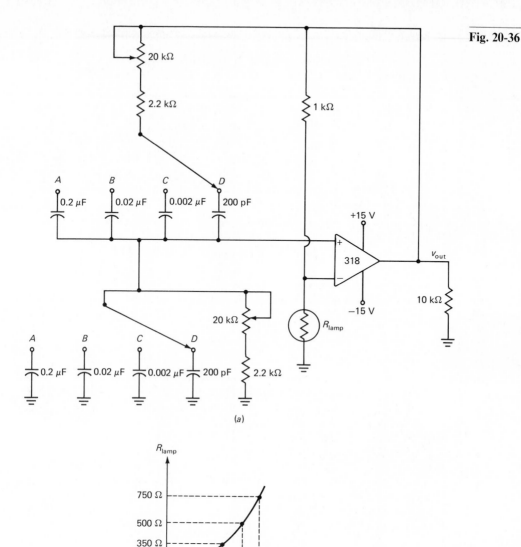

(a)

(b)

20-2. Position D in Fig. 20-36a is the highest frequency range of the oscillator. We can vary the frequency by ganged rheostats. What are the minimum and maximum frequencies of oscillation on this range?

20-3. Calculate the minimum and maximum frequency of oscillation for each position of the ganged switch of Fig. 20-36a.

20-4. To change the output voltage of Fig. 20-36a to a value of 6 V rms, what change can you make?

20-5. In Fig. 20-36a the cutoff frequency of the amplifier with negative feedback is at least one decade above the highest frequency of oscillation. What is the cutoff frequency?

20-6. The twin-T oscillator of Fig. 20-9 has $R = 100$ kΩ and $C = 0.01$ μF. What is the frequency of oscillation?

20-7. What is the approximate value of the dc emitter current in Fig. 20-37? The dc voltage from collector to emitter?

Fig. 20-37

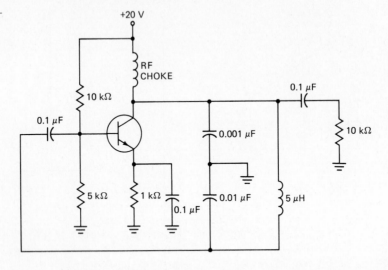

20-8. What is the approximate frequency of oscillation in Fig. 20-37? The value of *B*? For the oscillator to start, what is the minimum value of *A*?

20-9. If the oscillator of Fig. 20-37 is redesigned to get a CB amplifier, what is the feedback fraction?

20-10. A Hartley oscillator like Fig. 20-17*b* has $L_1 = 1$ μH and $L_2 = 0.2$ μH. What is the feedback fraction?

20-11. A crystal has a fundamental frequency of 5 MHz. What is the approximate value of the first overtone frequency? The second overtone? The third?

20-12. A crystal has a thickness of *t*. If you reduce *t* by 1 percent, what happens to the frequency?

20-13. The ac equivalent circuit of a crystal has these values: $L = 1$ H, $C_s = 0.01$ pF, $R = 1000$ Ω, and $C_m = 20$ pF.

    a. What is the series resonant frequency?

    b. What is the *Q* at this frequency?

20-14. A 555 timer is connected for monostable operation. If $R = 10$ kΩ and $C = 0.022$ μF, what is the width of the output pulse?

20-15. An astable 555 timer has $R_A = 10$ kΩ, $R_B = 2$ kΩ, and $C = 0.0047$ μF. What are the output frequency and duty cycle?

Troubleshooting

20-16. Does the output voltage of the Wien-bridge oscillator (Fig. 20-36*a*) increase, decrease, or stay the same for each of these troubles:

    a. Lamp open

    b. Lamp shorted

    c. Upper 20-kΩ potentiometer shorted

    d. Supply voltages are 20 percent low

    e. 10 kΩ open

20-17. The Colpitts oscillator of Fig. 20-37 will not start. Name at least three possible troubles.

20-18. You have designed and built an amplifier. It does amplify an input signal, but the output looks fuzzy on an oscilloscope. When you touch the circuit, the fuzz disappears, leaving a perfect signal. What do you think the trouble is, and how would you try to cure it?

20-19. The VCO of Fig. 20-34 is producing an output with a frequency of 20 kHz. When you move the wiper of Fig. 20-34a, the output frequency stays at 20 kHz. Which of these is the trouble:

a. Supply voltage low
b. Pin 1 open
c. Pin 4 grounded
d. Defective IC
e. Potentiometer is shorting V_{CC} to ground

Design

20-20. Design a Wien-bridge oscillator similar to Fig. 20-36a that meets these specifications: three decade frequency ranges covering 20 Hz to 20 kHz; output voltage of 5 V rms.

20-21. Select a value of L in Fig. 20-37 to get an oscillation frequency of 2.5 MHz.

20-22. Design a 555 timer circuit that free-runs at a frequency of 1 kHz and a duty cycle of 75 percent.

20-23. In Fig. 20-35a, select values for the resistors and the capacitor that will produce a ramp of 10 V in 5 μs. Assume $V_{CC} = 15$ V.

Challenging

20-24. Figure 20-38 shows an op-amp phase-shift oscillator. If $f_{2(CL)} = 1$ kHz, what is the phase shift through the two lag networks? What is the frequency of oscillation?

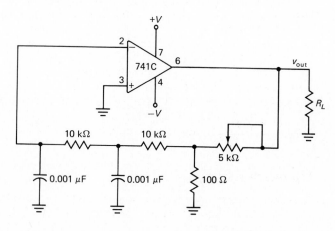

Fig. 20-38

20-25. The tuning capacitor of Fig. 20-39 is used to pull the crystal frequency. What do you think this means? Describe how the tuning capacitor pulls the frequency.

Fig. 20-39

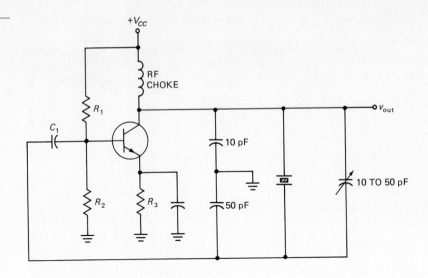

20-26. In Fig. 20-40, calculate the output frequency and pulse width for each of these values of R: 33 kΩ, 47 kΩ, and 68 kΩ.

Fig. 20-40

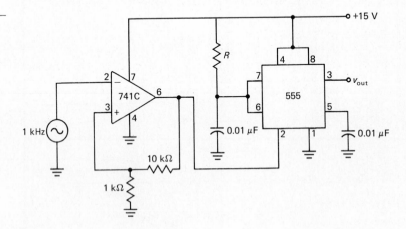

20-27. In Fig. 20-41, the level of the 1-kHz signal is adjusted to get a duty cycle of 90 percent out of the Schmitt trigger. Calculate the output frequency and slope for each of these values of R: 10 kΩ, 22 kΩ, 33 kΩ.

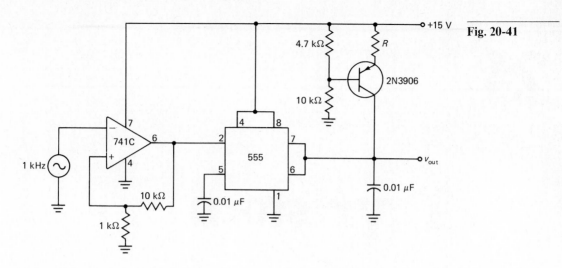

Fig. 20-41

Computer

20-28. The arithmetic function ATN(x) returns the arctangent (in radians) of x. To get the arctangent in degrees, multiply ATN(x) by 57.29578. Here is a program:

```
10 PRINT "ENTER R": INPUT R
20 PRINT "ENTER XC": INPUT XC
30 Y = 57.29578 * ATN(XC/R)
40 PRINT Y;" DEGREES"
```

What will the computer print out if $X_C/R = 1$?

20-29. Write a program that prints out 200 values of ϕ in Eq. (20-2). The first 100 values should correspond to X_C/R varying from 0.01 to 1 in steps of 0.01; the next 100 values should be for $R/X_C = 0.01$ to 1 in steps of 0.01.

20-30. Write a program that calculates the oscillation frequency for a Wien-bridge oscillator, a twin-T oscillator, a Colpitts oscillator, and a Clapp oscillator. Use a menu.

20-31. Write a program that calculates the frequency and duty cycle of an astable 555 timer.

21

Thyristors

A *thyristor* is a four-layer semiconductor device that uses internal feedback to produce latching action. Unlike bipolar transistors and FETs, which can operate either as linear amplifiers or as switches, thyristors can only operate as switches. Their main application is in controlling large amounts of load current for motors, heaters, lighting systems, and other such devices. Incidentally, the word "thyristor" is from Greek and means "door."

21-1 THE IDEAL LATCH

All thyristors can be explained in terms of the ideal *latch* shown in Fig. 21-1*a*. Notice that the upper transistor Q_1 is a *pnp* device and the lower transistor Q_2 is an *npn* device. As you can see, the collector of Q_1 drives the base of Q_2, and the collector of Q_2 drives the base of Q_1.

REGENERATION

Because of the unusual connection of Fig. 21-1*a*, we have positive feedback, also called *regeneration*. A change in current at any point in the loop is amplified and returned to the starting point with the same phase. For instance, if the Q_2 base current increases, the Q_2 collector current increases. This forces more base current through Q_1. In turn, this produces a larger Q_1 collector current, which drives the Q_2 base harder. This build-up in currents will continue until both transistors are driven into saturation. In this case, the latch acts like a closed switch (Fig. 21-1*b*).

On the other hand, if something causes the Q_2 base current to decrease, the Q_2 collector current will decrease. This reduces the Q_1 base current. In turn, there is less Q_1 collector current, which reduces the Q_2 base current even more. This regeneration continues until both transistors are driven into cutoff. At this time, the latch acts like an open switch (Fig. 21-1*c*).

The latch can be in either of two states, closed or open. It will remain in either state indefinitely. If closed, it stays closed until something causes the currents to decrease. If open, it stays open until something else forces the currents to increase.

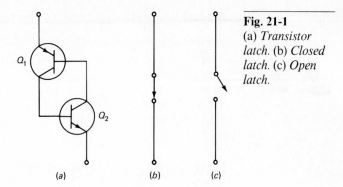

TRIGGERING

One way to close a latch is by *triggering*, applying a forward bias voltage to either base. For example, Fig. 21-2a shows a trigger (sharp pulse) hitting the Q_2 base. Suppose the latch is open before point A in time. Then the supply voltage appears across the open latch (Fig. 21-2b), and the operating point is at the lower end of the load line (Fig. 21-2d).

At point A in time, the trigger momentarily forward-biases the Q_2 base. The Q_2 collector current suddenly comes on and forces base current through Q_1. In turn, the Q_1 collector current comes on and drives the Q_2 base harder. Since the Q_1 collector now supplies the Q_2 base current, the trigger pulse is no longer needed. Once the regeneration starts, it will sustain itself and drive both transistors into saturation. The minimum input current needed to start the regenerative switching action is called the *trigger current*.

When saturated, both transistors ideally look like short circuits, and the latch is closed (Fig. 21-2c). Ideally, the latch has zero voltage across it when it is closed, and the operating point is at the upper end of the load line (Fig. 21-2d).

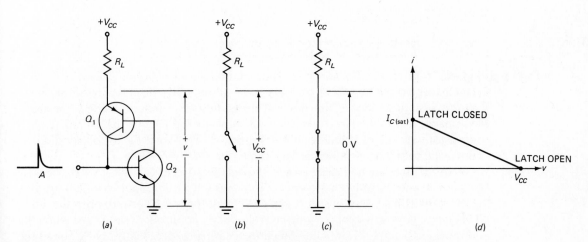

Fig. 21-2 (a) *Forward-bias trigger closed latch.* (b) *Supply voltage appears across open latch.* (c) *Zero voltage across closed latch.* (d) *Load line.*

BREAKOVER

Another way to close a latch is by *breakover*. This means using a large enough supply voltage V_{CC} to break down either collector diode. Once the breakdown begins, current comes out of one of the collectors and drives the other base. The effect is the same as if the base had received a trigger. Although breakover starts with a breakdown of one of the collector diodes, it ends with both transistors in the saturated state. This is why the term breakover is used instead of breakdown to describe this kind of latch closing.

LOW-CURRENT DROPOUT

How do we open an ideal latch? One way is to reduce the load current to zero. This forces the transistors to come out of saturation and return to the open state. For instance, in Fig. 21-2a we can open the load resistor. Alternatively, we can reduce the V_{CC} supply to zero. In either case, a closed latch will be forced to open. We call this type of opening *low-current dropout* because it depends on reducing the latch current to a low value.

REVERSE-BIAS TRIGGERING

Another way to open the latch is to apply a *reverse-bias trigger* in Fig. 21-2a. When a negative trigger is used instead of a positive one, the Q_2 base current decreases. This forces the Q_1 base current to decrease. Since the Q_1 collector current also decreases, the regeneration will rapidly drive both transistors into cutoff, which opens the latch.

CONCLUSION

Here is a summary that will help you to understand how different thyristors work:

1. We can close an ideal latch by forward-bias triggering or by breakover.
2. We can open an ideal latch by reverse-bias triggering or by low-current dropout.

21-2 THE FOUR-LAYER DIODE

Figure 21-3a is called a *four-layer* diode (also known as a Shockley diode). It is classified as a diode because it has only two external leads. Because of its four doped regions, it's often called a *pnpn* diode. The easiest way to understand how it works is to visualize it separated into two halves, as shown in Fig. 21-3b. The left half is a *pnp* transistor, and the right half is an *npn* transistor. Therefore, the four-layer diode is equivalent to the latch shown in Fig. 21-3c.

Because there are no trigger inputs, the only way to close a four-layer diode is by breakover, and the only way to open it is by low-current dropout. With a four-layer diode it is not necessary to reduce the current all the way to zero to open the latch. The internal transistors of the four-layer diode will come out of saturation when the current is reduced to a low value called the *holding current*. Figure 21-3d shows the schematic symbol for a four-layer diode.

After a four-layer diode breaks over, the voltage across it drops to a low value,

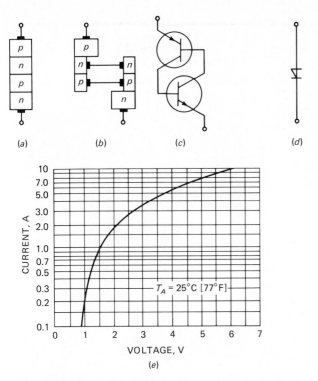

(a)   (b)   (c)   (d)

(e)

Fig. 21-3 (a) *Four-layer diode.* (b) *Equivalent structure.* (c) *Equivalent circuit.* (d) *Schematic symbol.* (e) *Example of current versus voltage in conducting four-layer diode.*

depending on how much current there is. For instance, Fig. 21-3*e* shows the current versus voltage for a 1N5158. Notice that the voltage increases with the current through the device: 1 V at 0.2 A, 1.5 V at 0.95 A, 2 V at 1.8 A, and so forth.

EXAMPLE 21-1

The 1N5158 of Fig. 21-4*a* has a breakover voltage of 10 V. If the diode is nonconducting and the input voltage is increased to +5 V, what is the diode current? If the input voltage is increased to +15 V, what is the diode current? (Use an approximate diode drop of 1 V.)

SOLUTION

Since the diode is initially off, the only way to make it conduct is to exceed its breakover voltage. Raising the input to +5 V has no effect on the diode, which remains open. When the input voltage is increased to +15 V, the four-layer diode breaks over and closes. Since it has a drop of approximately 1 V, the diode current is

$$I \cong \frac{15\text{ V} - 1\text{ V}}{100\ \Omega} = 140\text{ mA}$$

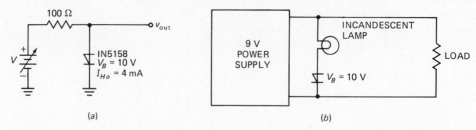

Fig. 21-4 (a) *Circuit with 1N5158.* (b) *Four-layer diode serves as overvoltage detector.*

EXAMPLE 21-2

In Fig. 21-4*a*, the diode has a holding current of 4 mA. If the voltage drop is 0.5 V at the dropout point, what is the input voltage that just produces low-current dropout?

SOLUTION

To open the four-layer diode, we have to reduce the current below the holding current of 4 mA. This means reducing the input voltage to slightly less than

$$V = 0.5 \text{ V} + (4 \text{ mA})(100 \ \Omega) = 0.9 \text{ V}$$

EXAMPLE 21-3

Describe the action in Fig. 21-4*b*.

SOLUTION

The four-layer diode has a breakover voltage of 10 V. As long as the power supply puts out 9 V, the four-layer diode is open and the lamp is dark. But if something goes wrong with the power supply and its voltage rises above 10 V, the four-layer diode latches and the lamp comes on. Even if the supply should return to 9 V, the diode remains latched as a record of the overvoltage that occurred. The only way to make the lamp go out is to turn off the supply.

The circuit is an example of an *overvoltage detector*. As long as the supply stays within normal limits, nothing happens. But if there is an overvoltage, even temporarily, the lamp comes on and stays on.

EXAMPLE 21-4

Figure 21-5*a* shows a *sawtooth* generator. Describe the circuit action.

SOLUTION

If the four-layer diode were not in the circuit, the capacitor would charge exponentially, and its voltage would follow the dashed curve of Fig. 21-5*b*. But

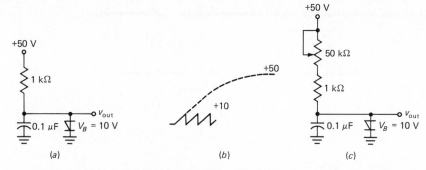

Fig. 21-5 (a) *Sawtooth generator.* (b) *Output waveform.* (c) *Adjustable sawtooth generator.*

the four-layer diode is in the circuit; therefore, as soon as the capacitor voltage reaches 10 V, the diode breaks over and the latch closes. This discharges the capacitor, producing the *flyback* (sudden decrease) of capacitor voltage. At some point on the flyback, the current drops below the holding current and the four-layer diode opens. The next cycle then begins.

Figure 21-5*a* is another example of a relaxation oscillator, a circuit whose output depends on the charging and discharging of a capacitor (or inductor). If we increase the *RC* time constant, the capacitor takes longer to charge to 10 V, and the frequency of the sawtooth wave is lower. For instance, with the potentiometer of Fig. 21-5*c* we can get a 50:1 range in frequency.

21-3 THE SILICON CONTROLLED RECTIFIER

The *silicon controlled rectifier* (SCR) is more useful than a four-layer diode because it has an extra lead connected to the base of the *npn* section, as shown in Fig. 21-6*a*. Again, you can visualize the four doped regions separated into two transistors, as shown in Fig. 21-6*b*. Therefore, the SCR is equivalent to a latch with a trigger input (Fig. 21-6*c*). Schematic diagrams use the symbol of Fig. 21-6*d*. Whenever you see this, remember that it is equivalent to a latch with a trigger input.

GATE TRIGGERING

The gate of an SCR is approximately equivalent to a diode (see Fig. 21-6*c*). For this reason, it takes at least 0.7 V to trigger an SCR. Furthermore, to get the regeneration started, a minimum input current is required. Data sheets list the *trigger voltage* and *trigger current* for SCRs. For example, the data sheet of a 2N4441 gives a typical trigger voltage of 0.7 V and a trigger current of 10 mA. The source driving the gate of a 2N4441 has to be able to supply at least 10 mA at 0.7 V; otherwise, the SCR will not latch shut.

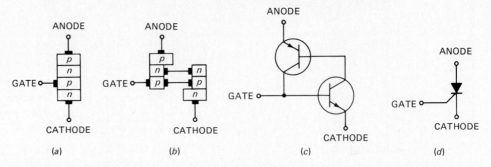

Fig. 21-6 *SCR.* (a) *Structure.* (b) *Equivalent structure.* (c) *Equivalent circuit.* (d) *Schematic symbol.*

BLOCKING VOLTAGE

SCRs are not intended for breakover operation. Breakover voltages range from around 50 V to more than 2500 V, depending on the SCR type number. Most SCRs are designed for trigger closing and low-current opening. In other words, an SCR stays open until a trigger drives its gate (Fig. 21-6*d*). Then the SCR latches and remains closed, even though the trigger disappears. The only way to open an SCR is with low-current dropout.

Most people think of an SCR as a device that blocks voltage until the trigger closes it. For this reason, the breakover voltage is often called the *forward blocking voltage* on data sheets. For instance, the 2N4441 has a forward blocking voltage of 50 V. As long as the supply voltage is less than 50 V, the SCR cannot break over. The only way to close it is with a gate trigger.

HIGH CURRENTS

Almost all SCRs are industrial devices that can handle large currents ranging from less than 1 A to more than 2500 A, depending on the type number. Because they are high-current devices, SCRs have relatively large trigger and holding currents. The 2N4441 can conduct up to 8 A continuously; its trigger current is 10 mA, and so is its holding current. This means that you have to supply the gate with at least 10 mA to control up to 8 A of anode current. (The anode and cathode are shown in Fig. 21-6*d*.) As another example, the C701 is an SCR that can conduct up to 1250 A with a trigger current of 500 mA and a holding current of 500 mA.

CRITICAL RATE OF RISE

In many applications, an ac supply voltage is used with the SCR. By triggering the gate at a certain point in the cycle, we can control large amounts of ac power to a load such as a motor, a heater, or some other load. Because of junction capacitances inside the SCR, it is possible for a rapidly changing supply voltage to trigger the SCR. Put another way, if the rate of rise of forward voltage is high enough, capacitive charging current can initiate the regeneration.

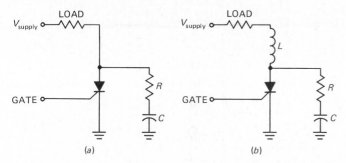

Fig. 21-7 (a) *RC snubber protects SCR against voltage transients on supply voltage.*
(b) *Inductor protects SCR against rapid current changes.*

To avoid false triggering of an SCR, the anode rate of voltage change must not exceed the *critical rate of voltage rise* listed on the data sheet. For instance, a 2N4441 has a critical rate of voltage rise of 50 V/μs. To avoid a false triggering, the anode voltage must not rise faster than 50 V/μs. As another example, the C701 has a critical rate of voltage rise of 200 V/μs. To avoid a false closure, the anode voltage must not increase faster than 200 V/μs.

Switching transients are the main cause of exceeding the critical rate of voltage rise. One way to reduce the effects of switching transients is with an *RC snubber,* shown in Fig. 21-7a. If a high-speed switching transient does appear on the supply voltage, its rate of rise is reduced at the anode because of the *RC* circuit. The rate of anode voltage rise depends on the load resistance, as well as on the R and C values.

Larger SCRs also have a *critical rate of current rise*. For instance, the C701 has a critical rate of current rise of 150 A/μs. If the anode current tries to rise faster than this, the SCR may be destroyed. Including an inductor in series, as shown in Fig. 21-7b, reduces the rate of current rise, as well as helping the *RC* snubber decrease the rate of voltage rise.

SCR CROWBAR

One of the applications of an SCR is to protect a load against an overvoltage from a power supply. For instance, Fig. 21-8a shows a positive supply of 20 V connected to a zener diode in series with a resistor. The voltage across the resistor drives the gate of an SCR, which is in parallel with the load. The zener voltage is 21 V. As long as the supply voltage is 20 V, the zener diode is open and the SCR gate voltage is zero. Since the SCR is open, 20 V appears across the load resistor.

If something goes wrong with the power supply and the voltage tries to rise above 21 V, the zener diode breaks down, and a voltage appears at the gate of the SCR. When this voltage exceeds approximately 0.7 V, the SCR latches and shorts out the load voltage. The action is similar to throwing a *crowbar* across the load terminals. Because the SCR turnon is very fast (1 μs for a 2N4441), the load is quickly protected against the damaging effects of a large overvoltage.

Crowbarring, though a drastic form of protection, is necessary with many digital ICs; they can't take much overvoltage. Rather than destroy expensive ICs, therefore, we can use an SCR crowbar to short the load terminals at the first sign of overvoltage.

Fig. 21-8
SCR crowbar.
(a) Soft turnon
circuit. (b)
Practical circuit.

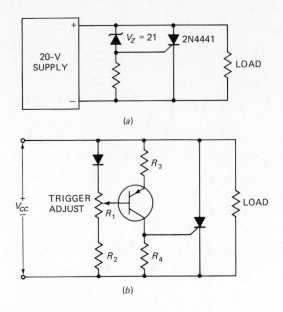

(a)

(b)

Power supplies with an SCR crowbar need current limiting to prevent excessive current when the SCR closes.

The crowbar of Fig. 21-8*a* has a soft turnon because of the zener diode. Figure 21-8*b* shows a more practical example of an SCR crowbar. The transistor provides voltage gain, which produces a much sharper turnon. When the voltage across R_4 exceeds approximately 0.7 V, the SCR turns on. An ordinary diode is included for temperature compensation of the transistor's base-emitter diode. The trigger adjust allows us to set the trip point of the circuit, typically around 10 to 15 percent above the normal voltage.

EXAMPLE 21-5

Figure 21-9 shows another SCR crowbar. How does the circuit work?

Fig. 21-9
SCR crowbar
using comparator
for fast turnon.

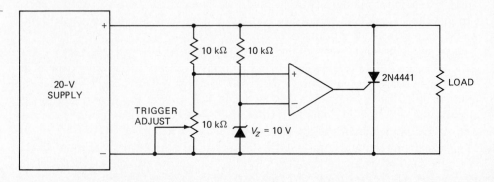

SOLUTION

The SCR is normally open because the supply voltage is only 20 V. Because of the zener diode, 10 V goes to the inverting input of the comparator. The trigger adjust produces slightly less than 10 V for the noninverting input. As a result, the error voltage is negative, and the comparator has a negative output. This output does nothing to the SCR.

If the supply voltage tries to rise above 20 V, the noninverting input becomes greater than 10 V. Since the error voltage is positive, the output of the comparator drives the SCR into conduction. This rapidly shuts down the supply by crowbarring the load terminals.

21-4 VARIATIONS OF THE SCR

There are other *pnpn* devices whose action is similar to that of the SCR. What follows is a brief description of these SCR variations. The devices to be discussed are for low-power applications.

PHOTO-SCR

Figure 21-10*a* shows a *photo-SCR,* also known as a light-activated SCR (LASCR). The arrows represent incoming light that passes through a window and hits the depletion layers. When the light is strong enough, valence electrons are dislodged from their orbits and become free electrons. When these free electrons flow out of the collector of one transistor into the base of the other, the regeneration starts and the photo-SCR closes.

After a light trigger has closed the photo-SCR, it remains closed, even though the light disappears. For maximum sensitivity to light, the gate is left open, as shown in Fig. 21-10*a.* If you want an adjustable trip point, you can include the trigger adjust shown in Fig. 21-10*b.* The gate resistor diverts some of the light-produced electrons and changes the sensitivity of the circuit to the incoming light.

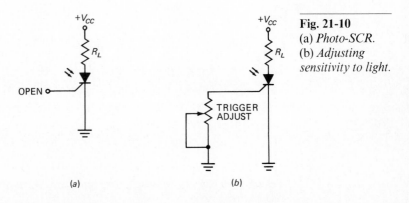

Fig. 21-10
(a) *Photo-SCR.*
(b) *Adjusting sensitivity to light.*

Fig. 21-11
GCS circuit.

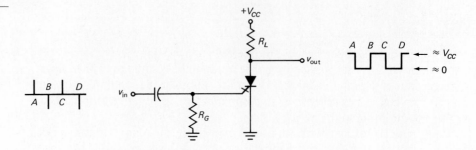

GCS

As mentioned earlier, low-current dropout is the normal way to open an SCR. But the *gate-controlled switch* (GCS) is designed for easy opening with a reverse-biased trigger. A GCS is closed by a positive trigger and opened by a negative trigger (or by low-current dropout). Figure 21-11 shows a GCS circuit. Each positive trigger closes the GCS, and each negative trigger opens it. Because of this, we get the square-wave output shown. The GCS is useful in counters, digital circuits, and other applications in which a negative trigger is available for turnoff.

SCS

Figure 21-12*a* shows the doped regions of a *silicon controlled switch* (SCS). Now an external lead is connected to each doped region. Visualize the device separated into two halves (Fig. 21-12*b*). Therefore, it's equivalent to a latch with access to both bases (Fig. 21-12*c*). A forward-bias trigger on either base will close the SCS. Likewise, a reverse-bias trigger on either base will open the device.

Figure 21-12*d* shows the schematic symbol for an SCS. The lower gate is called the cathode gate; the upper gate is the anode gate. The SCS is a low-power device compared with the SCR. It handles currents in milliamperes rather than amperes.

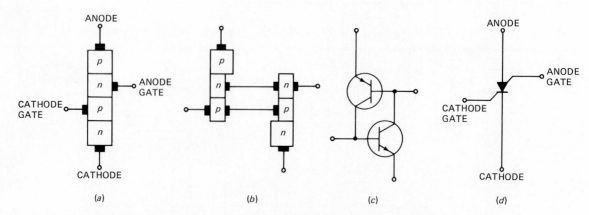

Fig. 21-12   *SCS. (a) Structure. (b) Equivalent structure. (c) Equivalent circuit. (d) Schematic symbol.*

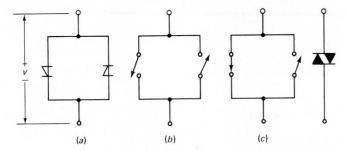

Fig. 21-13 *Diac. (a) Equivalent to parallel back-to-back four-layer diodes. (b) Equivalent circuit. (c) Left latch closed. (d) Schematic symbol.*

21-5 BIDIRECTIONAL THYRISTORS

Up until now, all devices have been unidirectional; current was only in one direction. This section discusses *bidirectional* thyristors, devices in which the current can flow in either direction.

DIAC

The *diac* can have latch current in either direction. The equivalent circuit of a diac is a pair of four-layer diodes in parallel, as shown in Fig. 21-13a, ideally the same as the latches in Fig. 21-13b. The diac is nonconducting until the voltage across it tries to exceed the breakover voltage in either direction.

For instance, if v has the polarity indicated in Fig. 21-13a, the left diode conducts when v tries to exceed the breakover voltage. In this case, the left latch closes, as shown in Fig. 21-13c. On the other hand, if the polarity of v is opposite to that of Fig. 21-13a, the right latch closes when v tries to exceed the breakover voltage.

Once the diac is conducting, the only way to open it is by low-current dropout. This means reducing the current below the rated holding current of the device. Figure 21-13d shows the schematic symbol for a diac.

TRIAC

The *triac* acts like two SCRs in parallel (Fig. 21-14a), equivalent to the two latches of Fig. 21-14b. Because of this, the triac can control current in either direction. The

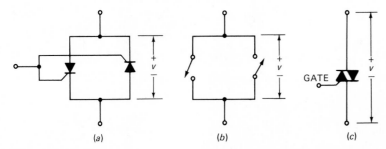

Fig. 21-14 *Triac. (a) Equivalent to parallel back-to-back SCRs. (b) Equivalent circuit. (c) Schematic symbol.*

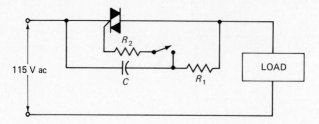

Fig. 21-15 *Low-power switch controls triac, which acts like high-power switch.*

breakover voltage is usually high, so that the normal way to turn on a triac is by applying a forward-bias trigger. Data sheets list the trigger voltage and trigger current needed to turn on a triac. If v has the polarity shown in Fig. 21-14a, we have to apply a positive trigger; this closes the left latch. When v has opposite polarity, a negative trigger is needed; it will close the right latch. Figure 21-14c is the schematic symbol for a triac.

Figure 21-15 is an example of a triac circuit. When the low-power switch is open, the triac is nonconducting and no ac power reaches the load. But when the switch is closed, current through R_2 turns on the triac during each half cycle. Resistor R_1 and capacitor C also act like an RC snubber to prevent switching transients from damaging the triac.

21-6 THE UNIJUNCTION TRANSISTOR

The *unijunction* transistor (UJT) has two doped regions with three external leads (Fig. 21-16a). It has one emitter and two bases. The emitter is heavily doped, having many holes. The n region, however, is lightly doped. For this reason, the resistance between the bases is relatively high, typically 5 to 10 kΩ when the emitter is open. We call this the *interbase resistance,* symbolized by R_{BB}.

INTRINSIC STANDOFF RATIO

Figure 21-16b shows the equivalent circuit of a UJT. The emitter diode drives the junction of two internal resistances, R_1 and R_2. When the emitter diode is nonconducting, R_{BB} is the sum of R_1 and R_2. When a supply voltage is between the two bases, as shown in Fig. 21-16c, the voltage across R_1 is given by

$$V_1 = \frac{R_1}{R_1 + R_2} V = \frac{R_1}{R_{BB}} V$$

or

$$V_1 = \eta V \qquad\qquad (21\text{-}1)$$

where

$$\eta = \frac{R_1}{R_{BB}}$$

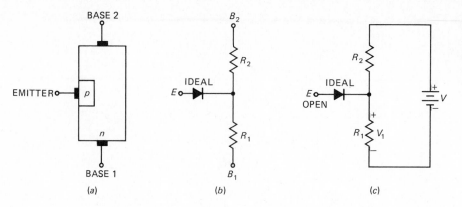

Fig. 21-16
UJT. (a) *Struc-
ture.* (b) *Equiva-
lent circuit.* (c)
Standoff voltage.

(The Greek letter η is pronounced eta, e as the a in face and a as the a in about.)

The quantity η is called the *intrinsic standoff ratio,* which is nothing more than the voltage-divider factor. The typical range of η is from 0.5 to 0.8. For instance, a 2N2646 has an η of 0.65. If this UJT is used in Fig. 21-16c with a supply voltage of 10 V,

$$V_1 = \eta V = 0.65(10 \text{ V}) = 6.5 \text{ V}$$

In Fig. 21-16c, V_1 is called the intrinsic standoff voltage because it keeps the emitter diode reverse-biased for all emitter voltages less than V_1. If V_1 equals 6.5 V, then ideally we have to apply slightly more than 6.5 V to the emitter to turn on the emitter diode.

HOW A UJT WORKS

In Fig. 21-17a, imagine that the emitter supply voltage is turned down to zero. Then the intrinsic standoff voltage reverse-biases the emitter diode. When we increase the emitter supply voltage, v_E increases until it is slightly greater than V_1. This turns on the emitter diode. Since the p region is heavily doped compared with the n region, holes are injected into the lower half of the UJT. The light doping of the n region gives these holes a long lifetime. These holes create a conducting path between the emitter and the lower base.

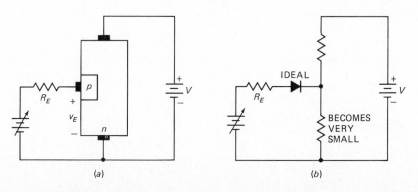

Fig. 21-17
(a) *UJT circuit.*
(b) R_1 *approaches
zero after emitter
diode turns on.*

Fig. 21-18
*UJT. (a) Struc-
ture. (b) Latch
equivalent circuit.
(c) Schematic
symbol.*

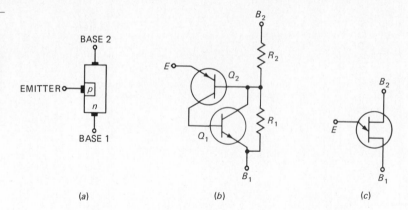

(a)   (b)   (c)

The flooding of the lower half of the UJT with holes drastically lowers resistance R_1 (Fig. 21-17b). Because R_1 is suddenly much lower in value, v_E suddenly drops to a low value, and the emitter current increases.

LATCH EQUIVALENT CIRCUIT

One way to remember how the UJT of Fig. 21-18a works is by relating it to the latch of Fig. 21-18b. With a positive voltage from B_2 to B_1, a standoff voltage V_1 appears across R_1. This keeps the emitter diode of Q_2 reverse-biased as long as the emitter input voltage is less than the standoff voltage. When the emitter input voltage is slightly greater than the standoff voltage, however, Q_2 turns on and regeneration takes over. This drives both transistors into saturation, ideally shorting the emitter and the lower base.

Figure 21-18c is the schematic symbol for a UJT. The emitter arrow reminds us of the upper emitter in a latch. When the emitter voltage exceeds the standoff voltage, the latch between the emitter and the lower base closes. Ideally, you can visualize a short between E and B_1. To a second approximation, a low voltage called the *emitter saturation* voltage, $V_{E(sat)}$, appears between E and B_1.

The latch stays closed as long as latch current (emitter current) is greater than holding current. Data sheets specify a *valley* current I_V, which is equivalent to holding current. For instance, a 2N2646 has an I_V of 6 mA; to hold the latch closed, the emitter current must be greater than 6 mA.

Fig. 21-19

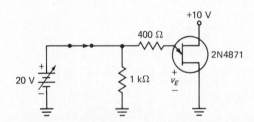

EXAMPLE 21-6

The 2N4871 of Fig. 21-19 has an η of 0.85. What is the ideal emitter current?

SOLUTION

The standoff voltage is

$$V_1 = 0.85(10 \text{ V}) = 8.5 \text{ V}$$

Ideally, v_E must be slightly greater than 8.5 V to turn on the emitter diode and close the latch. With the input switch closed, 20 V drives the 400-Ω resistor. This is more than enough voltage to overcome the standoff voltage. Therefore, the latch is closed and the emitter current equals

$$I_E \cong \frac{20 \text{ V}}{400 \text{ }\Omega} = 50 \text{ mA}$$

EXAMPLE 21-7

The valley current of a 2N4871 equals 7 mA, and the emitter voltage is 1 V at this point. At what value of emitter-supply voltage does the UJT in Fig. 21-19 open?

SOLUTION

As we reduce the emitter-supply voltage, the emitter current decreases. At the point at which it equals 7 mA, v_E is 1 V and the latch is about to open. The emitter-supply voltage at this time is

$$V = 1 \text{ V} + (7 \text{ mA})(400 \text{ }\Omega) = 3.8 \text{ V}$$

When V is less than 3.8 V, the UJT opens. Then it is necessary to raise V above 8.5 V to close the UJT.

21-7 THYRISTOR APPLICATIONS

Thyristors have become increasingly popular for controlling ac power to resistive and inductive loads, such as motors, solenoids, and heating elements. Compared with competing devices like relays, thyristors offer lower cost and better reliability. This section discusses some applications of thyristors to give you an idea of how they can be used.

UJT RELAXATION OSCILLATOR

Figure 21-20*a* shows a UJT *relaxation oscillator*. The action is similar to that of the relaxation oscillator of Example 21-4. The capacitor charges toward V_{CC}, but as soon as its voltage exceeds the standoff voltage, the UJT closes. This discharges the

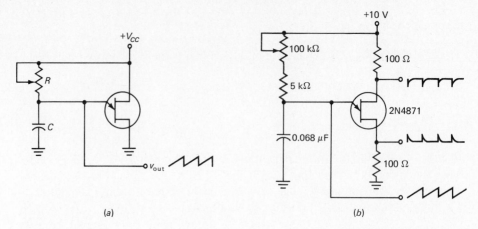

Fig. 21-20 *UJT circuits.* (a) *Sawtooth generator.* (b) *Sawtooth and trigger outputs.*

capacitor until low-current dropout occurs. As soon as the UJT opens, the next cycle begins. As a result, we get a sawtooth output.

If we add a small resistor to each base circuit, we can get three useful outputs: sawtooth waves, positive triggers, and negative triggers, as shown in Fig. 21-20*b*. The triggers appear during the flyback of the sawtooth because the UJT conducts heavily at this time. With the values of Fig. 21-20*b*, the frequency can be adjusted between 50 Hz and 1 kHz (approximately).

AUTOMOBILE IGNITION

Sharp trigger pulses out of a UJT relaxation oscillator can be used to trigger an SCR. For instance, Fig. 21-21 shows part of an automobile ignition system. With the distributor points open, the capacitor charges exponentially toward +12 V. As soon as the capacitor voltage exceeds the intrinsic standoff voltage, the UJT conducts heavily through the primary winding. The secondary voltage then triggers the SCR.

Fig. 21-21
UJT triggers SCR to produce spark for automobile ignition.

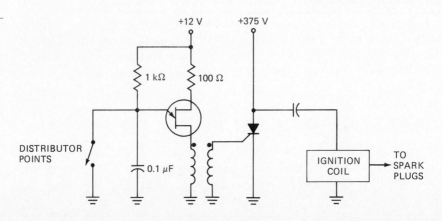

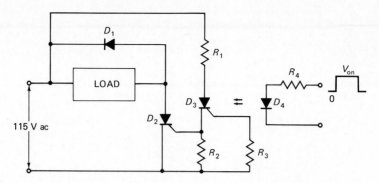

Fig. 21-22
*Optocoupler
control of SCR.*

When the SCR latches shut, the positive end of the output capacitor is suddenly grounded. As the output capacitor discharges through the ignition coil, a high-voltage pulse drives one of the spark plugs. When the points close, the circuit resets itself in preparation for the next cycle.

OPTOCOUPLER CONTROL

Figure 21-22 is an example of *optocoupler control*. When an input pulse turns on the LED (D_4), its light activates the photo-SCR (D_3). In turn, this produces a trigger voltage for the main SCR (D_2). In this way, we get isolated control of the positive half cycles of line voltage. An ordinary diode D_1 is needed to protect the SCR from inductive kickback and transients that may occur during the reverse half cycle.

DIAC-TRIGGERED SCR

In Fig. 21-23, the full-wave output from a bridge rectifier drives an SCR that is controlled by a diac and an *RC* charging circuit. By adjusting R_1, we can change the time constant and control the point at which the diac fires. Circuits like these can easily control several hundred watts of power to a lamp, heater, or other load. A four-layer diode could be used instead of a diac.

UJT-TRIGGERED SCR

Figure 21-24 shows another way to control an SCR, this time with a UJT relaxation oscillator. The load may be a motor, a lamp, a heater, or some other device. By

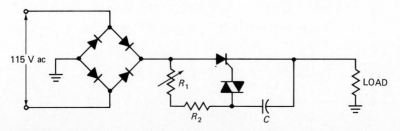

Fig. 21-23
*Controlling the
conduction angle
of an SCR.*

Fig. 21-24
*UJT relaxation
oscillator controls
conduction angle
of SCR.*

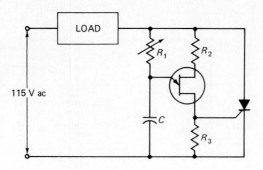

varying R_1, we can change the RC time constant and alter the point at which the UJT fires. This allows us to control the conduction angle of the SCR, which means that we are controlling the load current. A circuit like this represents half-wave control because the SCR is off during the negative half cycle.

FULL-WAVE CONTROL

The diac of Fig. 21-25 can trigger the triac on either half cycle of line voltage. Variable resistance R_1 controls the RC time constant of the diac control circuit. Since this changes the point in the cycle at which the diac fires, we have control over the triac condition angle. In this way, we can change the large load current.

MICROPROCESSOR-CONTROLLED SCR

In robotic systems a microprocessor controls motors and other loads. Figure 21-26 is a simple example of how it is done. A rectangular pulse from a microprocessor drives an emitter follower, whose output controls the gate of an SCR. While the rectangular control pulse is high, the SCR latches during the positive half cycles and shuts off during the negative half cycles. The duration of the rectangular pulse from the microprocessor determines the number of positive half cycles during which the load receives power.

Fig. 21-25
*Controlling the
conduction angle
of a triac.*

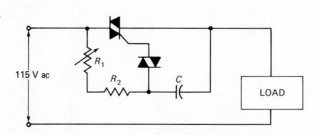

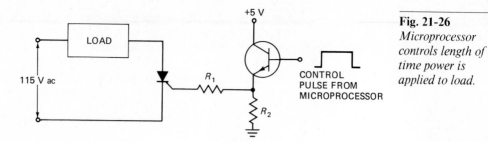

Fig. 21-26
*Microprocessor
controls length of
time power is
applied to load.*

PROBLEMS

Straightforward

21-1. The 1N5160 of Fig. 21-27a is conducting. If we allow 0.7 V across the diode at the dropout point, what is the value of V at which low-current dropout occurs?

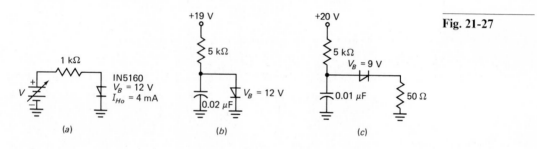

Fig. 21-27

21-2. With a supply of 19 V, it takes the capacitor of Fig. 21-27b exactly one time constant to charge to 12 V, the breakover voltage of the diode. If we neglect the voltage across the diode when conducting, what is the frequency of the sawtooth output?

21-3. The current through the 50-Ω resistor of Fig. 21-27c is maximum just after the diode latches. If we allow 1 V across the latched diode, what is the maximum current?

21-4. The 2N4216 of Fig. 21-28a has a trigger current of 0.1 mA. If we allow 0.8 V for the gate voltage, what is the value of V that turns on the SCR?

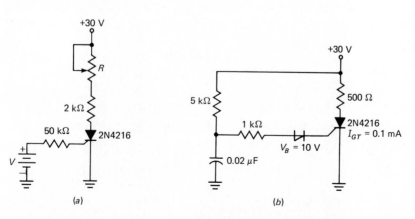

Fig. 21-28

21-5. The four-layer diode of Fig. 21-28b has a breakover voltage of 10 V. The SCR has a trigger current of 0.1 mA and a trigger voltage of 0.8 V. If the four-layer diode has a forward drop of approximately 0.7 V, what is the current through the diode just after it breaks over? The current through the 500-Ω resistor after the SCR turns on?

21-6. Figure 21-29a shows an alternative symbol for a diac. The MPT32 diac breaks over when the capacitor voltage reaches 32 V. It takes exactly one time constant for the capacitor to reach this voltage. How long after the switch is closed does the triac turn on? What is the ideal value of gate current when the diac breaks over? The load current after the triac has closed?

Fig. 21-29

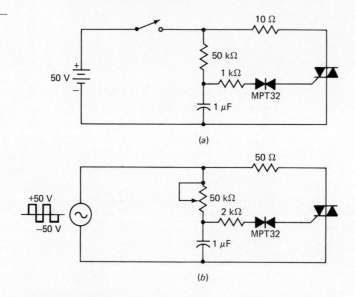

(a)

(b)

21-7. The frequency of the square wave in Fig. 21-29b is 10 kHz. It takes exactly one time constant for the capacitor to reach the breakover voltage of the diac. If the MPT32 breaks over at 32 V, what is the ideal value of gate current at the instant the diac breaks over? The ideal load current?

21-8. The UJT of Fig. 21-30a has an η of 0.63. Allowing 0.7 V across the emitter diode, what value of V just turns on the UJT?

Fig. 21-30

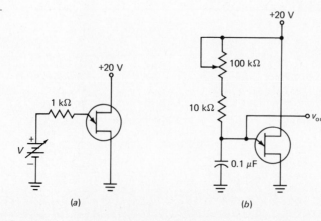

(a)

(b)

21-9. The valley current of the UJT in Fig. 21-30*a* is 2 mA. If the UJT is latched, we have to reduce V to get low-current dropout. Allowing 0.7 V across the emitter diode, what is the value of V that just opens the UJT?

21-10. The intrinsic standoff ratio of the UJT in Fig. 21-30*b* is 0.63. Ignoring the drop across the emitter diode, what are the minimum and maximum output frequencies?

Troubleshooting

21-11. Will the output frequency in Fig. 21-27*b* increase, decrease, or stay the same for each of these troubles:

a. Supply voltage at +15 V

b. Resistor is 20 percent high

c. Capacitor is 0.01 μF

d. Breakover voltage only 10 V

21-12. The SCR crowbar of Fig. 21-8*b* keeps a constant short across the load terminals even when the voltage is normal. Which of these is the trouble:

a. Compensating diode is open

b. R_3 is shorted

c. R_4 is shorted

d. SCR is open

21-13. The SCR of Fig. 21-26 will not come on. Name at least three possible causes.

Design

21-14. Select a value of C in Fig. 21-27*b* that produces an output frequency of approximately 20 kHz.

21-15. The UJT of Fig. 21-20*a* has an η of 0.63. The potentiometer has a maximum value of 10 kΩ. Select a value of C that produces an output frequency of approximately 50 kHz.

Challenging

21-16. In Fig. 21-8*b*, R_4 is 180 Ω. If the SCR trigger voltage is 1 V and the trigger current is 10 mA, what is the minimum collector current that triggers the SCR?

21-17. The UJT of Fig. 21-20*b* has an η of 0.75. What are the minimum and maximum output frequencies?

Computer

21-18. The arithmetic function EXP(x) produces the natural exponential of x. For instance,

```
10 Y = EXP(-2)
20 PRINT Y
30 STOP
```

When run, this program prints out e^{-2}. Rewrite this program so that it prints out the values of e^{-x} for all integer values of x between 1 and 10.

CHAPTER 22

Frequency Domain

In our previous work, we have emphasized time-domain analysis by working out voltages from one instant to the next. But time-domain analysis is not the only method. This chapter discusses frequency-domain analysis.

22-1 THE FOURIER SERIES

Figure 22-1a shows a sine wave with a peak V_P and a period T. Ac-dc books concentrate on the sine wave because it is the most fundamental. Earlier chapters have likewise emphasized the sine wave. Now we are ready to examine nonsinusoidal waves.

PERIODIC WAVES

The triangular wave of Fig. 22-1b traces its basic pattern during period T; after this, each cycle is a repetition of the first cycle. It is the same with the sawtooth of Fig. 22-1c and the half-wave signal of Fig. 22-1d; all cycles are copies of the first cycle. Waveforms with repeating cycles are called *periodic*; they have a period T in which the size and shape of every cycle is determined.

HARMONICS

The sine wave is extraordinary. By adding sine waves with the right amplitude and phase, we can produce the triangular wave of Fig. 22-1b. With a different combination of sine waves, we can get the sawtooth wave of Fig. 22-1c. And with still another combination of sine waves, we can produce the half-wave signal of Fig. 22-1d. In other words, any periodic wave is a superposition of sine waves.

These sine waves are harmonically related, meaning that their frequencies are *harmonics* (multiples) of a *fundamental* (lowest frequency). Given a periodic wave, you can measure its period T on an oscilloscope. The reciprocal of T equals the fundamental frequency. Symbolically,

$$f_1 = \frac{1}{T} \qquad \text{(fundamental)} \qquad (22\text{-}1)$$

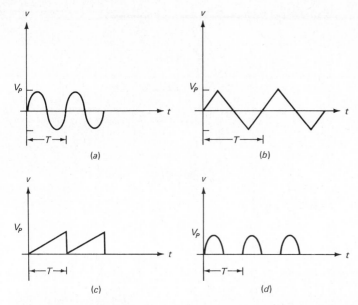

Fig. 22-1 *Periodic waveforms.* (a) *Sine wave.* (b) *Triangular wave.* (c) *Sawtooth wave.*
(d) *Half-wave rectified sine wave.*

The second harmonic has a frequency of

$$f_2 = 2f_1$$

The third harmonic has a frequency of

$$f_3 = 3f_1$$

In general, the nth harmonic has a frequency of

$$f_n = nf_1 \qquad (22\text{-}2)$$

As an example, Fig. 22-2 shows a sawtooth wave on the left. This periodic wave is equivalent to the sum of the harmonically related sine waves on the right. Also included is a battery to account for the average or dc value of the sawtooth wave. Since period T equals 2 ms, the harmonic frequencies equal

$$f_1 = \frac{1}{2 \text{ ms}} = 500 \text{ Hz}$$

$$f_2 = 2(500 \text{ Hz}) = 1000 \text{ Hz}$$
$$f_3 = 3(500 \text{ Hz}) = 1500 \text{ Hz}$$

and so on.

FOURIER SERIES

As a word formula, here is what Fig. 22-2 says:

Periodic wave = dc component + first harmonic
+ second harmonic + third harmonic + . . . + nth harmonic

Fig. 22-2
*Periodic wave is
equivalent to the
sum of sine waves.*

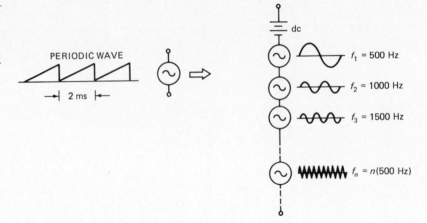

In precise mathematical terms,

$$v = V_0 + V_1 \sin(\omega t + \phi_1) + V_2 \sin(2\omega t + \phi_2)$$
$$+ V_3 \sin(3\omega t + \phi_3) + \ldots + V_n \sin(n\omega t + \phi_n)$$

This famous equation is called the *Fourier series*. It says that a periodic wave is a superposition of harmonically related sine waves. Voltage v is the value of the periodic wave at any instant in time; we can calculate this value by adding the dc component and the instantaneous values of the harmonics.

Theoretically, the harmonics continue to infinity; that is, n has no upper limit. In the laboratory, however, five to ten harmonics are enough to synthesize a periodic wave to within 5 percent. With the right combination of amplitudes (V_1, V_2, V_3, . . . , V_n) and angles (ϕ_1, ϕ_2, ϕ_3, . . . , ϕ_n), we can produce any periodic waveform.

22-2 THE SPECTRUM OF A SIGNAL

The Fourier theorem is the key to frequency-domain analysis. Since we already know a great deal about sine waves, we can reduce a periodic wave to its sine-wave components; then, by analyzing these sine waves, we are indirectly analyzing the periodic wave. In other words, there are two approaches in nonsinusoidal circuit analysis. We can figure out what a periodic wave does at each instant in time, or we can figure out what each harmonic does. Sometimes, the first approach (time-domain analysis) is faster. But often, the second approach (frequency-domain analysis) is superior.

SPECTRAL COMPONENTS

Suppose A represents the peak-to-peak value of a *sawtooth* wave. Using advanced mathematics, we can prove that

$$V_n = \frac{A}{n\pi} \tag{22-3}$$

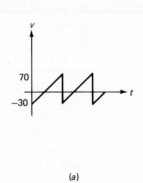

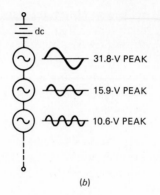

Fig. 22-3
*Sawtooth wave
and its harmonics.*

(a)

(b)

This says that the peak value of the nth harmonic equals A divided by $n\pi$. For example, Fig. 22-3a shows a sawtooth wave with a peak-to-peak value of 100 V; the harmonics have these peak values:

$$V_1 = \frac{100}{\pi} = 31.8 \text{ V}$$

$$V_2 = \frac{100}{2\pi} = 15.9 \text{ V}$$

$$V_3 = \frac{100}{3\pi} = 10.6 \text{ V}$$

and so on. Figure 22-3b shows the first three harmonics for the sawtooth wave of Fig. 22-3a.

With an oscilloscope (a time-domain instrument), you see the periodic signal as a function of time (Fig. 22-3a). The vertical axis represents voltage and the horizontal axis stands for time. In effect, the oscilloscope displays the instantaneous value v of the periodic wave.

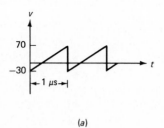

(a)

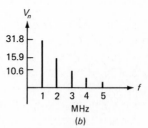

(b)

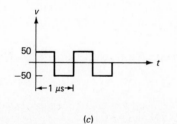

(c)

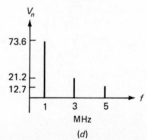

(d)

Fig. 22-4
*Waves and
spectra.*

The *spectrum analyzer* differs from the oscilloscope. To begin with, a spectrum analyzer is a frequency-domain instrument; its horizontal axis represents frequency. With a spectrum analyzer, we see the harmonic peak values versus frequency. For instance, if the sawtooth wave of Fig. 22-4a drives a spectrum analyzer, we will see the display of Fig. 22-4b. We call this kind of display a *spectrum*; the height of each line represents the harmonic peak value, and the horizontal location gives the frequency.

Every periodic wave has a spectrum, or set of vertical lines representing the harmonics. The spectrum normally differs from one periodic signal to the next. For instance, a square wave like Fig. 22-4c has a spectrum like Fig. 22-4d; this is different from the spectrum of the sawtooth wave of Fig. 22-4b.

FOUR BASIC SPECTRA

For future reference, Fig. 22-5 shows four periodic waves and their spectra. In each of these, A is the peak-to-peak value of the periodic wave. For convenience, we have shown only the harmonics to $n = 5$. In each case, the formula for harmonic peak

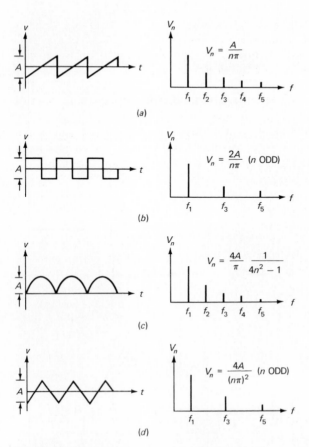

Fig. 22-5 *Some common waves and their spectra.* (a) *Sawtooth wave.* (b) *Square wave.* (c) *Full-wave rectified sine wave.* (d) *Triangular wave.*

values is given. For example, with a full-wave signal like Fig. 22-5c, we can calculate the peak value of any harmonic by using

$$V_n = \frac{4A}{\pi} \frac{1}{4n^2 - 1} \tag{22-4a}$$

Or if we want the peak values of harmonics for the square wave of Fig. 22-5b, we use

$$V_n = \frac{2A}{n\pi} \quad (n \text{ odd only}) \tag{22-4b}$$

HALF-WAVE SYMMETRY

Of the waveforms in Fig. 22-5, two have only odd harmonics. Many other waveforms have this odd-harmonic property, and it helps to know a quick method for identifying such waveforms.

Any waveform with *half-wave symmetry* has the odd-harmonic property. Half-wave symmetry means that you can invert the negative half cycle and get an exact duplicate of the positive half cycle. For example, Fig. 22-6a shows one period of a triangular wave. After you invert the negative half cycle, you have the *abc* half cycle (dashed line). This inverted half cycle is an exact duplicate of the positive half cycle. As a result, the triangular wave contains only odd harmonics. Similarly, the square wave of Fig. 22-6b has half-wave symmetry because the inverted negative cycle exactly duplicates the positive half cycle; this means that a square wave contains only odd harmonics.

In case you have any doubt, it helps to shift the inverted half cycle to the left; half-wave symmetry exists only if the shifted half cycle superimposes the positive half cycle. For instance, inverting the negative half cycle of a sawtooth wave produces the *abc* half cycle of Fig. 22-6c. When it is shifted to the left, the inverted half cycle exactly superimposes the positive half cycle; therefore, we are assured of half-wave symmetry and the odd-harmonic property.

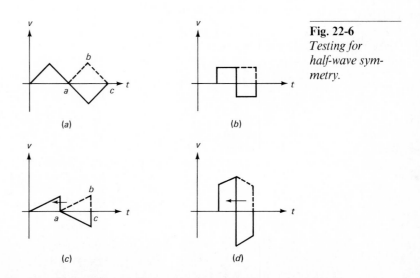

Fig. 22-6
Testing for half-wave symmetry.

The waveform of Fig. 22-6*d* does not have half-wave symmetry. When the negative half cycle is inverted and shifted to the left, it does not superimpose the positive half cycle. Therefore, Fig. 22-6*d* does not have the odd-harmonic property; it must contain at least one even harmonic.

THE DC COMPONENT

The *dc component* is the average value of the periodic wave, defined as

$$V_0 = \frac{\text{area under one cycle}}{\text{period}} \qquad (22\text{-}5)$$

As an example, Fig. 22-7*a* shows a sawtooth wave with a peak of 10 V and a period of 2 s. The area under one cycle is shaded and equals

$$\text{Area} = \tfrac{1}{2}(\text{base})(\text{height})$$
$$= \tfrac{1}{2}(2 \text{ s})(10 \text{ V}) = 10 \text{ V} \cdot \text{s}$$

Dividing by the period gives the average value of the sawtooth:

$$V_0 = \frac{10 \text{ V} \cdot \text{s}}{2 \text{ s}} = 5 \text{ V}$$

The area is positive above the horizontal axis, but negative below. If part of a cycle is above and part below the horizontal axis, you add the positive and negative areas algebraically to get the net area under the cycle. Figure 22-7*b* shows a square wave that swings from +70 V to −30 V. The first half cycle has an area above the horizontal axis; therefore, the area is positive. But the second half cycle is below the horizontal axis, so that the area is negative. Here is how to find the average value over the entire cycle

$$\text{Positive area} = (3 \text{ s})(70 \text{ V}) = 210 \text{ V} \cdot \text{s}$$
$$\text{Negative area} = (3 \text{ s})(-30 \text{ V}) = -90 \text{ V} \cdot \text{s}$$
$$\text{Net area under one cycle} = 210 - 90 = 120 \text{ V} \cdot \text{s}$$

Fig. 22-7
Areas and average values.

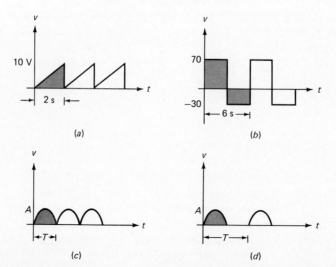

(a)

(b)

(c)

(d)

Dividing the net area by the period gives

$$V_0 = \frac{120 \text{ V} \cdot \text{s}}{6 \text{ s}} = 20 \text{ V}$$

This is the average value for the entire cycle of Fig. 22-7b; it is the value a dc voltmeter would read.

Using the area-over-period formula, you can calculate the average value of other period waves, provided the waves have linear segments. For instance, sawtooth, square, and triangular waves are made up of straight lines. Because of this, you can use well-known geometry formulas to calculate areas; after dividing by period, you have the average value.

But what do you do with half-wave and full-wave signals like Fig. 22-7c and d? The waveforms are nonlinear, and no simple geometry formulas are available for their areas. The only exact way to calculate the areas is with calculus; using calculus, we can derive these average values:

$$V_0 = 0.636A \qquad \text{(full wave)} \qquad (22\text{-}6)$$
$$\text{and} \qquad V_0 = 0.318A \qquad \text{(half wave)} \qquad (22\text{-}7)$$

EFFECT OF DC COMPONENT ON SPECTRUM

If we add a dc component to any waveform, the only change in the spectrum is the appearance of a line at zero frequency. The height of this line represents the dc voltage. In general, adding a dc component to a waveform has no effect on the harmonics; the only spectral change is a new line at zero frequency.

22-3 HARMONIC DISTORTION

When an amplified signal is small, only a small part of the transconductance curve is used. Because of this, the operation takes place over an almost linear arc of the curve. Operation like this is called *linear* because changes in output current are proportional to changes in input voltage. Linear operation means that the shape of the amplified waveform is the same as the shape of the input waveform. That is, we get no distortion when the operation is linear or small-signal.

But when the signal is large, we can no longer treat the operation as linear; changes in output current are no longer proportional to changes in input voltage. Because of this, we get *nonlinear distortion*. This section examines distortion from viewpoint of the frequency domain.

LARGE-SIGNAL OPERATION

When the signal swing is large, the operation becomes nonlinear. Figure 22-8 is an example of this. A sinusoidal V_{BE} voltage produces large swings along the transconductance curve. Because of the nonlinearity of the curve, the resulting current is no longer sinusoidal. In other words, the shape of the output current is no longer a true duplication of the input shape. Since the output current flows through a load resistance, the output voltage will also have nonlinear distortion.

Fig. 22-8
Nonlinear
distortion.

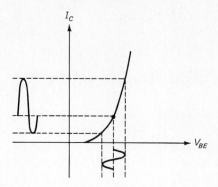

Figure 22-9*a* shows nonlinear distortion from the time-domain viewpoint. The input sine wave drives an amplifier. If the operation is large-signal, the amplified output voltage is no longer a pure sine wave. Arbitrarily, we have shown more gain on one half cycle than on the other; this kind of distortion is sometimes called *amplitude distortion*.

The frequency domain gives us insight into amplitude distortion. Figure 22-9*b* shows how to visualize the same situation in the frequency domain. The input spectrum is a single line at f_1, the frequency of the input sine wave. The output signal is distorted but still periodic; therefore, it contains the dc component and the harmonics shown. Arbitrarily, we have stopped with the fourth harmonic. The point is that a waveform with amplitude distortion contains a fundamental and harmonics; the strength of the higher harmonics is a clue to how bad the distortion is.

In fact, an alternative name for amplitude distortion is *harmonic distortion*. We use the term amplitude distortion when we visualize a signal in the time domain, and we use the term harmonic distortion when we are thinking of the signal in the frequency domain. When we are interested in the cause of the distortion, we use the term nonlinear distortion. All are synonyms for the kind of distortion that occurs

Fig. 22-9
Amplitude
distortion produces
harmonics in the
spectrum.

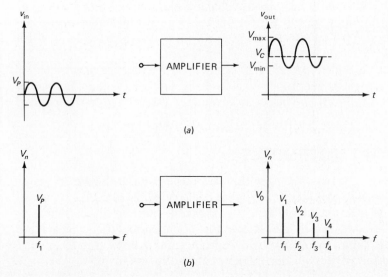

with one input sine wave. (The next chapter talks about distortion with two or more input sine waves.)

FORMULA FOR INDIVIDUAL HARMONIC DISTORTION

How are we going to compare the harmonic distortion of one amplifier with that of another? The larger the peak values of the harmonics, the larger the harmonic distortion. The simplest way to compare different amplifiers is to take the ratio of the harmonics to the fundamental. Specifically, we define the percent of second-harmonic distortion as

$$\text{Percent second-harmonic distortion} = \frac{V_2}{V_1} \times 100\% \qquad (22\text{-}8a)$$

The percent of third-harmonic distortion is

$$\text{Percent third-harmonic distortion} = \frac{V_3}{V_1} \times 100\% \qquad (22\text{-}8b)$$

and so on for any higher harmonic. In general, the percent of nth-harmonic distortion is

$$\text{Percent } n\text{th-harmonic distortion} = \frac{V_n}{V_1} \times 100\% \qquad (22\text{-}8c)$$

As an example, suppose the output spectrum of Fig. 22-9b has $V_1 = 2$ V, $V_2 = 0.2$ V, $V_3 = 0.1$ V, and $V_4 = 0.05$ V. Then, the harmonic distortions are

Percent second-harmonic distortion = 10%
Percent third-harmonic distortion = 5%
Percent fourth-harmonic distortion = 2.5%

FORMULA FOR TOTAL HARMONIC DISTORTION

Data sheets usually give total harmonic distortion, all harmonics lumped together and compared with the fundamental. The formula for total harmonic distortion is

$$\text{Percent total harmonic distortion} = \sqrt{(\text{percent second})^2 + (\text{percent third})^2 + \dots} \qquad (22\text{-}9)$$

As an example, for $V_1 = 2$ V, $V_2 = 0.2$ V, $V_3 = 0.1$ V, and $V_4 = 0.05$ V, the individual harmonic distortions are 10, 5, and 2.5 percent; the total harmonic distortion is

$$\text{Percent total harmonic distortion} = \sqrt{(10\%)^2 + (5\%)^2 + (2.5\%)^2} = 11.5\%$$

NEGATIVE FEEDBACK REDUCES HARMONIC DISTORTION

A final point: As discussed in Chap. 16, negative feedback reduces harmonic or nonlinear distortion by $1 + AB$, the desensitivity. For instance, suppose an amplifier has an open-loop harmonic distortion of 10 percent. If this amplifier is used with negative feedback where the desensitivity is 100, then the closed-loop harmonic distortion is only (10%)/100, or 0.1 percent.

22-4 OTHER KINDS OF DISTORTION

For nonlinear operation of an amplifier, an input sine wave produces a distorted output signal in the time domain; in the frequency domain, nonlinear operation is equivalent to a single-line spectrum producing an output spectrum with many lines (see Fig. 22-10*a*). This is what we mean by harmonic distortion; a pure input sine wave produces a fundamental and harmonics.

FREQUENCY DISTORTION

Frequency distortion is different. It has nothing to do with nonlinear distortion; frequency distortion can occur even with small-signal operation. The cause of frequency distortion is a change in amplifier gain with frequency. Figure 22-10*b* illustrates this kind of distortion. Arbitrarily, the input spectrum contains many equal-amplitude sinusoidal components. If the cutoff frequency of the amplifier is less than the highest sinusoidal frequency, the higher frequencies in the output spectrum are attenuated as shown.

Frequency distortion, therefore, is nothing more than a change in the spectrum of the signal caused by amplifier cutoff frequencies. This can affect the quality of speech or music signals. Speech and music are complex signals with many components in the spectrum. Unless all these components are in the bandwidth of the amplifier, the components in the output spectrum will not have the correct amplitudes. Because of this, the speech or music will sound different from the original input signal.

Fig. 22-10
(a) *Harmonic distortion.* (b) *Frequency distortion.*

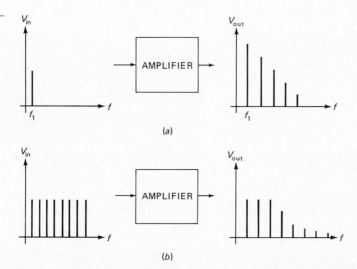

(a)

(b)

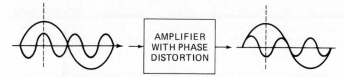

Fig. 22-11 *Phase distortion means harmonics shift with respect to the fundamental.*

PHASE DISTORTION

Phase distortion takes place when the phase of a harmonic is shifted with respect to the fundamental. As an example, the input signal of Fig. 22-11 shows the third harmonic peak in phase with the peak of the fundamental. If there is phase distortion, the third harmonic changes its phase with the fundamental at the output.

Frequency and phase distortion almost always occur together. In the midband of an amplifier, the voltage gain and phase shift are constant (either 0° or 180°). Because of this, no frequency or phase distortion can occur if all sinusoidal components are in the midband of the amplifier. Outside the midband, the voltage gain drops off and the phase angle changes; therefore, we get frequency and phase distortion if the spectrum contains components outside the midband.

PROBLEMS

Straightforward

22-1. A half-wave signal like Fig. 22-1*d* has a peak value of 20 V and a period of 25 μs. What are the frequencies of the first three harmonics?

22-2. Work out the first four odd-harmonic frequencies for a square wave with a peak-to-peak value of 30 V and a period of 5 ms.

22-3. A triangular wave like Fig. 22-1*b* has $V_P = 25$ V and $T = 2$ ms. What is the fundamental frequency? The frequency of the 25th harmonic?

22-4. A sawtooth wave like Fig. 22-1*c* has $V_P = 10$ V and $T = 20$ μs. What frequencies do the first three harmonics have?

22-5. Calculate the peak values of the first three odd harmonics in Prob. 22-2.

22-6. What is the peak value of the fundamental in Prob. 22-3? The peak value of the 10th harmonic?

22-7. Draw the spectrum for the sawtooth wave of Prob. 22-4. Include the dc component and the first four harmonics.

22-8. A square wave has a positive peak of +80 V and a negative peak of −20 V. What is the average voltage?

22-9. A sine wave drives an amplifier. The output spectrum has lines at f_1, f_2, and f_3. The corresponding peak values are 5 V, 0.4 V, and 0.2 V. Calculate the second-harmonic distortion, the third-harmonic distortion, and the total harmonic distortion.

22-10. An amplifier with a voltage gain of 500 has a total harmonic distortion of 0.2 percent for an input sine wave of 10 mV rms. If only the fundamental and the third harmonic are present in the output, what is the rms value of the third harmonic?

22-11. The input voltage to an amplifier is changed in decade steps as follows: $v_{in} = 0.1$ mV, 1 mV, 10 mV, 100 mV, and 1 V. The corresponding output voltages are

v_{out} = 10 mV, 100 mV, 1 V, 9 V, and 12 V. If you want to avoid harmonic distortion, what is the largest input voltage you should use?

22-12. A spectrum has four lines at f_1, f_2, f_3, and f_4 with corresponding peak voltages of 10 V, 10 V, 5 V, and 5 V. Calculate the total harmonic distortion.

22-13. A data sheet for an op amp says that the total harmonic distortion equals 3 percent when the output signal is 20 V. If you use this op amp in a noninverting negative-voltage-feedback system with a desensitivity of 500, what will the total harmonic distortion equal when the output signal has a peak-to-peak value of 20 V?

Troubleshooting

22-14. In Fig. 22-10*b*, all output spectral lines suddenly drop to one-tenth of the original value. Suggest one possible trouble inside the amplifier that would cause this.

Design

22-15. Some nearby equipment is radiating an interference signal at 1.25 MHz. You have a circuit that is picking up this unwanted signal. Design a notch filter to reject the interference signal. Use an *R* of 1 kΩ.

Challenging

22-16. Calculate the total harmonic distortion of a square wave. (Include up to the ninth harmonic in your calculations.)

Computer

22-17. Write a program that calculates the first 10 harmonics for a sawtooth wave, a square wave, a full-wave rectified sine wave, and a triangular wave. Your program should include a menu.

Frequency Mixing

When a sine wave drives a nonlinear circuit, harmonics of this sine wave appear in the output. If two sine waves drive a nonlinear circuit, we get harmonics of each sine wave, plus new frequencies called the sum and difference frequencies. This chapter describes the theory and application of these new output frequencies.

23-1 NONLINEARITY

Figure 23-1 shows a nonlinear graph of output voltage versus input voltage. If the signal is small, the instantaneous operating point swings over a small part of the curve. In a case like this, the operation is linear and

$$v_{\text{out}} = Av_{\text{in}} \tag{23-1a}$$

where A is a constant that is equal to the voltage gain.

POWER SERIES

When the signal swing is large, we must use a more complicated equation that accounts for the nonlinearity. It can be shown that a power series is needed for large-signal operation:

$$v_{\text{out}} = Av_{\text{in}} + Bv_{\text{in}}^2 + Cv_{\text{in}}^3 + Dv_{\text{in}}^4 + \cdots \tag{23-1b}$$

A power series like this applies to any nonlinear curve like Fig. 23-1. Each distinct curve has its own set of coefficients (A, B, C, . . .). Notice that the power series contains linear operation as a special case; when v_{in} is small, all higher-order terms drop out, leaving $v_{\text{out}} = Av_{\text{in}}$.

SMALL-SIGNAL OPERATION WITH TWO INPUT SINE WAVES

What happens if two sine waves drive an amplifier? We can express the first sine wave by

$$v_x = V_x \sin \omega_x t$$

Fig. 23-1
*Nonlinear
input-output
curve.*

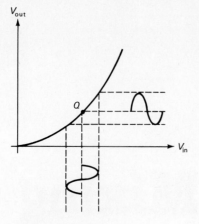

and the second by

$$v_y = V_y \sin \omega_y t$$

Figure 23-2*a* shows these sine waves; arbitrarily, v_x has the higher frequency. If the sinusoidal voltages are in series, as shown in Fig. 23-2*b*, the total input voltage is

$$v_{in} = v_x + v_y = V_x \sin \omega_x t + V_y \sin \omega_y t \qquad (23\text{-}1c)$$

The sum of these two sine waves looks like Fig. 23-2*c* and the spectrum like Fig. 23-2*d*. In other words, if you look at the input voltage with an oscilloscope, you see Fig. 23-2*c*. If you look at the input voltage with a spectrum analyzer, you see Fig. 23-2*d*.

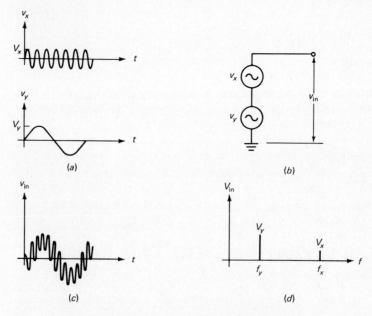

Fig. 23-2 (a) *Two input signals.* (b) *Series connection of two sine-wave sources.* (c) *Additive waveform.* (d) *Spectrum of additive waveform.*

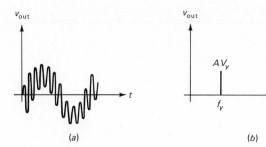

Fig. 23-3
(a) *Amplified additive signal.*
(b) *Spectrum.*

When both signals are small, the operation is linear and Eq. (23-1a) applies:

$$v_{out} = Av_{in} = A(v_x + v_y)$$
$$= AV_x \sin \omega_x t + AV_y \sin \omega_y t$$

What does this equation say? It says that the ac output voltage is the sum of each input sine wave amplified by A. Figure 23-3a shows how this output signal looks in the time domain; it is nothing more than the signal of Fig. 23-2c amplified by A. Figure 23-3b shows the spectrum of the amplified signal; it is the original input spectrum with each line amplified by A. Especially important, no harmonics or other lines appear in the output spectrum for small-signal operation.

ADDITIVE WAVEFORMS

A waveform like Fig. 23-3a is called an *additive waveform* because it is the sum or superposition of two sine waves. You often see waveforms like this. As an example, if you are amplifying a sine wave with frequency f_y, an interference signal with frequency f_x may somehow get into the amplifier and be added to the desired signal. In this case, you will see the additive waveform of Fig. 23-3a at the output of the amplifier.

The interference signal can get into the amplifier in a number of ways. It may be excessive power supply ripple being added to the desired signal. Or it may be a signal induced by nearby electrical apparatus, or possibly a transmitted radio signal. In any event, remember that whenever you see an additive waveform like Fig. 23-3a, it represents the sum of two sine waves.

LINEAR OPERATION WITH MANY SINUSOIDAL COMPONENTS

What we have derived for two input sine waves applies to any number of input sine waves. In other words, if we have 10 input sine waves, linear amplification results in 10 output sine waves. The output spectrum will contain 10 spectral lines with the same frequencies as the input spectrum; each component is amplified by A.

A good example of a multi-sine-wave input is the signal produced by middle C of a piano. This signal contains a fundamental with a frequency of 256 Hz plus harmonics to around 10,000 Hz (approximately 40 harmonics); the strength of the harmonics compared with the fundamental makes this note sound different from all others. If we linearly amplify this middle-C signal, the output spectrum contains only the original harmonics amplified by factor A. Because of this, the output is louder, but it still has the distinct sound of middle C. A good high-fidelity amplifier, therefore, is a small-signal or linear amplifier; it does not add or subtract any spectral components; furthermore, it amplifies each component by the same amount.

23-2 MEDIUM-SIGNAL OPERATION WITH ONE SINE WAVE

For typical values of A, B, C, etc., in Eq. (23-1b), the first higher-power term to become important is the quadratic term. That is, there is a level of input signal between the small-signal and large-signal cases at which only the first two terms of Eq. (23-1b) are important:

$$v_{out} = Av_{in} + Bv_{in}^2 \qquad (23\text{-}2)$$

Because this special case lies between small-signal and large-signal operation, we call it the *medium-signal* case. Any transistor amplifier can operate medium-signal, with only a linear and a quadratic term in the equation for ac output voltage. With a further increase in the signal, a bipolar amplifier goes into the large-signal case, where the cubic and higher-power terms become important.

THE FET AMPLIFIER

Unlike a bipolar amplifier, a FET amplifier operates medium-signal all the way to saturation and cutoff. In other words, when we increase the input signal, the FET amplifier continues to operate medium-signal; only the linear and quadratic terms appear in the expression for output voltage. Equation (23-2) applies to a FET amplifier for any signal level, provided the output signal is not clipped. This is a direct consequence of the parabolic or square-law transconductance curve. We do not have the time to prove it, but it is possible to derive Eq. (23-2) from the transconductance equation of a FET. Because of this, as long as the instantaneous operating point remains along the transconductance curve, Eq. (23-2) applies to a FET amplifier.

QUADRATIC TERM PRODUCES SECOND HARMONIC

The quadratic term Bv_{in}^2 causes second-harmonic distortion. Suppose the input voltage is a sine wave expressed by

$$v_{in} = V_x \sin \omega_x t$$

In the medium-signal case, the output voltage equals

$$\begin{aligned} v_{out} &= Av_{in} + Bv_{in}^2 \\ &= AV_x \sin \omega_x t + BV_x^2 \sin^2 \omega_x t \end{aligned} \qquad (23\text{-}3a)$$

A useful expansion formula proved in trigonometry is

$$\sin^2 A = \tfrac{1}{2} - \tfrac{1}{2} \cos 2A \qquad (23\text{-}3b)$$

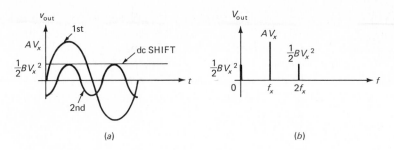

Fig. 23-4
(a) *x output in
time domain.* (b)
*Spectrum of x
output.*

where A represents angle. If we let $A = \omega_x t$, we can use Eq. (23-3b) to rearrange Eq. (23-3a); the result is

$$v_{out} = \tfrac{1}{2}BV_x^2 + AV_x \sin \omega_x t - \tfrac{1}{2}BV_x^2 \cos 2\omega_x t \qquad (23\text{-}4)$$

Each term in this expression is important. Briefly, here is what each means:

$\tfrac{1}{2}BV_x^2$—this has a constant value and represents a dc shift.

$AV_x \sin \omega_x t$—the linear output term; this is the amplified input sine wave.

$\tfrac{1}{2}BV_x^2 \cos 2\omega_x t$—this term has a radian frequency of $2\omega_x$, equivalent to a cycle frequency of $2f_x$; therefore, this term represents the second harmonic of the input sine wave.

Figure 23-4a shows each of these components in the time domain; if we add these components, we get the total waveform as it would appear on an oscilloscope. Figure 23-4b shows the spectrum; note the dc component, the fundamental with a frequency f_x, and the second harmonic with a frequency $2f_x$.

SUMMARY

For the medium-signal case, an input sine wave v_x produces a dc shift, an amplified fundamental, and a second harmonic. In later discussion, we will refer to the right-hand member of Eq. (23-4) as the *x output*. We can visualize the *x* output either in the time domain (Fig. 23-4a) or in the frequency domain (Fig. 23-4b).

23-3 MEDIUM-SIGNAL OPERATION WITH TWO SINE WAVES

When two input sine waves drive a medium-signal amplifier, something remarkable happens; in addition to harmonics, *new frequencies* appear in the output. To be specific, here is what we will prove jn this section. When two input sine waves with frequencies f_x and f_y drive an amplifier in the medium-signal case, the output spectrum contains sinusoidal components with these frequencies:

f_x and f_y: the two input frequencies

$2f_x$ and $2f_y$: the second harmonics of the input frequencies

$f_x + f_y$: a new frequency equal to the sum of the input frequencies

$f_x - f_y$: a new frequency equal to the difference of the input frequencies

For instance, if the two input frequencies are 1 kHz and 20 kHz, the output spectrum contains sinusoidal frequencies of 1 kHz and 20 kHz (the two input frequencies), 2 kHz and 40 kHz (second harmonics), 21 kHz (the sum), and 19 kHz (the difference).

THE CROSS PRODUCT

When two input sine waves drive an amplifier, we can express the input voltage as

$$v_{in} = v_x + v_y$$

where v_x and v_y are sine waves. For the medium-signal case, the output voltage equals

$$
\begin{aligned}
v_{out} &= Av_{in} + Bv_{in}^2 \\
&= A(v_x + v_y) + B(v_x + v_y)^2 \\
&= Av_x + Av_y + Bv_x^2 + 2Bv_xv_y + Bv_y^2
\end{aligned}
$$

By rearranging, we get

$$v_{out} = \underbrace{Av_x + Bv_x^2}_{x \text{ output}} + \underbrace{Av_y + Bv_y^2}_{y \text{ output}} + \underbrace{2Bv_xv_y}_{\text{cross product}} \qquad (23\text{-}5)$$

The first two terms in this equation are the x output discussed in the preceding section, that is, the output from a medium-signal amplifier when only v_x drives the amplifier; therefore, we already know that these two terms result in a dc shift, an amplified fundamental f_x, and a second harmonic $2f_x$.

The next pair of terms, $Av_y + Bv_y^2$ is called the *y output;* because they are identical to the x output except for the y subscript, the y-output terms result in a dc shift, an amplified fundamental f_y, and a second harmonic $2f_y$. In other words, the y output is what we get out of a medium-signal amplifier when a sine wave of frequency f_y is the only input to the amplifier.

The unusual thing about Eq. (23-5) is the *cross product* $2Bv_xv_y$. If the cross product were *not* present in this equation, the output would be the superposition of the x output and the y output. This would allow us to take each input separately, find how the amplifier responds to it, and sum the x and y outputs to get the total output. But the cross product is there, and because of it the superposition theorem gives us only part of the total output.

SUM AND DIFFERENCE FREQUENCIES

In Eq. (23-5), we already know that the x-output terms represent a dc shift, an amplified fundamental f_x, and a second harmonic $2f_x$. Likewise, the y-output terms represent another dc shift, an amplified fundamental f_y, and a second harmonic $2f_y$. All that remains now is the meaning of the cross product.

When we substitute

$$v_x = V_x \sin \omega_x t$$

and

$$v_y = V_y \sin \omega_y t$$

into the cross product, we get

$$
\begin{aligned}
2Bv_xv_y &= 2B(V_x \sin \omega_x t)(V_y \sin \omega_y t) \\
&= 2BV_xV_y(\sin \omega_x t)(\sin \omega_y t)
\end{aligned}
\tag{23-6}
$$

The product of two sine waves can be expanded by the trigonometric identity

$$\sin A \sin B = \tfrac{1}{2}\cos(A - B) - \tfrac{1}{2}\cos(A + B)$$

By letting $A = \omega_x t$, and $B = \omega_y t$, we can expand and rearrange Eq. (23-6) to get

$$2Bv_xv_y = BV_xV_y \cos(\omega_x - \omega_y)t - BV_xV_y \cos(\omega_x + \omega_y)t \tag{23-7}$$

The first term on the right side is a sinusoidal component with a radian frequency of $\omega_x - \omega_y$, equivalent to a cycle of frequency of $f_x - f_y$; therefore, this first term represents the difference frequency. The second term is also a sinusoidal component, but it has a cycle frequency of $f_x + f_y$; so, this second term represents the sum frequency.

OUTPUT SPECTRUM

Equation (23-5) gives the output voltage from a medium-signal amplifier driven by two input sine waves. The first two terms are the x output; these terms represent a dc component, an amplified fundamental f_x, and a second harmonic $2f_x$; Fig. 23-5a shows the spectrum for this x output. In Eq. (23-5) the y-output terms represent another dc component, an amplified fundamental f_y, and a second harmonic $2f_y$; Fig. 23-5b illustrates the spectrum for the y-output terms. In Eq. (23-5) the cross product represents the sum and difference terms; Fig. 23-5c shows the spectrum of this cross product.

The total output signal is the sum of the x output, the y output, and the cross product. The signal is too complicated to draw in the time domain, but it is easy to show in the frequency domain. Figure 23-5d is the output spectrum of an amplifier driven by two sine waves in the medium-signal case. As we can see, the zero-frequency line is the sum of the dc components. More important, the remaining lines represent sinusoidal signals: we have each input frequency and its harmonic, plus sum and difference frequencies.

SUMMARY

We have found something new and useful. Put two sinusoidal signals into a medium-signal amplifier, and out come six sinusoidal signals with frequencies f_x, $2f_x$, f_y, $2f_y$, a sum frequency of

$$\text{Sum} = f_x + f_y$$

and a difference frequency of

$$\text{Difference} = f_x - f_y$$

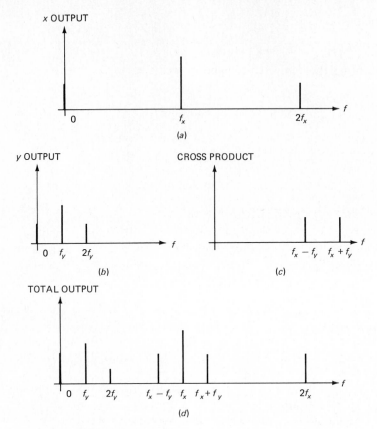

Fig. 23-5 *Spectra.* (a) *x output.* (b) *y output.* (c) *Sum and difference components.* (d) *Total output.*

23-4 LARGE-SIGNAL OPERATION WITH TWO SINE WAVES

What happens when two sine waves drive a bipolar amplifier in the large-signal mode? We get an output spectrum containing each input frequency, all harmonics of these frequencies, and sum and difference frequencies produced by every combination of harmonics.

DERIVATION

For large-signal operation we need Eq. (23-1*b*), which says

$$v_{\text{out}} = Av_{\text{in}} + Bv_{\text{in}}^2 + Cv_{\text{in}}^3 + \cdots$$

This is an infinite series; there is no limit to the number of terms; each new term we add to the expression is the next higher power of v_{in}. With two input sine waves,

$$v_{\text{in}} = v_x + v_y$$

where v_x and v_y are sine waves. After substitution into the infinite series,

$$v_{\text{out}} = A(v_x + v_y) + B(v_x + v_y)^2 + C(v_x + v_y)^3 + \cdots$$

By applying the binominal theorem to each higher-power term, we can rearrange the expression to get

$$v_{\text{out}} = (Av_x + Bv_x^2 + Cv_x^3 + \cdots) + (Av_y + Bv_y^2 + Cv_y^3 + \cdots)$$
$$+ (2Bv_xv_y + 3Cv_x^2v_y + 3Cv_xv_y^2 + \cdots) \quad (23\text{-}8)$$

The first parenthesis in this equation is the *x output:*

$$x \text{ output} = Av_x + Bv_x^2 + Cv_x^3 + \cdots$$

This is the output we would get if v_x alone were driving the amplifier. With trigonometry we can prove that v_x^n produces the nth harmonic of f_x. Therefore, the x output contains an amplified fundamental of frequency f_x, a second harmonic of frequency $2f_x$, a third harmonic of frequency $3f_x$, and so forth.

The second parenthesis in Eq. (23-8) is the *y output:*

$$y \text{ output} = Av_y + Bv_y^2 + Cv_y^3 + \cdots$$

This is the output we would get if only v_y were driving the amplifier. Again, the nth-power term introduces the nth harmonic; therefore, the spectrum of the y output contains spectral lines at f_y, $2f_y$, $3f_y$, and so forth.

The third parenthesis in Eq. (23-8) is a collection of cross products from the binomial expansion of each power of $v_x + v_y$:

$$\text{Cross products} = 2Bv_xv_y + 3Cv_x^2v_y + 3Cv_xv_y^2 + \cdots$$

Following the last term are more terms of the form

$$Kv_x^mv_y^n$$

By advanced mathematics, we can prove that each cross product produces a sum frequency of $mf_x + nf_y$ and a difference frequency of $mf_x - nf_y$. Or,

$$\text{New } f\text{'s} = mf_x \pm nf_y \quad (23\text{-}9)$$

where m and n can be any positive integers.

ORDERLY CALCULATIONS

Equation (23-9) is the formula for every possible sum and difference frequency generated in a large-signal amplifier driven by two sine waves. Here is an orderly way to use this important equation:

Group 1. First harmonic of v_x combining with each harmonic of v_y gives frequencies of
$$f_x \pm f_y$$
$$f_x \pm 2f_y$$
$$f_x \pm 3f_y$$
etc.

Group 2. Second harmonic of v_x combining with each harmonic of v_y gives frequencies of

$2f_x \pm f_y$
$2f_x \pm 2f_y$
$2f_x \pm 3f_y$
etc.

Group 3. Third harmonic of v_x combining with each harmonic of v_y gives frequencies of

$3f_x \pm f_y$
$3f_x \pm 2f_y$
$3f_x \pm 3f_y$
etc.

and so on. In this way, we can calculate any higher sum and difference frequencies of interest.

A concrete example will help. Suppose the two input sine waves to a large-signal amplifier have frequencies of $f_x = 100$ kHz and $f_y = 1$ kHz. Then we can calculate sum and difference frequencies as follows:

Group 1.

100 kHz $\pm$ 1 kHz = 99 and 101 kHz
100 kHz $\pm$ 2 kHz = 98 and 102 kHz
100 kHz $\pm$ 3 kHz = 97 and 103 kHz
etc.

Group 2.

200 kHz $\pm$ 1 kHz = 199 and 201 kHz
200 kHz $\pm$ 2 kHz = 198 and 202 kHz
200 kHz $\pm$ 3 kHz = 197 and 203 kHz
etc.

Group 3.

300 kHz $\pm$ 1 kHz = 299 and 301 kHz
300 kHz $\pm$ 2 kHz = 298 and 302 kHz
300 kHz $\pm$ 3 kHz = 297 and 303 kHz
etc.

Continuing like this, we can find any sum and difference frequencies of interest.

NEGATIVE FREQUENCIES

When calculating difference frequencies, you may come up with a negative frequency. For instance, suppose the two input frequencies are 10 kHz and 3 kHz. Then group 1 has these frequencies:

10 kHz $\pm$ 3 kHz = 7 and 13 kHz
10 kHz $\pm$ 6 kHz = 4 and 16 kHz
10 kHz $\pm$ 9 kHz = 1 and 19 kHz
10 kHz $\pm$ 12 kHz = -2 and 22 kHz

Whenever a negative frequency turns up, take the absolute value. In this example, −2 kHz becomes 2 kHz. The reason you can disregard the negative sign is because the cosine terms of Eq. (23-7) and other cross products have the same value for positive and negative frequency. That is,

$$\cos (A - B) = \cos (B - A)$$

SUMMARY

Unquestionably, the situation in a large-signal amplifier is complex. Put two sine waves into an amplifier, and out come two amplified sine waves, their harmonics, and every conceivable sum and difference frequency produced by the input frequencies and their harmonics. Theoretically, an infinite number of spectral components exist. As a practical matter, the size of these components decreases for higher values of m and n.

23-5 INTERMODULATION DISTORTION

If we amplify speech or music in a nonlinear amplifier, the sum and difference frequencies in the output make the speech or music sound radically different. *Intermodulation distortion* is the change in the spectrum caused by the sum and difference frequencies.

CHORD C-E-G

For reasons not yet understood, music sounds good when the notes have a mathematical relation. In Fig. 23-6a, middle C (point 1) has a fundamental frequency of 256 Hz. The next C on the right is an *octave* higher (point 2); this C has a fundamental frequency of 512 Hz, exactly double that of middle C. Play two notes that are an octave apart, and the combination sounds pleasant.

Chords may sound even better. Play chord C-E-G (point 4), and it sounds very pleasant. Here is something interesting. Chord C-E-G has these mathematical properties:

C	E	G	Notes in Chord
256	320	384	Fundamental frequency, Hz
1	$1\frac{1}{4}$	$1\frac{1}{2}$	Ratio to middle C

The ratios suggest that a chord sounds good when the fundamentals are quarter multiples of the lowest frequency.

Besides the fundamental frequencies shown, chord C-E-G contains harmonics to around 10 kHz. The amplitudes of these harmonics give the chord its distinct sound.

Figure 23-6b shows part of the spectrum for chord C-E-G. There are fundamental frequencies of 256-320-384, second harmonics of 512-640-768, third harmonics, fourth harmonics, and so on. The relative amplitudes of these harmonics must remain the same with respect to the fundamentals if we want to retain the exact

Fig. 23-6
(a) *Piano keyboard.* (b) *C-E-G spectrum.*

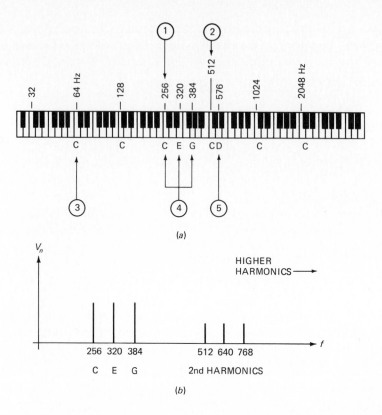

(a)

(b)

sound. If this spectrum drives a *linear amplifier,* each spectral component receives the same gain, assuming that all components are in the midband of the amplifier. In terms of the superposition theorem, the output is the sum of each amplified spectral component.

SUM AND DIFFERENCE FREQUENCIES

If the spectrum of chord C-E-G drives a *nonlinear amplifier,* the output spectrum contains all the input spectral lines plus the sums and differences of every possible combination of these components. How will this sound? Terrible! The ear immediately detects sum and difference frequencies because these frequencies sound like mistakes.

For instance, the first two spectral lines of Fig. 23-6b have frequencies of 256 and 320 Hz; these components produce a difference frequency of 64 Hz and a sum frequency of 576 Hz. The 64-Hz component corresponds to the C note two octaves below middle C (point 3 in Fig. 23-6a). This extra C note is not unpleasant because it is harmonically related to middle C; however, it does represent a new component not in the original sound. Much worse is the sum frequency 576 Hz; this corresponds to the D note one octave above middle C (point 5 in Fig. 23-6a). This D note does not belong in chord C-E-G and sounds *discordant;* the effect is the same as if the pianist made a mistake.

Besides the sum and difference frequencies of the first two fundamentals, we get many other sum and difference frequencies produced by other combinations. As a result, nonlinear amplification produces many discordant notes.

RELATION TO NONLINEAR DISTORTION

Nonlinear distortion causes harmonic and intermodulation distortion. Any device or circuit with a nonlinear input-output relation results in nonlinear distortion of the signal. In the time domain this means that the shape of the periodic signal changes as it passes through the nonlinear circuit. In the frequency domain the result is a change in the spectrum of the signal. If only one input sine wave is present, only harmonic distortion occurs; if two or more input sine waves are involved, both harmonic and intermodulation distortion occur.

23-6 FREQUENCY MIXERS

A *frequency mixer* is used in almost every radio and television receiver; it is also used in many other electronic systems.

THE BASIC IDEA

Figure 23-7 shows all the key ideas behind a frequency mixer. Two input sine waves drive a nonlinear circuit. As before, this results in all harmonics and intermodulation components. The bandpass filter then passes one of the intermodulation components, usually the difference frequency $f_x - f_y$. Therefore, the final output of a typical mixer is a sine wave with frequency $f_x - f_y$. In terms of spectra, the frequency mixer is a circuit that produces an output spectrum with a single line at $f_x - f_y$ when the input spectrum is a pair of lines at f_x and f_y.

A low-pass filter may be used in place of a bandpass filter, provided that $f_x - f_y$ is less than f_x or f_y. For instance, if f_x is 2 MHz and f_y is 1.8 MHz, then

$$f_x - f_y = 2 \text{ MHz} - 1.8 \text{ MHz} = 0.2 \text{ MHz}$$

In this case, the difference is lower than either input frequency, so we can use a low-pass filter if we wish. (Low-pass filters are usually easier to build than bandpass filters.)

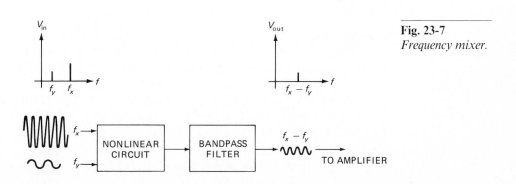

Fig. 23-7
Frequency mixer.

But in applications in which $f_x - f_y$ is between f_x and f_y, we must use a bandpass filter. As an example, if $f_x = 2$ MHz and $f_y = 0.5$ MHz, then

$$f_x - f_y = 2 \text{ MHz} - 0.5 \text{ MHz} = 1.5 \text{ MHz}$$

To pass only the difference frequency, we are forced to use a bandpass filter.

USUAL SIZE OF INPUT SIGNALS

In most applications, one of the input signals to the mixer will be *large*. This is necessary to ensure nonlinear operation; unless one of the signals is large, we cannot get intermodulation components. This large input signal is often supplied by an oscillator or signal generator.

The other input signal is usually *small*. By itself, this signal produces only small-signal operation of the mixer. Often one of the reasons this signal is small is because it is a weak signal coming from an antenna.

The normal inputs to a mixer, therefore, are:

1. A large signal adequate to produce medium- or large-signal operation of the mixer

2. A small signal that by itself can produce only small-signal operation

TRANSISTOR MIXER

Figure 23-8 is an example of a transistor mixer. One signal drives the base and the other signal drives the emitter. The resulting collector current contains harmonics and intermodulation components. With the *LC* tank tuned to the difference frequency, the output signal has a frequency of $f_x - f_y$.

DIODE MIXERS

Instead of using a transistor for the nonlinear device, we can use a diode. We drive the diode with an additive waveform containing frequencies f_x and f_y. The diode current then contains harmonics and intermodulation components. With a filter, we can remove the difference frequency.

Incidentally, *heterodyne* is another word for mix, and *beat frequency* is synonymous with difference frequency. In Fig. 23-8, we are heterodyning two input signals to get a beat frequency of $f_x - f_y$.

CONVERSION GAIN

Conversion gain refers to the power gain of the mixer. Specifically,

$$\text{Conversion gain} = \frac{P_{\text{out}}}{P_{\text{in}}} \tag{23-10}$$

where P_{out} is the output power of the difference signal and P_{in} is the input power of the smaller input signal. As an example, suppose the smaller signal in Fig. 23-8 delivers 10 μW to the emitter; if the output power of the difference signal is 40 μW, we have a

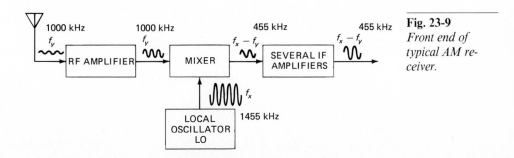

Fig. 23-8
Transistor mixer.

conversion gain of

$$\text{Conversion gain} = \frac{40\ \mu\text{W}}{10\ \mu\text{W}} = 4$$

This is equivalent to a decibel conversion gain of 6 dB.

Also useful is the conversion voltage gain, defined as

$$\text{Conversion voltage gain} = \frac{v_{\text{out}}}{v_{\text{in}}} \qquad (23\text{-}11)$$

MIXERS IN AM RECEIVERS

Figure 23-9 shows the front end of a typical AM broadcast receiver. The antenna delivers a weak signal to the radio-frequency (RF) amplifier. Although this increases the signal strength, the f_y signal driving the mixer is still small. The other mixer input comes from a local oscillator (LO); this signal is large enough to produce nonlinear operation. As a result, the output of the mixer is a signal with a frequency $f_x - f_y$.

The difference signal drives several stages called the intermediate-frequency (IF) amplifiers. These IF amplifiers provide most of the gain in the receiver. The large signal coming out of the IF amplifiers then drives other circuits.

Fig. 23-9
Front end of typical AM receiver.

Here is a numerical example. When you tune to a station with a frequency of 1000 kHz, you are doing the following:

1. Adjusting a capacitor to tune the RF amplifier to 1000 kHz

2. Adjusting another capacitor to set the LO frequency to 1455 kHz

As shown in Fig. 23-9, the output of the mixer is 455 kHz. The IF amplifiers are fixed-tuned to this frequency; therefore, the 455-kHz signal receives maximum gain from the IF strip (group of amplifiers).

If you tune to a station with a frequency of 1200 kHz, the RF stage is tuned to 1200 kHz and the LO stage to 1655 kHz. The difference frequency is still 455 kHz. In other words, by deliberate design the frequency out of the mixer always equals 455 kHz, no matter what station you tune to. We will discuss the reason for this in the next chapter. Briefly, this is the reason: It is easier to design a group of amplifier stages tuned to a constant frequency than to try to gang-tune these stages to each station frequency.

Another point: We used an IF of 455 kHz because it is one of the common IFs in commercial receivers. Other typical values are 456 kHz and 465 kHz. Many automobile radios use IFs of 175 kHz or 262 kHz. Regardless of the exact value, any modern receiver has a group of IF amplifiers fixed-tuned to the same frequency. The incoming signal frequency is converted to this IF.

23-7 SPURIOUS SIGNALS

A serious problem may arise in mixer applications. This section describes the problem and its cure.

UNWANTED SIGNALS

The output filter of a mixer passes any signal whose frequency is in the passband. It is possible for small unwanted signals to get through the filter along with the desired difference frequency. As we saw in Sec. 23-4, many difference frequencies are generated in a mixer; these are given by

$$\text{Difference frequencies} = mf_x - nf_y$$

where m and n take on all integer values. It is possible to find values of m and n that produce difference frequencies close to $f_x - f_y$. Such difference frequencies are called spurious signals because they can pass through the filter along with the desired difference frequency.

Here is an example. Suppose $f_x = 10.5$ MHz and $f_y = 2.5$ MHz. The desired difference is

$$f_x - f_y = 10.5 \text{ MHz} - 2.5 \text{ MHz} = 8 \text{ MHz}$$

There are many values of m and n that produce difference frequencies near this. For instance, if $m = 2$ and $n = 5$,

$$mf_x - nf_y = 2(10.5 \text{ MHz}) - 5(2.5 \text{ MHz}) = 8.5 \text{ MHz}$$

This frequency is close to the desired frequency; if enough of this 8.5-MHz signal reaches the output, it may interfere with the desired signal. We can also find other spurious frequencies close to the desired frequency. Because of this, the mixer output contains the desired signal plus many spurious signals.

Spurious signals occur for values of *m* and *n* greater than unity. For this reason, they are weak compared with the desired signal. Therefore, in some applications the bipolar mixer is acceptable even though it produces small spurious output signals.

THE FET MIXER

There are many applications in which spurious signals out of the bipolar mixer cause problems. The simplest way to eliminate spurious signals is to operate the mixer medium-signal; the output voltage for this type of operation is

$$v_{out} = Av_{in} + Bv_{in}^2$$

As we proved in Sec. 23-3, medium-signal operation produces only the input frequencies, their second harmonics, the sum frequency, and the difference frequency. There are no other difference frequencies; therefore, there can be no spurious signals with medium-signal operation.

Because the FET operates medium-signal all the way to cutoff and saturation, it is an ideal device to use in a mixer. With a FET mixer, spurious signals are virtually eliminated. For this reason, the FET is superior to a bipolar transistor when it comes to mixer applications.

For the FET to operate medium-signal, the large signal must not drive the FET into saturation or cutoff. If this should happen, cubic and higher-power terms will creep into the expression for output voltage. In other words, if the LO signal is large enough to cause clipping, the spurious signal content will increase. Ideally, the LO signal should cause the FET to swing over as much of the transconductance curve as possible without clipping; this results in the highest conversion gain with minimum spurious signals.

23-8 PHASE-LOCKED LOOPS

A *phase-locked loop* (PLL) is a feedback loop with a phase detector, a low-pass filter, an amplifier, and a voltage-controlled oscillator (VCO). Rather than feeding back a voltage, the PLL feeds back a frequency and compares it with the incoming frequency. This allows the VCO to lock on to the incoming frequency.

PHASE DETECTOR

Suppose we have a mixer with input frequencies of 50 kHz and 50 kHz. Then the difference frequency is zero, which represents dc. This means that a dc voltage comes out of a mixer when the input frequencies are equal.

A *phase detector* is a mixer optimized for use with equal input frequencies. It is called a phase detector (or phase comparator) because the amount of dc voltage depends on the phase angle ϕ between the input signals. As the phase angle changes, so does the dc voltage.

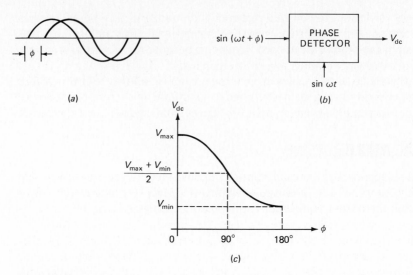

Fig. 23-10   (a) *Two sine waves with phase difference.* (b) *Phase detector driven by equal-frequency signals.* (c) *Dc output voltage decreases as phase angle increases.*

Figure 23-10*a* illustrates the phase angle between two sinusoidal signals. When these signals drive the phase detector of Fig. 23-10*b*, a dc voltage comes out. One type of phase detector has a dc output voltage that varies as shown in Fig. 23-10*c*. When the phase angle ϕ is 0, the dc voltage is maximum. As the phase angle increases from 0° to 180°, the dc voltage decreases to a minimum value. When ϕ is 90°, the dc output is the average of the maximum and minimum outputs.

For example, suppose a phase detector has a maximum output of 10 V and a minimum output of 5 V. When the two inputs are in phase, the dc output is 10 V. When the inputs are 90° out of phase, the dc output is 7.5 V. When the inputs are 180° out of phase, the dc output is 5 V. The key idea is that the dc output decreases when the phase angle increases.

VCO

Chapter 20 showed how a 555 timer could be operated as a *voltage-controlled oscillator* (VCO) by applying a dc voltage to the control input. Recall the idea. When the dc voltage in Fig. 23-11*a* increases, the frequency of the output signal decreases.

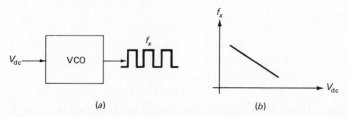

Fig. 23-11   (a) *VCO generates a rectangular wave.* (b) *Output frequency is inversely proportional to dc input voltage.*

In other words, a dc voltage controls the oscillator frequency. Typically, the frequency decreases linearly with an increase in dc voltage (see Fig. 23-11b).

Many other designs for VCOs are possible. For example, one approach uses an LC oscillator with a varactor (voltage-controlled capacitor). By varying the dc voltage applied to the varactor, we can change the capacitance and control the resonant frequency. The important thing to remember about a VCO is this: An input dc voltage controls the output frequency.

PHASE-LOCKED LOOP

Figure 23-12a is a block diagram for a phase-locked loop (PLL). An input signal with a frequency of f_x is one of the inputs to a phase detector. The other input comes from a VCO. The output of the phase detector is filtered by a low-pass filter. This removes the original frequencies, their harmonics, and the sum frequency. Only the difference frequency (a dc voltage) comes out of the low-pass filter. This dc voltage then controls the frequency of the VCO.

The feedback system of Fig. 23-12a *locks* the VCO frequency on to the input frequency. When the system is working correctly, the VCO frequency equals f_x, the same as the input signal. Therefore, the phase detector has two inputs with equal frequency. The phase angle between these inputs determines the amount of dc output. Figure 23-12b shows the phasors for the input signal and the VCO.

If the input frequency changes, the VCO frequency will track it. For instance, if the input frequency f_x increases slightly, its phasor rotates faster, and the phase angle increases, as shown in Fig. 23-12c. This means that less dc voltage will come out of the phase detector. The lower dc voltage forces the VCO frequency to increase until it equals f_x.

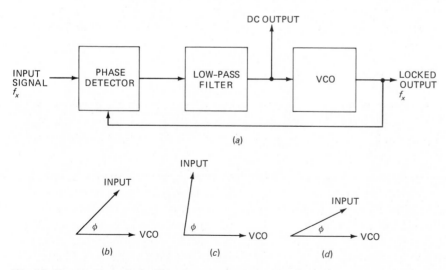

Fig. 23-12   (a) *Phase-locked loop.* (b) *Phase angle between input signal and VCO signal.* (c) *An increase in input frequency is equivalent to an increase in phase angle.* (d) *A decrease in input frequency is equivalent to a decrease in phase angle.*

On the other hand, if the input frequency decreases, its phasor slows down, and the phase angle decreases, as shown in Fig. 23-12d. Now more dc voltage will come out of the phase detector. This causes the VCO frequency to decrease until it equals the input frequency. In other words, the PLL automatically corrects the VCO frequency and phase angle.

Here is a numerical example. Suppose the VCO is locked on to an input frequency of 50 kHz. If the input frequency increases to 51 kHz, the phase detector immediately sends less voltage to the VCO and increases its frequency to 51 kHz. If the input frequency later decreases to 49 kHz, the phase detector sends more dc voltage to the VCO and decreases its frequency to 49 kHz. In either case, the feedback automatically adjusts the phase angle to produce a dc voltage that locks the VCO frequency on to the input frequency.

The *lock range* B_L is the range of frequencies the VCO can produce; it is given by

$$B_L = f_{max} - f_{min} \qquad (23-12)$$

where f_{max} = maximum VCO frequency
f_{min} = minimum VCO frequency

For example, if the VCO frequency can vary from 40 to 60 kHz, the lock range equals

$$B_L = 60 \text{ kHz} - 40 \text{ kHz} = 20 \text{ kHz}$$

Once the PLL is locked on, the input frequency f_x can vary from 40 to 60 kHz; the VCO will track the input frequency, and the locked output will equal f_x.

FREE-RUNNING MODE

Recall the astable 555 timer with no control voltage. It oscillated at a natural frequency determined by the circuit components. The same is true of the VCO in Fig. 23-12a. If the input signal is disconnected, the VCO oscillates in a free-running mode, with its frequency determined by its circuit elements.

CAPTURE AND LOCK

Assume that the PLL is free-running or unlocked. The PLL can lock on to the input frequency if it lies within the capture range, a band of frequencies centered on the free-running frequency. The formula for *capture range* is

$$B_C = f_2 - f_1 \qquad (23-13)$$

where f_2 and f_1 are the highest and lowest frequencies that the PLL can lock onto. The capture range is always less than or equal to the lock range and is related to the cutoff frequency of the low-pass filter. The lower the cutoff frequency, the smaller the capture range.

Here is an example. Suppose the PLL can initially lock into a frequency as high as 52 kHz or as low as 48 kHz. Then the capture range is 4 kHz, with a center frequency of 50 kHz. If the lock range is 20 kHz and lock has been acquired, then the input frequency can vary gradually from 40 to 60 kHz without losing lock.

LOCKED OUTPUT

One use for the locked output f_x of a PLL is to synchronize the horizontal and vertical oscillators of TV receivers to the incoming sync pulses. PLLs can also automatically tune each TV channel by locking on to the channel frequency. Still another use for PLLs is to lock on to weak signals from satellites and other distant sources; this improves the signal-to-noise ratio.

In general, the locked output is a signal with the same frequency as the input signal. Even though the input signal may drift over a rather large frequency range, the output frequency will remain locked on. This eliminates the need to tune a resonant circuit for maximum output.

FM OUTPUT

Figure 23-13*a* shows an *LC* oscillator with a variable tuning capacitor. If the capacitance is varied, the oscillation frequency changes. Figure 23-13*b* illustrates the output signal. This is an example of *frequency modulation* (FM). If the capacitance of Fig. 23-13*a* varies sinusoidally at a rate of 1 kHz, the modulation frequency is 1 kHz.

When an FM signal like Fig. 23-13*b* is the input to a PLL like Fig. 23-12*a*, the VCO will track the input frequency as it changes. As a result, a fluctuating voltage comes out of the low-pass filter. This voltage has the same frequency as the modulating signal. In other words, the varying dc output now represents a demodulated FM output. This is useful in FM receivers. If the modulating signal is music, the signal out of the FM output will be the same music.

THE 565

The NE565 from Signetics is a 14-pin IC that can be connected to external components to form a PLL. Figure 23-14 shows a simplified block diagram. Pins 2 and 3 are a differential input to the phase detector. If a single-ended input is preferred, pin 3 is grounded and the input signal is applied to pin 2. Pins 4 and 5 are usually connected together. In this way, the VCO output becomes an input to the phase detector. In those applications in which the locked output is desired, pin 4 is the output pin.

An external timing resistor is connected to pin 8, and an external timing capacitor to pin 9. These two components determine the free-running frequency of the VCO, given by

$$f = \frac{0.3}{R_T C_T} \tag{23-14}$$

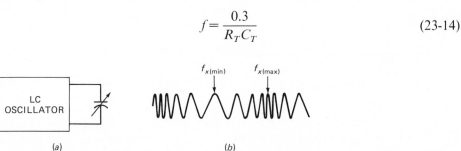

(a) (b)

Fig. 23-13   (a) *Variable capacitance on LC oscillator.* (b) *Frequency modulation.*

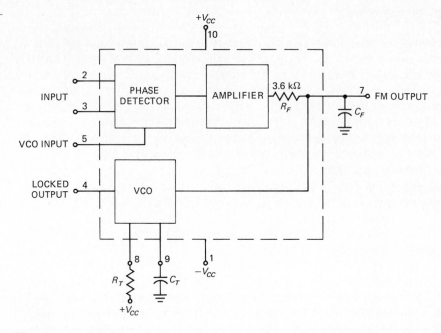

Fig. 23-14
*Block diagram of
NE565.*

R_T and C_T are selected to produce a free-running VCO frequency at the center of the input-frequency range. If you want to lock on to an input frequency between 40 and 60 kHz, you choose R_T and C_T to produce a free-running VCO frequency of 50 kHz.

Pin 7 is the FM output, used only when an FM signal is driving the phase detector. In FM receivers, a demodulated signal comes out of this pin. This signal then goes to other amplifiers and eventually comes out the loudspeaker.

Notice the filter capacitor C_F between pin 7 and ground. This capacitor and the internal 3.6-kΩ resistor form a low-pass *RC* filter that removes the original frequencies, their harmonics, and the sum frequency. The cutoff frequency of this filter is given by

$$f = \frac{1}{2\pi R_F C_F} \tag{23-15}$$

The lower the cutoff frequency of this filter, the smaller the capture range. In some applications, the filter capacitor is omitted and the capture range almost equals the lock range.

23-9 NOISE

As mentioned earlier, noise contains sinusoidal components at all frequencies. Some of these noise components will mix with the LO signal and produce difference frequencies in the output of the mixer. Because of this, the remainder of the system amplifies both the desired signal and unwanted noise out of the mixer. In general, noise is any kind of unwanted signal that is not derived from or related to the input signal.

SOME TYPES OF NOISE

Where does noise come from? Electric motors, neon signs, power lines, ignition systems, lightning, and so on, set up electric and magnetic fields. These fields can induce noise voltage in electronic circuits. To reduce noise of this type, we can shield the circuit and its connecting cables.

Power supply ripple is also classified as noise because it is independent of the desired signal. As we know, ripple can get into signal paths through biasing resistors (and also by induction). By using regulated power supplies and shielding, we can reduce ripple to an acceptably low level.

If you bump or jar a circuit, the vibrations may move capacitor plates, inductor windings, and so on. This results in noise called *microphonics*. In this respect, the transistor is far superior to the vacuum tube; because of its solid-state construction, a transistor has negligible microphonics.

THERMAL NOISE

We can eliminate or at least minimize the effects of ripple, microphonics, and external field noise. But there is little we can do about *thermal noise*. Figure 23-15a shows the idea behind this kind of noise. Inside any resistor are conduction-band electrons. Since these electrons are loosely held by the atoms, they tend to move randomly in different directions, as shown. The energy for this motion comes from the thermal energy of the surrounding air; the higher the ambient temperature, the more active the electrons.

The motion of billions of electrons is pure chaos. At some instants in time, more move up than down, producing a small negative voltage across the resistor. At other instants, more move down than up, producing a positive voltage. If it were amplified and viewed on an oscilloscope, this noise voltage would resemble Fig. 23-15b. Like any voltage, noise has an rms or heating value. As an approximation, the highest noise peaks are about four times the rms value.

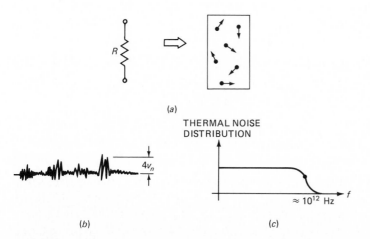

(a)

THERMAL NOISE DISTRIBUTION

$4v_n$

$\approx 10^{12}$ Hz

(b)   *(c)*

Fig. 23-15   *Thermal noise.* (a) *Random electron motion.* (b) *Appearance on oscilloscope.* (c) *Spectral distribution.*

The changing size and shape of the noise voltage implies that it includes components of many different frequencies. An advanced derivation shows that the noise spectrum is like Fig. 23-15c. As we can see, noise is uniformly distributed throughout the practical frequency range. The cutoff frequency of approximately 10^{12} Hz is far beyond the capability of electronic circuits. For this reason, most people say that noise contains sinusoidal components at all frequencies.

How much noise voltage does a resistor produce? This depends on the temperature, the bandwidth of the system, and the size of the resistance. Specifically,

$$v_n = \sqrt{4kTBR} \tag{23-16}$$

where v_n = rms noise voltage
$\quad k$ = Boltzmann's constant (1.37×10^{-23})
$\quad T$ = absolute temperature, °C + 273
$\quad B$ = noise bandwidth, Hz
$\quad R$ = resistance, Ω

In Eq. (23-16), the noise voltage increases with temperature, bandwidth, and resistance.

At room temperature (25°C), the formula reduces to

$$v_n = 1.28(10^{-10})\sqrt{BR} \tag{23-17}$$

The noise bandwidth B is approximately equal to the 3-dB bandwidth of the amplifier, mixer, or system being analyzed. As an example, a transistor radio has an overall bandwidth of about 10 kHz. This is an approximate value of B in Eq. (23-17). A television receiver, on the other hand, has a bandwidth of about 4 MHz; this is the value of B to use for noise calculations in a television receiver.

Using Eq. (23-16) or (23-17), we can estimate the amount of noise produced by any resistor in an amplifier. Those resistors near the input of the amplifier are most important because their noise will be amplified most and will dominate the final output noise. We can get a rough estimate of the input noise level by calculating the noise generated by the Thevenin resistance driving the amplifier or system.

Here is an example. Suppose the source driving an amplifier has a Thevenin resistance of 5 kΩ. If the amplifier has a bandwidth of 100 kHz and the ambient temperature is 25°C, then the noise at the input to the amplifier is

$$v_n = 1.28(10^{-10})\sqrt{10^5(5000)} = 2.86 \ \mu V$$

This gives us an estimate of the input noise level. If a received signal is smaller than this, it will be masked or covered by the noise.

PROBLEMS

Straightforward

23-1. Two sine waves drive a medium-signal amplifier. If the input frequencies are 56 kHz and 84 kHz, what frequencies does the output contain?

23-2. A large-signal amplifier has input frequencies of 1 kHz and 4 kHz. What harmonic frequencies does the output contain? What are the sum and difference frequencies in the first group?

23-3. The output spectrum of an amplifier contains 1 kHz, 2 kHz, 3 kHz, 4 kHz, and 5 kHz. If the amplifier is linear, how many input signals does it have?

23-4. On the piano keyboard of Fig. 23-6a, what is the fundamental frequency of the lowest C note? The highest C note? The highest E note?

23-5. The input spectrum of a medium-signal amplifier contains frequencies of 384 Hz, 480 Hz, and 576 Hz. What frequencies does the output spectrum contain?

23-6. The sum and difference frequencies out of a medium-signal amplifier are 2 kHz and 10 kHz. What are the input frequencies?

23-7. When tuned to channel 3, the mixer in a television receiver has a small input signal with a frequency of 63 MHz and an LO signal with a frequency of 107 MHz. What is the difference frequency out of the mixer?

23-8. AM station frequencies are from 535 to 1605 kHz. In an AM radio, the received signal is one input to a mixer, and the LO signal is the other input. The frequency of the LO signal is 455 kHz greater than that of the received signal. What are the minimum and maximum LO frequencies? If a Hartley oscillator is used to generate the LO signal, what is the ratio of maximum to minimum tuning capacitance?

23-9. A bipolar mixer has an input signal power of 3 μW and an outut signal power of 1.5 μW. What is the conversion power gain in decibels?

23-10. A FET mixer has an input signal voltage of 10 mV and an output signal voltage of 40 mW. Calculate the conversion voltage gain in decibels.

23-11. Figure 23-16 shows a 565 connected as a phase-locked loop. What is the VCO frequency when the wiper of the potentiometer is all the way up? When it is all the way down?

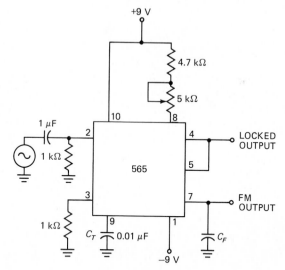

Fig. 23-16

23-12. Calculate the cutoff frequency of the low-pass filter in Fig. 23-16 when C_F is 0.1 μF. (Refer to Fig. 23-14 if necessary.)

23-13. What is the thermal noise voltage at room temperature for a bandwidth of 1 MHz and a resistance of
a 1 MΩ
b. 10 kΩ
c. 100 Ω

23-14. How much thermal noise voltage is there at room temperature for a resistance of 10 kΩ and a bandwidth of 100 kHz? For a bandwidth of 1 kHz? For 10 Hz?

23-15. A noninverting voltage amplifier has a closed-loop voltage gain of 1000 and a bandwidth of 150 kHz. If the Thevenin source resistance is 1 kΩ, how much noise voltage is there at the output of the amplifier?

Troubleshooting

23-16. In Fig. 23-8, all dc voltages are normal, but no difference frequency appears across R_L. Name at least three possible troubles.

23-17. A FET mixer has excessive spurious signals at the output. Which of these is a possible trouble:
   a. FET open
   b FET not being driven hard enough
   c. Large signal drives FET into cutoff
   d. No supply voltage

23-18. The PLL of Fig. 23-16 cannot lock on to the incoming frequency. Name at least five possible troubles.

Design

23-19. In Fig. 23-14, R_T is 10 kΩ. Select a value of C_T to get a free-running frequency of 5 kHz.

23-20. Select a value of C_F in Fig. 23-14 to get a cutoff frequency of 500 Hz.

Challenging

23-21. Two sinusoidal signals with frequencies of 1 MHz and 7.4 MHz drive a large-signal amplifier. Calculate the sum and difference frequencies produced by the first five harmonics of each input signal.

Computer

23-22. Write a program that calculates the sum and difference frequencies in Prob. 23-21.

Amplitude Modulation

Radio, television, and many other electronic systems would be impossible without modulation; it refers to a low-frequency signal (typically audio) controlling the amplitude, frequency, or phase of a high-frequency signal (usually radio frequency).

24-1 BASIC IDEA

When a low-frequency signal controls the amplitude of a high-frequency signal, we get amplitude modulation (AM). Figure 24-1a shows a simple modulator. A high-frequency signal v_x is the input to a potentiometer. Therefore, the amplitude of v_{out} depends on the position of the wiper. If we move the wiper up and down sinusoidally, v_{out} resembles the AM waveform of Fig. 24-1b. Notice that the amplitude or peak values of the high-frequency signal are varying at a low-frequency rate.

The high-frequency input signal is called the *carrier,* and the low-frequency signal is called the *modulating signal*. Hundreds of carrier cycles normally occur during one cycle of the modulating signal. For this reason, the positive peaks of the carrier are so closely spaced that they form a solid upper envelope (Fig. 24-1c). Similarly, the negative peaks form a lower envelope.

A MODULATED RF AMPLIFIER

Figure 24-2 is an example of a modulated RF stage. Here is how it works. The carrier signal v_x is the input to a CE amplifier. The circuit amplifies the carrier by a factor of A, so that the output is Av_x. The modulating signal is part of the biasing. Therefore, it produces low-frequency variations in emitter current. In turn, this produces variations in r'_e and A. For this reason, the amplified output signal looks like the AM waveform shown, where the peaks of the output vary sinusoidally with the modulating signal. Stated another way, the upper and lower envelopes have the shape of the modulating signal.

INPUT VOLTAGES

For normal operation, Fig. 24-2 should have a small carrier. We do not want the carrier to influence voltage gain; only the modulating signal should do this. There-

Fig. 24-1
*Amplitude
modulation.*

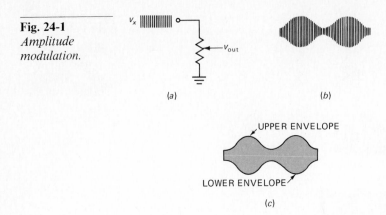

(a)

(b)

UPPER ENVELOPE

LOWER ENVELOPE

(c)

fore, the operation should be small-signal with respect to the carrier. On the other hand, the modulating signal affects the Q point. To produce noticeable changes in voltage gain, the modulating signal has to be large. For this reason, the operation is large-signal with respect to the modulating signal.

INPUT FREQUENCIES

Usually, the carrier frequency f_x is much greater than the modulating frequency f_y. In Fig. 24-2, we need f_x at least 100 times greater than f_y. Here is the reason. The capacitors should look like low impedances to the carrier and like high impedances to the modulating signal. If they do, the carrier is coupled into and out of the circuit, but the modulating signal is blocked from the output.

Fig. 24-2
*A modulated RF
stage.*

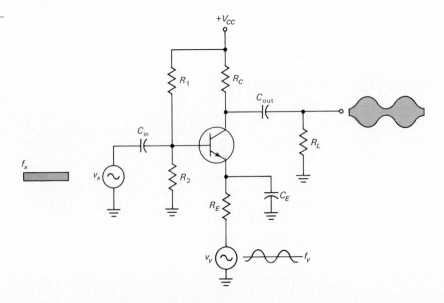

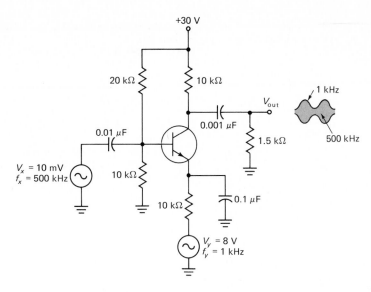

Fig. 24-3

EXAMPLE 24-1

In Fig. 24-3, the input peak of the carrier is 10 mV. The input peak of the modulating signal is 8 V. Calculate the minimum, quiescent, and maximum voltage gain.

SOLUTION

When the modulating voltage is zero, the voltage across the emitter resistor is 9.3 V. So, the emitter current is 0.93 mA, and r'_e is approximately 26.9 Ω. The quiescent voltage gain therefore equals

$$A = \frac{r_C}{r'_e} = \frac{10{,}000 \parallel 1500}{26.9 \ \Omega} = 48.5 \qquad \text{(quiescent)}$$

At the instant the modulating voltage reaches a positive peak of 8 V, only 1.3 V is across the emitter resistor. At this instant, the emitter current is 0.13 mA and r'_e is 192 Ω. The corresponding voltage gain is

$$A = \frac{10{,}000 \parallel 1500}{192 \ \Omega} = 6.79 \qquad \text{(minimum)}$$

At the negative peak of the modulating signal, the voltage across the emitter resistor is 17.3 V. The emitter current becomes 1.73 mA, r'_e decreases to 14.5 Ω, and

$$A = \frac{10{,}000 \parallel 1500}{14.5} = 90 \qquad \text{(maximum)}$$

As far as the carrier is concerned, the CE amplifier changes its voltage gain from a low of 6.79 to a high of 90. Therefore, the output signal is an AM waveform, as shown in Fig. 24-3. In following sections, the quiescent voltage gain is designated A_0, the minimum voltage gain A_{min}, and the maximum voltage gain A_{max}. In this example, $A_0 = 48.5$, $A_{min} = 6.79$, and $A_{max} = 90$.

24-2 PERCENT MODULATION

Ideally, a sinusoidal modulating signal produces a sinusoidal variation in voltage gain, expressed by

$$A = A_0(1 + m \sin \omega_y t) \tag{24-1}$$

where A = instantaneous voltage gain
  A_0 = quiescent voltage gain
  m = modulation coefficient

As the sine function varies between -1 and 1, the voltage gain varies sinusoidally between $A_0(1 - m)$ and $A_0(1 + m)$. For example, if $A_0 = 100$ and $m = 0.5$, then the voltage gain varies sinusoidally between a minimum voltage gain of

$$A_{min} = 100(1 - 0.5) = 50$$

and a maximum voltage gain of

$$A_{max} = 100(1 + 0.5) = 150$$

In Eq. (24-1), m controls the amount of modulation. The larger m is, the greater the change in voltage gain. Percent modulation is typically used to measure the amount of amplitude modulation:

$$\text{Percent modulation} = m \times 100\% \tag{24-2}$$

If m is 0.5, then the percent modulation is 50 percent. When m is 0.9, the percent modulation is 90 percent.

You can measure m as follows. Given an AM signal like Fig. 24-4a, the maximum peak-to-peak voltage is $2V_{max}$, and the minimum is $2V_{min}$. These peak-to-peak values are related to m by the following formula:

$$m = \frac{2V_{max} - 2V_{min}}{2V_{max} + 2V_{min}} \tag{24-3}$$

Fig. 24-4

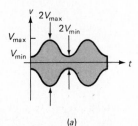

(a)

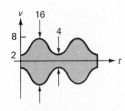

(b)

As an example, a waveform like Fig. 24-4b has a modulation coefficient of

$$m = \frac{16 - 4}{16 + 4} = 0.6$$

which is equivalent to 60 percent.

24-3 AM SPECTRUM

The output voltage of a modulated RF stage looks like Fig. 24-5a and equals

$$v_{out} = Av_x$$

If the carrier is sinusoidal, we may write

$$v_{out} = AV_x \sin \omega_x t$$

where V_x is the peak value of the input carrier. From Eq. (24-1), the output voltage becomes

$$v_{out} = A_0(1 + m \sin \omega_y t)(V_x \sin \omega_x t)$$

or

$$v_{out} = A_0V_x \sin \omega_x t + mA_0V_x \sin \omega_y t \sin \omega_x t \qquad (24\text{-}4)$$

UNMODULATED CARRIER

The first term in Eq. (24-4) represents a sinusoidal component with a peak of A_0V_x and a frequency of f_x. Figure 24-5b is the graph of the first term. We call this the unmodulated carrier because it is the output voltage when m equals zero.

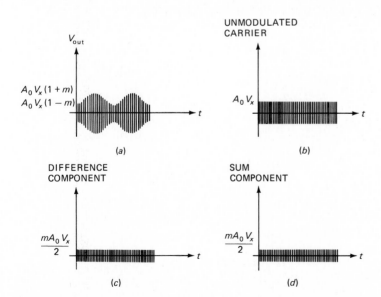

Fig. 24-5 (a) *AM signal.* (b) *Unmodulated carrier.* (c) *Difference component.* (d) *Sum component.*

CROSS PRODUCT

The second term in Eq. (24-4) is a cross product of two sine waves, similar to the cross products that occur in a mixer. As derived in the previous chapter, the product of two sine waves results in two new frequencies: a sum and a difference. Specifically, the second term of Eq. (24-4) equals

$$mA_0V_x \sin \omega_y t \sin \omega_x t = \frac{mA_0V_x}{2}\cos{(\omega_x - \omega_y)t} - \frac{mA_0V_x}{2}\cos{(\omega_x + \omega_y)t}$$

The first term on the right is a sinusoid with a peak value of $mA_0V_x/2$ and a difference frequency of $f_x - f_y$. The second term is also a sinusoid with a peak of $mA_0V_x/2$ but a sum frequency $f_x + f_y$. Figure 24-5c and d shows these sinusoidal components.

SPECTRAL COMPONENTS

In the time domain an AM signal like Fig. 24-5a is the the superposition of three sine waves (Fig. 24-5b through d). One sine wave has the same frequency as the carrier, another has the difference frequency, and the third has the sum frequency.

In terms of spectra, here is what AM means. Figure 24-6a is the input spectrum to a modulated RF stage. The first line represents the large modulating signal with frequency f_y. The second line is for the small carrier with frequency f_x. Figure 24-6b is the output spectrum. Here we see the amplified carrier between the difference and sum components. The difference component is sometimes called the lower side frequency and the sum the upper side frequency.

The circuit implication of an AM signal is this: An AM signal is equivalent to three sine-wave sources in series, as shown in Fig. 24-6c. This equivalence is not a mathematical fiction; the side frequencies really exist. In fact, with narrowband filters we can separate the side frequencies from the carrier.

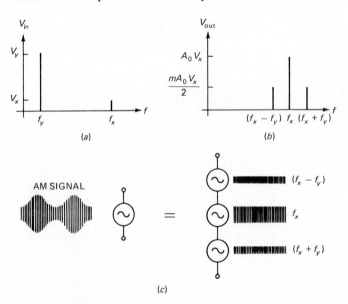

Fig. 24-6 (a) *Input spectrum.* (b) *Output spectrum.* (c) *AM is the sum of three sine waves.*

24-4 THE ENVELOPE DETECTOR

Once the AM signal is received, the carrier's work is over. Somewhere in the receiver there is a special circuit that separates the modulating signal from the carrier. We call this circuit a *demodulator* or *detector*.

DIODE DETECTOR

Figure 24-7*a* shows one type of demodulator. Basically, it is a peak detector. Ideally, the peaks of the input signal are detected, so that the output is the upper envelope. For this reason, the circuit is called an *envelope detector*.

During each carrier cycle, the diode turns on briefly and charges the capacitor to the peak voltage of the particular carrier cycle. Between peaks, the capacitor discharges through the resistor. If we make the *RC* time constant much greater than the period of the carrier, we get only a slight discharge between cycles. This removes most of the carrier signal. The output then looks like the upper envelope with a small ripple, as shown in Fig. 24-7*b*.

REQUIRED *RC* TIME CONSTANT

But here is a crucial idea. Between points *A* and *C* in Fig. 24-7*b*, each carrier peak is smaller than the preceding one. If the *RC* time constant is too long, the circuit cannot detect the next carrier peak (see Fig. 24-7*c*). The hardest part of the envelope to follow occurs at *B* in Fig. 24-7*b*. At this point, the envelope is decreasing at its fastest rate. With calculus, we can equate the rate of change of the envelope and the capacitor discharge to prove that

$$f_{y(\text{max})} = \frac{1}{2\pi RCm} \tag{24-5}$$

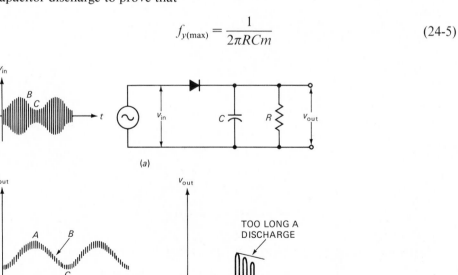

Fig. 24-7   *Envelope detector.*

where m is the modulation coefficient. With this equation, we can calculate the highest envelope frequency that the detector can follow without attenuation. If the envelope frequency is greater than $f_{y(max)}$, the detected output drops 20 dB per decade.

The stages following the envelope detector are usually audio amplifiers with an upper cutoff frequency less than the carrier frequency. For this reason, the small carrier ripple in Fig. 24-7b is reduced by these audio stages. Occasionaly, we may add a low-pass filter on the output of the detector to remove the small carrier ripple.

24-5 THE SUPERHETERODYNE RECEIVER

The superheterodyne receiver (superhet) has a constant selectivity and is easier to tune over the frequency range. Figure 24-8 is the block diagram of a superhet. Section 23-6 explained the mixing action near the front end of the superhet. Recall that the received signal mixes with the LO signal to produce the IF signal. Because the LO of Fig. 24-8 operates 455 kHz above the RF signal, the IF spectrum has a center frequency of 455 kHz.

The IF signal is amplified by several IF stages (see Fig. 24-8). The output of the last IF stage is envelope-detected to retrieve the modulating signal. This signal then drives the audio amplifiers and the loudspeaker. The output of the detector also feeds an AGC voltage back to the IF amplifiers. (AGC was discussed in Secs. 17-1 and 17-3.)

As mentioned earlier, it's impractical to fabricate inductors and large capacitors on a chip. This is why radio and television ICs contain only transistors and resistors. For these chips to work properly, you have to connect external L's and C's to tune the RF and IF stages.

The LM1820 is an example of an integrated AM receiver. It contains an RF amplifier, an oscillator, a mixer, IF amplifiers, and an AGC detector. If we connect external tank circuits, an audio amplifier like the LM386, and a loudspeaker, we get a complete AM radio.

Fig. 24-8
Superheterodyne receiver.

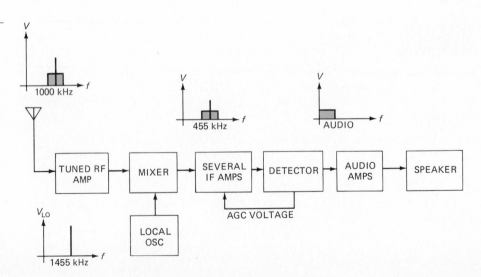

In summary, the superhet is the standard approach in most communication systems. The key idea is to use a mixer to shift the received spectrum down to an intermediate frequency. At this constant and lower frequency, the IF stages can efficiently amplify the signal prior to envelope detection and audio amplification.

PROBLEMS

Straightforward

24-1. In the modulated RF stage of Fig. 24-3, R_E is changed to 20 kΩ and V_y to 4 V. Calculate A_0, A_{min}, and A_{max}.

24-2. If $A_{min} = 40$ and $A_{max} = 60$, what does the percent modulation equal?

24-3. An AM signal has $V_{max} = 1.8$ V and $V_{min} = 1.2$ V. What does m equal? The percent modulation?

24-4. The modulating frequency is 250 Hz and the carrier frequency is 400 kHz. What are the side frequencies?

24-5. A modulated RF stage has $m = 0.3$ and $A_0 = 100$. If the input carrier has a peak voltage of 10 mV, what is the peak value of the output carrier? Of each side frequency?

24-6. An AM signal drives an envelope detector with an R of 10 kΩ and C of 1000 pF. If the percent modulation is 30 percent, what is the highest modulating frequency detectable without attenuation?

24-7. A superhet has an intermediate frequency of 455 kHz. What is the LO frequency when the received frequency is 540 kHz? When it is 1600 kHz?

Troubleshooting

24-8. In Fig. 24-3, the output signal is unmodulated and has a peak value of only 1.3 mV. Which of the following is a possible trouble:
  a. Input coupling capacitor open
  b. Output coupling capacitor open
  c. Emitter bypass capacitor open

Design

24-9. Select a value of R_E in Fig. 24-3 to produce a quiescent voltage gain of 40.

Challenging

24-10. An RF carrier is modulated by a sinusoidal signal. Prove that the maximum power in the side frequencies is one-half the power in the carrier.

Computer

24-11. Write a program that calculates the cutoff frequency of an envelope detector.

Appendix 1
Mathematical Derivations

Proof of Eq. (7-5)

The starting point for this derivation is the rectangular *pn* junction equation derived by Shockley:

$$I = I_s(\epsilon^{Vq/kT} - 1) \qquad\qquad \text{(A-1)}$$

where I = total diode current
I_s = reverse saturation current
V = total voltage across the depletion layer
q = charge on an electron
k = Boltzmann's constant
T = absolute temperature, $^\circ$C + 273

Equation (A-1) does *not* include the bulk resistance on either side of the junction. For this reason, the equation applies to the total diode only when the voltage across the bulk resistance is negligible.

At room temperature, q/kT equals approximately 40, and Eq. (A-1) becomes

$$I = I_s(\epsilon^{40V} - 1) \qquad\qquad \text{(A-2)}$$

(Some books use $39V$, but this is a small difference.) To get r_e', we differentiate I with respect to V.

$$\frac{dI}{dV} = 40 I_s \epsilon^{40V}$$

Using Eq. (A-2), we can rewrite this as

$$\frac{dI}{dV} = 40\,(I + I_s)$$

Taking the reciprocal gives r'_e:

$$r'_e = \frac{dV}{dI} = \frac{1}{40(I + I_s)} = \frac{25 \text{ mV}}{I + I_s} \tag{A-3}$$

Equation (A-3) includes the effect of reverse saturation current. In a practical linear amplifier, I is much greater than I_s (otherwise, the bias is unstable). For this reason, the practical value of r'_e is

$$r'_e = \frac{25 \text{ mV}}{I}$$

Since we are talking about the emitter depletion layer, we add the subscript E to get

$$r'_e = \frac{25 \text{ mV}}{I_E}$$

Proof of Eq. (10-42)

In Fig. 10-13a, the instantaneous power dissipation during the *on* time of the transistor is

$$p = V_{CE}I_C$$
$$= V_{CEQ}(1 - \sin\theta)I_{C(\text{sat})}\sin\theta$$

This is for the half-cycle when the transistor is conducting; during the *off* half-cycle, $p = 0$ ideally.

The average power dissipation equals

$$p_{\text{av}} = \frac{\text{area}}{\text{period}} = \frac{1}{2\pi}\int_0^\pi V_{CEQ}(1 - \sin\theta)I_{C(\text{sat})}\sin\theta\, d\theta$$

After evaluating the definite integral over the half-cycle limits of 0 to π, and dividing by the period 2π, we have the average power over the *entire cycle* for one transistor:

$$p_{\text{av}} = \frac{1}{2\pi}V_{CEQ}I_{C(\text{sat})}\left[-\cos\theta - \frac{\theta}{2}\right]_0^\pi$$

$$= 0.068\,V_{CEQ}I_{C(\text{sat})} \tag{A-4}$$

This is power dissipation in each transistor over the entire cycle, assuming 100 percent swing over the ac load line.

If the signal does not swing over the entire load line, the instantaneous power equals

$$p = V_{CE}I_C = V_{CEQ}(1 - k\sin\theta)I_{C(\text{sat})}k\sin\theta$$

where k is a constant between 0 and 1; k represents the fraction of the load line being used. After integrating

$$p_{\text{av}} = \frac{1}{2\pi}\int_0^\pi p\, d\theta$$

you get

$$p_{\text{av}} = \frac{V_{CEQ}I_{C(\text{sat})}}{2\pi}\left(2k - \frac{\pi k^2}{2}\right) \tag{A-5}$$

Since p_{av} is a function of k, we can differentiate and set dp_{av}/dk equal to zero to find the maximizing value of k:

$$\frac{dp_{av}}{dk} = \frac{V_{CEQ}I_{C(\text{sat})}}{2\pi}\,(2 - k\pi) = 0$$

Solving for k gives

$$k = \frac{2}{\pi} = 0.636$$

With this value of k, Eq. (A-5) reduces to

$$p_{av} = 0.107V_{CEQ}I_{C(\text{sat})} \cong 0.1V_{CEQ}I_{C(\text{sat})}$$

Since $I_{C(\text{sat})} = V_{CEQ}/R_L$ and $V_{CEQ} = PP/2$, the foregoing equation can be written as

$$P_{D(\text{max})} = \frac{PP^2}{40R_L}$$

Proof of Eqs. (12-14) and (12-15)

Start with the transconductance equation

$$I_D = I_{DSS}\left[1 - \frac{V_{GS}}{V_{GS(\text{off})}}\right]^2 \tag{A-6}$$

The derivative of this is

$$\frac{dI_D}{dV_{GS}} = g_m = 2I_{DSS}\left[1 - \frac{V_{GS}}{V_{GS(\text{off})}}\right]\left[-\frac{1}{V_{GS(\text{off})}}\right]$$

or

$$g_m = -\frac{2I_{DSS}}{V_{GS(\text{off})}}\left[1 - \frac{V_{GS}}{V_{GS(\text{off})}}\right] \tag{A-7}$$

When $V_{GS} = 0$, we get

$$g_{m0} = -\frac{2I_{DSS}}{V_{GS(\text{off})}} \tag{A-8}$$

or by rearranging,

$$V_{GS(\text{off})} = -\frac{2I_{DSS}}{g_{m0}}$$

This proves Eq. (12-15). By substituting the left member of Eq. (A-8) into Eq. (A-7),

$$g_m = g_{m0}\left[1 - \frac{V_{GS}}{V_{GS(\text{off})}}\right]$$

This is the proof of Eq. (12-14).

Proof of Eq. (14-35)

Current gain is measured under special conditions of r_c equals zero and R_S equals infinity (Fig. 14-20c), in other words, an ac short across the output and a current source driving the input. For these conditions, the Thevenin resistance in the base lag network becomes

$$R = \beta r'_e$$

and the capacitance is

$$C = C'_e + C'_c$$

The break frequency of the base lag network under these special conditions is called the *beta-cutoff frequency* f_β. It equals

$$f_\beta = \frac{1}{2\pi\beta r'_e(C'_e + C'_c)} \tag{A-9}$$

By a derivation which we will discuss in a moment, f_β and f_T are related as follows:

$$f_\beta = \frac{f_T}{\beta} \tag{A-10}$$

With this relation, Eq. (A-9) can be rearranged to get

$$C'_e + C'_c = \frac{1}{2\pi f_T r'_e}$$

or, since C'_e is much greater than C'_c,

$$C'_e \cong \frac{1}{2\pi f_T r'_e}$$

A correct derivation for Eq. (A-10) is not easy. You have to include the delay time of carriers diffusing through the base. If interested, see Pettit, J. M., and M. M. McWhorter: *Electronic Amplifier Circuits*, McGraw-Hill Book Company, New York, 1961, pp. 26–35.

Proof of Eq. (14-52a)

The overall voltage gain of two identical cascaded dc-coupled stages is

$$A_1 A_2 = \frac{A_{1(\text{mid})}}{\sqrt{1 + (f/f_c)^2}} \frac{A_{2(\text{mid})}}{\sqrt{1 + (f/f_c)^2}}$$

The 3-dB cutoff frequency occurs when the product of the denominators is

$$\sqrt{1 + (f/f_c)^2} \quad \sqrt{1 + (f/f_c)^2} = \sqrt{2}$$

which simplifies to

$$1 + (f/f_c)^2 = \sqrt{2}$$

Solving for the overall cutoff frequency gives

$$f = f_c \sqrt{2^{1/2} - 1}$$

A similar derivation applies to n stages with a final result of

$$f = f_c \sqrt{2^{1/n} - 1}$$

Likewise, a similar derivation can be applied to the lower cutoff frequency and the overall bandwidth to get

$$B_n = B \sqrt{2^{1/n} - 1}$$

Proof of Eq. (15-22)

The equation of a sinusoidal voltage is

$$v = V_P \sin \omega t$$

The derivative with respect to time is

$$\frac{dv}{dt} = \omega V_P \cos \omega t$$

The maximum rate of change occurs for $t = 0$. Furthermore, as the frequency increases, we reach the point where the maximum rate of change just equals the slew rate. At this critical point,

$$S_R = \left(\frac{dv}{dt}\right)_{\text{max}} = \omega_{\text{max}} V_P = 2\pi f_{\text{max}} V_P$$

Solving for f_{max} in terms of S_R, we get

$$f_{\text{max}} = \frac{S_R}{2\pi V_P}$$

Proof of Eq. (16-8)

Here is the derivation for closed-loop output impedance. Begin with

$$A_{CL} = \frac{A}{1 + AB}$$

Substitute

$$A = A_u \frac{R_L}{r_{\text{out}} + R_L}$$

where A is the loaded gain (R_L connected), A_u is the unloaded gain (R_L disconnected). After substitution for A, the closed-loop gain simplifies to

$$A_{CL} = \frac{A_u}{1 + A_u B + r_{\text{out}}/R_L}$$

When

$$1 + A_u B = \frac{r_{\text{out}}}{R_L}$$

A_{CL} will drop in half, implying the load resistance matches the Thevenin output resistance of the feedback amplifier. Solving for R_L gives

$$R_L = \frac{r_{out}}{1 + A_u B}$$

This is the value of load resistance that forces the closed-loop voltage gain to drop in half, equivalent to saying it equals the closed-loop output impedance:

$$r_{out(CL)} = \frac{r_{out}}{1 + A_u B}$$

In any practical feedback amplifier, r_{out} is much smaller than R_L, so that A approximately equals A_u. This is why you almost see the following expression for output impedance:

$$r_{out(CL)} = \frac{r_{out}}{1 + AB}$$

where $r_{out(CL)}$ = closed-loop output impedance

r_{out} = open-loop output impedance

AB = open-loop gain

Proof of Eq. (16-20)

Assume an inverting amplifier has a feedback resistor R_F between its input and output terminals. Designate the input voltage as V_1 and the output voltage as V_2. Then the current through the feedback resistor is

$$I = \frac{V_1 + V_2}{R_F} \tag{A-11}$$

which can be written as

$$I = \frac{V_1 + A V_1}{R_F} = \frac{V_1 (1 + A)}{R_F}$$

or

$$\frac{V_1}{I} = \frac{R_F}{1 + A}$$

Since this is the ratio of input voltage to input current, we can write

$$r_{in(CL)} = \frac{R_F}{1 + A}$$

This is part of Miller's theorem for a feedback resistor. Notice that the equivalent Miller input resistance equals R_F divided by $1 + A$, which means a very small input resistance.

To prove the second half of the Miller theorem for a feedback resistor, proceed as follows. Since $A = V_2/V_1$, Eq. (A-11) becomes

$$I = \frac{V_2/A + V_2}{R_F} = \frac{V_2(1/A + 1)}{R_F}$$

or

$$\frac{V_2}{I} = \frac{A R_F}{1 + A}$$

This is the output Miller equivalent resistance.

Proof of Eq. (16-22)

Because of the virtual ground in Fig. 16-17, essentially all of the input current flows through R_1. Summing voltages around the circuit gives

$$-v_{error} + i_{in}R_1 - (i_{out} - i_{in})R_2 = 0 \qquad \text{(A-12)}$$

With the following substitutions,

$$v_{error} = \frac{v_{out}}{A}$$

and

$$v_{out} = i_{out}R_L + (i_{out} - i_{in})R_2$$

Eq. (A-12) can be rearranged as

$$\frac{i_{out}}{i_{in}} = \frac{AR_1 + (1+A)R_2}{R_L + (1+A)R_2}$$

Since A is usually much greater than 1, this reduces to

$$\frac{i_{out}}{i_{in}} = \frac{A(R_1 + R_2)}{R_L + AR_2}$$

Furthermore, AR_2 is usually much greater than R_L, and the foregoing simplifies to

$$\frac{i_{out}}{i_{in}} = \frac{R_1}{R_2} + 1$$

Proof of Eq. (16-24)

If the open-loop voltage gain rolls off at a rate of 20 dB per decade, then we can write

$$A = \frac{A_{mid}}{1 + jf/f_2} \qquad \text{(A-13)}$$

where A = open-loop voltage gain at any frequency
$\quad A_{mid}$ = open-loop voltage gain in the midband
$\qquad f$ = frequency of input signal
$\qquad f_2$ = open-loop cutoff frequency
By substituting Eq. (A-13) into

$$A_{CL} = \frac{A}{1 + AB}$$

and simplifying, we can get

$$A_{CL} = \frac{A_{mid}}{1 + A_{mid}B + jf/f_2}$$

The closed-loop cutoff frequency occurs when the real part of the denominator equals the imaginary part:

$$1 + A_{mid}B = \frac{f}{f_2}$$

or
$$f = (1 + A_{mid}B)f_2$$

Since this is the closed-loop cutoff frequency, we can write it as

$$f_{2(CL)} = (1 + A_{mid}B)f_2$$

where $f_{2(CL)}$ = closed-loop upper cutoff frequency
A_{mid} = open-loop gain in midband
B = feedback fraction
f_2 = open-loop upper cutoff frequency

Proof of Eqs. (17-9) and (17-10)

In Fig. 17-9, we can sum voltages to get

$$-v_{in} + i_{in}R_S + i_{in}R_F + v_{out} = 0$$

or
$$i_{in} = \frac{v_{in} - v_{out}}{R_S + R_F} \tag{A-14}$$

Summing voltages around a second loop gives

$$-v_{in} + i_{in}R_S + v_{error} = 0$$

or
$$-v_{in} + i_{in}R_S - \frac{v_{out}}{A} = 0 \tag{A-15}$$

By substituting Eq. (A-14) into (A-15) and rearranging,

$$A_{CL} = \frac{-R_F}{R_S} \frac{AB}{1 + AB} \tag{A-16}$$

The factor A is frequency-sensitive because it equals

$$A = \frac{A_{mid}}{1 + jf/f_2}$$

Substituting this into Eq. (A-16) and rearranging gives

$$A_{CL} = \frac{-R_F}{R_S} \frac{A_{mid}B}{1 + A_{mid}B + jf/f_2}$$

The denominator $1 + A_{mid}B + jf/f_2$ is the key. Equating the real and imaginary parts gives the cutoff frequency:

$$f_{2(CL)} = (1 + A_{mid}B)f_2 \tag{A-17}$$

This says the closed-loop cutoff frequency is increased by a factor of $1 + A_{mid}B$.

The closed-loop gain-bandwidth product is $A_{CL}f_{2(CL)}$. With Eq. (A-17), we can write

$$A_{CL}f_{2(CL)} = A_{CL}(1 + A_{mid}B)f_2$$

With Eq. (A-16), the foregoing can be arranged as

$$A_{CL}f_{2(CL)} = \frac{-R_F}{R_F + R_S} f_{unity} \tag{A-18}$$

Proof of Eq. (18-21)

The change in capacitor voltage is given by

$$\Delta V = \frac{IT}{C} \tag{A-19}$$

In the positive half-cycle of input voltage (Fig. 18-23*a*), the capacitor charging current is ideally

$$I = \frac{V_P}{R}$$

Since T is the rundown time of the output ramp, it represents half of the output period. If f is the frequency of the input square wave, then $T = 1/2f$. Substituting for I and T in Eq. (A-19) gives

$$\Delta V = \frac{V_P}{2fRC}$$

The input voltage has a peak-to-peak value of $2V_P$, while the output voltage has a peak-to-peak value of ΔV. Therefore, the equation can be written as

$$v_{\text{out(p-p)}} = \frac{v_{\text{in(p-p)}}}{4fRC}$$

Proof of Eq. (18-22)

The UTP has a value of $+BV_{\text{sat}}$ and the LTP a value of $-BV_{\text{sat}}$. Start with the basic switching equation that applies to any RC circuit:

$$v = v_i + (v_f - v_i)(1 - e^{-t/RC}) \tag{A-20}$$

where v = instantaneous capacitor voltage
$\quad v_i$ = initial capacitor voltage
$\quad v_f$ = target capacitor voltage
$\quad t$ = charging time
$\quad RC$ = time constant

In Fig. 18-25*b*, the capacitor charge starts with an initial value of $-BV_{\text{sat}}$ and ends with a value of $+BV_{\text{sat}}$. The target voltage for the capacitor voltage is $+V_{\text{sat}}$ and the capacitor charging time is half a period, $T/2$. Substitute into Eq. (A-20) to get

$$BV_{\text{sat}} = -BV_{\text{sat}} + (V_{\text{sat}} + BV_{\text{sat}})(1 - e^{-T/2RC})$$

This simplifies to

$$\frac{2B}{1 + B} = 1 - e^{-T/2RC}$$

By rearranging and taking the antilog, the foregoing becomes

$$T = 2RC \ln \frac{1 + B}{1 - B}$$

Proof of Eqs. (19-6) and (19-7)

In Fig. 19-4a, the load current I flows through R_4. Therefore, the voltage at the input to the voltage divider is

$$V = IR_4 + V_{out}$$

The voltage reaching the base of Q_3 is

$$V_B = K(IR_4 + V_{out})$$

Since the emitter voltage is

$$V_E = V_{out}$$

the base-emitter voltage is

$$V_{BE} = K(IR_4 + V_{out}) - V_{out}$$

or

$$V_{BE} = KIR_4 + (K - 1) V_{out}$$

Solving for I gives

$$I = \frac{V_{BE} + (1 - K)V_{out}}{KR_4} \qquad \text{(A-21)}$$

When the load terminals are shorted, the load current equals I_{SL} and Eq. (A-21) reduces to

$$I_{SL} = \frac{V_{BE}}{KR_4}$$

where I_{SL} = shorted-load current
$V_{BE} = Q_3$ base-emitter voltage, 0.6 to 0.7 V
K = feedback fraction of R_5-R_6 voltage divider
R_4 = current-sensing resistance

When the regulator is operating normally at its maximum load current, Eq. (A-21) gives

$$I_{max} = I_{SL} + \frac{(1 - K) V_{out}}{KR_4}$$

where I_{max} = maximum load current with regulated output
V_{out} = regulated output voltage

Proof of Eq. (20-20)

Start with Eq. (A-20), the switching equation for any RC circuit. In Fig. 20-30b, the initial capacitor voltage is zero, the target capacitor voltage is $+V_{CC}$, and the final capacitor voltage is $+2V_{CC}/3$. Substitute into Eq. (A-20) to get

$$\frac{2V_{CC}}{3} = V_{CC}(1 - e^{-W/RC})$$

This simplifies to

$$e^{-W/RC} = \frac{1}{3}$$

Solving for W gives

$$W = 1.0986RC \cong 1.1RC$$

Proof of Eqs. (20-22) and (20-23)

In Fig. 20-32b, the capacitor upward charge takes time W. The capacitor voltage starts at $+V_{CC}/3$ and ends at $+2V_{CC}/3$ with a target voltage of $+V_{CC}$. Substitute into Eq. (A-20) to get

$$\frac{2V_{CC}}{3} = \frac{V_{CC}}{3} + \left(V_{CC} - \frac{V_{CC}}{3} \right) (1 - e^{-W/RC})$$

This simplifies to

$$e^{-W/RC} = 0.5$$

or $\qquad\qquad W = 0.693RC = 0.693(R_A + R_B)C$

The discharge equation is similar, except that R_B is used instead of $R_A + R_B$. In Fig. 20-32b, the discharge time is $T - W$, which leads to

$$T - W = 0.693R_B C$$

Therefore, the period is

$$T = 0.693(R_A + R_B)C + 0.693R_B C$$

and the duty cycle is

$$D = \frac{0.693(R_A + R_B)C}{0.693(R_A + R_B)C + 0.693R_B C} \times 100\%$$

or $\qquad\qquad\qquad D = \frac{R_A + R_B}{R_A + 2R_B} \times 100\%$

To get the frequency, take the reciprocal of the period T:

$$f = \frac{1}{T} = \frac{1}{0.693(R_A + R_B)C + 0.693R_B C}$$

or $\qquad\qquad\qquad f = \frac{1.44}{(R_A + 2R_B)C}$

Appendix 2
Standard Resistance Values ± 5%

Ω	Ω	Ω	kΩ	kΩ	kΩ	MΩ	MΩ
1.0	10	100	1.0	10	100	1.0	10
1.1	11	110	1.1	11	110	1.1	
1.2	12	120	1.2	12	120	1.2	
1.3	13	130	1.3	13	130	1.3	
1.5	15	150	1.5	15	150	1.5	
1.6	16	160	1.6	16	160	1.6	
1.8	18	180	1.8	18	180	1.8	
2.0	20	200	2.0	20	220	2.0	
2.2	22	220	2.2	22	220	2.2	
2.4	24	240	2.4	24	240	2.4	
2.7	27	270	2.7	27	270	2.7	
3.0	30	300	3.0	30	300	3.0	
3.3	33	330	3.3	33	330	3.3	
3.6	36	360	3.6	36	360	3.6	
3.9	39	390	3.9	39	390	3.9	
4.3	43	430	4.3	43	430	4.3	
4.7	47	470	4.7	47	470	4.7	
5.1	51	510	5.1	51	510	5.1	
5.6	56	560	5.6	56	560	5.6	
6.2	62	620	6.2	62	620	6.2	
6.8	68	680	6.8	68	680	6.8	
7.5	75	750	7.5	75	750	7.5	
8.2	82	820	8.2	82	820	8.2	
9.1	91	910	9.1	91	910	9.1	

Appendix 3
Capacitance Values

pF	μF	μF	μF	μF
10	0.001	0.1	10	1000
12	0.0012			
13	0.0013			
15	0.0015	0.15	15	
18	0.0018			
20	0.002			
22	0.0022	0.22	22	2200
24				
27				
30				
33	0.0033	0.33	33	3300
36				
43				
47	0.0047	0.47	47	4700
51				
56				
62				
68	0.0068	0.68	68	6800
75				
82				
100	0.01	1.0	100	10,000
110				
120				
130				
150	0.015	1.5		

Capacitance Values *(Continued)*

pF	μF	μF	μF	μF
180				
200				
220	0.022	2.2	220	22,000
240				
270				
300				
330	0.033	3.3	330	
360				
390				
430				
470	0.047	4.7	470	47,000
510				
560				
620				
680	0.068	6.8		
750				
820				82,000
910				

Answers to Selected Odd-Numbered Problems

Chapter 1

1-1. 0.1 V   **1-3.** 22.5 A   **1-5.** 0.218 A, 10.9 V   **1-7.** 200 Ω
1-9. R_S is less than or equal to 0.2 Ω **1-11.** 4.81 mA; no   **1-13.** Disconnect load resistor and use voltmeter to measure open-circuit voltage.
1-15. 6 mA, 4 mA, 3 mA, 2.4 mA, 2 mA, 1.71 mA, and 1.5 mA
1-17. Disconnect load resistor, measure voltage between A and B; replace voltage source by short and use ohmmeter to measure the resistance between A and B **1-19.** Draw current source of 6 mA in parallel with a resistance of 2 kΩ; 5.35 mA   **1-21.** R_1 open, R_2 shorted (typically solder bridge)
1-23. Load open   **1-25.** V_S = 30 V, R_1 = 4 kΩ, and R_2 = 4 kΩ
1-27. Measure open-circuit voltage across battery; this is V_{TH}. Next, measure short-loaded current. Third, calculate V_{TH}/I_{SL} to get R_{TH}.   **1-29.**
24.7 μA, 21.1 μA, 18.5 μA, 16.4 μA, 14.8 μA, 13.5 μA, and 12.3 μA
1-31. TODAY IS SUNNY on line 1, PROGRAMMING IS EASY on line 2, MAYBE I CAN LEARN THIS STUFF on line 3.   **1-33.** 10 PRINT "FOOTHILL COLLEGE"; 20 PRINT "12345 EL MONTE ROAD"; 30 PRINT "LOS ALTOS HILLS, CA 94022"; 40 END

Chapter 2

2-1. 64 nA, 2.05 μA   **2-3.** 2.6(10^6)   **2-5.** Draw straight line through origin and (4 mA, 8 V)   **2-7.** 40 mW, 85 mW   **2-9.** 40 mA, 8 V, 36 mA, 0.8 V, 32 mW, load line shifts downward by factor of 2
2-11. 8.3 mA   **2-13.** 44.4 μA, no   **2-15.** 1N914, 1N4001
2-17. 70 Ω, 15.6 Ω, 8.2 Ω **2-19.** Open   **2-21.** Shorted diode, open 100 kΩ **2-23.** No supply voltage, R_1 open, or R_2 shorted.
2-25. 23.4 kΩ **2-27.** Nearest standard sizes: R_1 = 2.4 kΩ, R_2 = 620 Ω
2-29. I_S = 0.528 μA, I_{SL} = 4.47 μA   **2-33.** Prints out 11500

Chapter 3

3-1. 103.5 V, 126.5 V   **3-3.** 1N4002, 1N1183   **3-5.** 28.3 V, 18 V, 60 mA   **3-7.** 60 mA, 56.6 V, 30 mA   **3-9.** 135 mA, 67.5 mA, 84.9 V
3-11. Yes   **3-13.** Approximately 1.9 V   **3-15.** 12.5 V, 62.5 mA, 1.04 V, 31.3 mA, 25 V   **3-17.** 10.7 A   **3-19.** 1953 V, 1302 V
3-21. 2546 V, 25.4 mA, 212 V   **3-23.** 0.1 V; protect meter against excessive current   **3-25.** Open filter capacitor, filter capacitor too small, load resistor too small   **3-27.** Defective transformer, open diodes, shorted filter capacitor, shorted load resistor, no ground on left diodes.
3-29. 10.6 V, 184 μF (nearest standard is 220 μF), 11 mA, 21.2 V
3-31. 416 μF (nearest standard is 470 μF)   **3-33.** Average the values of a sine wave over small interval like 10°   **3-35.** 64.3 A

Chapter 4

4.1. 300 mW   **4-3.** 0.05 V   **4-5.** 751 Ω **4-7.** 132 mA, 0.384 V
4-9. 0.05 V   **4-11.** 26.7 mV   **4-13.** 3 kΩ **4-15.** 12 V, 51 mA, 29.2 mV   **4-17.** 160 Ω **4-21.** 23 mA   **4-23.** 16.4 mA, 18.1 mA
4-25. 0.38 V   **4-27.** 6.5 MHz, 11.3 MHz   **4-29.** 125 MHz
4-31. Zener diode   **4-33.** b, d, f   **4-35.** V_Z = 6.8 V, R_S = 204 Ω (nearest standard is 200 Ω)   **4-37.** 525 Ω (next higher standard is 560 Ω)   **4-39.** 0.438 V   **4-41.** Prints maximum zener current.

Chapter 5

5-1. 20,000 electrons, 980,000 electrons   **5-3.** 0.995, 200   **5-5.** 10 mV, 0.1 V, 0.5 V   **5-7.** 0.125 mA   **5-11.** 0.1 μA, 9.999 V   **5-13.** 120 mW
5-15. 2 mA, 20 V   **5-17.** 0.198 mA, 0, yes   **5-19.** 0.915 mA, 10.6 mA, 0
5-21. 1 mA, 10 V, 8.2 V   **5-23.** Ideally zero   **5-25.** a. 5 mA b. 10 mA, 0
c. 10 V   **5-27.** Troubled d   **5-29.** a. On b. Off c. Off d. Off
5-31. Change R_E to 120 Ω **5-33.** Use an R_E of 470 Ω **5-35.** a. 3.6 V
b. 0.24 mA c. 2.4 μA

Chapter 6

6-1. 16.9 V   **6-3.** 220   **6-5.** 4.93 V, 6.9 V, 5.77 V   **6-7.** 0.73 V, 0.72 V, 0.83 V   **6-9.** First: 2.86 V, 2.16 V, 9 mA, 11 V, 99 mW; second: 2.83 V, 2.13 V, 17.8 mA, 10.9 V, 156 mW; third: 3.05 V, 2.35 V, 15.7 mA, 10.3 V, 125 mW   **6-11.** 6.11 mW   **6-13.** −2.18 V, −1.48 V, −9.11 V
6-15. 20 mA   **6-17.** a. high b. high c. low d. high e. low f. high g. low

h. high i. high j. low   **6-19.** Shorted 1.8 kΩ, open transistor, open 470 Ω, shorted 510 Ω, open 2.4 kΩ, and no supply voltage   **6-21.** One possible design is $R_1 = 7.5$ kΩ, $R_2 = 1.2$ kΩ, $R_E = 1$ kΩ, $R_C = 3.9$ kΩ   **6-23.** Use $V_Z = 7.5$ V and $R_E = 180$ Ω   **6-25.** 9 V, 8.3 V   **6-27.** a. Goes to line 6000 b. line 7000

Chapter 7

7-1. 3.98 μF (next higher value is 4.7 μF)   **7-3.** 159 μF (220 μF)
7-9. 2.5 kΩ, 500 Ω, 250 Ω, 50 Ω, 25 Ω, 5 Ω, 2.5 Ω   **7-11.** 19.2 Ω
7-13. –261 mV   **7-15.** 208, 322   **7-17.** –226 mV   **7-19.** –162 mV
7-21. 67.6 mV   **7-23.** 238 mV   **7-25.** 134 mV   **7-27.** a. All dc voltage normal, no ac input voltage to the first base. b. Dc collector voltage of first stage is low. c. Dc voltages normal, but no ac signal at final output. d. Dc voltages normal, but voltage gain of first stage is low. e. Dc voltages of second stage low because transistor is saturated. f. Dc collector voltage low in first stage, no ac output voltage. g. Dc collector voltage high in second stage, no ac output signal.   **7-29.** One design is $R_1 = 2.7$ kΩ, $R_2 = 470$ Ω, $R_E = 130$ Ω, $R_C = 620$ Ω, $r_E = 20$ Ω   **7-33.** The program calculates and prints out the Thevenin voltage gain of a grounded-emitter stage.

Chapter 8

8-1. 7.5 V, 6.8 V, 15 V, 0.567 mA, 0.567 mA, 3.54 μA   **8-3.** 28.7 mV
8-5. First: 2.14 V, 1.44 V, 10.2 V, 1.44 mA, 1.44 mA, 18 μA; second: 7.5 V, 6.8 V, 15 V, 0.829 mA, 0.829 mA, 10.4 μA   **8-7.** –268 mV
8-9. 13 mA, 108 μA, 0.9 μA   **8-11.** 9.32 mV   **8-13.** –251 mV
8-15. 5.5 V, 8.62 mA, 0.15 Ω   **8-17.** 2.9 V, 2.2 V, 8.95 V, 1.83 mA, 1.83 mA, 24.4 μA   **8-19.** The common lead grounds the emitter
8-21. Approximately 6.4 mV   **8-23.** Either trouble c or d.   **8-25.** One solution: $R_1 = 220$ kΩ, $R_2 = 220$ kΩ, and $R_E = 1.8$ kΩ   **8-27.** 4.48 V, 3.78 V, 11.2 V, 3.78 mA, 3.78 mA, (3.78 mA)/h_{FE} **8-29.** 1.79 V, approximately zero

Chapter 9

9-1. 215, –144, 2.98 kΩ, 49.5 kΩ   **9-3.** 342 Ω, 165 Ω, 62 Ω, 30 Ω, 14.8 Ω, 5.87 Ω, 3.12 Ω   **9-5.** 119, –41   **9-7.** –122, 0.973
9-9. 2 kΩ, 1, – 131, 14 μS   **9-11.** 5.83, 1.49(10^{-4}), –0.994, 0.199 μS
9-13. C_2 open   **9-15.** One solution is $R_1 = 4.3$ kΩ, $R_2 = 750$ Ω, $R_E =$

750 Ω, R_C = 3 kΩ, and R_L = 3.3 kΩ **9-17.** 7544   **9-19.** 25 kΩ,
0.999, –73, 4 μS   **9-21.** a. no b. yes c. no

Chapter 10

10-1. $I_{C(\text{sat})}$ = 1.18 mA, $V_{CE(\text{cut})}$ = 8.84 V, PP = 6.98 V   **10-3.** 5.2 V
10-5. 7.14 V, ac load line passes 7.28 mA and 9.25 V   **10-7.** –16.2, 125,
2027, 9.27 mW, 280 mW, 45.4 mA, 441 mW, 2.1%   **10-9.** 2.04%
10-13. 39.8 mA, 338 mA, 74%   **10-15.** 11 mA, 19 mA   **10-17.** 74.7 Ω
10-19. 1 W, 0.2 W   **10-21.** 103 mA   **10-23.** 2.75 W   **10-25.** 184°C
10-27. c   **10-29.** d   **10-31.** Change R_1 to 2 kΩ **10-37.** All except b

Chapter 11

11-1. 4.98 MHz, 12.8 mA, 15 V, 200 kHz, 30 V   **11-3.** –4.3 V
11-5. 14.94 MHz   **11-7.** 62.6 kΩ **11-9.** 5.01 MHz   **11-11.** 19.1 V,
11.7 W   **11-13.** 5.15 V   **11-15.** 50 mA   **11-17.** b, d, e
11-19. 5 kΩ **11-21.** V_{CC} = 10 V, L = 23.9 μH, C = 1060 pF.
11-23. 5 kΩ, 25   **11-25.** 2000, 3000

Chapter 12

12-1. 1.5(10^{11})   **12-3.** I_D = 0.032(1 + V_{GS}/8)^2, 8 mA, 18 mA
12-5. 5.12 mA, –1.38 V, 5.78 V   **12-7.** 1.98 mA, 7.5 V, 8.5 V, 11.4 V
12-9. 1.98 mA, 8.47 V   **12-11.** 3000 μS   **12-13.** –2 V
12-15. –16.1 mV   **12-17.** 200 mV   **12-19.** 1.03 mV, 20 mV
12-21. 200 Ω, increase   **12-23.** b   **12-25.** 333 Ω (nearest standard is
330 Ω)   **12-27.** R_E = 6.2 kΩ, R_D = 8.2 kΩ **12-29.** 3.9 mA, –1.05 V
12-31. 16 mA, 30 V; 2.94 mA

Chapter 13

13-1. 4.5 mA, 12.5 mA   **13-3.** 100 MΩ, –8, 1 kΩ **13-5.** 2.25 mA
13-7. –7.74   **13-9.** +5 V, 0   **13-11.** a   **13-13.** b   **13-15.** 7.5 kΩ
13-17. 976 mV   **13-19.** 25 μs, 1 kHz   **13-21.** Calculates K, 7 V, 6

Chapter 14

14-1. 345 Hz, 260 Hz   **14-3.** 41.2 Hz, 740 Hz   **14-5.** 42.9 Hz

14-7. –25, 637 Hz   **14-9.** 194 kHz, 10.6 MHz   **14-11.** 1273 pF
14-13. 11.5 Hz, 7.8 MHz   **14-15.** 250 W   **14-19.** 40 dB, 0 dB
14-21. 0.44 μs   **14-23.** 350 kHz   **14-25.** c   **14-27.** b
14-29. 0.0398 μF (use 0.022 μF and 0.018 μF)   **14-31.** 5 Ω, 4 kΩ

Chapter 15

15-1. 0.715 mA, 1.43 mA, 7.85 V   **15-3.** 7.15 μA, 5.96 μA, 1.19 μA
6.56 μA   **15-5.** 1.83 μA, 1.53 μA, 0.3 μA, 1.68 μA   **15-7.** 143,
10.5 kΩ **15-9.** 173, –0.603, 49.1 dB   **15-11.** 3906, –1, 71.8 dB
15-13. 0.1 V   **15-15.** 9.85 V   **15-17.** 9 V   **15-19.** a. Approximately
36 dB b. 21 V c. 1000   **15-21.** 1.5 V/μs   **15-23.** 3.33 V/μs
15-25. 2.36 V/μs   **15-27.** a. 800 kHz b. 2 MHz c. 4 MHz
15-29. –0.7 V   **15-31.** 7.15 kΩ (6.8 kΩ)   **15-33.** 293 kΩ (300 kΩ)
15-35. 10.7 Ω, 187   **15-37.** 0.93 mA, 186

Chapter 16

16-1. 1,000,000, 0.02, 50   **16-3.** 0.15 μV, 1 mV, 1.5 μV, 1 mV, 150
16-5. 334 MΩ, 0.3 Ω **16-7.** 105 mV   **16-9.** 20,000 MΩ
16-11. 100 mV, 1 V, 10 V   **16-13.** 1 Ω, 10 Ω, 100 Ω **16-15.** 1 mA
16-17. 10 kHz, 15.9 V   . **16-19.** 100 and 10 kHz, 5 and 200 kHz, 2 and
500 kHz   **16-21.** 3 MHz, 3 MHz   **16-23.** 2 MHz   **16-25.** c
16-27. Defective LF355, pot not connected to pin 1 or pin 5, +15 V not
connected to wiper, either supply voltage missing   **16-29.** R_1 = 240 kΩ
and R_2 = 1 kΩ **16-33.** 41 mV

Chapter 17

17-1. 10   **17-3.** A_1 = 194, A_2 = 48, $A_1 A_2$ = 9312   **17-5.** 7, 10
17-7. –100, 10 kHz   **17-9.** –1, –0.5, –0.25, –0.125   **17-11.** 10 mA,
500 Ω **17-13.** 1 mA, –0.5 V   **17-15.** 15.9 kHz, 40 dB per decade.
17-17. 796 Hz, 15.9 kHz   . **17-19.** V_{out} goes into positive or negative
saturation   **17-21.** One solution: R_1 = 20 kΩ and R_2 = 1 kΩ
17-23. One solution: R = 1 kΩ

Chapter 18

18-1. 7 μV, 0.4 mA, 0.127 mA   **18-3.** 0, +600 mV, 20 kHz
18-5. +3.72 V, –13 to –14 V, +13 to +14 V   **18-7.** +0.271 V and –0.271 V

18-9. 0.638 pF   **18-11.** –1 V   **18-13.** 90.9 mV   **18-15.** 12.8 kHz,
13V   **18-17.** 1 kΩ shorted, 47 kΩ open   **18-19.** C_2 open, R_5 shorted,
noninverting and inverting connections reversed   **18-21.** One approximate
solution: 2.7 kΩ and 1 kΩ for upper, 6.8 kΩ and 1 kΩ for lower
18-23. One solution: R_1 = 1 kΩ, R_2 = 5.1 kΩ   **18-25.** 141 μs

Chapter 19

19-1. 34.5 V   **19-3.** 0.63 A   **19-5.** 8.81 to 15.3 V   **19-7.** 6.66 A,
2.59 A, 59.9 W, 23.3 W   **19-9.** 0.1%   **19-11.** 0.211%   **19-13.** 5 to
15.8 V   **19-15.** 0.508 A   **19-17.** 1.25 to 27.3 V, 1.25 V   **19-19.** 8 V
19-21. a. increase. b. decrease. c. decrease. d. increase   **19-23.** a. Same
b. decrease c. decrease (most likely) d. decrease   **19-25.** 130 Ω
19-27. 3333 μF (3300 μF)   **19-29.** 331, 208, 16.9 Ω   **19-31.** 18.6 mA

Chapter 20

20-1. 9 V   **20-3.** A: 35.8 to 362 Hz; B: 358 Hz to 3.62 kHz; C: 3.58 kHz
to 36.2 kHz; D: 35.8 kHz to 362 kHz   **20-5.** 3.62 MHz   **20-7.** 5.97
mA, 14 V   **20-9.** 0.2   **20-11.** 10 MHz, 15 MHz, 20 MHz
20-13. a. 1.59 MHz b. Approximately 10,000   **20-15.** 21.9 kHz, 85.7%
20-17. Open 10-kΩ biasing resistor, shorted 5 kΩ, open 1 kΩ, open choke,
open 5 μH, open 0.001 μF, etc.   **20-19.** d   **20-21.** 4.46 μH
20-23. One solution: R_1 = 1 kΩ, R_2 = 2.2 kΩ, R_E = 2 kΩ, and C = 0.001 μF
20-25. Pulling the frequency means changing it slightly; the tuning capacitor
adds to the mounting capacitance   **20-27.** 41 V/ms, 18.6 V/ms, and
12.4 V/ms

Chapter 21

21-1. 4.7 V   **21-3.** 0.16 A   **21-5.** 9.3 mA, 60 mA   **21-7.** 16 mA,
1 A   **21-9.** 2.7 V   **21-11.** a. Same. b. Decrease. c. Increase. d. Increase.
21-13. Load is open, R_1 is open, R_2 is shorted, transistor is open, no +5 V,
etc.   **21-15.** 0.002 μF   **21-17.** 82 Hz, 1.72 kHz

Chapter 22

22-1. 40 kHz, 80 kHz, 120 kHz   **22-3.** 500 Hz, 12.5 kHz   **22-5.** 19.1 V,
6.37 V, 3.82 V   **22-7.** The spectrum has five lines: one at zero frequency
and four more lines at 50 kHz, 100 kHz, 150 kHz, and 200 kHz; the

corresponding peak values are 5 V, 3.18 V, 1.59 V, 1.06 V, and 0.796 V
22-9. 8%, 4%, 8.94%   **22-11.** 10 mV   **22-13.** 0.006%   **22-15.** Use
either a Wien bridge or a twin-T filter; one design is R = 1.27 kΩ and C =
100 pF

Chapter 23

23-1. Original: 56 kHz and 84 kHz; second harmonics: 112 kHz and
168 kHz; sum: 140 kHz; difference: 28 kHz   **23-3.** 5   **23-5.** Input
frequencies: 384, 480, and 576 Hz; second harmonics: 768, 960, and
1152 Hz; sum and differences: 864, 960, 1056, 96, 192 Hz   **23-7.** 44 MHz
23-9. –3 dB   **23-11.** 3.09 kHz, 6.38 kHz   **23-13.** a. 128 μV b. 12.8 μV
c. 1.28 μV   **23-15.** 1.57 μV   **23-17.** c   **23-19.** 0.006 μF
23-21. Group 1: 6.4 and 8.4, 5.4 and 9.4, 4.4 and 10.4, 3.4 and 11.4, 2.4
and 12.4; group 2: 13.8 and 15.8, 12.8 and 16.8, 11.8 and 17.8, 10.8 and
18.8, 9.8 and 19.8; group 3: 21.2 and 23.2; 20.2 and 24.2; 19.2 and 25.2;
18.2 and 26.2; 17.2 and 27.2; group 4: 28.6 and 30.6, 27.6 and 31.6, 26.6
and 32.6, 25.6 and 33.6, 24.6 and 34.6; group 5: 36 and 38, 35 and 39, 34
and 40, 33 and 41, 32 and 42

Chapter 24

24-1. 24.3, 13.8, 34.7   **24-3.** 0.2, 20%   **24-5.** 1 V, 0.15 V, 0.15 V
24-7. 995 kHz, 2055 kHz   **24-9.** 12 kΩ

Index

A

B

P

Q

U

UJT (unijunction transistor), 664–667
UJT-triggered silicon controlled rectifier, 669
Unbalanced load, 78–79
Unbiased diode, 27–28
Unijunction transistor (UJT), 664–667, 669
Unity-gain frequency, 458, 498
Universal bias, 154–161
Unwanted oscillations, 633–638
Upper trip point, 558
Upside-down *pnp* circuits, 164–167

V

Valence orbit, 19
Valley current, 666
Varactor, 104–106
Varicap, 104–106
Varistor, 108
VCO (voltage-controlled oscillator), 644, 704–705
Vertical MOS (VMOS), 378–382
Video amplifier, 464
Video automatic gain control, 526–527
Virtual ground, 491–492
VMOS (vertical MOS), 378–382
Voltage amplifiers, 512–528
  inverting, 517–528
  noninverting, 512–517
Voltage compliance, 106, 124
Voltage-controlled current sources, 530–533
Voltage-controlled oscillator (VCO), 644, 704–705
Voltage-divider bias, 154–161, 333–334
Voltage doubler, 71–72
Voltage feedback:
  inverting, 488–494
  noninverting, 471–484

Voltage-feedback regulation, 582–587
Voltage follower, 483–484
Voltage gain, 181–182, 203
Voltage multiplier, 71–73
Voltage quadrupler, 72–73
Voltage regulation, 221–225, 466
Voltage-regulator diode, 88–91
Voltage-sourcing the base, 136
Voltage-to-current converter, 486, 530–533
Voltage tripler, 72
Voltage-variable capacitance, 104–106
Voltage-variable resistance, 354
VVR (voltage-variable resistance), 354

W

Wafer, 436
Waveform conversion, 572–575
Waveform generation, 575–578
Wideband amplifier, 464
Wien-bridge oscillator, 610–616
Window, 558
  of IC, 437
Window comparator, 557–559

Y

y parameters, 233

Z

z parameters, 232
Zener diode, 86–91
Zener dropout point, 95–96
Zener follower, 221–225
Zener regulator, 92–97
Zener resistance, 88
Zero bias of MOSFET, 368
Zero-crossing detector, 140–141, 552